AF781804

Biochemistry: Modern Theory, Techniques and Applications

Biochemistry: Modern Theory, Techniques and Applications

Edited by Jim Wang

SYRAWOOD
PUBLISHING HOUSE
New York

Published by Syrawood Publishing House,
750 Third Avenue, 9th Floor,
New York, NY 10017, USA
www.syrawoodpublishinghouse.com

Biochemistry: Modern Theory, Techniques and Applications
Edited by Jim Wang

© 2018 Syrawood Publishing House

International Standard Book Number: 978-1-68286-514-9 (Hardback)

This book contains information obtained from authentic and highly regarded sources. Copyright for all individual chapters remain with the respective authors as indicated. All chapters are published with permission under the Creative Commons Attribution License or equivalent. A wide variety of references are listed. Permission and sources are indicated; for detailed attributions, please refer to the permissions page and list of contributors. Reasonable efforts have been made to publish reliable data and information, but the authors, editors and publisher cannot assume any responsibility for the validity of all materials or the consequences of their use.

The publisher's policy is to use permanent paper from mills that operate a sustainable forestry policy. Furthermore, the publisher ensures that the text paper and cover boards used have met acceptable environmental accreditation standards.

Trademark Notice: Registered trademark of products or corporate names are used only for explanation and identification without intent to infringe.

Cataloging-in-Publication Data

Biochemistry : modern theory, techniques and applications / edited by Jim Wang.
p. cm.
Includes bibliographical references and index.
ISBN 978-1-68286-514-9
1. Biochemistry. I. Wang, Jim.
QH345 .B56 2018
572--dc23

Printed in China.

TABLE OF CONTENTS

PREFACE

Biochemistry studies the processes that occur within an organism. It has evolved as a field over the past few decades and has successfully been able to explain the complexities of life. It studies structures and functions of biological macromolecules. This book traces the progress of this field and highlights some of its key concepts and applications. For all the readers who are interested in biochemistry, the case studies included herein will serve as an excellent guide to develop a comprehensive understanding. The aim of this book is to present researches that have transformed this discipline and aided its advancement.

This book is the end result of constructive efforts and intensive research done by experts in this field. The aim of this book is to enlighten the readers with recent information in this area of research. The information provided in this profound book would serve as a valuable reference to students and researchers in this field.

At the end, I would like to thank all the authors for devoting their precious time and providing their valuable contribution to this book. I would also like to express my gratitude to my fellow colleagues who encouraged me throughout the process.

Editor

1

Pronounced Metabolic Changes in Adaptation to Biofilm Growth by *Streptococcus pneumoniae*

Raymond N. Allan[1,2]*, Paul Skipp[3,4], Johanna Jefferies[1,5], Stuart C. Clarke[1,5,6], Saul N. Faust[1,2,6], Luanne Hall-Stoodley[1,2,7]◔, Jeremy Webb[3,6]◔

1 Academic Unit of Clinical and Experimental Sciences, Faculty of Medicine and Institute for Life Sciences, University of Southampton, Southampton, United Kingdom, **2** Southampton NIHR Wellcome Trust Clinical Research Facility, University Hospital Southampton NHS Foundation Trust, Southampton, United Kingdom, **3** Centre for Biological Sciences, University of Southampton, Southampton, United Kingdom, **4** Centre for Proteomic Research, Institute for Life Sciences, University of Southampton, Southampton, United Kingdom, **5** Public Health England, Southampton, United Kingdom, **6** Southampton NIHR Respiratory Biomedical Research Unit, University Hospital Southampton NHS Foundation Trust, Southampton, United Kingdom, **7** Microbial Infection and Immunity, Centre for Microbial Interface Biology, The Ohio State University, Columbus, Ohio, United States of America

Abstract

Streptococcus pneumoniae accounts for a significant global burden of morbidity and mortality and biofilm development is increasingly recognised as important for colonization and infection. Analysis of protein expression patterns during biofilm development may therefore provide valuable insights to the understanding of pneumococcal persistence strategies and to improve vaccines. iTRAQ (isobaric tagging for relative and absolute quantification), a high-throughput gel-free proteomic approach which allows high resolution quantitative comparisons of protein profiles between multiple phenotypes, was used to interrogate planktonic and biofilm growth in a clinical serotype 14 strain. Comparative analyses of protein expression between log-phase planktonic and 1-day and 7-day biofilm cultures representing nascent and late phase biofilm growth were carried out. Overall, 244 proteins were identified, of which >80% were differentially expressed during biofilm development. Quantitatively and qualitatively, metabolic regulation appeared to play a central role in the adaptation from the planktonic to biofilm phenotype. Pneumococci adapted to biofilm growth by decreasing enzymes involved in the glycolytic pathway, as well as proteins involved in translation, transcription, and virulence. In contrast, proteins with a role in pyruvate, carbohydrate, and arginine metabolism were significantly increased during biofilm development. Downregulation of glycolytic and translational proteins suggests that pneumococcus adopts a covert phenotype whilst adapting to an adherent lifestyle, while utilization of alternative metabolic pathways highlights the resourcefulness of pneumococcus to facilitate survival in diverse environmental conditions. These metabolic proteins, conserved across both the planktonic and biofilm phenotypes, may also represent target candidates for future vaccine development and treatment strategies. Data are available via ProteomeXchange with identifier PXD001182.

Editor: Jens Kreth, University of Oklahoma Health Sciences Center, United States of America

Funding: This work was funded by The Bill and Melinda Gates Foundation grant GCE 1006884 (http://www.gatesfoundation.org). Funding was received by JW, SCC, SNF, LHS, and JJ. The funders had no role in study design, data collection and analysis, decision to publish, or preparation of the manuscript.

Competing Interests: The authors have declared that no competing interests exist.

* Email: r.allan@soton.ac.uk

◔ These authors contributed equally to this work.

Introduction

Pneumococcal disease accounts for a significant global burden of morbidity and mortality, particularly in younger children and older adults. *Streptococcus pneumoniae* is a Gram-positive facultative anaerobe that colonises the majority of children in the first year of life and is associated with mostly asymptomatic carriage in the human respiratory tract. However, multifactorial and incompletely understood mechanisms allow *S. pneumoniae* to cause localised infection such as otitis media and sinusitis, and/or invasive disease (meningitis, bacteremia and pneumonia). The transition from asymptomatic colonisation of the respiratory mucosal epithelium without inflammation to the initiation of infection is thought to correlate with the expression of specific cell surface molecular properties such as adhesins and capsule polysaccharides under different conditions [1]. Although pneumococcal conjugate vaccines have reduced carriage and disease caused by capsular strains included in these vaccines, there has been a rise in the prevalence of capsular strains not included in the vaccines (serotype replacement). With over 90 different pneumococcal serotypes, there is a need for more effective vaccines that are conserved across serotypes, immunogenic in all age groups, and effective against both colonisation and infection by *S. pneumoniae* [2].

Aggregated pneumococci adherent to human middle ear mucosae of children undergoing treatment for chronic otitis media (COM) indicate that *S. pneumoniae* is present in biofilms on the mucosal epithelium [3], an observation supported by animal models [4,5]. Pneumococcal biofilms were also observed on adenoid mucosal epithelium from children undergoing adenoidectomy for the treatment of obstructive sleep apnea, suggesting that pneumococcal biofilms are present in the nasopharynx in both the presence and absence of overt infection [6,7]. The ability of *S. pneumoniae* to develop biofilms has been shown to affect both colonisation and disease states [8–10], and understanding *S. pneumoniae* biofilm development is important for studying the mechanisms of persistence.

S. pneumoniae adapts to different environments in the human host by modulating proteins and carbohydrates that may function as virulence factors. In addition to different capsule serotypes, multiple proteins modulate cell attachment and contribute to both colonisation and invasion. *In vitro* studies with pneumococcal clinical strains have demonstrated that biofilm growth downregulates or disrupts capsule expression, and induces the production a complex extracellular matrix consisting of carbohydrates, proteins and DNA [11–14] and differential expression of a broad spectrum of proteins [15,16]. We hypothesized that protein expression patterns during biofilm growth might highlight important phenotypic changes facilitating pneumococcal persistence or virulence. On the other hand, conserved protein expression in both planktonic and biofilm growth, might inform better vaccine strategies since these proteins would be present regardless of the mode of growth or type of infection.

Serotype 14 is a common serotype in pneumococcal infection worldwide, causing both localised and invasive disease. It exhibits high levels of antibiotic resistance [17] and is prevalent in paediatric disease such as otitis media [18]. Serotype 14 was chosen in this study as a model strain because of its high biofilm-forming index based on multiparametric analysis of different clinical *S. pneumoniae* strains [13]. Previous proteomic analyses of *S. pneumoniae* growth have used conventional 1DE and 2DE gel-based methodologies [15,16,19–21]. iTRAQ (isobaric tag for relative and absolute quantitation) is a high-throughput gel-free proteomic profiling approach enabling quantitative high resolution comparison of protein profiles between multiple phenotypes or growth conditions within a single experiment. To investigate the pneumococcal serotype 14 biofilm proteome, we compared nascent 1-day biofilms lacking complex structure, highly structured 7-day biofilms, and log-phase planktonic growth to determine: 1) distinct patterns of differential expression comparing planktonic, nascent and late-phase biofilm growth, and 2) conserved patterns of protein regulation during planktonic and biofilm growth.

Methods

Bacterial strain and growth conditions

A Serotype 14 (ST124) blood sample isolate provided by the UK Public Health England (Health Protection Agency) Microbiology Services laboratory, Southampton, was selected based on multi-parametric, statistically-based criteria for biofilm development [13] [personal observation RNA/LHS]. The strain was subcultured from a frozen stock onto Columbia blood agar (CBA) plates (Oxoid, U.K.) and incubated for 14 hours at 37°C/5% CO_2. Colonies were re-suspended in 2 ml Brain Heart Infusion (BHI) broth (Oxoid, U.K.), centrifuged at 3,200×*g* for 5 min to remove cell debris and secreted lytic enzymes (personal communication, Public Health England), and the supernatant used to inoculate fresh dilute BHI broth. For planktonic growth, cultures were grown to mid-exponential phase ($OD_{600} = 0.3$; $\sim 1\times10^8$ cells).

Biofilm Formation

For biofilm formation, mid-exponential planktonic cultures (2 ml×10^8 cells) were used to inoculate individual wells of polystyrene 6-well plates (Corning Incorporated, Costar, U.S.A.), and all wells supplemented with 2 ml BHI diluted 1:5 with non-pyrogenic sterile dH_2O. Cultures were incubated under static conditions at 37°C/5% CO_2 with replacement of warm, fresh diluted BHI daily.

Colony Forming Unit (CFU) Enumeration and Total Cell Counts

All medium was removed and biofilms washed twice using diluted BHI. Biofilms were resuspended in 1 ml Hank's balanced salt solution (HBSS) as described [13]. Briefly, bacterial suspensions (n = 3) underwent 10-fold dilutions in HBSS and 5×20 µl of each dilution spot plated onto CBA plates and incubated at 37°C/5% CO_2. For total cell counts (n = 3) bacterial suspensions were diluted 10-fold, stained with Syto 9 (Life Technologies, U.S.A.) according to manufacturer instructions, and individual cells counted using a haemocytometer (Marienfeld-Superior, Germany) and a Leica epifluorescent microscope using a 100x oil immersion lens.

Scanning Electron Microscopy (SEM)

Biofilms (n = 3) were grown in 6-well plates containing ethanol-sterilized 13 mm glass coverslips (V.W.R., U.K.) under static conditions at 37°C/5% CO_2. All medium was removed and biofilms washed twice using diluted BHI. Coverslips were removed using sterile forceps, placed in primary fixative solution (3% gluteraldehyde, 0.1 M sodium cacodylate (pH 7.2), 0.15% Alcian blue), and incubated at 4°C overnight. The primary fixative was replaced with 0.1 M sodium cacodylate (pH 7.2), followed by the secondary fixative solution (0.1 M osmium tetroxide, 0.1 M sodium cacodlyate; pH 7.2), and finally with 0.1 M sodium cacodylate (pH 7.2). Each treatment was incubated for 1 h at room temperature. Coverslips were processed through an ethanol series to 100%, critical point dried, and sputter coated and biofilms imaged using FEI Quanta 200 scanning electron microscope.

Confocal Laser Scanning Microscopy (CLSM)

Biofilms (n = 6) were grown in 35 mm glass bottom microwell dishes (MatTek Corporation, U.S.A.) under static conditions at 37°C with 5% CO_2, washed twice with HBSS and stained with LIVE/DEAD BacLight Bacterial Viability Kit (Life Technologies, U.S.A.) according to manufacturer instructions. Biofilms were examined immediately using a Leica SP5 LSCM with inverted stand using a 63x oil immersion lens, performing sequential scanning using 0.5 µm sections. Three random fields of view were captured and images were analysed using Leica LCS Software.

Protein Extraction

For planktonic protein samples, cultures were grown to mid-exponential phase and centrifuged. The supernatant was discarded and the pellet washed twice in 1 ml Hanks' Balanced Salt Solution (HBSS; Gibco, Life Technologies, U.K.) at 9,500×*g*. For biofilm protein samples 1 and 7 day cultured biofilms were washed twice with HBSS and resuspended in 1 ml HBSS using sterile cell scrapers. The bacterial suspension was centrifuged (9,500×*g*), the supernatant removed and the pellet resuspended in 0.1 M triethylammonium bicarbonate (TEAB) buffer (Sigma-Aldrich, U.K.) with 0.1% Rapigest SF surfactant (Waters, U.K.). Both the planktonic and biofilm samples were lysed in lysing matrix B (MP

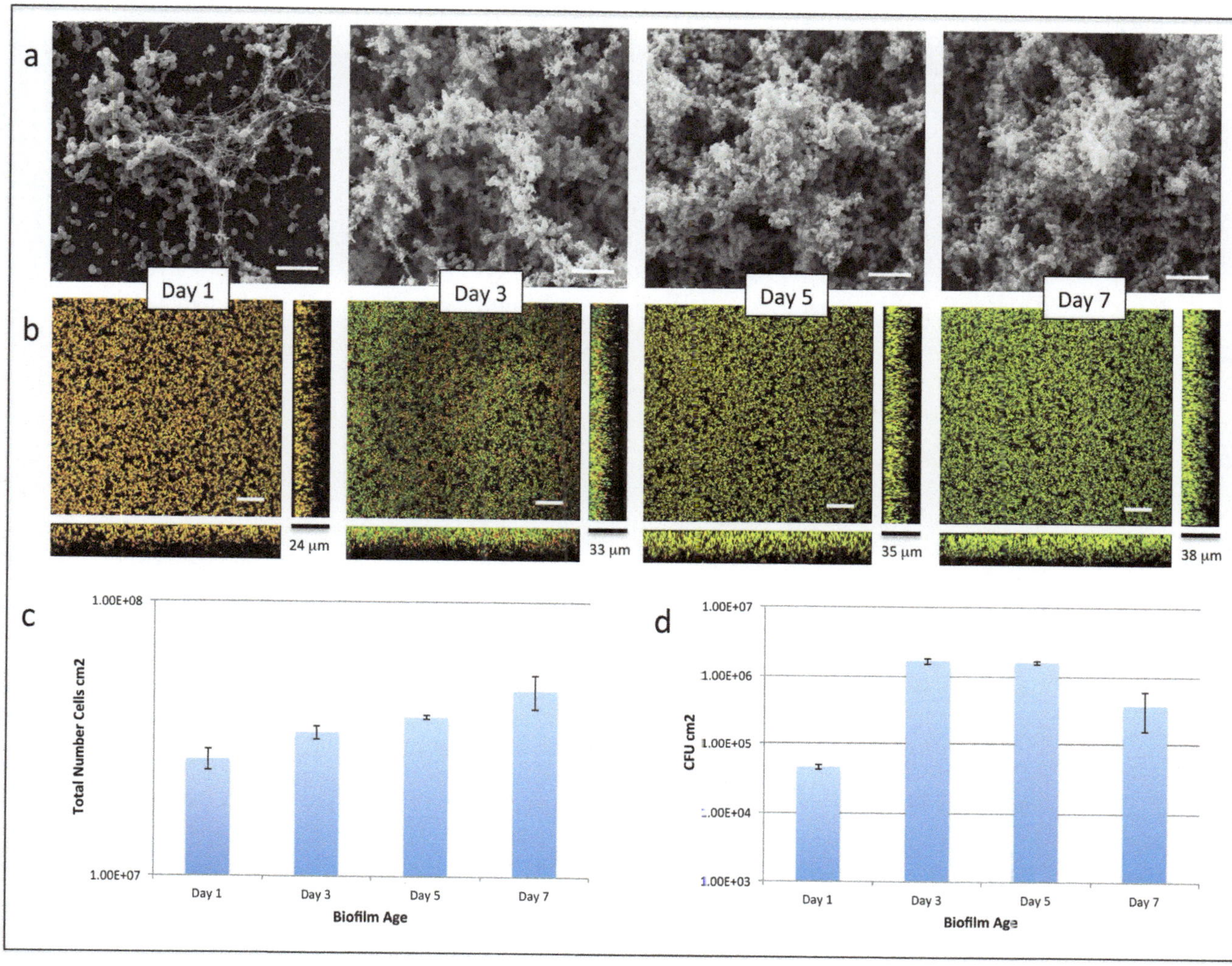

Figure 1. ***S. pneumoniae*** **serotype 14 day 1, 3, 5 and 7 biofilms imaged using (a) scanning electron microscopy with Alcian Blue staining (8,000x magnification; scale bar: 5 μm), and (b) confocal microscopy using Live/Dead staining, with sideviews (YZ - right) and (XZ - bottom) sagittal sections of the biofilm.** Scale bar beside YZ sagittal sections represents average biofilm thickness. Scale bar in xy pane: 30 μm. (c) Bar chart comparing the total number of cells between day 1, 3, 5 and 7 biofilms. (d) Bar chart comparing number of viable cells between day 1, 3, 5 and 7 biofilms through CFU cm^{-2} measurement.

Bioscience, U.K.) using a Hybaid Ribolyser Homogenizer (Hybaid, U.K.) in six 30 second sessions with 30 second storage on ice between sessions. The lysates were centrifuged at 855×*g*/5 min and the supernatant retained (soluble protein fraction). Protein quantification of the soluble protein fraction was performed using Coomassie Protein Assay Reagent (Thermo Fisher Scientific, U.K.), measuring optical density (595 nm) and quantified against albumin standards (Thermo Fisher Scientific, U.K.).

iTRAQ labelling

Comparative analyses of protein expression between planktonic log-phase culture and 1-day and 7-day biofilms were performed on 3 technical replicates of 3 biological replicates. iTRAQ labelling of samples was performed according to the manufacturer's instructions using an iTRAQ Reagent-8 plex Multiplex Kit (AB Sciex U.K. Limited). Protein samples were normalized to ~60 μg and 2 μl of the reducing agent and 50 mM (tris-(2-carboxyethyl) phosphine were added and mixed. Samples were incubated at 60°C for 1 h. Methyl methane-thiosulfonate in isopropanol (1 μl of 200 mM) was then added and incubated for a further 10 min at room temperature. Proteins were digested by adding 10 μl of 1 mg/ml trypsin in 80 mM $CaCl_2$. Samples were incubated for 16 h at 37°C. Labelling of tryptic peptides with iTRAQ 8-plex tags was achieved by mixing the appropriate iTRAQ labeling reagent with the relevant sample, and incubation at room temperature for 2 h.

Mass Spectrometry

Two dimensional separations were performed using a nanoAcquity 2D UPLC system (Waters). For the first dimension separation, 2.0 μl of the prepared protein lysates containing 3.2 μg of iTRAQ labeled peptides were injected onto a 5 μm Xbridge BEH130 C18, 300 μm ID×50 mm (Waters) column equilibrated in 20 mM ammonium formate, pH 10 (buffer A). The first dimension separation was achieved by increasing the concentration of acetonitrile (buffer B) in 7 steps consisting of 11, 14, 16, 20, 25, 50, 65%. At each step the programmed percentage composition was held for 1 min at a low rate of 1000 nl/min and the eluant diluted by buffer C (H_20+0.1% formic acid)

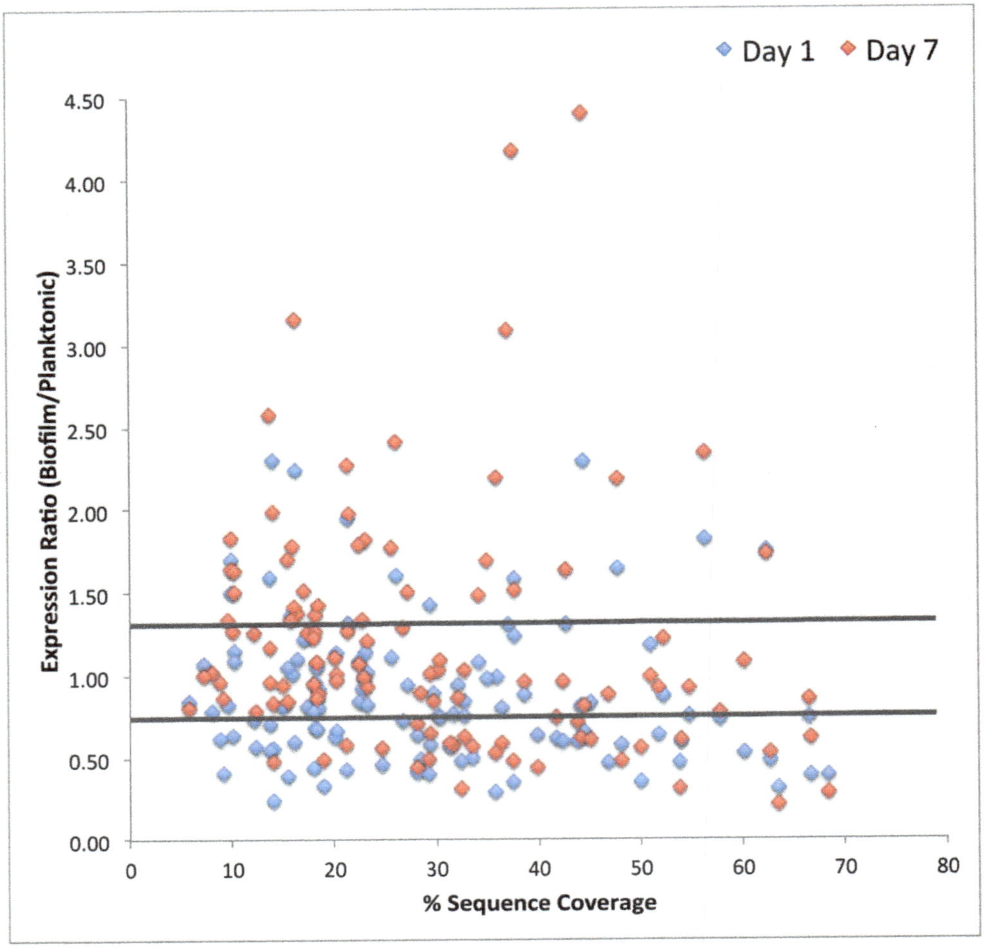

Figure 2. Volcano plot of the complete iTRAQ proteomic dataset comparing the fold change in protein expression between mid-exponential planktonic *S. pneumoniae* and day 1 and day 7 biofilms. Proteins represented met inclusion criteria of ≥3 peptide matches, ≥50 protein score, and ≥5% sequence coverage ($p<0.05$). Comparative protein data with >1.3 and <0.77 ratios (marked by horizontal lines) were identified as having differential expression.

from the second dimension pump at a flow rate of 20 µl/min, effectively diluting the ammonium formate and acetonitrile, allowing trapping of the eluting peptides onto a Symmetry C18, 180 µm×20 mm trapping cartridge (Waters). After 20 min washing of the trap column, peptides were separated using an in-line second dimension analytical separation performed on a 75 µm ID×200 mm, 1.7 µm BEH130 C18, column (Waters) using a linear gradient of 5 to 40% B (buffer A = water containing 0.1% formic acid, buffer B = 95% acetonitrile containing 0.1% formic acid) over 120 min with a wash to 85% B at a flow rate of 300 nl/min. All separations were automated and performed on-line to the mass spectrometer.

All data were acquired using a Q-tof Global Ultima (Waters Ltd, Manchester, UK) fitted with a nanoLockSpray source. A survey scan was acquired from *m/z* 350 to 1700 with the switching criteria for MS to MS/MS including ion intensity and charge state. The collision energy used to perform MS/MS was varied according to the mass and charge state of the eluting peptide. Peak lists were generated using ProteinLynx Global Server 2.4 (Waters Ltd, Manchester, UK). The following parameters were used for processing the MS/MS spectra; normal background subtraction with a 25% background threshold and medium de-isotoping with a threshold of 1%, no smoothing was performed.

Peak list generation and database searching

Peak lists from the MS/MS analysis of were submitted to the MASCOT search engine version 2.2.1 (Matrix Science, London, UK) and the data searched against a protein translation of the CGSP14 NCBI genome (http://www.ncbi.nlm.nih.gov/Taxonomy/Browser/wwwtax.cgi?id = 516950). The corresponding quantitative information using the iTRAQ reporter ions was also obtained via MASCOT.

A maximum of one missed cleavage for tryptic digestion and fixed modifications for methyl methane-thiosulfonation of cysteine and the N-terminus and lysine side chains using the 8-plex iTRAQ label were allowed (Applied Biosystems, Warrington, UK). Variable modification for the oxidation of methionine and iTRAQ modification of tyrosine were also allowed. Precursor ion and sequence ion mass tolerances were set at 100 ppm and 0.3 Da respectively. Protein identifications required the assignment of ≥2 different peptides with a significance threshold for accepting a match of $p<0.05$. The protein ratios were calculated using MASCOT version 2.2.1, where the peptide ratios were weighted and median normalisation was performed, automatic outlier removal was chosen and the peptide threshold was set to 'at least homology'. False discovery rates were calculated by running all spectra against a decoy database using the MASCOT software.

	Protein	Accession No.	Protein Expression Ratios: 1-day Biofilm/ Planktonic	Protein Expression Ratios: 7-day Biofilm/ Planktonic	Peptide Matches	% Sequence Coverage
Translation	elongation factor Tu	YP_001836191	0.76	0.92	14	54.7
	elongation factor EF-2	YP_001835001	1.07	1.49	15	34.1
	elongation factor Ts	YP_001836897	0.82	0.62	10	44.2
	30S ribosomal protein S1	YP_001835523	0.61	0.74	13	41.8
	30S ribosomal protein S7	YP_001835000	0.39	0.62	4	66.7
	50S ribosomal protein L4	YP_001834936	0.63	0.92	8	51.7
	30S ribosomal protein S5	YP_001834952	0.83	0.60	5	45.1
	50S ribosomal protein L1	YP_001835309	0.66	0.81	8	44.5
	50S ribosomal protein L3	YP_001834935	0.43	0.57	3	21.2
	50S ribosomal protein L23	YP_001834937	0.64	0.44	4	39.8
	50S ribosomal protein L5	YP_001834947	0.59	0.61	7	53.9
	30S ribosomal protein S9	YP_001835026	0.46	0.56	3	24.6
	30S ribosomal protein S3	YP_001834941	0.41	0.71	4	28.1
	30S ribosomal protein S2	YP_001836898	0.35	0.48	4	37.5
	50S ribosomal protein L22	YP_001834940	0.38	0.28	4	68.4
	30S ribosomal protein S8	YP_001834949	0.80	0.59	3	36.4
	50S ribosomal protein L10	YP_001836061	0.77	0.58	4	31.6
	50S ribosomal protein L17	YP_001834961	0.49	0.57	5	33.6
	50S ribosomal protein L13	YP_001835025	0.29	0.52	3	35.8
	translation initiation factor IF-2	YP_001835236	0.84	0.81	5	5.9
	elongation factor Tu family protein	YP_001835355	0.68	1.37	5	18.2
	50S ribosomal protein L14	YP_001834945	0.35	0.56	4	50
	30S ribosomal protein S18	YP_001836242	0.57	0.48	4	48.1
	50S ribosomal protein L18	YP_001834951	0.31	0.22	3	63.4
	50S ribosomal protein L15	YP_001834954	0.63	0.44	3	28.1
	30S ribosomal protein S10	YP_001834934	0.73	0.77	4	57.8
	30S ribosomal protein S4	YP_001834800	0.59	0.96	5	42.4
	50S ribosomal protein L6	YP_001834950	0.60	0.71	4	43.8
	50S ribosomal protein L2	YP_001834938	0.94	0.86	4	32.1
	30S ribosomal protein S13	YP_001834958	0.57	0.58	3	31.4
	50S ribosomal protein L29	YP_001834943	0.48	0.31	3	32.4
	30S ribosomal protein S15	YP_001836318	0.40	0.49	3	29.2
	ribosomal subunit interface protein	YP_001836890	1.94	2.27	3	21.4
Transcription	DNA-directed RNA polymerase alpha subunit	YP_001834960	0.75	1.03	4	30.2
	DNA-binding protein HU	YP_001835885	0.47	0.31	3	53.8
	transcription elongation factor NusA	YP_001835233	0.55	0.84	4	14
	DNA-directed RNA polymerase beta~ subunit	YP_001836642	0.82	1.33	3	9.6
	ribose-phosphate pyrophosphokinase	YP_001835901	1.31	1.27	3	21.3
Metabolism – Glycolysis	enolase (phosphopyruvate hydratase)	YP_001835867	0.47	0.52	18	62.7
	pyruvate kinase	YP_001835590	0.87	1.21	16	52.3
	phosphoglycerate kinase	YP_001835194	1.17	0.99	14	51
	glyceraldehyde-3-phosphate dehydrogenase	YP_001836696	0.58	0.65	8	29.5
	phosphoglyceromutase	YP_001836345	0.74	0.85	8	66.5
	glucose-6-phosphate isomerase	YP_001836754	0.33	0.48	3	18.9
	6-phosphofructokinase	YP_001835589	0.84	1.02	5	32.8
	fructose-bisphosphate aldolase	YP_001835283	0.56	0.78	3	12.3
	phosphoglucomutase	YP_001836198	0.64	1.26	3	10.1
Metabolism – Pyruvate Metabolism	L-lactate dehydrogenase	YP_001835798	1.23	1.51	10	37.7
	formate acetyltransferase	YP_001835173	1.30	3.09	14	37
	alcohol dehydrogenase	YP_001835014	1.30	1.63	9	42.8
	alcohol dehydrogenase, iron-containing	YP_001836708	1.51	1.63	6	9.9
	alcohol dehydrogenase, zinc-containing	YP_001836739	1.70	1.82	3	9.9
Metabolism – Amino-acid Metabolism	arginine deiminase	YP_001836835	1.58	4.17	8	37.7
	ornithine carbamoyltransferase	YP_001836836	2.24	3.16	4	16.3
	aspartate aminotransferase	YP_001836677	0.62	0.96	3	8.9
	asparagine synthetase AsnA	YP_001836652	1.31	1.97	4	21.5
	adenylosuccinate synthetase	YP_001834736	0.24	0.48	3	14
	glutamate dehydrogenase	YP_001835989	0.78	1.02	3	8
Metabolism – Carbohydrate Metabolism	tagatose 1,6-diphosphate aldolase	YP_001835825	2.29	4.39	11	44.5
	6-phosphogluconate dehydrogenase	YP_001835088	0.94	1.50	6	27.2
	sugar ABC transporter, ATP-binding protein	YP_001836282	1.60	2.41	5	26.1
	PTS system, IIB component, putative	YP_001835318	1.82	2.34	3	56.4
	UDP-glucose 4-epimerase	YP_001836305	0.81	1.25	5	17.4
	phosphoenolpyruvate-protein phosphotransferase	YP_001835838	0.96	1.26	7	18.2
	PTS system, mannose-specific IIAB components	YP_001835012	1.42	1.01	7	29.5
	sugar ABC transporter, sugar-binding protein	YP_001836373	1.59	2.57	5	13.8
	PTS system, mannose-specific IID component	YP_001835010	1.13	1.82	6	23.1
	maltose/maltodextrin ABC transporter, maltose/maltodextrin-binding protein	YP_001836789	1.21	1.51	4	17.1
	PTS system, fructose specific IIABC components	YP_001835543	0.55	0.96	4	13.8
Metabolism – Other Metabolic	glycogen phosphorylase family protein	YP_001836787	1.07	1.00	4	7.2
	uracil phosphoribosyltransferase	YP_001835412	0.52	1.08	7	60.2
	pyridine nucleotide-disulfide oxidoreductase	YP_001836287	0.62	1.11	5	20.1
	ATP synthase subunit B	YP_001836209	0.75	1.09	9	30.3
	ATP synthase subunit A	YP_001836211	0.73	1.26	4	12.2
	3-ketoacyl-(acyl-carrier-protein) reductase	YP_001835136	0.48	0.89	4	28.4
	3-oxoacyl-(acyl carrier protein) synthase	YP_001835137	0.81	0.94	4	15
Virulence / Competence / Stress	pyruvate oxidase	YP_0018[illegible]5396	0.99	2.19	17	35.9
	NADH oxidase	YP_0018[illegible]6175	0.39	0.84	4	15.5
	general stress protein 24, putative	YP_0018[illegible]6498	1.74	1.73	10	62.4
	molecular chaperone DnaK	YP_0018[illegible]5207	0.75	0.63	16	32.8
	chaperonin GroEL	YP_0018[illegible]6598	1.14	1.01	7	20.2
	superoxide dismutase, manganese-dependent	YP_0018[illegible]5432	0.84	1.07	3	22.4
Protein Biosynthesis / Alteration / Degradation	ATP-dependent Clp protease, ATP-binding subunit, putative	YP_0018[illegible]5059	2.30	1.99	3	14.1
	endopeptidase O	YP_0018[illegible]5337	0.81	0.89	6	18.5
	glutamyl-tRNA amidotransferase subunit A	YP_0018[illegible]5148	1.02	1.21	3	23.2
	signal recognition particle protein	YP_0018[illegible]5967	1.01	1.78	3	15.9
	glycyl-tRNA synthetase beta subunit	YP_0018[illegible]5180	1.04	1.70	4	15.5
	arginyl-tRNA synthetase	YP_0018[illegible]5762	1.15	1.50	3	10.3
	phenylalanyl-tRNA synthetase beta subunit	YP_00183[illegible]264	1.09	1.63	3	10.2
	lysyl-tRNA synthetase	YP_00183[illegible]379	0.68	1.08	3	18.3
	aspartyl-tRNA synthetase	YP_00183[illegible]796	0.78	1.22	3	18.1
	trigger factor	YP_00183[illegible]116	0.88	0.96	9	38.6
Cell Division	cell division protein FtsZ	YP_00183[illegible]355	0.66	0.97	3	20.3
	cell division protein DivIVA	YP_00183[illegible]350	0.82	0.93	3	23.3
	septation ring formation regulator EzrA	YP_00183[illegible]473	1.09	1.37	4	16.4
	cell division protein FtsH	YP_00183[illegible]729	1.09	1.78	3	22.5
Transport	ABC transporter, substrate-binding protein	YP_00183[illegible]869	1.11	1.77	4	25.7
	bacteriocin transport accessory protein	YP_00183[illegible]201	1.64	2.18	4	47.8
	translocase	YP_00183[illegible]391	0.41	0.86	3	9.1
Miscellaneous	polyribonucleotide nucleotidyltransferase	YP_001835[illegible]270	0.71	1.17	6	13.7
	thioredoxin	YP_001836[illegible]71	0.89	0.84	3	29.8
	lipoate-protein ligase, putative	YP_001835[illegible]54	0.92	1.42	4	18.5
	putative manganese-dependent inorganic pyrophosphatase	YP_001836[illegible]37	0.87	0.86	3	18.3
	flavodoxin	YP_001835[illegible]81	0.47	0.88	3	46.9
	lipoprotein	YP_001835[illegible]04	0.99	0.99	5	23
	SPFH domain-containing protein/band 7 family protein	YP_001836[illegible]41	0.44	0.95	3	18.1
Unknown Function	hypothetical protein SPCG_1532	YP_001836[illegible]49	0.73	1.29	5	26.8
	hypothetical protein SPCG_0438	YP_001835[illegible]55	1.05	1.07	7	18.4
	hypothetical protein SPCG_0428	YP_001835[illegible]45	0.98	1.69	4	34.9
	hypothetical protein SPCG_1503	YP_001836[illegible]20	1.36	1.34	4	15.8
	hypothetical protein SPCG_2153	YP_001836[illegible]70	0.59	1.41	4	16.1
	hypothetical protein SPCG_1901	YP_001836[illegible]18	0.92	1.33	3	22.8

> 2.0	> 2 fold upregulation
1.3 - 1.99	< 2 fold upregulation
0.77 - 1.29	no change in expression
0.5 - 0.76	< 2 fold downregulation
< 0.5	> 2 fold downregulation

Figure 3. Comparative iTRAQ analyses of *S. pneumoniae* serotype 14 1-day and 7-day old *in vitro* biofilms to mid-exponential planktonic population protein expression. Inclusion criteria: ≥3 peptide matches, ≥50 protein score, ≥5% sequence coverage ($p<0.05$). Comparative protein data with >1.3 and <0.77 ratios identified as having differential expression.

Group	Category	Protein	Accession No.	Protein Expression Ratios: 1-day Biofilm/ Planktonic	Protein Expression Ratios: 7-day Biofilm/ Planktonic	Score	% Sequence Coverage
	Translation	30S ribosomal protein S6	YP_001836244	0.19	0.24	126	26
		50S ribosomal protein L19	YP_001835977	0.50	1.05	107	40
		50S ribosomal protein L11	YP_001835308	0.27	0.14	51	37.6
		30S ribosomal protein S17	YP_001834944	0.55	0.74	762	34.9
		50S ribosomal protein L9	YP_001836888	0.72	0.51	197	15.3
		peptide chain release factor 1	YP_001835715	0.89	1.19	225	10
		30S ribosomal protein S12	YP_001834999	1.44	0.92	46	24.1
	Transcription	transcription antitermination protein NusG	YP_001836690	0.78	0.96	62	14.2
		RNA polymerase sigma factor	YP_001835924	0.67	0.94	77	18.4
Metabolism	Pyruvate Metabolism	acetoin dehydrogenase complex, E3 component, dihydrolipoamide dehydrogenase, putative	YP_001835853	0.49	0.72	162	14.3
		lactate oxidase	YP_001835380	0.70	1.40	85	8.2
	Amino-acid Metabolism	methionine--tRNA ligase	YP_001835454	0.25	0.50	38	7.1
		2,3,4,5-tetrahydropyridine-2-carboxylate N-succinyltransferase, putative	YP_001836779	0.60	0.73	159	13.8
		3-deoxy-7-phosphoheptulonate synthase	YP_001836389	1.55	1.46	198	12
		serine hydroxymethyltransferase	YP_001835719	1.39	1.39	124	11.2
		glutamine amidotransferase, class I	YP_001835906	0.97	1.39	87	13.1
		cysteine synthase	YP_001836894	1.92	2.57	159	9.5
	Sugar Metabolism	glucose-6-phosphate 1-dehydrogenase	YP_001835777	0.58	0.83	200	7.5
		galactose-6-phosphate isomerase	YP_001835822	1.23	1.87	263	22.7
		D-fructose-6-phosphate amidotransferase	YP_001834993	0.90	1.65	263	9.8
		transketolase	YP_001836712	1.30	1.67	240	6.5
		UTP-glucose-1-phosphate uridylyltransferase	YP_001836775	1.34	1.14	221	17.4
		galactose-6-phosphate isomerase	YP_001835823	1.60	1.77	137	26.3
		PTS system, IIC component	YP_001834780	0.90	2.38	58	10.3
	Other Metabolic	acyl carrier protein	YP_001835133	0.54	0.54	1620	40.5
		acetyl-CoA carboxylase alpha subunit	YP_001835142	0.76	1.22	201	15.7
		aminopeptidase N	YP_001835463	0.84	1.08	269	8.7
		UDP-N-acetylglucosamine 1-carboxyvinyltransferase	YP_001836648	1.17	0.77	143	15.5
		UDP-N-acetylmuramoyl-L-alanyl-D-glutamate synthetase	YP_001835360	1.14	1.26	98	9.6
		purine nucleoside phosphorylase	YP_001835495	1.16	1.19	77	14.8
		ATP synthase subunit B	YP_001836213	1.45	1.20	120	7.9
		enoyl-(acyl-carrier-protein) reductase	YP_001835134	1.60	1.23	78	11.7
		bifunctional GMP synthase/glutamine amidotransferase protein	YP_001836150	1.25	1.94	43	11.5
		catabolite control protein A	YP_001836682	0.78	1.37	22	5.2
		glycerol kinase	YP_001836871	0.90	3.54	195	5.6
		acetyl-CoA carboxylase beta subunit	YP_001835141	1.44	2.27	53	14.6
	Virulence / Competence / Stress	universal stress protein	YP_001836679	0.84	1.86	200	31
		pneumolysin	YP_001836615	1.10	1.36	170	15.5
		DNA-binding response regulator CiaR	YP_001835464	1.51	1.75	59	19.1
	Protein Biosynthesis / Alteration / Degradation	glutamyl-tRNA synthetase	YP_001836753	0.76	1.03	210	14.8
		ATP-dependent protease ATP-binding subunit	YP_001836271	0.54	0.89	195	9.5
		asparaginyl-tRNA synthetase	YP_001836245	0.58	1.15	50	11.4
		aminopeptidase C	YP_001835009	0.79	1.21	164	10.6
		seryl-tRNA synthetase	YP_001835127	1.11	1.14	135	8.8
		prolyl-tRNA synthetase	YP_001834991	0.83	1.34	151	13.8
		peptidase T	YP_001835701	1.07	1.51	63	10.6
		peptidylprolyl isomerase	YP_001835673	1.22	1.60	46	8
	Transport	oligopeptide ABC transporter, ATP-binding protein AmiE	YP_001836579	0.47	0.62	72	10.1
		oligopeptide ABC transporter, ATP-binding protein AmiF	YP_001836578	0.87	1.56	54	10
		ABC transporter, ATP-binding protein	YP_001835529	1.69	2.26	164	16.3
	Miscellaneous	FeS assembly protein SufB	YP_001835533	0.60	1.13	47	11.5
		lipoprotein	YP_001834870	0.81	1.10	95	16.2
		cell wall surface anchor family protein	YP_001834797	1.12	0.88	46	2.2
		thiol peroxidase	YP_001836341	0.82	1.43	129	31.4
		nitroreductase family protein	YP_001835300	1.06	1.32	94	12.4
		inositol-5-monophosphate dehydrogenase	YP_001836911	1.15	1.33	72	11.8
		GTP-binding protein EngA	YP_001836398	1.48	3.28	154	15.8
		glucose-inhibited division protein A	YP_001834840	1.37	4.01	64	8
		Gfo/Idh/MocA family oxidoreductase	YP_001836376	1.28	2.86	42	7.6
	Unknown	hypothetical protein SPCG_1897	YP_001836614	0.79	0.66	401	11.8
		hypothetical protein SPCG_0966	YP_001835683	0.83	0.94	343	20.9
		hypothetical protein SPCG_0759	YP_001835476	1.06	1.26	61	22.2

> 2.0	> 2 fold upregulation
1.3 - 1.99	< 2 fold upregulation
0.77 - 1.29	no change in expression
0.5 - 0.76	< 2 fold downregulation
< 0.5	> 2 fold downregulation

Figure 4. Qualitative data listing proteins identified with 2 peptide matches using comparative iTRAQ analysis of *S. pneumoniae* serotype 14 1-day and 7-day old *in vitro* biofilms to planktonic population protein expression. Comparative protein data with >1.3 and <0.77 ratios identified as having differential expression.

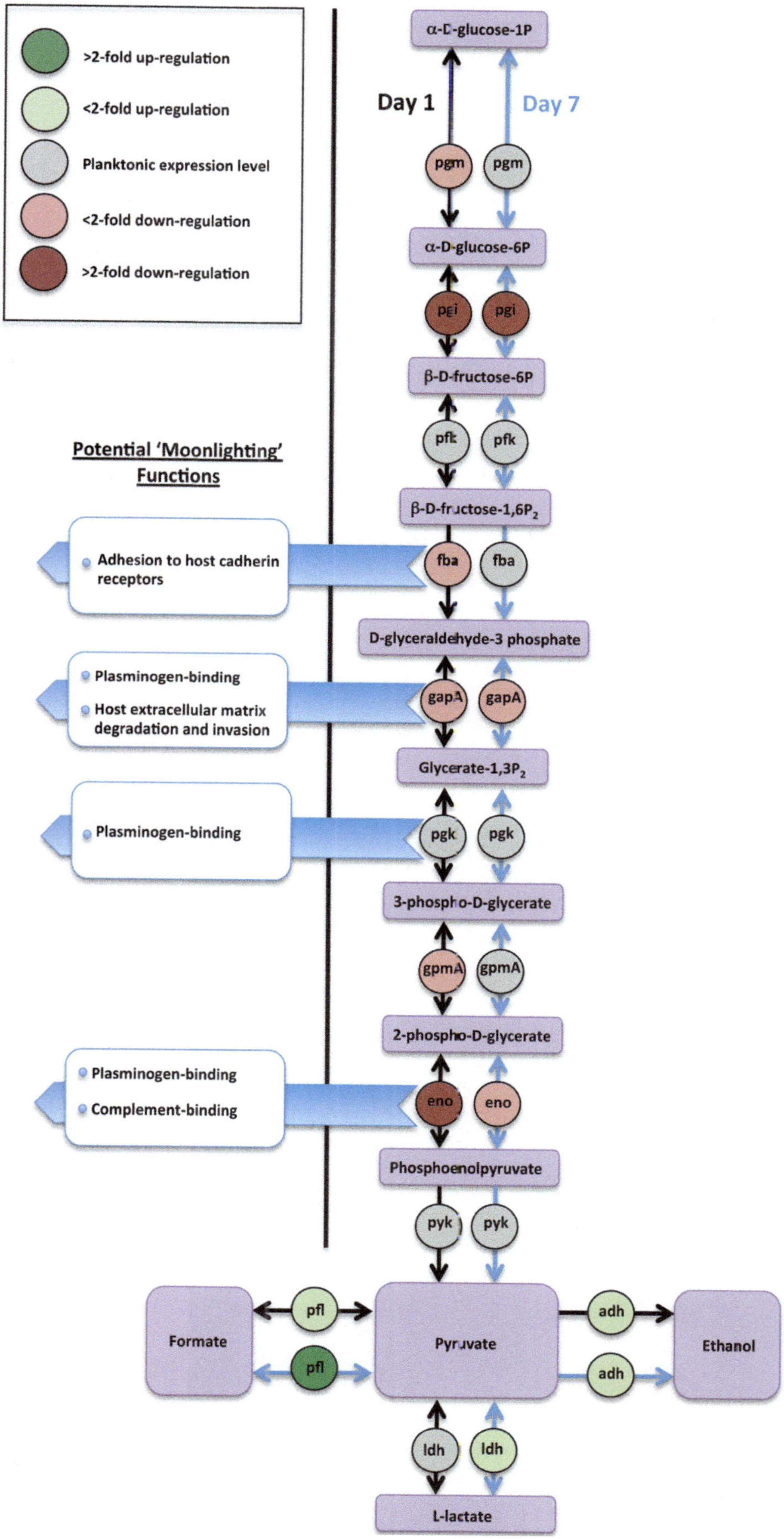

>2-fold up-regulation
<2-fold up-regulation
Planktonic expression level
<2-fold down-regulation
>2-fold down-regulation
Potential 'Moonlighting' Functions
Adhesion to host cadherin receptors
Plasminogen-binding
Host extracellular matrix degradation and invasion
Plasminogen-binding
Plasminogen-binding
Complement-binding
Day 1
Day 7
α-D-glucose-1P
pgm
α-D-glucose-6P
pgi
β-D-fructose-6P
pfk
β-D-fructose-1,6P2
fba
D-glyceraldehyde-3 phosphate
gapA
Glycerate-1,3P2
pgk
3-phospho-D-glycerate
gpmA
2-phospho-D-glycerate
eno
Phosphoenolpyruvate
pyk
Formate
pfl
Pyruvate
adh
Ethanol
ldh
L-lactate

Figure 5. Change in expression of enzymes involved in the glycolysis/gluconeogenesis pathway and pyruvate metabolism of *S. pneumoniae* serotype 14 during biofilm formation. Based on iTRAQ expression data comparing mid-exponential planktonic cultures to day 1 and 7 biofilms. Proteins with potential moonlighting capability highlighted with their putative alternative functions. Pgm - phosphoglucomutase; pgi – glucose-6-phosphate isomerase; pfk–6-phosphofructokinase; fba – fructose biphosphate aldolase; gapA – glyceraldehyde-3-phosphate dehydrogenase; pgk – phosphoglycerate kinase; gpmA – phosphoglyceromutase; eno – enolase; pyk – pyruvate kinase, ldh – lactate dehydrogenase; adh – alcohol dehydrogenase, pfl – formate acetyltransferase.

Inclusion criteria for quantitative analysis were set at ≥3 peptide matches, ≥50 protein score, ≥5% sequence coverage ($p< 0.05$). Comparative protein data with >1.3 and <0.77 ratios were identified as having differential expression. For qualitative identification the inclusion criteria were ≥2 peptide matches, 50 protein score and 10% sequence coverage.

Results

Pneumococcus grew over 7 days from unstructured cell clusters to a structurally complex biofilm comprised of metabolically active, viable cells

SEM demonstrated that day 1 biofilms were characterized by the presence of small clusters of cells interspersed amongst individual cells and diplococci (Fig. 1a) surrounded by extracellular material

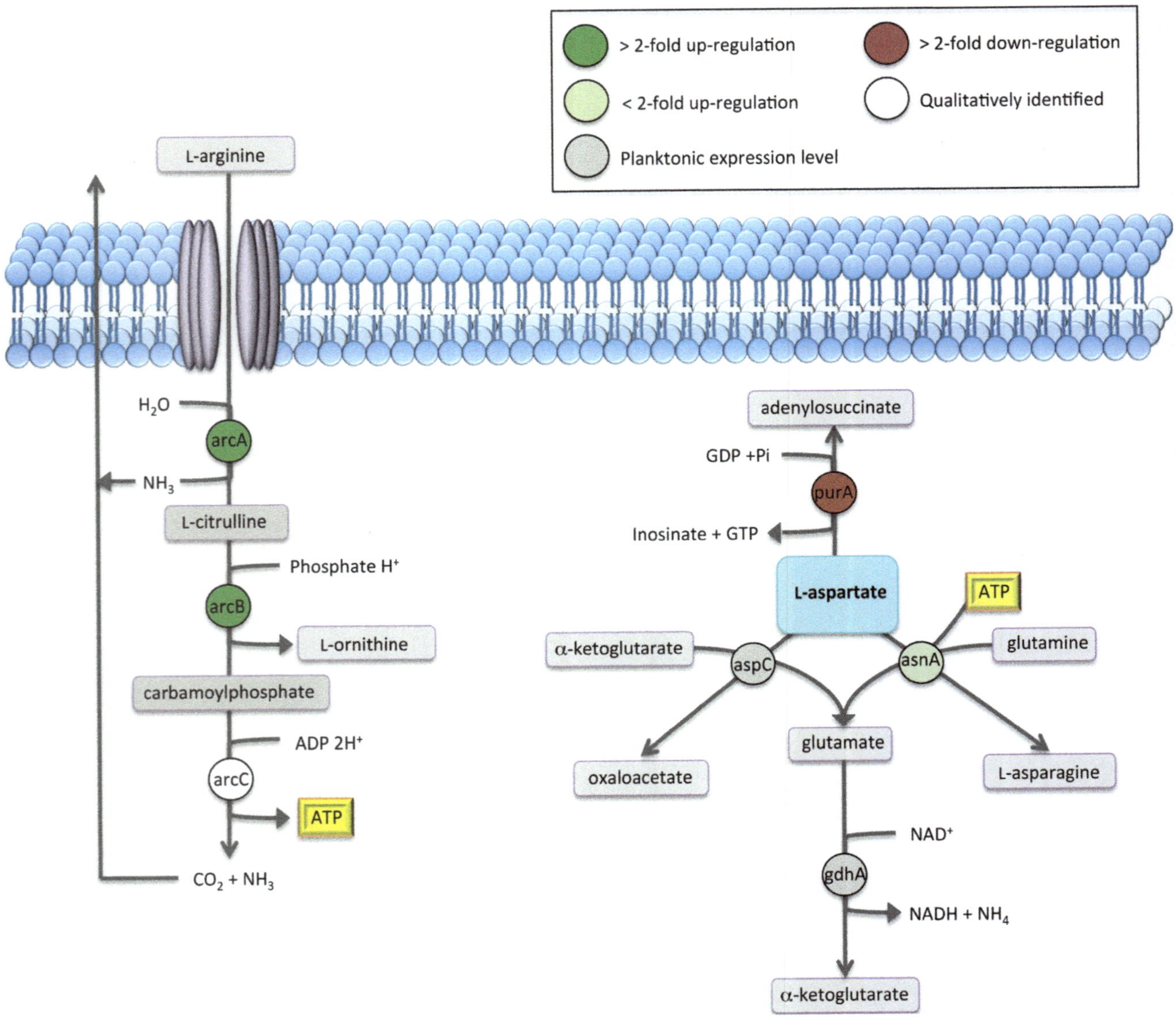

Figure 6. Change in expression of enzymes involved in amino-acid metabolism of day 7 *S. pneumoniae* serotype 14 biofilms. Based on iTRAQ expression data comparing mid-exponential planktonic cultures to day 7 biofilms. Highlighting the arginine deiminase (ADI) pathway consisting of arginine deiminase (arcA), ornithine carbamoyltransferase (arcB) and carbamate kinase (arcC), and components of L-aspartate metabolism consisting of adenylosuccinate synthetase (purA), aspartate aminotransferase (aspC), asparagine synthetase (asnA), and glutamate dehydrogenase (gdhA).

linking numerous clusters together. Adherent growth after 7 days resulted in structurally complex biofilms characterized by large towers of cells connected by chains of cells. CLSM following Live/Dead staining demonstrated few live cells in day 1 biofilms, whereas day 7 biofilms were comprised of live cells adjacent to copious amounts of extracellular matrix (Fig. 1b). Day 7 biofilms also demonstrated increased numbers of metabolically active cells following Calcein violet staining (data not shown), and a ten-fold increase in CFU/cm^2 compared with nascent (1-day) biofilms (Fig. 1c & 1d). These results agree with other reports showing that Live/Dead correlates with viability in pneumococcal biofilms [12–14].

Pneumococcus exhibited extensive proteomic modulation during biofilm development

Overall, 244 individual proteins were identified using iTRAQ profiling of which 112 met inclusion criteria for quantitative analysis (≥3 peptide matches; >5% sequence coverage and a 50+ score; P<0.05) (Fig. 2 & 3). Additionally, 62 proteins with 2 peptide matches were qualitatively identified (Fig. 4). The majority (>80%) of proteins were differentially expressed during biofilm development. Nascent (1-day) pneumococcal biofilms differed significantly from log-phase planktonic culture with 47% (53/112) of identified proteins down-regulated and only 16% (18/112) up-regulated at this time point (Fig. 2).

Adherent growth for 7 days resulted in further changes in protein expression compared with early biofilm growth with 44% of proteins (49/112) overall differentially expressed. Nearly a quarter (24%; 27/112) of the proteins differentially expressed in the nascent 1-day old biofilm returned to levels similar to log-phase planktonic growth in the late-phase 7-day old biofilm, and a further 16% (18/112) proteins were up-regulated; only 3/112 (2.7%) were down-regulated in late compared with early biofilm growth. When late biofilm growth was compared with log-phase planktonic protein expression 35/112 proteins (31%) were up-regulated, 48/112 (43%) were similarly expressed, and 29/112 (26%) were down-regulated. Nearly 20% (21/112) of proteins were quantitatively unchanged in all profiles. iTRAQ profiling suggests the transition to adherent growth was a dynamic adaptive process consistent with previous proteomic analyses demonstrating remarkable changes in expression profiles during pneumococcal biofilm development. Pneumococcal translational capacity was substantially reduced during biofilm development. Of the 33 identified proteins specific to translation, 29 (88%) were down-regulated with over half (15/29) more than two-fold, together with increased levels of a ribosomal subunit interface protein, which facilitates arrest of translation [22].

Glycolytic proteins were significantly decreased during nascent pneumococcal biofilm development

Previous proteomic analyses of *S. pneumoniae* indicate that biofilm and planktonic pneumococci substantially differ in a number of proteins involved in biosynthesis and metabolism [15,16,21], and iTRAQ profiling broadly supported these. Thirty-eight metabolic proteins were quantitatively identified and divided into 4 subsets specific to glycolysis, amino acid, carbohydrate and pyruvate metabolism, with a final group comprised of proteins with miscellaneous metabolic roles (Fig. 3). Of all metabolic proteins quantified, 37% (14/38) were decreased in the nascent biofilm, and 32% (12/38) increased. After 7 days of adherent growth however, all but 4 (11%) proteins initially down-regulated in the nascent biofilm, returned to levels similar to log-phase planktonic growth. Additionally, 40% (15/38) of metabolic proteins were increased in late biofilm growth suggesting that pneumococcal biofilms adapted to adherent growth, exhibiting protein levels observed in log-phase growth or up-regulating specific proteins associated with pyruvate, sugar and amino acid metabolism (Fig. 3).

Compared with previous pneumococcal proteomic analyses, iTRAQ identified all 9 proteins mediating the glycolytic (Embden-Meyerhof-Parnas) pathway allowing us, for the first time, to fully and quantitatively characterize changes in the pathway during biofilm formation. The majority of these proteins were decreased in nascent biofilm development consistent with other reports demonstrating down-regulation of glycolytic proteins in early TIGR4 biofilms and following epithelial cell contact [23,24]. However, after 7 days of adherent growth, half of these proteins were quantitatively commensurate with levels observed in log-phase planktonic growth (Fig. 3 & 5). All proteins associated with pyruvate metabolism were up-regulated in adherent pneumococci (Fig. 3 & 5).

Proteins involved with amino-acid metabolism and carbohydrate utilization were up-regulated in adherent pneumococci

In contrast to the decrease in glycolytic proteins in adherent bacteria, there was a significant increase in proteins involved in amino acid and sugar metabolism. Analysis of these proteins revealed that 47% (8/17) were increased in nascent biofilms, and 3/17 (18%) were decreased. By day 7 59% (10/17) were significantly increased, and only one decreased, suggesting that adherent pneumococci utilized alternative metabolic pathways during biofilm development. Six proteins associated with amino acid metabolism were identified, 3 of which were up-regulated in both nascent and 7 day biofilms (Fig. 3 & 6). Arginine deiminase (ADI) and ornithine carbamoyltransferase, enzymes which facilitate the degradation of L-arginine to generate 1 mol ATP, were up-regulated >3-fold (Fig. 3 & 6). Aspartate aminotransferase and adenyolsuccinate synthetase were down-regulated in adherent pneumococci, while glutamate dehydrogenase corresponded with planktonic levels (Fig. 3 & 6).

Our data suggest *S. pneumoniae* utilised a range of carbohydrates during biofilm development, since 90% (10/11) of the proteins associated with sugar metabolism were either up-regulated or unchanged during nascent biofilm growth relative to log-phase planktonic growth. Furthermore, (64%; 7/11) proteins were significantly increased in day 7 biofilms with the remainder consistent with planktonic *S. pneumoniae* (Fig. 3). Five PTS system proteins were identified, with mannose-specific IIAB and IID components, and a putative IIB galactitol-specific component up-regulated. The phosphoenolpyruvate-protein (PEP) PTS, an active-transport system specific for carbohydrates, was unchanged between log-phase planktonic and adherent bacteria. A fructose II ABC component, was down-regulated in nascent biofilms, but commensurate with planktonic growth at day 7 biofilms (Fig. 3).

Additionally, three ABC transporter system proteins were significantly up-regulated in biofilm pneumococci (Fig. 3). An ATP-binding protein and a sugar-binding protein from a sugar ABC transporter were both increased >2 fold in day 7 adherent *S. pneumoniae* and correspond to *msmK* and the sialic acid ABC transporter protein (*SatABC*). A multi-substrate ABC transporter (maltose/maltodextrin-binding protein of the maltose/maltodextrin ABC transporter) corresponding to *MalX* was also up-regulated in the pneumococcal biofilm profile. Tagatose 1,6-diphosphate aldolase demonstrated the highest levels of expression in the biofilm phenotype (>4-fold at day 7). Several additional sugar metabolism proteins were identified qualitatively (Fig. 4), including a PTS system IIC component and another ABC transporter.

Proteins associated with pneumococcal virulence and stress response were quantitatively different in biofilms

There was a >2-fold decrease in the level of NADH oxidase (nox) in nascent *S. pneumoniae* biofilms similar to a previous report [9], however, in day 7 biofilms, levels were similar to planktonic levels (Fig. 3). Superoxide dismutase (SOD) was unchanged between log-phase planktonic and biofilm pneumococci, indicating a requirement for protection from molecular oxygen in both conditions. Pyruvate oxidase, required for most stages of infection was also quantitatively unchanged between nascent and planktonic culture, but was >2-fold higher during day 7 biofilms. Bacterocin transport accessory protein (Bta) was also significantly higher (>2-fold) at this timepoint, suggesting increased production of bacteriocins by biofilm pneumococci. Stress proteins GroEL and DnaK, chaperone proteins facilitating appropriate protein conformation, and trigger factor were either decreased or unchanged compared with planktonic bacteria, however the pneumococcal general stress protein 24 (gls24) was increased in both nascent and late phase biofilms.

Discussion

Biofilm growth by S. *pneumoniae* is incompletely understood, but mounting evidence suggests that it plays a role in colonisation and the ability to persist in the host [3,4,7–16,23,24]. We used a static biofilm system to model conditions present in chronic otitis media where there is an absence of flow [3,13] to interrogate the hard-wired, intrinsic ability of *S. pneumoniae* to develop biofilms and obtain sufficient quantities of protein for analysis, without the presence of proteins from cell lines or animal cells to interfere with quantitative proteomic analysis.

Our data (to our knowledge the first study of *S. pneumoniae* using a quantitative gel-free approach and the first proteomic analysis of biofilms using a serotype 14 clinical strain) identify proteins whose role in biofilm development has previously been uncharacterized. iTRAQ analyses suggest that the regulation of metabolism plays a pivotal role during pneumococcal adaptation to aggregated biofilm growth. The glycolytic pathway, for which a quantitative profile of all enzyme components has been characterized for the first time, was generally down-regulated. In contrast, there was significant up-regulation of proteins involved in carbohydrate, pyruvate and arginine metabolism. These findings suggest that targeting proteins essential for pneumococcal adaption to biofilm growth, and particularly those involved in sugar and arginine metabolism, may offer new targets for treating pneumococcal-associated infections. Several proteins were conserved in both the planktonic and biofilm phenotypes, and may represent candidates for protein-based vaccines.

Protein profiling highlighted pronounced changes in the pneumococcal proteome during adaptation to the biofilm phenotype consistent with previous studies [15,24]. Proteins related to translation, transcription, virulence and glycolysis were significantly decreased, particularly in nascent biofilms. Together with experiments showing many dead and metabolically inactive cells in nascent biofilms, pneumococcus appeared to adopt a quiescent phase while it adapted to biofilm growth. These results also suggest this time point may be characterized by increased pneumococcal lysis with lysed bacteria contributing substantially to the extracellular matrix, since *LytA* mutants failed to produce biofilms with an extracellular matrix [14]. These results are also similar to reports showing that pneumococcal biofilms are suggestive of a quiescent phase that may not be strongly immunogenic [16,23]. After 7 days however, 2/3 of identified proteins quantitatively increased, with most of the remaining proteins statistically similar to log-phase planktonic growth. These data correlated with increased CFU/cm^2, and viable, metabolically-active biofilm pneumococci *in situ* suggesting that biofilms were comprised of both viable and dead cells during adaption to biofilm growth (Fig. 1).

This nascent phase may be similar to the prioritization of carbon source utilization recently observed in planktonic lag phase growth in *E. coli* [25]. Moreover, the decrease in proteins associated with cell division and translation may be important in the decreased susceptibility of biofilm pneumococci to antibiotics that target the replication machinery and protein and cell wall synthesis [13,14,23,26]. Even a small population of metabolically active cells in structurally complex biofilms would comprise a reservoir of viable pneumococci with the potential for propagation to new sites within the host [1]. Notably, other reports indicate that biofilm *S. pneumoniae* are hyperadhesive [23,27]. However, biofilm development is multifactorial with disease potential related to clinical isolate, extracellular matrix production and previous cultivation with epithelial cells [12–14,28,29].

In our in vitro study, a key difference in planktonic and biofilm proteomic profiles was the expression of metabolic proteins. This is the first study to provide quantitative analysis of the complete set of glycolytic enzymes and indicated reduced levels of the majority of these proteins during the adaptation to biofilm growth, consistent with previous data showing decreased expression of 6 glycolytic proteins [16]. However, our analysis at a later time point in biofilm development suggests that further biofilm development resulted in half of these proteins returning to levels similar to planktonic pneumococci by day 7. Interestingly, 4 glycolytic proteins represent potential "moonlighting" proteins comprising a class of anchorless cell membrane proteins with dual protein binding and metabolic functions [30,31]. Phoshopyruvate hydratase (α-enolase), glyceraldehyde-3-phosphate dehydrogenase (GapDH) and fructose bisphosphate aldolase (Fba) were significantly decreased in pneumococcal biofilms, whereas phosphoglycerate kinase (Pgk) was unchanged. Decreased expression of these proteins in *in vitro* biofilms suggests that *S. pneumoniae* modulated these bifunctional proteins upon adapting to biofilm growth and may promote a non-invasive phenotype capable of reducing an inflammatory response that facilitates persistence [1]. Given the streamlined pneumococcal genome, the ability of proteins to perform different functions depending on their localization in the cell would be a resourceful adaptation. Consistent with this possibility an SPFH domain-containing protein/band 7 family protein was decreased in nascent biofilms, but increased >2-fold in day 7 biofilms. These lipid raft-associated proteins aggregate to form membrane micro-domains that function in the recruitment of multi-protein complexes.

Evidence of increased pyruvate metabolism, and particularly lactate dehydrogenase, suggest that pneumococcal biofilm cells respond to a changing oxygen environment over time. The marked increase in arginine deiminase and ornithine carbamoyltransferase, two of three enzymes comprising the ADI system in pneumococcal biofilms, warrant further study. The ADI system has been shown to play a role in pathogenesis, energy metabolism and the protection of streptococcal biofilms from acidic conditions [32], and is widely distributed among bacteria with conservation of structure, but diversification of operons and regulation [33]. In particular, ADI has been characterized in oral streptococci where it plays a role in protection from lethal acidic conditions, production of ATP and survival. Recently, extracellular *Streptococcus intermedius* ADI was shown to inhibit *Porphyromonas gingivalis* biofilm formation by an interspecies signaling mechanism [33]. ADI in other streptococci has been shown to inhibit PBMC proliferation to be up-regulated in response to human serum [34,35].

iTRAQ quantitative data indicate that carbohydrate metabolism is especially important during pneumococcal adaptation to biofilm growth. Proteins associated with carbohydrate utilization, including phosphotransferase system (PTS) and ABC transporter proteins that determine carbohydrate substrate selection and fermentation were significantly increased in biofilm bacteria [36–39] suggesting that *S. pneumoniae* was capable of metabolising a wide range of carbohydrates during biofilm development. Several PTS systems and ABC transporters, many of which have been shown to play a role in pathogenesis [36], were upregulated in the biofilm including a mannose-specific IIAB component up-regulated in 1 day adherent bacteria and mannose-specific IID significantly up-regulated at 7-days (Fig. 3). These membrane spanning proteins, corresponding to the ManL and ManN nomenclature in D39 and TIGR 4, are widely conserved and have been shown to transport glucose and galactose, GlcNAc and GlcN, in addition to mannose [37,40]. The phosphoenolpyruvate-protein (PEP) PTS, an active-transport system specific for carbohydrates phosphorylating incoming sugar substrates during translocation across the cell membrane, was unchanged between planktonic and biofilm pneumococci suggesting this system was active in both growth conditions consistent with PEP's role in pyruvate metabolism. Inhibition of PTS systems demonstrated attenuated virulence in *Staphylococcus aureus* and *Haemophilus influenzae* [41].

Several ABC transporter system proteins were significantly up-regulated in pneumococcal biofilms. A multi-substrate ABC transporter (maltose/maltodextrin-binding protein of the maltose/maltodextrin ABC transporter) corresponding to *MalX* was identified along with several additional sugar metabolism proteins. Pneumococcal ABC transporters may utilise multiple carbohydrates as nutrient substrates [36,37,40], since free carbohydrates are scarce in the human nasopharynx whereas complex glycans are abundant. Specifically, an ATP-binding protein corresponding to *msmK* was increased over two-fold in day 7 biofilms. MsmK acts as a common ATPase for multiple carbohydrate transporters in pneumococcus and mutation of *msmK* reduced colonization in a mouse model [38]. Similarly, in a screen of pneumococcal biofilm mutants, several PTS and ABC transporters were shown to be defective in colonization [42].

Pneumococcal biofilm development also resulted in a quantitative increase in several proteins associated with a transparent phenotype, which plays an important role in colonization and adherence [5,43]. Tagatose-1, 6-diphosphate aldolase, ABC transporter sugar-binding protein, ADI and ornithine carbamoyl-transferase were similarly upregulated in a transcriptional profiling study of *S. pneumoniae* comparing transparent and opaque colony variants [43] and have been linked to pneumococcal colonisation of the airway in several studies [9,16,44].

Pneumococcal virulence proteins associated with infection, persistence and competitive fitness were mostly downregulated during biofilm development. NADH oxidase (*nox*) decreased over two-fold in nascent biofilms. NADH oxidase regulates competence, virulence, and pneumococcal persistence through its actions as an oxygen sensor, in detoxifying oxygen, and in improving the efficiency of glucose catabolism [45] and plays an important role in pneumococcal infection in animal models of both pneumonia and otitis media [46]. However, pyruvate oxidase (*spxB*) was significantly upregulated in day 7 biofilms. SpxB appears to be important in carbohydrate selection and capsule production [47], and in reducing competition by other organisms, since pneumococcal extracellular H_2O_2 inhibits the growth of competing bacteria [48,49]. Recently pyruvate oxidase was shown to be essential for pneumococcal aggregate formation and biofilm growth [27] and mutation of *spxB* resulted in changes in colonization and transparent/opaque phenotype expression [47,50]. Pneumococcal biofilm growth resulted in increased levels of bacteriocin transport accessory protein (Bta). Since small antimicrobial peptides play a role in pneumococcal competition with closely related bacterial species [51] biofilm growth may enhance *S. pneumoniae* fitness [1].

Finally quantitative proteomic analyses of pneumococcal biofilms may be useful in identifying cross-protective protein antigen candidates that target colonization across serotypes. The iTRAQ dataset highlighted both proteins that were modulated between planktonic and biofilm modes of growth, and those stably expressed in both conditions and may be useful in screening potential antigens for mucosal or transcutaneously administered vaccines [52,53]. Several prominent proteins quantified in iTRAQ proteomic profiles appear to be immunogenic. For example, nearly two-thirds of the 17 immunogenic proteins identified by MALDI-TOF analysis are present in iTRAQ biofilm profiles, including 2 of 2 proteins that were highly immunogenic across all age groups: pyruvate oxidase and α-enolase [54]. Fba and GapDH were other strongly immunogenic glycolytic enzymes and 6PGD, a surface-associated protein with putative adhesin-like activity, was significantly increased in day 7 pneumococcal biofilms. Present in multiple pneumococcal strains and highly immunogenic in adults, 6PGD was also immunoprotective in a mouse model of infection [55]. Although NADH oxidase (nox) was recently evaluated as a potential vaccine candidate that resulted in decreased pneumococcal colonisation in the nasopharynx and lungs in mice [56] iTRAQ data suggest this protein was downregulated in nascent biofilms. The maltose/maltodextrin ABC transporter binding protein MalX, however, also identified as a vaccine candidate, was quantitatively unchanged in nascent biofilms and increased in day 7 biofilm pneumococci [40]. Notably, this protein has also been identified through proteomic analysis to be a widely conserved pneumococcal antigen capable of eliciting mucosal T_H17 responses that abrogated colonization [53].

S. pneumoniae metabolic regulation appears to be of key importance in biofilm growth. Although marked downregulation of glycolysis, was observed, alternative metabolic pathways likely enhance the ability of pneumococcus to adapt to biofilm growth and facilitate survival. Such metabolic proteins may provide novel cross protective target antigens underlying pneumococcal persistence in multiple serotypes.

Acknowledgments

We would like to thank David Johnston and the Southampton Biomedical Imaging Unit for their support in imaging the biofilms. We acknowledge the PRIDE team for the deposition of our mass spectrometry data to the ProteomeXchange Consortium (http://proteomecentral.proteomexchange.org) via the PRIDE partner repository with the dataset identifier PXD001182.

Author Contributions

Conceived and designed the experiments: RNA PS JJ SCC SNF LHS JW. Performed the experiments: RNA. Analyzed the data: RNA LHS PS. Contributed reagents/materials/analysis tools: PS JJ SCC SNF LHS JW. Contributed to the writing of the manuscript: RNA PS JJ SCC SNF LHS JW.

References

1. Weiser JN (2010) The pneumococcus: why a commensal misbehaves. J Mol Med (Berl) 88: 97–102.
2. Kadioglu A, Weiser JN, Paton JC, Andrew PW (2008) The role of Streptococcus pneumoniae virulence factors in host respiratory colonization and disease. Nat Rev Microbiol 6: 288–301.
3. Hall-Stoodley L, Hu FZ, Gieseke A, Nistico L, Nguyen D, et al. (2006) Direct detection of bacterial biofilms on the middle-ear mucosa of children with chronic otitis media. JAMA 296: 202–211.
4. Reid SD, Hong W, Dew KE, Winn DR, Pang B, et al. (2009) Streptococcus pneumoniae forms surface-attached communities in the middle ear of experimentally infected chinchillas. J Infect Dis 199: 786–794.
5. Weiser JN, Markiewicz Z, Tuomanen EI, Wani JH (1996) Relationship between phase variation in colony morphology, intrastrain variation in cell wall physiology, and nasopharyngeal colonization by Streptococcus pneumoniae. Infect Immun 64: 2240–2245.
6. Hoa M, Tomovic S, Nistico L, Hall-Stoodley L, Stoodley P, et al. (2009) Identification of adenoid biofilms with middle ear pathogens in otitis-prone children utilizing SEM and FISH. Int J Pediatr Otorhinolaryngol 73: 1242–1248.
7. Nistico L, Kreft R, Gieseke A, Coticchia JM, Burrows A, et al. (2011) Adenoid reservoir for pathogenic biofilm bacteria. J Clin Microbiol 49: 1411–1420.
8. Camilli R, Pantosti A, Baldassarri L (2011) Contribution of serotype and genetic background to biofilm formation by Streptococcus pneumoniae. Eur J Clin Microbiol Infect Dis 30: 97–102.
9. Oggioni MR, Trappetti C, Kadioglu A, Cassone M, Iannelli F, et al. (2006) Switch from planktonic to sessile life: a major event in pneumococcal pathogenesis. Mol Microbiol 61: 1196–1210.
10. Marks LR, Reddinger RM, Hakansson AP (2012) High levels of genetic recombination during nasopharyngeal carriage and biofilm formation in Streptococcus pneumoniae. MBio 3.
11. Moscoso M, Garcia E, Lopez R (2006) Biofilm formation by Streptococcus pneumoniae: role of choline, extracellular DNA, and capsular polysaccharide in microbial accretion. J Bacteriol 188: 7785–7795.
12. Allegrucci M, Sauer K (2007) Characterization of colony morphology variants isolated from Streptococcus pneumoniae biofilms. J Bacteriol 189: 2030–2038.
13. Hall-Stoodley L, Nistico L, Sambanthamoorthy K, Dice B, Nguyen D, et al. (2008) Characterization of biofilm matrix, degradation by DNase treatment and evidence of capsule downregulation in Streptococcus pneumoniae clinical isolates. BMC Microbiol 8: 173.
14. Marks LR, Parameswaran GI, Hakansson AP (2012) Pneumococcal interactions with epithelial cells are crucial for optimal biofilm formation and colonization in vitro and in vivo. Infect Immun 80: 2744–2760.
15. Allegrucci M, Hu FZ, Shen K, Hayes J, Ehrlich GD, et al. (2006) Phenotypic characterization of Streptococcus pneumoniae biofilm development. J Bacteriol 188: 2325–2335.
16. Sanchez CJ, Hurtgen BJ, Lizcano A, Shivshankar P, Cole GT, et al. (2011) Biofilm and planktonic pneumococci demonstrate disparate immunoreactivity to human convalescent sera. BMC Microbiol 11: 245.
17. Ding F, Tang P, Hsu MH, Cui P, Hu S, et al. (2009) Genome evolution driven by host adaptations results in a more virulent and antimicrobial-resistant Streptococcus pneumoniae serotype 14. BMC Genomics 10: 158.
18. Hiller NL, Janto B, Hogg JS, Boissy R, Yu S, et al. (2007) Comparative genomic analyses of seventeen Streptococcus pneumoniae strains: insights into the pneumococcal supragenome. J Bacteriol 189: 8186–8195.
19. Bae SM, Yeon SM, Kim TS, Lee KJ (2006) The effect of protein expression of Streptococcus pneumoniae by blood. J Biochem Mol Biol 39: 703–708.
20. Encheva V, Gharbia SE, Wait R, Begum S, Shah HN (2006) Comparison of extraction procedures for proteome analysis of Streptococcus pneumoniae and a basic reference map. Proteomics 6: 3306–3317.
21. Lee KJ, Bae SM, Lee MR, Yeon SM, Lee YH, et al. (2006) Proteomic analysis of growth phase-dependent proteins of Streptococcus pneumoniae. Proteomics 6: 1274–1282.
22. Agafonov DE, Kolb VA, Spirin AS (2001) Ribosome-associated protein that inhibits translation at the aminoacyl-tRNA binding stage. EMBO Rep 2: 399–402.
23. Sanchez CJ, Kumar N, Lizcano A, Shivshankar P, Dunning Hotopp JC, et al. (2011) Streptococcus pneumoniae in biofilms are unable to cause invasive disease due to altered virulence determinant production. PLoS One 6: e28738.
24. Orihuela CJ, Radin JN, Sublett JE, Gao G, Kaushal D, et al. (2004) Microarray analysis of pneumococcal gene expression during invasive disease. Infect Immun 72: 5582–5596.
25. Schultz D, Kishony R (2013) Optimization and control in bacterial Lag phase. BMC Biol 11: 120.
26. McCoy LS, Xie Y, Tor Y (2011) Antibiotics that target protein synthesis. Wiley Interdiscip Rev RNA 2: 209–232.
27. Blanchette-Cain K, Hinojosa CA, Akula Suresh Babu R, Lizcano A, Gonzalez-Juarbe N, et al. (2013) Streptococcus pneumoniae Biofilm Formation Is Strain Dependent, Multifactorial, and Associated with Reduced Invasiveness and Immunoreactivity during Colonization. MBio 4.
28. Trappetti C, van der Maten E, Amin Z, Potter AJ, Chen AY, et al. (2013) Site of isolation determines biofilm formation and virulence phenotypes of Streptococcus pneumoniae serotype 3 clinical isolates. Infect Immun 81: 505–513.
29. Hakansson AP (2014) Pneumococcal Adaptive Responses to Changing Host Environments. J Infect Dis.
30. Fulde M, Bernardo-Garcia N, Rohde M, Nachtigall N, Frank R, et al. (2013) Pneumococcal phosphoglycerate kinase interacts with plasminogen and its tissue activator. Thromb Haemost 111.
31. Chhatwal GS (2002) Anchorless adhesins and invasins of Gram-positive bacteria: a new class of virulence factors. Trends Microbiol 10: 205–208.
32. Casiano-Colon A, Marquis RE (1988) Role of the arginine deiminase system in protecting oral bacteria and an enzymatic basis for acid tolerance. Appl Environ Microbiol 54: 1318–1324.
33. Cugini C, Stephens DN, Nguyen D, Kantarci A, Davey ME (2013) Arginine deiminase inhibits Porphyromonas gingivalis surface attachment. Microbiology 159: 275–285.
34. Degnan BA, Palmer JM, Robson T, Jones CE, Fischer M, et al. (1998) Inhibition of human peripheral blood mononuclear cell proliferation by Streptococcus pyogenes cell extract is associated with arginine deiminase activity. Infect Immun 66: 3050–3058.
35. Yang Q, Zhang M, Harrington DJ, Black GW, Sutcliffe IC (2011) A proteomic investigation of Streptococcus agalactiae reveals that human serum induces the C protein beta antigen and arginine deiminase. Microbes Infect 13: 757–760.
36. Hartel T, Eylert E, Schulz C, Petruschka L, Gierok P, et al. (2012) Characterization of central carbon metabolism of Streptococcus pneumoniae by isotopologue profiling. J Biol Chem 287: 4260–4274.
37. Bidossi A, Mulas L, Decorosi F, Colomba L, Ricci S, et al. (2012) A functional genomics approach to establish the complement of carbohydrate transporters in Streptococcus pneumoniae. PLoS One 7: e33320.
38. Marion C, Aten AE, Woodiga SA, King SJ (2011) Identification of an ATPase, MsmK, which energizes multiple carbohydrate ABC transporters in Streptococcus pneumoniae. Infect Immun 79: 4193–4200.
39. Hardy GG, Magee AD, Ventura CL, Caimano MJ, Yother J (2001) Essential role for cellular phosphoglucomutase in virulence of type 3 Streptococcus pneumoniae. Infect Immun 69: 2309–2317.
40. Buckwalter CM, King SJ (2012) Pneumococcal carbohydrate transport: food for thought. Trends Microbiol 20: 517–522.
41. Kok M, Bron G, Erni B, Mukhija S (2003) Effect of enzyme I of the bacterial phosphoenolpyruvate: sugar phosphotransferase system (PTS) on virulence in a murine model. Microbiology 149: 2645–2652.
42. Munoz-Elias EJ, Marcano J, Camilli A (2008) Isolation of Streptococcus pneumoniae biofilm mutants and their characterization during nasopharyngeal colonization. Infect Immun 76: 5049–5061.
43. King SJ, Hippe KR, Gould JM, Bae D, Peterson S, et al. (2004) Phase variable desialylation of host proteins that bind to Streptococcus pneumoniae in vivo and protect the airway. Mol Microbiol 54: 159–171.
44. Marion C, Stewart JM, Tazi MF, Burnaugh AM, Linke CM, et al. (2012) Streptococcus pneumoniae can utilize multiple sources of hyaluronic acid for growth. Infect Immun 80: 1390–1398.
45. Auzat I, Chapuy-Regaud S, Le Bras G, Dos Santos D, Ogunniyi AD, et al. (1999) The NADH oxidase of Streptococcus pneumoniae: its involvement in competence and virulence. Mol Microbiol 34: 1018–1028.
46. Yu J, Bryant AP, Marra A, Lonetto MA, Ingraham KA, et al. (2001) Characterization of the Streptococcus pneumoniae NADH oxidase that is required for infection. Microbiology 147: 431–438.
47. Carvalho SM, Farshchi Andisi V, Gradstedt H, Neef J, Kuipers OP, et al. (2013) Pyruvate oxidase influences the sugar utilization pattern and capsule production in Streptococcus pneumoniae. PLoS One 8: e68277.
48. Regev-Yochay G, Trzcinski K, Thompson CM, Malley R, Lipsitch M (2006) Interference between Streptococcus pneumoniae and Staphylococcus aureus: In vitro hydrogen peroxide-mediated killing by Streptococcus pneumoniae. J Bacteriol 188: 4996–5001.
49. Regev-Yochay G, Trzcinski K, Thompson CM, Lipsitch M, Malley R (2007) SpxB is a suicide gene of Streptococcus pneumoniae and confers a selective advantage in an in vivo competitive colonization model. J Bacteriol 189: 6532–6539.
50. Syk A, Norman M, Fernebro J, Gallotta M, Farmand S, et al. (2014) Emergence of Hypervirulent Mutants Resistant to Early Clearance During Systemic Serotype 1 Pneumococcal Infection in Mice and Humans. J Infect Dis.
51. Dawid S, Roche AM, Weiser JN (2007) The blp bacteriocins of Streptococcus pneumoniae mediate intraspecies competition both in vitro and in vivo. Infect Immun 75: 443–451.
52. Novotny LA, Clements JD, Bakaletz LO (2013) Kinetic analysis and evaluation of the mechanisms involved in the resolution of experimental nontypeable Haemophilus influenzae-induced otitis media after transcutaneous immunization. Vaccine 31: 3417–3426.
53. Moffitt KL, Gierahn TM, Lu YJ, Gouveia P, Alderson M, et al. (2011) T(H) 17-based vaccine design for prevention of Streptococcus pneumoniae colonization. Cell Host Microbe 9: 158–165.
54. Ling E, Feldman G, Portnoi M, Dagan R, Overweg K, et al. (2004) Glycolytic enzymes associated with the cell surface of Streptococcus pneumoniae are

antigenic in humans and elicit protective immune responses in the mouse. Clin Exp Immunol 138: 290–298.
55. Daniely D, Portnoi M, Shagan M, Porgador A, Givon-Lavi N, et al. (2006) Pneumococcal 6-phosphogluconate-dehydrogenase, a putative adhesin, induces protective immune response in mice. Clin Exp Immunol 144: 254–263.
56. Muchnik L, Adawi A, Ohayon A, Dotan S, Malka I, et al. (2013) NADH oxidase functions as an adhesin in Streptococcus pneumoniae and elicits a protective immune response in mice. PLoS One 8: e61128.

2

Protein Aggregation in *E. coli*: Short Term and Long Term Effects of Nutrient Density

Ulfat I. Baig[1,2], Bharati J. Bhadbhade[1], Dincy Mariyam[2], Milind G. Watve[2*]

1 Department of Microbiology, Abasaheb Garware College, Pune, Maharashtra, India, 2 Department of Biology, Indian Institute of Science Education and Research, Pune, Maharashtra, India

Abstract

During exponential growth some cells of *E. coli* undergo senescence mediated by asymmetric segregation of damaged components, particularly protein aggregates. We showed previously that functional cell division asymmetry in *E. coli* was responsive to the nutritional environment. Short term exposure as well as long term selection in low calorie environments led to greater cell division symmetry and decreased frequency of senescent cells as compared to high calorie environments. We show here that long term selection in low nutrient environment decreased protein aggregation as revealed by fluorescence microscopy and proportion of insoluble proteins. Across selection lines protein aggregation was correlated significantly positively with the RNA content, presumably indicating metabolic rate. This suggests that the effects of caloric restriction on cell division symmetry and aging in *E. coli* may work via altered protein handling mechanisms. The demonstrable effects of long term selection on protein aggregation suggest that protein aggregation is an evolvable phenomenon rather than being a passive inevitable process. The aggregated proteins progressively disappeared on facing starvation indicating degradation and recycling demonstrating that protein aggregation is a reversible process in *E. coli*.

Editor: Aamir Nazir, CSIR-Central Drug Research Institute, India

Funding: The authors would like to thank Rajiv Gandhi Science and Technology Commission's Maharashtra Gene Bank Project for the partial funding. The funders had no role in study design, data collection and analysis, decision to publish, or preparation of the manuscript.

Competing Interests: The authors have declared that no competing interests exist.

* Email: milind@iiserpune.ac.in

Introduction

Senescence in bacteria in spite of conditions favoring growth became a focus of active interest only recently. In *Caulobacter* [1], [2] there is a clear distinction between mother cell and daughter cell and the mother cell has been shown to slow down and stop reproducing after a certain number of division cycles. The division in such a species is morphologically and functionally asymmetric. For some time it was thought that there is no senescence or natural death in bacteria that undergo a morphologically symmetric division. On this background the demonstration by Stewart *et al* [3] that even in *E. coli* the cell division was functionally asymmetric and that the "old pole" cell gradually slowed down its growth and stopped dividing came as a surprise. This finding triggered both theoretical and empirical work and gave rise to many new possibilities.

Aging in *E. coli* is thought to happen because of asymmetric segregation of damaged components to the daughter cells. The damaged components were later shown to be mainly, if not exclusively, protein aggregates (PAs) [4], [5], [6]. There appear to be two possible alternative evolutionary interpretations of the asymmetric segregation phenomenon. According to one senescence is inevitable and cellular polarization is suggested to have evolved to restrict senescence to one daughter lineage [7], [8]. According to the other, there are two possible ways of handling damaged proteins. One is to repair or recycle the damaged components and the other is to dump them asymmetrically in one cell to leave the other free of damage. Mathematical models show that repair or recycle could be a better strategy in severely nutrient limited environments and asymmetric dumping may work better in rich nutrient environments [9]. These models expect that the extent of protein aggregation and/or the asymmetry of segregation would be responsive to different nutritional environments.

In the initial concept of bacterial aging the distinction of old pole and new pole was thought to be critical, with the old pole cell always retaining the aggregate and senescing [3], [4], [8]. However there is a possibility that there can be pole flipping, i. e. the damaged components may drift away from the morphologically old pole with some probability. Even if the probability is small the old pole cell may rejuvenate and maintain its growth rate over a long time [10]. Some studies have indicated that the old pole cell is not always the slower of the two in terms of growth and next division [11]. Therefore the role of morphologically old and new poles may be less important than assumed earlier, nevertheless the importance of asymmetric segregation of protein aggregates remains more robust theoretically as well as empirically.

Protein aggregation is associated with aging and related disorders across wide diversity of taxa including humans [12], [13], [14]. Because all organisms face the important challenges of protein misfolding and aggregation, strategies to cope with the problem are expected to have evolved and be conserved [15], [16]. However our understanding of the evolution of protein handling

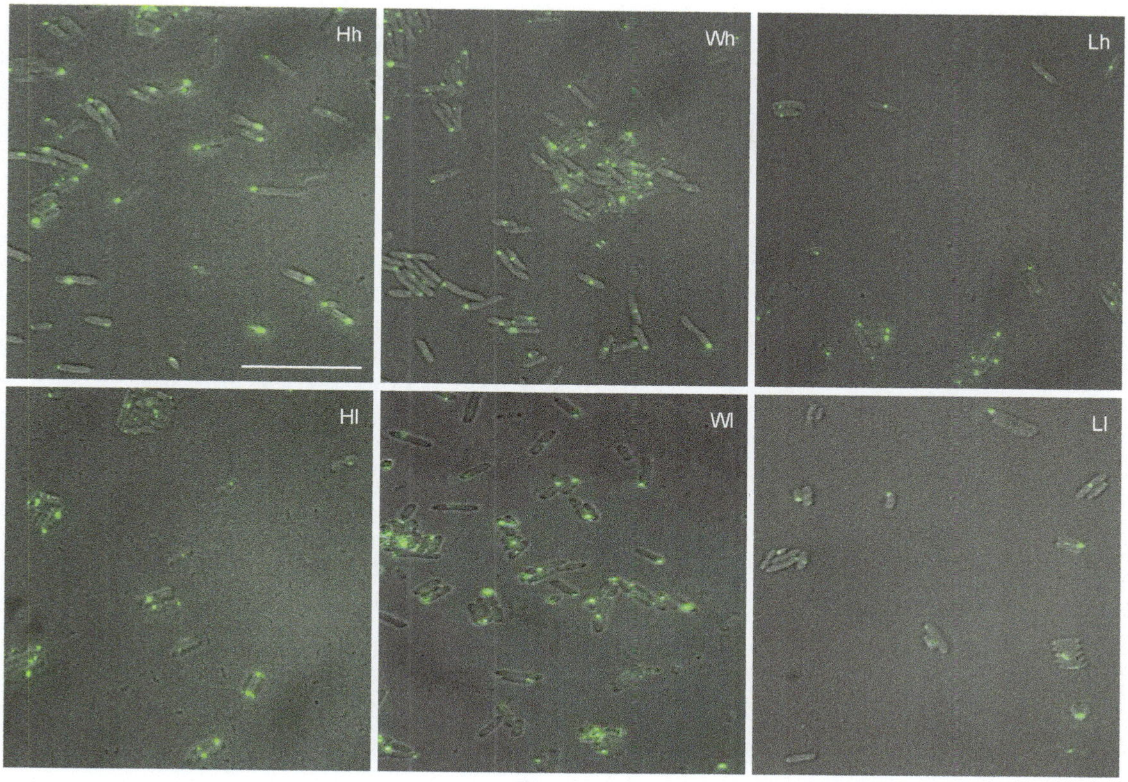

Figure 1. Superimposed DIC and fluorescence images of the three selection lines generated from MGAY in two current nutrient conditions. *W* stands for the wild type or ancestral strain. Selection was carried out for an estimated 1800 generation under constant low (*L*) and high (*H*) nutrient environments. Before microscopy all the three lines were grown in high (*h*) or low (*l*) glucose concentration for 16 hours. (e.g. *Hl:* a lineage selected under high nutrient concentration but currently being grown in low concentration). The same notation is used in all figures. All lineages showed visible protein aggregates under stress free conditions but the proportion of cells with aggregates and mean intensities differed across selection and current nutrient conditions. The scale bar represents 10 μ.

mechanisms is as yet very poor. It is interesting to ask why components of proteostasis machinery are not constitutively expressed at high levels, or why it did not evolve to perfection; whether an efficient proteostasis is impossible or whether the cost benefits of a highly efficient system are not favorable. This question cannot be context independent and it is likely that the cost-benefits of efficient protein handling systems are more beneficial under certain conditions and costly under others. One of the suggested costs in multicellular organisms is the increased potentiation of cancerous growth [17]. Alternatively a trade-off between somatic maintenance and reproduction may decide an optimum level of investment in proteostasis mechanisms in somatic cells [18]. This optimum could be much short of perfection and therefore protein aggregation is commonly observed. Understanding protein folding and mechanisms of PA formation in bacteria can serve as a simpler model for the mechanics as well as the evolutionary dynamics of proteostasis mechanisms which can ultimately help us understand how and why human conformational diseases progress.

In bacteria aggregate formation and features are influenced by various growth conditions such as temperature, pH [19], culture phase [5] and glucose/oxygen availability [20]. All proteins in inclusion bodies were initially considered as misfolded, but it has been shown that some active polypeptides are present inside aggregates [21] presumably due to accidental trapping. Misfolded proteins are presumably deposited in the nucleoid free space because of molecular crowding and these aggregates get deposited more frequently at poles [4], [6]. This leads to asymmetric inheritance of denatured proteins which are a burden for old pole cell. Cells that are less burdened with aggregates divide faster as compared to the ones with a greater burden [16].

Since the association of protein aggregates with asymmetric damage segregation, asymmetry in cell growth and division time and aging has been shown by many studies, we take these associations for granted and investigate,

i. Whether protein aggregation is an inevitable phenomenon or the cell has some control over it. Is it possible to avoid or minimize protein aggregation and whether the cell executes these mechanisms all the time or strategically under certain conditions?
ii. Whether asymmetric segregation of protein aggregation is an inevitable phenomenon or is a strategic behaviour subject to differential natural selection under different environments?
iii. Whether protein aggregates are outright waste products or whether they have any beneficial role under some conditions?

A common approach to study protein aggregation in the context of aging is to follow individual cells and their subsequent daughters for a few generations [3], [4], [6], [10]. This is done using fluorescence microscopy to reveal the protein aggregates. A possible objection to this approach is that the cells under observation are continuously or repeatedly exposed to UV light during imaging which might alter the fate of proteins [22], [23]. Since the extent of alteration in protein handling during fluorescence microscopy has not been estimated we use a cross

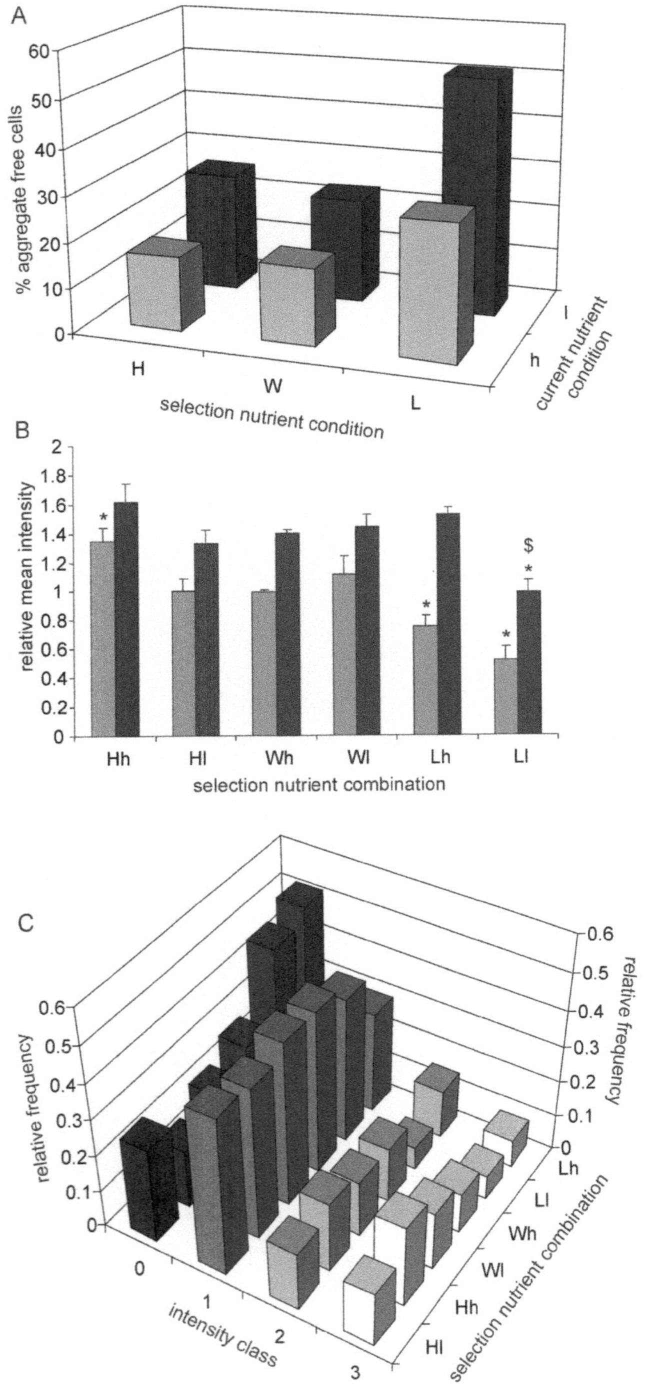

Figure 2. Effect of selection and current nutrient concentration on protein aggregation. (n = 200±20 each for data in Fig. 2A to C). A: The proportion of aggregate free cells: Both short term exposure and long term selection in low nutrient concentrations increased the proportion of aggregate free cells. (chi sq = 38.15, p<0.0001). B: The mean fluorescence intensities of cells (expressed relative to mean *Wh* including aggregate free cells) under the 6 selection nutrition combinations: The grey columns denote the average including cells without visible aggregates. The black columns denote average calculated by excluding the aggregate free cells. *denotes significant difference from Wh. $ denotes significant difference between corresponding *h* and *l* conditions (at p<0.05). C: Frequency distribution of cells with fluorescence intensity classes of aggregates: The modal class is 1 except in *Lh* and *Ll* where the modal class is 0. The proportion of cells in different intensity classes across the three selection lines differed significantly at high nutrient concentration (for *Wh, Hh* and *Lh* Chi square = 57.157, df = 6, P<0.0001) as well as low nutrient concentration (For *Wl, Hl* and *Ll* Chi square = 35.317, df = 6, P<0.0001). The effect of current nutrient concentration was not significant for *W* (Chi square = 2.35, df = 3, P = 0.503) but was significant for *H* (Chi square = 7.138, df = 3, P = 0.0676) and *L* (Chi square = 9.415, df = 3, P = 0.0243).

sectional approach where cells are not observed longitudinally but a growing population is sampled at a cross section in time. This does not reveal information of old or new poles and mother or daughter cells but inferences based on the statistical distribution of aggregates are nevertheless possible. The cross sectional data can therefore be cleaner and important despite the limitation that the inferences are necessarily statistical and indirect.

Materials and Methods

Strains used: We studied MGAY, obtained from A. Lindner (INSERM, France) which constitutively expresses yellow fluorescent protein (YFP) fused to the C terminus of Inclusion body protein (Ibp) A under the control of the endogenous chromosomal IbpA promoter. This allows visualization of protein aggregates in the cells under fluorescence microscopy [4] [24].

Selection regime

Similar to Lele *et al* [11] we grew the above strain continuously under two concentrations of glucose in a mineral medium. On agar media with high (10 mg/ml) or low (0.1 mg/ml) concentrations of glucose as the sole source of carbon and energy, cultures were serially transferred up to an estimated 1800 generations so that at the end of the prolonged sub-culturing we had three cell lineages, one each selected for either of the nutritional environments and one wild type. We denote the strain selected under high caloric environment as *H* and the one selected under low caloric environment as *L*. The current medium condition for the experiment is denoted by small letters *h* and *l* for high and low caloric conditions respectively (e.g. a lineage selected under high concentration but currently being grown in low concentration is denoted as *Hl*).

In addition we also used the lines selected by Lele *et al* [11] originating from *E. coli* KL16 and 2563 with an identical protocol for an estimated 3000 generations. These lines did not have the fluorescent tag and therefore could not be used for microscopic observation of protein aggregates but were subject to differential estimation of soluble and insoluble protein contents (see below) which is thought to be a surrogate of protein aggregation. Lele *et al* [11] used three ancestral strains of *E. coli,* but since some of the selected lines from one of them lost viability only two of the three were used for this study.

Growth conditions

All the selection lines were maintained on agar media with the respective nutrient concentration (*h* or *l*). Before preparing for microscopy or cell composition studies they were grown in respective liquid media on shake flask at 180 rpm for 16–18 hours at 37°C.

The starvation experiment was performed with two alternative starvation conditions. In one the cells were continued to be incubated in the same medium after entering the stationary phase and were observed with fluorescence microscopy at every 12 hours up to 48 hours. Alternatively the cells in early stationary phase were harvested, washed in distilled water three times and then suspended in distilled water again and incubated at 37°C.

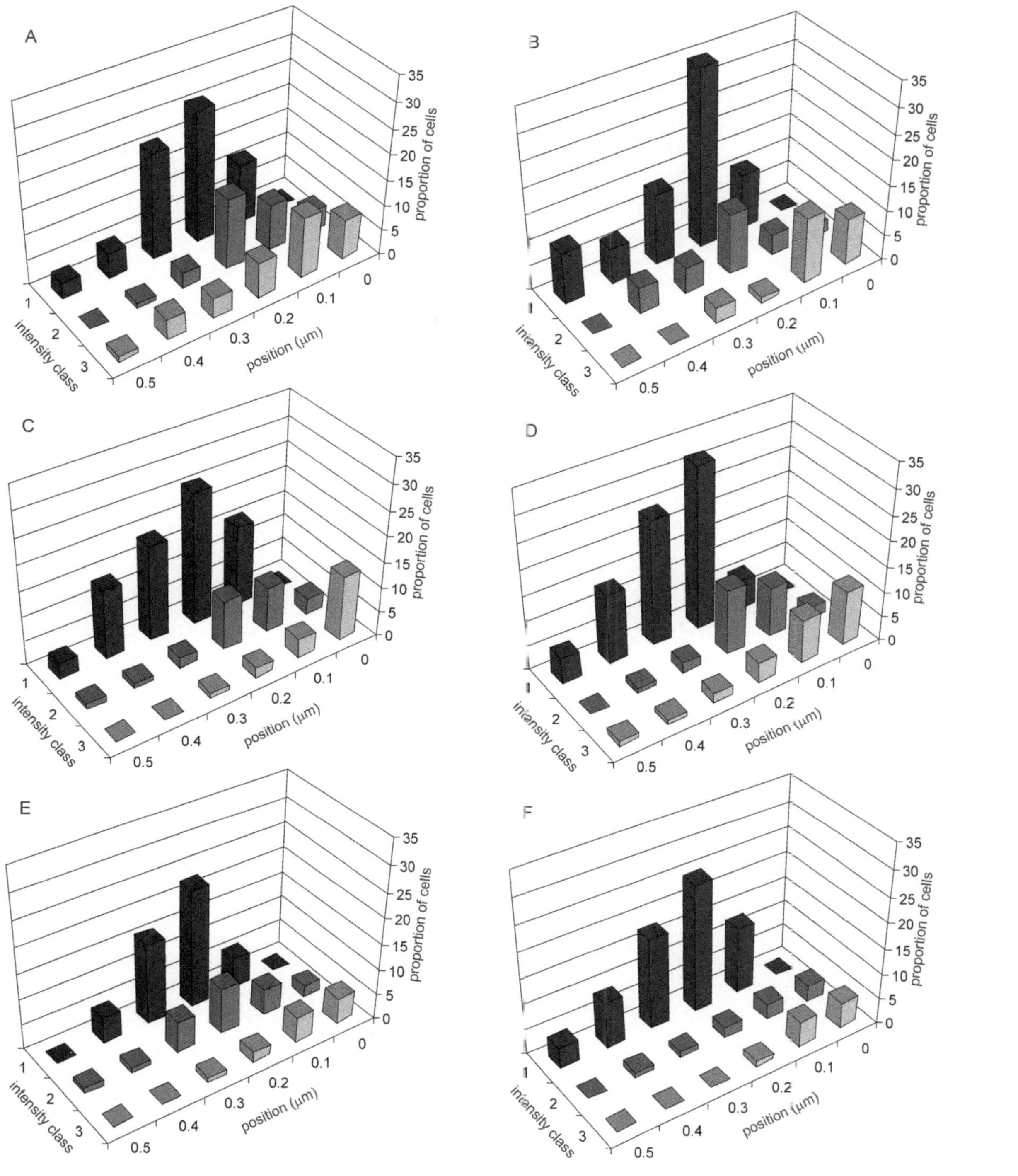

Figure 3. Spatial distribution of intensity classes under the six selection nutrition combinations. Only cells with visible aggregates are included. Positions 0 to 0.5 indicate the position of the aggregate along the cel length, 0 representing the nearest pole and 0.5 representing the centre of the cell. Note that the pattern is similar in all the lines. Aggregates in class 1 can be observed anywhere along the length with a subpolar peak. Older aggregates are more frequently polar. Similarity across selection demonstrates that although there is a difference in the rate of formation of the aggregates, once formed their further fate and spatial dynamics may be unaffected by selection or standing nutrient concentration. A–F represents *Hh*, *Hl*, *Wh*, *Wl*, *Lh* and *Ll* respectively.

Observations with fluorescence microscopy were made after every 12 hours up to 48 hours.

Fluorescence microscopy

For the selection lines derived from *E. coli* MGAY, when the cells reached an absorbance 0.2 (600 nm), they were placed on a slide that was layered with a 1.5% agarose pad and covered with a cover slip to immobilize the cells. Observations were made using

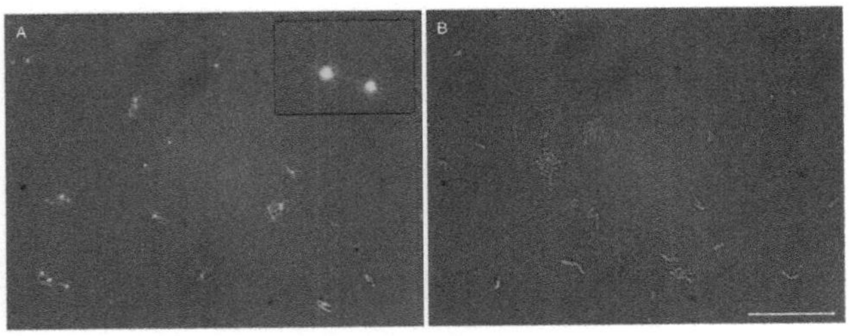

Figure 4. Comparative superimposed DIC and fluorescence images of protein aggregates in cells immediately after entering the stationary phase. (A) and after starvation for 4 days (B). The scale bar represents 20 μ. Inset: some of the larger aggregates did not disappear after prolonged starvation (30 days) but altered the morphology showing diffused fluorescence sometimes even exceeding the cell boundary.

differential interference contrast (DIC) and fluorescence imaging on Zeiss Axioimager M-1 upright microscope (100X/1.40 Oil DIC M27) with a color digital camera controlled by Axio software 4.8. Images both DIC/fluorescence were taken using exposure time of 30 millisecond illumination. The resulting images were processed and analyzed using Image J 1.44c software [25]. The fluorescence image alone was used for quantification of fluorescence intensity reflecting aggregate size whereas composite image formed by merging DIC and fluorescence was used to determine the location of protein aggregates in a cell.

Fluorescence intensity classification

Fluorescence given by the protein aggregates was quantified in arbitrary units using integrated fluorescence intensity corrected for the background. The aggregates were classified into four intensity classes namely 0 (cells with no visible aggregates), 1 with intensity <1500 units, 2 with intensities between 1500 and 3000 and 3 with intensities greater than 3000 units. In an experiment pertaining to the effect of starvation on PAs, classes were resolved at finer intervals with a larger sample size (500 to 800 cells per observation).

To characterize the position of protein aggregates in *E. coli* cells, we measured the distances of an aggregate from the nearest pole. The distance was normalized by the cell length. As a result the normalized positions range from 0 to 0.5 which were divided in to 5 classes. The class 0 to 0.1 represents the polar location and 0.4 to 0.5 represents central position.

Estimation of RNA and protein content of cells

The RNA and soluble and insoluble protein concentrations of cells were determined. A density of 10^7–10^8 cells/ml was used for the studies in which RNA extraction was done by hot phenol method [25], [26]. RNA quality was further checked by Agilent 2100 bioanalyzer and the RIN number for the same was 8 or above. Estimation of nucleic acids was done by Thermo scientific Nano drop 2000 [27]. For the quantification of soluble and insoluble proteins exponentially growing cells were harvested from 10-liter cultures and then soluble and insoluble fractions were extracted by Maisonneuve *et al* method [5] and estimated by bichinconic acid method [28].

Results

Low nutrient concentration reduces protein aggregation

In selection lines derived from MGAY strain fluorescent protein aggregates were visible and quantifiable in all cell lineages under both *h* and *l* conditions in the absence of heat or any other external stress (Fig. 1). The proportion of aggregate free cells showed a gradient along both current nutrient concentration and selection axes. The maximum proportion of aggregate free cells was found in *Ll* whereas the minimum was in *Hh* (Fig. 2A). The mean fluorescence intensities of the aggregates calculated including zero intensity cells (aggregate free cells) showed a similar trend where both the effects of current concentrations and selection could be seen (Fig. 2B). After removing zero intensity cells from the average calculation both the effects remained in the same direction but were weaker (Fig. 2B). When the aggregates were categorized in 4 intensity classes, the frequency distribution was positively skewed in all the lines at both nutrient concentrations with the modal frequency being at zero for *Lh* and *Ll* and at intensity class 1 for all others (Fig. 2C). The decreasing frequency of larger aggregates is expected by the dynamics of cell division and asymmetric segregation. This distribution is similar to the one expected by a Leslie matrix model of stable age structure [9]. Since in an exponentially growing population with stable age class distribution the largest frequency is that of the youngest class, maximum frequency would be in the aggregate free or smaller aggregate class. The frequency of zero aggregate class will be determined by the relative rates of cell growth and division versus protein aggregation. It appears from the frequencies (>50%) that in *W* and *H* lineages under both *h* and *l* conditions, a visible aggregate appears in a cell in the first generation itself whereas in *Lh* and *Ll* since the frequency of cells showing aggregates is less than 50%, the first generation cells are presumably all free of visible aggregates.

Separating the effects of current nutrient concentration and selection, we compared the selection lines at a given nutrient concentration and nutrient concentrations for a given selection. It can be seen (Fig. 2C) that both current nutrient concentration and long term selection altered the distribution of intensity classes significantly. The main difference was in the zero intensity class, i.e. aggregate free cells. Among the cells with aggregates the differences between selection and current nutrition conditions were marginal. This indicates that the main difference caused by nutrient concentration and selection is likely to be in the rate of formation or early development of the aggregates and the difference in the later development could be smaller or absent.

Apart from the rates of aggregate formation and growth, the fate and spatial dynamics of the aggregates does not appear to be affected by the different selection and current conditions. This is apparent in the spatial distribution of the intensity classes in cells having aggregates (Fig. 3). The patterns in this distribution are remarkably similar across groups with the possible exception of *Ll* which has very few aggregates in class 2 and 3. Because of their smaller number it is difficult to decide whether their spatial distribution is significantly different from other lineages. The picture generalizable across selection and treatments is that smaller aggregates may originate anywhere in the cell with a higher probability at sub central or sub polar region. Larger aggregates tend to be polar more frequently, but they can be seen at other positions occasionally.

With a low frequency, cells with more than one aggregate at a time could be observed. Since there is only one aggregate in majority of the cells, it is obvious that this aggregate will go to only one of the daughter cells so that the distribution of aggregated proteins will be asymmetric. Since theoretical models [9] expect more symmetric division in *l* type of environment, it may be expected that cells with more than one aggregates and distribution of these aggregates to both the daughter cells would be more common in *L* and/or *l*. However the total frequency of this

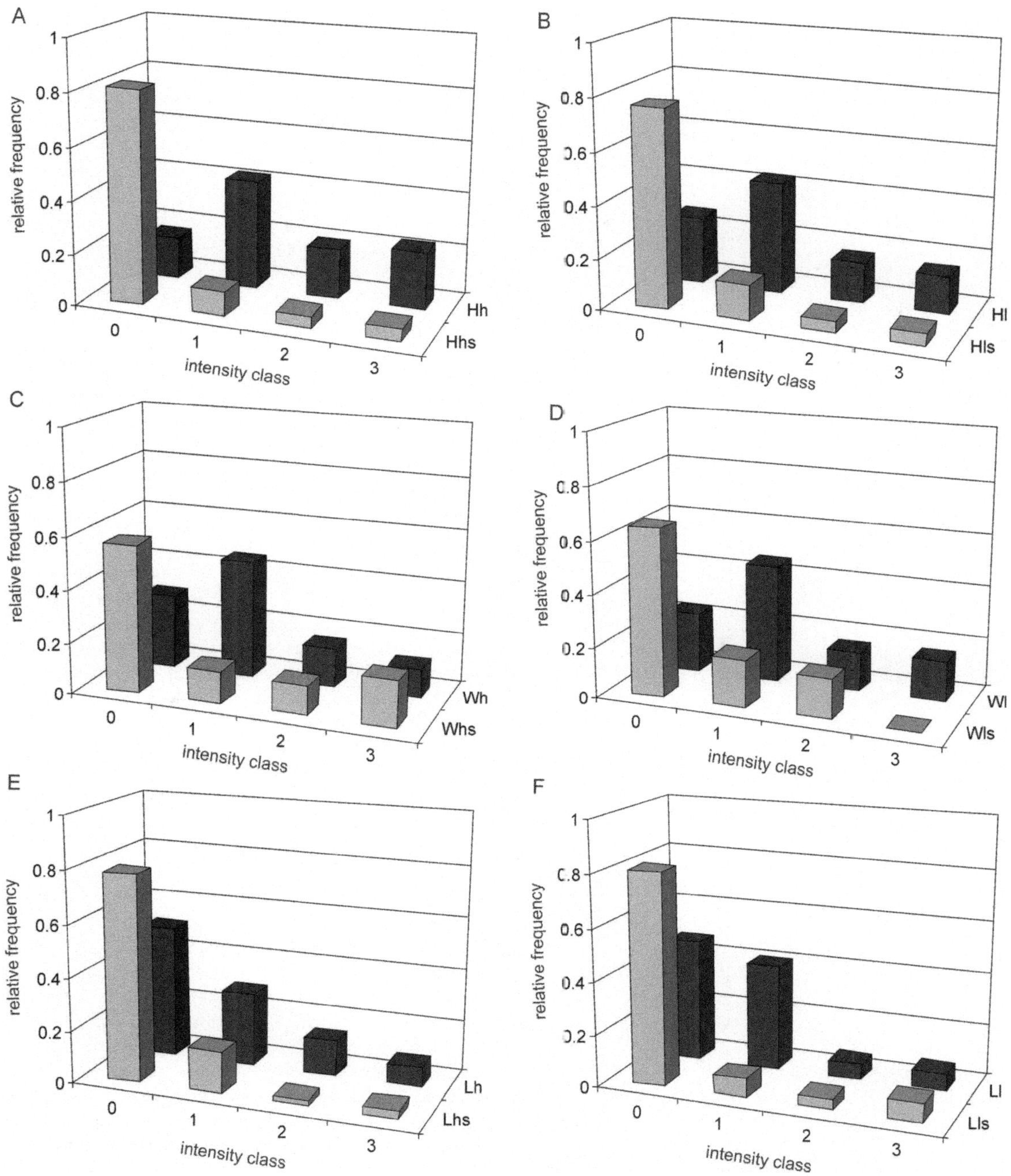

Figure 5. The distribution of intensity classes before and after starvation for 4 days. A–F represents *Hh*, *Hl*, *Wh*, *Wl*, *Lh* and *Ll* respectively. The suffix *s* stands for starvation. Although the distribution before starvation was different in all selection-concentration combinations, the distribution after starvation is similar. In all lines the aggregate free class increased significantly and class 1 reduced consistently. The trend in class 2 and 3 was less consistent.

phenomenon was so small that this could not be tested. Also at such a low frequency it is unlikely to make a difference to the population.

Protein aggregates progressively disappear on starvation

In order to test whether the aggregated proteins are degraded and recycled on facing starvation we compared aggregate distribution in cells at the end of exponential phase with the same culture continued in the stationary phase for 4, 8, 15 and 30 days. By the 4th day itself the proportion of aggregate free cells had increased in all lineages (Fig. 4). In the 8th, 15th and 30th day pictures, smaller aggregate classes were not observed, some large aggregates were still seen but with an altered morphology and diffused fluorescence around them, the fluorescence occasionally diffusing out of the cell (Fig. 4 inset). The distribution of intensity classes was substantially altered by 4 day starvation (Fig. 5) in all selection lines. The proportion of no aggregate cells increased and that in intensity class 1 decreased consistently. In *Hh* there was a 5 fold rise in the proportion of aggregate free cells and 80% decline in class 1. On the other hand since *Lh* already had over 50% cells in the aggregate free class there was a 1.54 fold rise and class 1

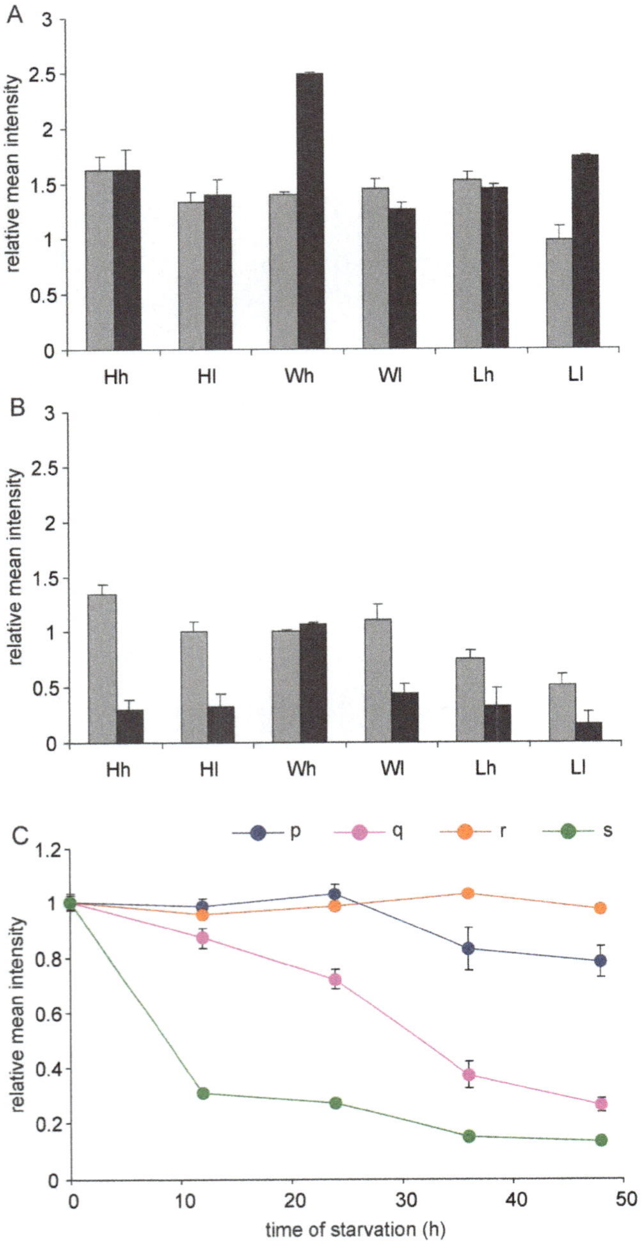

Figure 6. The effect of 4 day's starvation on visible protein aggregates. (n = 100±20 each for data in Fig. 2A to C). A and B: The mean fluorescence intensities before (grey bars) and after (black bars) starvation for 4 days. Means calculated excluding (A) and including (B) aggregate free cells. All columns are significantly reduced in B but not reduced or significantly increased in A. This suggests that all aggregate classes are not degraded proportionately. Many of the larger ones could be relatively resistant to degradation while the smaller ones degrade fast, thereby increasing the mean excluding aggregate free cells after degradation. *denotes significant difference from Wh. $ denotes significance between corresponding starved and unstarved populations (at $p<0.05$). C: The time course of decrease in the mean fluorescence intensity with starvation: The intensities are expressed relative to the mean intensity at zero hours. p: represents mean excluding aggregate free cells and q: represents including them in continued stationary phase. r: represents averages excluding aggregate free cells and s: represents including aggregate free cells in distilled water. Error bars show standard error of the mean.

declined by 46%. There was no consistent trend in classes 2 and 3. Although the trends in classes 2 and 3 are difficult to infer because of smaller frequencies, it is possible that the frequency in these classes does not decline as rapidly as class 1. The mean fluorescence intensity was reduced substantially on starvation when the aggregate free cells were included in the average calculation. However, when these cells were excluded the mean intensity did not decrease significantly or actually increased in some groups (Fig. 6). This indicates that the aggregates were not being degraded proportionately. The smaller aggregates were more likely to be degraded rapidly the larger ones being relatively resistant to degradation. The occasional appearance of large aggregates in 8, 15 and 30 day starved cells with diffusing fluorescence suggests that at least some of the large aggregates may not be degraded at all. This could be either because above a critical size aggregation might become irreversible, or cells containing very large aggregates may be dead. The dynamics of large aggregate degradation, if any, could be qualitatively different than the smaller ones.

In order to get a finer time as well size resolution we studied the fate of protein aggregates during stationary phase of *Wh* at every 12 hours using a larger sample size (500 to 800 cells each). The cells were either incubated in the same medium or washed and transferred to distilled water and both were observed under fluorescence microscopy after every 12 hours up to 48 hours. In the continued stationary phase experiment there was no significant reduction in the viable count whereas during and after washing and incubation in distilled water the viable count decreased by approximately one third. The pattern of change in the frequency distribution of fluorescence intensities with time can help us resolve between alternative hypotheses for the reduction in the mean intensities with time. Apart from the hypothesis of interest that the protein aggregates are degraded and recycled during starvation, the other possible explanations for the reduction in mean intensity and increase in the proportion of aggregate free cells are (i) the fluorescent protein has a short half life compared to the duration of the experiment and therefore fluorescence decreases with time (ii) senescent cells with large PAs die and disintegrate thereby reducing the mean fluorescence in the surviving population (iii) some replication continues in the stationary phase but there is little protein aggregation owing to low nutrient availability and therefore the proportion of aggregate free cells increases. The three alternative explanations make differential predictions that are testable by observing the changes in mean intensities and the frequency distribution of intensities with time. (i) GFP and YFP are known to be stable proteins [29] and therefore are unlikely to have disintegrated during the experimental period. We observed some large PAs retaining fluorescence up to 30 days (Fig. 4) indicating that fluorescence could remain substantially longer. Exponential reduction in fluorescence should give rise to a greater reduction in the high intensity classes than in the low intensity classes on a linear scale. This does not match with the observed pattern of decrease in the mean (Fig. 6 C) as well as frequency distributions (Fig. 7). Also we observed a significant decrease in the insoluble protein fraction (IPF) of the cells which reduced from 0.613 at the beginning of starvation to 0.267 at the end of 72 hours supporting the hypothesis that the aggregates actually degraded and half life of fluorescence is not an adequate explanation of the observed decline. (ii) If the reduction in mean is brought about by the death of cells with large PAs the right hand tail of the distribution would be curtailed first and the reduction in the mean excluding zero class should be comparable or faster than the reduction in the mean including zero class. This did not match with the observed

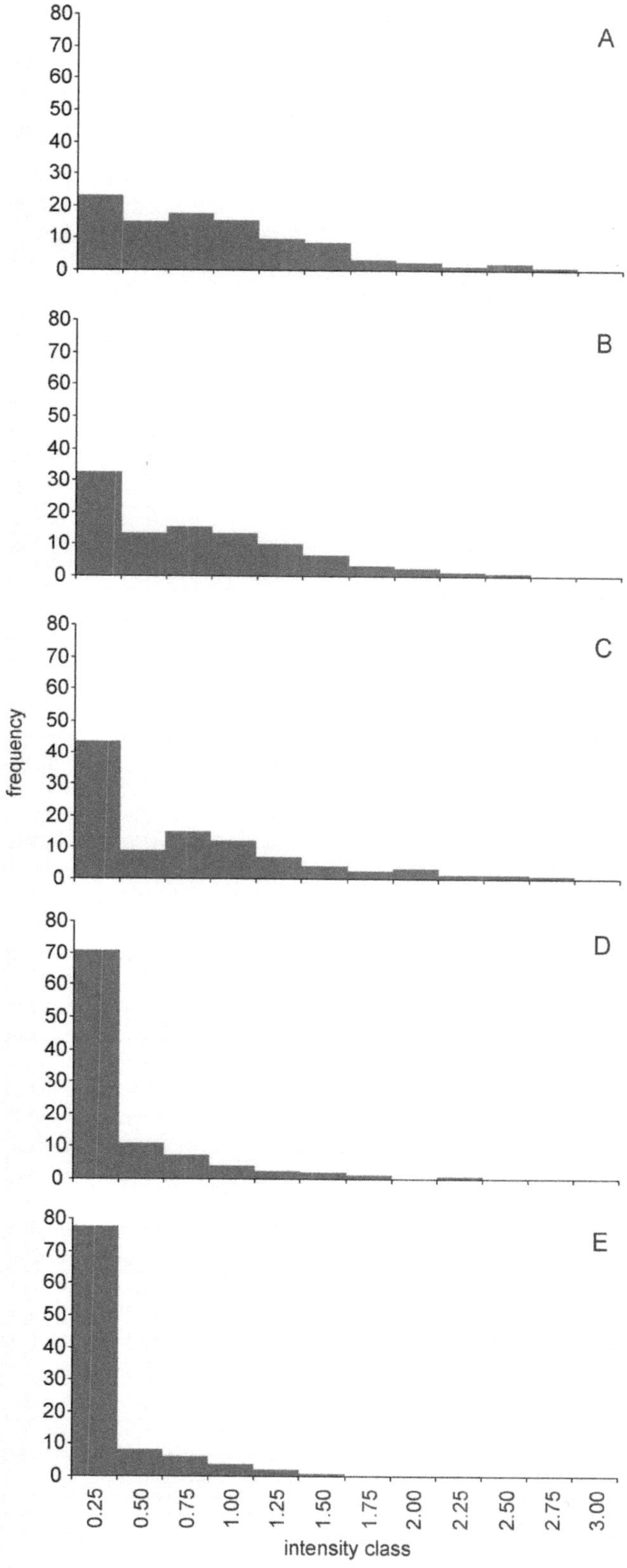

Figure 7. The progressively changing frequency distribution of fluorescence intensities on exposure to starvation. A: at the beginning of stationary phase B to E: 12, 24, 36, and 48 hours of starvation respectively. The intensities are classified at a higher resolution but the scale is matched with that in other figures. Note the transient bimodality created presumably due to faster degradation of smaller aggregate classes. Trends were similar but somewhat slower (not shown here) when cells were followed in the same medium after reaching stationary phase. The transient bimodality was consistently seen across both the conditions.

pattern (Fig. 7). Also when the cells were incubated in continued stationary phase we did not observe a decrease in the total viable count of cells in early stationary phase which makes this explanation unlikely. (iii) If new cells are born without aggregates, the frequency of zero class would increase but there would be no change in the mean calculated excluding zero class. The mean excluding aggregate free cells did not decrease in some and decreased marginally in other sets in this study which is compatible with this hypothesis. However, in order to explain the observed pattern in the decline of mean and the frequency distribution, the population should have increased between 1.5 to 1.8 fold. This is unlikely for a starving population. Also the total viable count did not increase either in continued stationary phase or in distilled water. In distilled water the viable count decreased whereas in continued stationary phase it did not. However, in spite of this difference the percent reduction in fluorescence in the two was not significantly different. Therefore we do not consider selective growth or death as adequate explanations of the observed decline.

A closer look at the pattern of change in the frequency distribution (Fig. 7) reveals that there appears a transient bimodality during early (12 and 24 hours) starvation. This is compatible with our earlier suggestion that the smaller aggregates disappeared at a faster rate than the rate at which the larger ones reduced their size.

Correlation between visible protein aggregates, estimates of insoluble protein fraction and RNA content of cells

The standing RNA content (Fig. 8A) of a cell is a direct indication of metabolic activity and therefore we expect the RNA levels to be different under different nutritional conditions and under differential selection. For all the three strains and in each of the three replicates we use the *Wh* as a reference and express the chemistry of all other lineages relative to it. Therefore throughout the results *Wh* has a mean of unity and variance of zero. Results show that the total RNA content of cells was substantially different. The RNA content of *l* was lower than that of *h* in 8 out of the 9 pairs. Selection had larger and significant effects. In all the three strains *L* selected lines had significantly lower RNA content under both *h* and *l* conditions as compared to *W* and *H* lines. The RNA contents were not consistently affected by *H* selection across all the three strains.

The total protein contents (Fig. 8B) of the cells also followed a similar pattern. In 7 out of the 9 pairs *l* conditions showed lower cell protein content than *h* conditions. Selection played a greater role, particularly *L* having a substantial reduction effect in all the three strains. More dramatic were the differences in soluble versus insoluble fractions in cell proteins (Fig. 8C). There was no consistent pattern across *l* and *h* conditions but *L* selection altered the ratio of insoluble to soluble proteins by almost an order of magnitude. Protein aggregates contribute to the insoluble fraction and therefore the two findings that *L* selection reduces visible protein aggregates and also reduces the fraction of insoluble proteins are mutually supportive. In MGAY there was good correlation between the mean fluorescence intensity of protein aggregates and IPF (Fig. 9). Therefore we can justifiably use IPF as a surrogate for protein aggregation in the other selection lines too. The total protein as well as insoluble protein fraction correlated

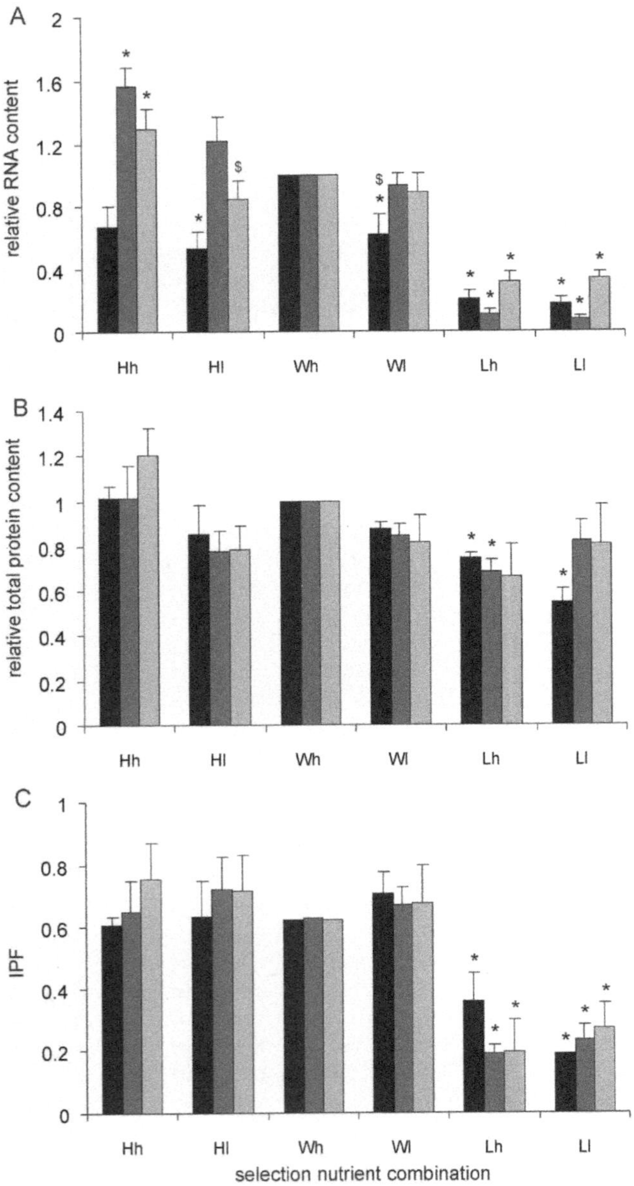

Figure 8. RNA and protein contents of cells. (A) Total RNA (ng/μl) (B) Total protein (μg/ml) and (C) The fraction of insoluble protein (IPF) content of the selection lines under high and low current nutrition. All are expressed relative to Wh. Black columns stand for KL16, dark grey for 2563 and light grey for MGAY. RNA contents differed over an order of magnitude in the *L* selected lines. *denote significant difference from *Wh*. $ denote significant difference between corresponding *h* and *l* conditions.

significantly positively with total RNA content in the pooled data over all the strains and their selection lines (Fig. 10).

The relation between nutritional environment and functional cell division time asymmetry has been shown earlier [11]. If protein aggregates are responsible for cell division asymmetry, we expect our data to correlate with earlier results. Correlations of the insoluble fraction with cell division time asymmetry, frequency of senescent cells, growth rates and growth yields of selection lines derived from KL16 and 2563 (obtained from Lele *et al.* 2011) [11] were not significant parametrically. However when separately ranked for the two strains under the two growth conditions, ranked correlations across the three selection regimes were significantly positive between IPF and asymmetry index ($r = 0.625$, $n = 12$, $p < 0.005$) and significantly negative between IPF and growth rate ($r = -0.625$, $n = 12$, $p < 0.05$). The proportion of senescent cells in the three selection regimes pooled over h and l and ranked separately for the two strains correlated positively with IPF ($r = 0.875$, $n = 6$, $p < 0.05$). This indicates that the two strains differ substantially in the pattern of quantitative response to h and l but the direction of response to selection is similar and therefore after correcting for the strain difference and current nutritional environment, protein aggregation is significantly positively correlated with cell division asymmetry and the frequency of senescent cells and negatively with growth rate in a non-parametric correlation. The correlation with growth yield nevertheless remained insignificant even after the correction. It is observed from the combined results of Lele *et al* [11] and this work that growth yield is affected significantly by h and l conditions whereas protein aggregation and cell division symmetry by H and L selection.

Discussion

Protein aggregation has an evolvable component

We showed previously that caloric restriction in short as well as long term reduces asymmetry in cell division time in *E. coli* and thereby the frequency of cells showing the signs of aging [11]. This study primarily demonstrates a relationship between caloric restriction and protein aggregation in the species supporting the hypothesis that the effect of caloric restriction on cell division symmetry and aging might be mediated through altered proteostasis. In all the three strains studied, in the lines selected under low nutrient environments there was less protein aggregation as compared to those selected in nutrient rich environments. The effect of nutrients on protein aggregation can potentially be a direct effect of nutrient concentration on the metabolic dynamics or it could be an evolved strategy of the cell to handle proteins differently under different nutrient conditions. The long term selection experiment gives support to the latter although the former is not eliminated. The quantitative effects of selection on protein aggregation suggest that there is some evolvable genetic control over protein aggregation although chemically and thermodynamically protein misfolding and aggregation may be an inevitable process. The possible cellular control over protein aggregation appears to operate more effectively at an early stage before visible protein aggregates appear since the different selection lines differ substantially in the proportion of visibly aggregate free cells but within cells showing aggregates, the frequency distribution of intensity or the spatial distribution along the cell length is only marginally different (Fig. 3). This has important implications for the inevitability versus evolved strategy debate. There appears to be substantial strategic control on protein aggregation, but the asymmetric distribution of the aggregates to the daughter cells is invariant and not responsive to nutrient based selection.

Lele *et al.* [11] demonstrated selection for symmetry in different strains of *E. coli* independently subject to the same selection regime. Protein aggregation could not be demonstrated in these strains by using fluorescence, but we showed here that the insoluble protein fraction showed response to selection in the same direction. Further in MGAY in which both IPF and visible aggregates can be monitored, there was a good correlation between fluorescence intensity of visible aggregates and IPF. Therefore IPF can be used as a reliable surrogate for protein aggregation. This allows us to test the generality of the results

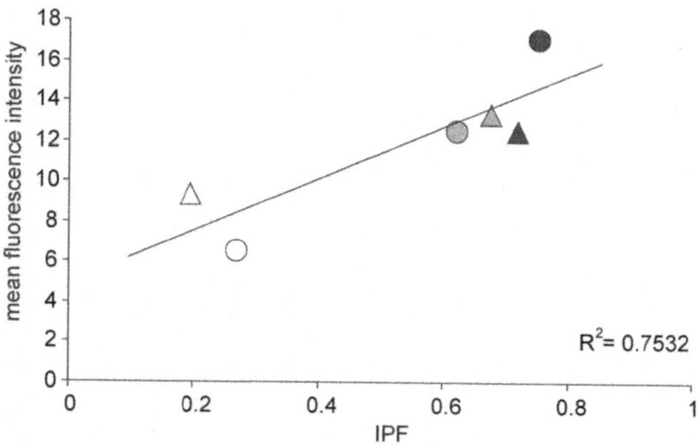

Figure 9. Correlation between mean fluorescence intensity and insoluble protein fraction in MGAY. The circles denote *h* and triangles denote *l* conditions (r = 0.868, n = 6, p<0.05). Open circles and triangles stand for *L* selection, grey for *W* and black for *H*. The correlation validates the use of IPF as a surrogate for protein aggregation in strains without a fluorescent marker.

across different strains. In all the three strains selection in low nutrient environment reduced the tendency to aggregate proteins and selection in high nutrient concentration increased it independent of the effects of standing nutrient concentration. This illustrates the generality and reproducibility of the phenomenon. Unlike some of the earlier studies that used heat or some other stress to induce protein aggregation [24], [30], [31] we observed the phenomenon here with no such external inducer of protein denaturation. Also we made a onetime cross sectional observation without repeatedly exposing the organisms to fluorescence. This is important in the light of the objection raised on the earlier studies that the observed protein aggregation and aging may be a response to the stress rather than being a natural phenomenon [23].

Selection on protein handling?

Since protein aggregation is related to aging in *E. coli* [4], [6], [32], [33], the demonstration that caloric restriction reduces protein aggregates has relevance to aging. It appears that the phenomenon of caloric restriction increasing the lifespan begins with bacteria along the hierarchy of life and here the mechanism is likely to be much simpler as compared to multicellular organisms. In bacteria this phenomenon can perhaps be more easily interpreted simultaneously at proximate and ultimate levels. Although many gaps in data still remain, it appears that when nutrient availability is good, the cells may adopt a use and throw strategy in handling proteins because it allows rapid growth [9] albeit with reduced energy efficiency. On the other hand in nutrient poor environments, energy efficiency could be more important for survival and therefore recycling of proteins is more adaptive than dumping them in the form of aggregates. It is easy to understand why such a strategy would be adaptive in a resource poor environment.

Is there any advantage of protein aggregates?

Our findings point to another dimension of protein aggregation that aggregation is not the end point. The aggregated proteins can be and seem to be reutilized when the cells face starvation. Within four days of starvation the proportion of aggregate free cells increases significantly in all selection lines. But interestingly within the cells that retain aggregates, the mean aggregate intensity is

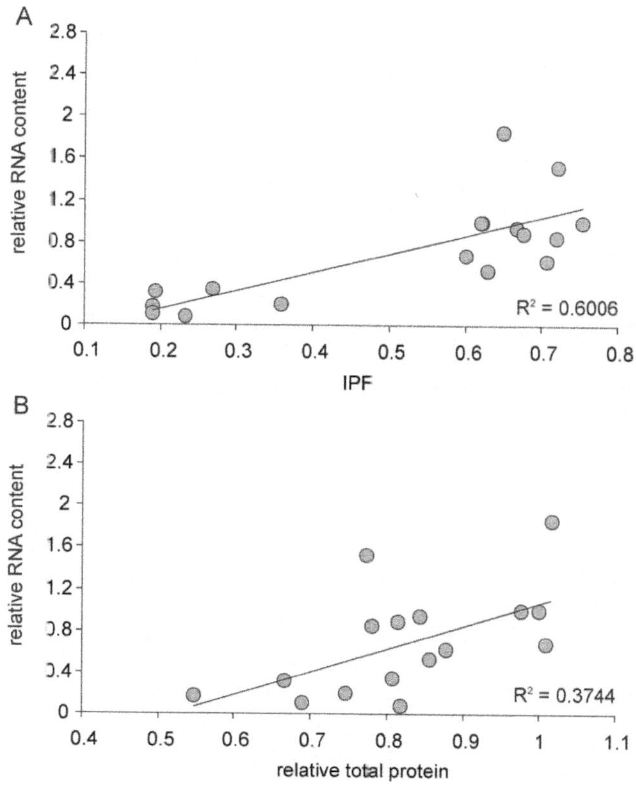

Figure 10. Correlation with total RNA content of insoluble protein fraction (r = 0.7749, n = 18, p<0.05) and the total protein (r = .612, n = 18, p<0.05) in the pooled data over all the strains and their selection lines.

only marginally decreased in some, not significantly decreased, or in fact increased in some cases. If all aggregates degraded with some rate, both the number of aggregate free cells and mean intensity of aggregates should have decreased proportionately. The more sluggish change in the later indicates that the larger aggregates do not degrade as efficiently. Some of them remain undegraded till 30 days. The rate of reutilization of aggregates is likely to be negatively associated with the size of protein aggregates. The smaller aggregates disappear faster on facing starvation and the larger ones appear to linger on disproportionately. This could be simply an effect of decrease in surface to volume ratio of the aggregates or aggregates above a threshold may interfere with some metabolic functions of the cell directly or indirectly making the cell non-viable. This has implications for the biology of aging. If protein aggregation is a mechanism of aging, the reutilization of aggregates points to the possibility of reversibility of aging. Not only that, the aggregated proteins may facilitate survival during starvation. These possibilities need to be tapped in near future.

Implications for the evolution of aging

Many issues in the evolution of aging are debated over a long time and bacterial aging is likely to provide some new insights into it [34]. The demonstration of aging in *E. coli* was taken as a convincing evidence for the universality and therefore inevitability of aging for the living world. Although many of the mechanisms leading to aging including oxidative damages, protein aggregation, telomere shortening, glycosylation etc are inevitable, the patterns of aging are not universal across species [35]. The probability of death and reproductive potential does not necessarily decrease with age in all life forms. The reutilization of protein aggregates raises a possibility that a senescent bacterial cell with a protein aggregate may have better chances of surviving starvation although this has not yet been directly demonstrated.

The reutilization of protein aggregates has a few other possible implications too. *E. coli* is not known to store fat as energy reserve which most multicellular animals do. In humans the tendency to store fat has been popularly known as "thrift" and is said to have evolved as a result of repeated exposure to starvation [36], [37], [38]. There are three close parallels between fat storage in higher animals and protein aggregates in *E. coli*. In both the cases high energy intake increases the storage of fat/proteins, under starvation both can be reutilized and excessive adiposity as well as protein aggregates appear to negatively affect at least some of the normal metabolic processes. This suggests that *E. coli* can be a good model to study the evolution of thrift. Although this was not the objective of the present study, it makes certain suggestions in this context. Continued exposure to low nutrient availability appears to increase the metabolic efficiency, compatible with the thrift hypothesis, but it does not result into excessive storage/aggregation. Instead growth at low nutrients increases growth yield and the efficiency of reutilization rather than the efficiency of storage. Thus there are similarities as well as differences with the prevalent concept of thrift in humans. This warrants further evolutionary experiments specifically designed to probe the evolution of thrift in a bacterial model.

The correlation of protein aggregates with metabolic rate is consistent with the Watve *et al.* [9] hypothesis prediction that asymmetric segregation should increase the net metabolic rate of the population although the cells with aggregates would slow down. Although we did not directly measure metabolic rate, the order of magnitude difference in the standing RNA content probably reflects the increased rate of transcription and thereby metabolism in general. It is important to realize that although we showed that the growth rate is negatively correlated with protein aggregation, this correlation is done after correcting for the main growth rate determining factor that is the currently available nutrient concentration. Without this correction the correlation was positive but non-significant. More surprising is the order of magnitude difference in the total DNA content which most probably reflects the standing chromosomal copy number and simultaneous replication forks [39]. This difference has many implications for the mutation dynamics, mutation tolerance, neutral space and evolvability of the organism. This is an intricate mathematical problem which we do not want to explore here. It is interesting to note that there is a substantial change in the overall cell chemistry indicating some fundamental alterations in metabolic regulation as a result of selection. The cells evolve to become more dilute in terms of protein and nucleic acid concentrations when they evolve in dilute environments.

There are likely to be other direct or indirect benefits of aging like processes in bacteria. Growth arrest accompanying protein aggregation is suggested to give temporary resistance to antibiotics [40]. For accruing this benefit too, the aging process needs to be reversible in bacteria. If such cells have a greater probability of surviving antibiosis or prokaryotic predation [41] and they can reutilize the protein aggregates, the population has a better chance of rejuvenating. Even if a senescent cell with large protein aggregate may have lost its ability to rejuvenate, the stored proteins may be available as a nutrient source to neighboring cells after the death and degeneration of the senescent cell. Thus there can be multiple possible advantages of senescence in bacteria which makes a good case for senescence being an adaptively evolved phenomenon rather than being an inevitable fate. Also since some of the advantages such as better chances of surviving starvation are possible individual advantages and others such as better growth rate are population advantages, both programmed and non-programmed aging are likely to be operative in bacteria. Alternatively the programmed versus non-programmed aging debate may in fact be irrelevant to bacteria [34].

Acknowledgments

The authors would like to extend their most sincere appreciation to Dr. Neelesh Dahanukar who worked with us for final set up of figures. Special mention ought to be made of Mr. Vijay Vital for helping with imaging and Ms. Uttara Lele for sharing the data.

Author Contributions

Conceived and designed the experiments: UB MW. Performed the experiments: UB DM. Analyzed the data: UB DM MW BB. Contributed to the writing of the manuscript: UB MW.

References

1. Ackermann M, Stearns S, Jenal U (2003) Senescence in a bacterium with asymmetric division. Science 300(5627): 1920–1920. doi:10.1126/science.1083532.
2. Ackermann M, Chao L, Bergstrom C, Doebeli M (2007) On the evolutionary origin of aging. Aging Cell 6(2): 235–244. doi:10.1111/j.1474-9726.2007.00281.x.
3. Stewart E, Madden R, Gregory P, Taddei F (2005) Aging and death in an organism that reproduces by morphologically symmetric division. PLoS Biol 3(2): 295–300. doi:10.1371/journal.pbio.0030045.
4. Linder A, Madden R, Demarez A, Stewart E, Taddei F (2008) Asymmetric segregation of proteins aggregates associated with cellular aging and rejuvenation. Proc Natl Acad Sci USA 105(8): 3076–3081. doi:10.1073/pnas.0708931105.
5. Maisonneuve E, Ezraty B, Dukan S (2008) Protein aggregates: an aging factor involved in cell death. Jour of Bact 190(18): 6070–6075. doi:10.1128/JB.00736-08.
6. Coquel A, Jacob J, Primet M, Demarez A, Dimiccoli M, et al. (2013) Localization of protein aggregation in *Escherichia coli* is governed by diffusion and nucleoid macromolecular crowding effect. PLoS Comp Biol 9(4): e1003038. doi:10.1371/journal.pcbi.1003038.
7. Macara I, Mili S (2008) Polarity and differential inheritance–universal attributes of life? Cell 135(5): 801–812. doi:10.1016/j.cell.2008.11.006.
8. Liu B, Larsson L, Caballero A, Hao X, Öling D, et al. (2010) The polarisome is required for segregation and retrograde transport of protein aggregates. Cell 140(2): 257–267. doi:10.1016/j.cell.2009.12.031.
9. Watve M, Parab S, Jogdand P, Keni S (2006) Aging may be a conditional strategic choice and not an inevitable outcome for bacteria. Proc Natl Acad Sci USA 103(40): 14831–14835. doi:10.1073/pnas.0606499103.
10. Wang P, Robert L, Pelletier J, Dang W, Taddei F, et al. (2010) Robust growth of *Escherichia coli*. Curr Biol 20(12): 1099–1103. doi:10.1016/j.cub.2010.04.045.
11. Lele U, Baig U, Watve M (2011) Phenotypic plasticity and effects of selection on cell division symmetry in *Escherichia coli*. PLoS ONE 6(1): e14516. doi:10.1371/journal.pone.0014516.
12. Mullins R, Russell S, Anderson D, Hageman G (2000) Drusen associated with aging and age-related macular degeneration contain proteins common to extracellular deposits associated with atherosclerosis, elastosis, amyloidosis, and dense deposit disease. The FASEB Journal 14(7): 835–846.
13. Butterfield D, Kanski J (2001) Brain protein oxidation in age-related neurodegenerative disorders that are associated with aggregated proteins. Mech Ageing Dev 122(9): 945–962. doi:10.1016/S0047-6374(01)00249-4.
14. Ballatore C, Lee V, Trojanowski J (2007) Tau-mediated neurodegeneration in Alzheimer's disease and related disorders. Nature Rev Neurosci 8(9): 663–672. doi:10.1038/nrn2194.
15. De Groot N, Ventura S (2010) Protein aggregation profile of the bacterial cytosol. PLoS ONE 5(2): e9383. doi:10.1371/journal.pone.0009383.
16. Bukau B, Weissman J, Horwich A (2006) Molecular chaperones and protein quality control. Cell 125(3): 443–451. doi:10.1016/j.cell.2006.04.014.
17. Whitesell L, Lindquist S (2005) HSP90 and the chaperoning of cancer. Nat Rev Cancer 5(10): 761–772. doi:10.1038/nrc1716.
18. Kirkwood T (1977) Evolution of ageing. Nature 270(5635): 301–304. doi:10.1038/270301a0.

19. Strandberg L, Enfors S (1991) Factors influencing inclusion body formation in the production of a fused protein in *Escherichia coli*. Appl Environ Microbiol 57(6): 1669–1674.
20. Kwiatkowska J, Matuszewska E, Kuczyńska-Wiśnik D, Laskowska E (2008) Aggregation of *Escherichia coli* proteins during stationary phase depends on glucose and oxygen availability. Res Microbiol 159: 651–657. doi:10.1016/j.resmic.2008.09.008.
21. Carrio M, Villaverde A (2005) Localization of chaperones DnaK and GroEL in bacterial inclusion bodies. J Bacteriol 187(10): 3599–3601. doi:10.1128/JB.187.10.3599-3601.2005.
22. Neumann M, Gabel D (2002) Simple method for reduction of autofluorescence in fluorescence microscopy. Journal of Histochem Cytochem 50(3): 437–439. doi:10.1177/002215540205000315.
23. Rang C, Peng A, Poon A, Chao L (2012) Ageing in *Escherichia coli* requires damage by an extrinsic agent. Microbiol 158(6): 1553–1559. doi:10.1099/mic.0.057240-0.
24. Laskowska E, Bohdanowicz J, Kuczyńska-Wiśnik D, Matuszewska E, Kędzierska S, et al (2004) Aggregation of heat-shock-denatured, endogenous proteins and distribution of the IbpA/B and Fda marker-proteins in *Escherichia coli* WT and grpE280 cells. Microbiol 150(1): 247–259. doi:10.1099/mic.0.26470-0.
25. Chomczynski P, Sacchi N (1987) Single-step method of RNA isolation by acid guanidinium thiocyanate-phenol-chloroform extraction. Anal Biochem 162(1): 156–159. doi:10.1016/0003-2697(87)90021-2.
26. Ares M (2012) Bacterial RNA isolation. Cold Spring Harbor Protocols 2012(9): 1024–1027. doi:10.1101/pdb.prot071068.
27. Gallagher S, Desjardins P (2001) Quantitation of DNA and RNA with absorption and fluorescence spectroscopy. Current Protocols in Human Genetics, A-3D. doi/10.1002/0471142727.mba03ds76/abstract.
28. Olson B, Markwell J (2007) Assays for Determination of Protein Concentration. Current Protocols in Protein Science. 48: 3.4: 3.4.1–3.4.29. doi:10.1002/0471140864.ps0304s48.
29. Lowder M, Unge A, Maraha N, Jansson J, Swiggett J, et al. (2000) Effect of starvation and the viable-but-nonculturable state on green fluorescent protein (GFP) fluorescence in GFP-tagged *Pseudomonas fluorescens* A506. Appl Environ Microb 66(8): 3160–3165. doi:10.1128/AEM.66.8.3160-3165.2000.
30. Zeng X, Bhasin S, Wu X, Lee J, Maffi S, et al. (2004) Hsp70 dynamics in vivo: effect of heat shock and protein aggregation. J Cell Sci 117(21): 4991–5000. doi:10.1242/jcs.01373.
31. Tyedmers J, Mogk A, Bukau B (2010) Cellular strategies for controlling protein aggregation. Nat Rev Mol Cell Biol 11(11): 777–788. doi:10.1038/nrm2993.
32. Winkler J, Seybert A, König L, Pruggnaller S, Haselmann U, et al. (2010) Quantitative and spatio-temporal features of protein aggregation in *Escherichia coli* and consequences on protein quality control and cellular ageing. The EMBO Journal 29(5): 910–923. doi:10.1038/emboj.2009.412.
33. Lloyd-Price J, Hakkinen A, Kandhavelu M, Marques I, Chowdhury S, et al. (2012) Asymmetric disposal of individual protein aggregates in *Escherichia coli*, One Aggregate at a Time. J of Bacteriol 194(7): 1747–1752. doi:10.1128/JB.06500-11.
34. Baig U, Bhadbhade B, Watve M (2014) Evolution of aging and death: what insights bacteria can give. Quart Rev Biol 89(3): 209–223.
35. Jones O, Scheuerlein A, Salguero-Gómez R, Camarda CG, Schaible R, et al. (2013) Diversity of ageing across the tree of life. Nature 505: 169–173. doi:10.1038/nature12789.
36. Neel J (1962) Diabetes mellitus: a "thrifty" genotype rendered detrimental by 'progress"? Am J Hum Genet 14(4): 353–362.
37. Baig U, Belsare P, Watve M, Jog M (2011) Can thrifty gene (s) or predictive fetal programming for thriftiness lead to obesity? 2011:861049 J of Obes doi:10.1155/2011/861049.
38. Watve M (2013) Doves, Diplomats, and Diabetes: A Darwinian Interpretation of Type 2 Diabetes and Related Disorders. Springer. doi:10.1007/978-01-4614-4409-1_1.
39. Fossum S, Crooke E, Skarstad K (2007) Organization of sister origins and replisomes during multifork DNA replication in *Escherichia coli*. The EMBO Journal 26(21): 4514–4522. doi:10.1038/sj.emboj.7601871.
40. Leszczynska D, Matuszewska E, Kuczynska-Wisnik D, Furmanek-Blaszk B, Laskowska E (2013) The formation of persister cells in stationary-phase cultures of *Escherichia Coli* is associated with the aggregation of endogenous proteins. PLoS ONE 8(1): e54737. doi:10.1371/journal.pone.0054737.
41. Kumbhar C, Watve M (2013) Why antibiotics: A comparative evaluation of different hypotheses for the natural role of antibiotics and an evolutionary synthesis. Nat Sci 05(04): 26–40. doi:10.4236/ns.2013.54A005.

3

Anomalies in Network Bridges Involved in Bile Acid Metabolism Predict Outcomes of Colorectal Cancer Patients

Sunjae Lee[1], **KiYoung Lee[2]**, **Seyeol Yoon[1]**, **Jae W. Lee[3]***, **Doheon Lee[1]***

1 Department of Bio and Brain Engineering, KAIST, Yuseong-gu, Daejeon, Republic of Korea, **2** Department of Medical Informatics, School of Medicine, Ajou University, Yeongtong-gu, Suwon-si, Republic of Korea, **3** Neuroscience Section, Papé Family Pediatric Research Institute, Department of Pediatrics, Oregon Health and Science University, Portland, Oregon, United States of America

Abstract

Biomarkers prognostic for colorectal cancer (CRC) would be highly desirable in clinical practice. Proteins that regulate bile acid (BA) homeostasis, by linking metabolic sensors and metabolic enzymes, also called bridge proteins, may be reliable prognostic biomarkers for CRC. Based on a devised metric, "bridgeness," we identified bridge proteins involved in the regulation of BA homeostasis and identified their prognostic potentials. The expression patterns of these bridge proteins could distinguish between normal and diseased tissues, suggesting that these proteins are associated with CRC pathogenesis. Using a supervised classification system, we found that these bridge proteins were reproducibly prognostic, with high prognostic ability compared to other known markers.

Editor: Antonio Moschetta, IRCCS Istituto Oncologico Giovanni Paolo II, Italy

Funding: This work was supported by grants from the NIH (DK064678 to J.W.L.). K.L. was supported by Basic Science Research Program through the National Research Foundation of Korea (NRF) funded by the Ministry of Education, Science and Technology (NRF-2013R1A2A2A04013317). D.L. was supported by grants of the Korea Health Technology R&D Project, Ministry of Health and Welfare, Republic of Korea (A112022); and also by the Bio-Synergy Research Project (NRF-2012M3A9C4048758) of the Ministry of Science, ICT and Future Planning through the National Research Foundation. The funders had no role in study design, data collection and analysis, decision to publish, or preparation of the manuscript.

Competing Interests: The authors have declared that no competing interests exist.

* Email: dhlee@kaist.ac.kr (DL); leejae@ohsu.edu (JWL)

These authors contributed equally to this work.

Introduction

Colorectal cancer (CRC) is the third leading cause of cancer deaths worldwide, with 746,000 persons dying from this disease in 2012 [1]. Prognostic biomarkers would improve treatment strategies through risk stratifications [2]. To date, however, few indicators of patient prognosis have been identified, impeding the selection and timing of adjuvant therapy for at-risk patients.

Prognostic biomarkers should be mechanistically relevant to disease pathogenesis. Although current data-driven expression-signatures, where gene expression patterns are highly correlated with patient prognosis, have shown substantial prognostic ability, they have not revealed underlying mechanism and thus obscured proper therapeutic interventions [3]. Biological hypotheses have provided a priori evidence of mechanistic relevance [4], but existing targeted hypothesis-driven approaches are likely to miss out numerous genes related to the biological hypotheses, requiring new alternative approaches to find many hypothesis-relevant genes.

Bile acids (BAs) are carcinogenic [5,6], with high-fat diets modulating BA homeostasis and altered levels of BAs leading to CRC pathogenesis. For example, a BA-supplemented diet in mice has been shown to induce CRCs directly, suggesting that BAs are carcinogenic [7]. However, although BAs lead to CRC pathogenesis, BAs were not utilized as practical markers. At in vivo levels, they were weak and indistinctive between patients with CRC and matched controls across studies [8] since changed BA levels by food intake are temporary and weak, thus difficult to detect. Anomalies in genes regulating cellular BA homeostasis are more of determinate factors to develop CRCs.

Proteins involved in the regulation of the homeostasis of not only BAs but all metabolites include metabolic sensors and metabolic enzymes. Metabolic sensors recognize the metabolic information during the regulation of homeostasis by detecting the levels of intracellular metabolites [9–11]. For example, the farnesoid X receptor (FXR, also known as NR1H4) detect the level of intracellular BAs, with this information utilized during the regulation of cellular BA homeostasis. Metabolic enzymes catalyze the reactions of metabolites, altering their intracellular levels. Anomalies in these sensors and enzymes would therefore alter BA homeostasis [12,13] and ultimately affect CRC pathogenesis. For example, genetic defects in BA regulating enzymes or sensor proteins were found to lead to CRC pathogenesis [14,15]. However, these genes also were not prognostic markers due to the low incidence of mutations in CRCs.

Interestingly, additional factors that are neither metabolic sensors nor enzymes were shown to modulate BA homeostasis [16]. As an alternative method of identifying reliable prognostic markers, we hypothesized that these factors may relay information on metabolic status between metabolic sensors and enzymes, functionally linking these two classes of molecules. These factors, called bridge proteins, may serve as reliable prognostic markers in patients with CRC, because anomalies in these proteins would disturb the delivery of metabolic information and the proper regulation of BA homeostasis. Current targeted approaches would be ineffective in probing relay proteins specifically between metabolic sensors and enzymes, due in large part to the lack of a method to quantify the relay degree of proteins. Systematic approaches, using information about known molecular interactions and the proteins connecting sensors and enzymes may identify and distinguish bridge proteins implicated in cellular signaling networks.

Here, we propose a network-based approach that identifies prognostic markers among proteins that play a critical role possibly linking sensors and enzymes of BA metabolism, relating to known biological hypothesis. These proteins, referred to as bridge proteins, can be assessed systematically based on information about molecular interactions recorded in several databases. To this end, we have defined a "bridgeness" metric, representing the degrees of connection between sensors and enzymes, and propose key bridge proteins as network markers for prognosis in patients with CRC. Using this "hypothesis-initiated" approach, we identified a set of markers that could better predict outcomes in patients with CRC than previously identified prognostic markers. A network-based investigation of biomarkers based on their bridgeness property may identify prognostic biomarkers implicated in cellular networks.

Results

Bridge networks and bridge proteins for bile acid metabolism

Our network-based approach identified 50 bridge proteins as reliable prognostic markers (**Table S1**). Top-ranked bridge proteins included peroxisome proliferator-activated receptor gamma, coactivator 1 alpha (PPARGC1A), hepatocyte nuclear factor 4 alpha (HNF4A), glycogen synthase kinase 3 beta (GSK3B), retinoid X receptor gamma (RXRG), caspase 8, apoptosis-related cysteine peptidase (CASP8), CREB binding protein (CBP), peroxisome proliferator-activated receptor alpha (PPARA), p53 (also known as TP53), E1A binding protein p300 (EP300) and retinoid X receptor alpha (RXRA). Notably, RXRA, forming a heterodimer with a BA sensor, FXR, participates in the regulation of BA homeostasis [17]. Also, p53 regulates BA homeostasis by linking between a BA sensor and BA enzymes, leading to abnormal BA accumulation by its defect [16,18]. Likewise, some bridge proteins that function in regulating BA homeostasis are summarized in **Table S2**, showing evidence that bridge proteins, though they are computationally selected, may participate in the regulation of BA homeostasis.

To investigate these bridge proteins, we constructed a reference network for BA metabolism (**Figure 1**), a network composed of metabolic sensors, metabolic enzymes and proteins linking sensors and enzymes. Pivotal bridge proteins that regulate given metabolic pathways were investigated by first integrating previous knowledge and interactome data. To date, 53 enzymes, including transporters, have been reported to be involved in BA metabolism and recorded in the EHMN database (**Table S3**) [19]. As detecting BAs and regulating their levels by altering downstream pathways for BAs, FXR has been found in vivo and in vitro to be a sensor for BAs [11]. Based on previous knowledge and the database, the sensor and enzymes were included in a BA bridge network. Large-scale interactome data from the databases, including HPRD [20] and TRANSFAC [21], were integrated to identify proteins that link sensors and enzymes (**Figure 1B**). We found that 10,805 genes or gene products were responsible for 110,741 interactions; of these gene products, we extracted only the sensors, enzymes and related intermediate proteins. All proteins responsible for direct and indirect interactions between sensors and enzymes were considered, with any intermediate protein being a possible bridge protein.

Constraints were subsequently imposed on both proteins and their interactions by considering the tissue-specific context of metabolism (**Figure 1C; Materials and Methods**). Despite abundant information on large-scale interactome data, there may be selection biases and tissue-specific variations. As a result of imposing constraints, we obtained a final reference network of 63,070 edges and 7,011 nodes, with sensors and enzymes constituting 23 nodes (**Figure 1D**, see **Figure S1** for the final reference network).

From the reference network, we selected bridge proteins, among intermediate proteins, that better link BA sensors and BA enzymes, using a "bridgeness" metric, assuming that the highly linking proteins critically regulate BA homeostasis through delivering metabolic information (**Figure 1E; Methods**). Compared with other existing centralities, including degree, closeness and betweenness centralities (see **Text S1**), our method was better able to focus on a particular protein's connections in specific paths between sensors and enzymes, regardless of the connections in other unrelated paths on the network. As expected, locally dense proteins among paths between BA sensors and BA enzymes contribute significantly to the regulation of BA metabolism; thus, these proteins may be associated with CRC carcinogenesis. We therefore focused on the prognostic potential of bridge proteins with high bridgeness scores.

Biological characteristics of bridge proteins

Before investigating their prognostic potentials, we examined the biological characteristics of bridge proteins that were computationally selected by bridgeness scores in CRCs. First, we identified expression patterns of bridge proteins embedded in CRCs; we examined discriminative patterns of bridge proteins at the transcriptomic level, using gene-expression profiles of CRC patients, as described previously [22]. Using univariate Student *t*-tests, we checked the ability of individual bridge proteins to distinguish between normal colon ($N = 54$) and primary CRC tissue samples ($N = 186$) at the transcriptomic level. Of the top-50 proteins, 42 (84%) were significantly discriminative (two sided $P < 0.01$). Gene ontology enrichment analysis of these 42 proteins revealed that most were enriched in terms such as "regulation of transcription from RNA polymerase II promoter" and "transcription regulator activity", which are related to regulatory roles in cellular processes (**Table S4**). They were also enriched in CRC pathogenic pathway-related terms, such as "canonical Wnt receptor signaling pathway" and "axin-APC-beta-catenin-GSK3B complex", suggesting the relevance of these bridge proteins to CRC pathogenesis.

Next, we compared the p-value distributions of i) bridge proteins, ii) a sensor and an enzyme, and iii) a combined group of i) and ii) (**Figure 2**). Compared with the background distribution

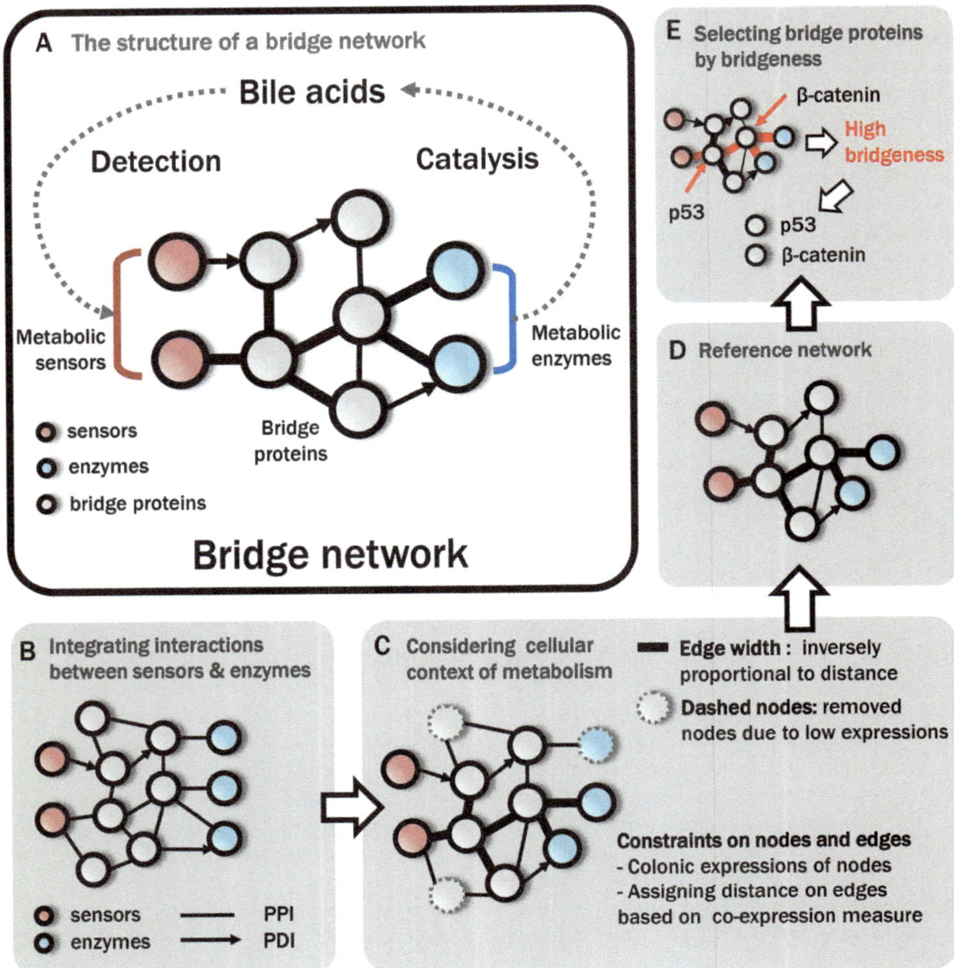

Figure 1. A bridge network for bile acid metabolism for determining bridge proteins. The overall process of the network construction is described in (**B–E**). (**A**) Structure of a bridge network, composed of a metabolic sensor (red), a metabolic enzyme (blue) and a bridge protein (gray). Metabolic enzymes catalyze the reactions of metabolites. Metabolic sensors detect the levels of intracellular metabolites. Bridge proteins link metabolic sensors and metabolic enzymes. (**B**) Integration of possible interactions between sensors and enzymes using protein-protein interactions (PPI) and protein-DNA interactions (PDI). Information on sensors and enzymes was collected from published studies and databases. (**C**) Imposing constraints on nodes and edges of an integrated network. (**D**) A final reference network to identify bridge proteins. (**E**) Selection of bridge proteins from the reference network by their bridgeness scores.

of p-values from overall gene products detected in a microarray ($N = 12{,}752$), the p-value distribution of the combined group was somewhat right-shifted (Kolmogorov-Smirnov (KS) test, one-sided $P = 7.89 \times 10^{-2}$). However, when we focused only on the bridge proteins, they showed high statistical significance in the KS test ($P = 2.93 \times 10^{-3}$), indicating that the discriminative power of bridge proteins, at the transcriptome level, was significantly greater than that of overall gene products in the microarray. Interestingly, sensor and enzyme proteins showed similar distributions relative to background ($P = 0.812$), indicating that sensor and enzyme proteins are less informative than bridge proteins in distinguishing between normal and diseased colon tissues.

We also investigated whether the top-50 bridge proteins are a feasible number of selections showing high statistical significance. We therefore compared the p-value distributions of selections with various numbers of bridge proteins, using the KS-test. The top-50 bridge proteins showed the lowest p-value on this comparison (**Figure S2**), with the statistical significance of selected bridge proteins being lower. Hence, we focused on the top-50 bridge proteins in further analysis. We also included other constraints used in network construction in a similar fashion (**Figure S3**).

We next compared the discriminative power of selected bridge proteins from different networks, through multivariate classification (**Figure 3A**) (**See detailed process in Materials and Methods**). The generated networks for comparisons were: (i) a bridge network developed from BA metabolism, (ii) a bridge network developed from glucose metabolism (i.e., glycolysis pathway) and (iii) a whole protein network without confining by sensors and enzymes in certain metabolic pathways. We also compared randomly selected proteins regardless of their interactions. Glycolysis was chosen for comparison to BA metabolism due to its relevance to common cancer progression [23,24]. As expected, the discriminative power of a BA bridge network at the transcriptome level exceeded that of a glycolysis bridge network because glycolysis is not specifically involved in CRCs. The ability of components of the BA bridge network to classify a sample as normal colon or primary CRC tissue (**Figure 3B**) largely exceeded that of randomly selected gene products. In contrast, components of other networks, including that involved in

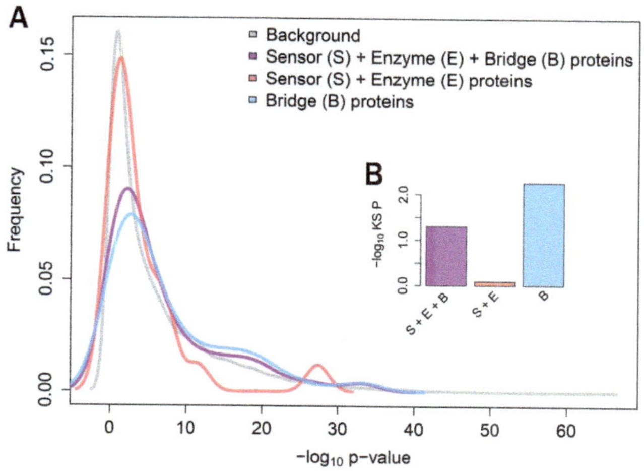

Figure 2. p-value distributions of components of a bridge network for bile acid metabolism. (**A**) p-value distributions of (i) sensor, enzyme and bridge proteins (S + E + B), (ii) sensor and enzyme proteins (S + E) and (iii) bridge proteins (B). (**B**) Comparisons of those p-value distributions with background p-value distribution. The statistical significance levels of shifted p-value distributions were determined by one-sided Kolmogorov Smirnov tests.

glycolysis, were equal to or barely exceeded randomly selected gene products in discriminative ability. That is, only gene expression levels of bridge proteins selected from a BA bridge network according to bridgeness were informative in distinguishing between normal colon and CRC.

We then examined CRC stage-specific expression patterns of selected bridge proteins. Most sporadic CRCs develop from normal colon via adenomatous polyps, with the sequence involving accumulated genetic anomalies in a stepwise manner [25]. To identify stage-specific variations in bridge proteins, we performed multivariate classifications between normal colons and adenomatous polyps and between polyps and primary CRCs. We found substantial variations in gene expressions of bridge proteins between normal colons and polyps (**Figure 3C and D**). Namely, bridge proteins associated with BA metabolism varied substantially during early stages of CRC pathogenesis, suggesting that these bridge proteins may be initiators of CRC tumorigenesis. We also found that bridge proteins from BA metabolism and glycolysis exhibited inverse patterns between polyps and primary CRCs, showing weaker, but substantial, variations during later stage of CRC pathogenesis, as if these changes were followers of CRC development (**Figure 3C and D**). Together, these findings showed that bridge proteins from BA metabolism and glycolysis behaved commutatively during CRC progression.

Furthermore, using pathway enrichment tests, we observed other meaningful biological characteristics of bridge proteins. Bridge proteins involved in BA metabolism were enriched in CRC-related pathways, including the Wnt (KEGG ID: hsa04310; false discovery rate-adjusted, hypergeometric $P = 4.47 \times 10^{-5}$), CRC (KEGG ID: hsa05210; $P = 2.80 \times 10^{-5}$) and common cancer (KEGG ID: hsa05200; $P = 6.94 \times 10^{-10}$) pathways (**Table S5**). This finding indicates that most bridge proteins are involved in CRC pathogenesis-related pathways and have the potential to promote CRCs through these pathways. Thus, characteristics determined from discriminative patterns and enrichment tests indicate that bridge proteins selected by bridgeness are associated with CRC pathogenesis.

Potential of bridge proteins as prognostic markers

To assess the prognostic ability of computationally-selected bridge proteins, we assessed their expression patterns in patients classified as having a good or poor prognosis. First, we clustered patients in an unsupervised way, based on similarities of expression patterns, and compared survival outcomes among patients in clusters. Total 178 patients from previous dataset [26] were clustered into three subgroups using a hierarchical clustering algorithm: BA-m1 ($N = 106$), BA-m2 ($N = 28$) and BA-m3 ($N = 44$) (**Figure 4A**). The Kaplan-Meier method with the log-rank test showed that among three subgroups of patients, the relapse-free survival was significantly different, indicating their substantial prognostic potential ($P = 2.37 \times 10^{-3}$) (**Figure 4B**). Then, we assessed the prognostic potential of other known expression-signature markers in the same way. Using expression patterns of genes selected in Wang et al [27] and ColoPrint [28], we classified patients into three subgroups and compared survival outcomes among their subgroups (ColoPrint's subgroups: col-m1 ($N = 20$), col-m2 ($N = 1$) and col-m3 ($N = 157$); Wang's subgroups: wang-m1 ($N = 19$), wang-m2 ($N = 3$) and wang-m3 ($N = 156$)). As a result, subgroups of patients clustered by ColoPrint's genes can distinguish between good and poor prognoses ($P = 2.75 \times 10^{-8}$), though just a single patient found in the poorest prognosis group (col-m2), but Wang's genes were not prognostic ($P = 0.258$) (**Figure 4C and D**). In addition, known molecular markers, including p53 mutations ($P = 0.233$), mismatch repair gene status ($P = 9.8 \times 10^{-2}$), KRAS mutations ($P = 5.75 \times 10^{-2}$), and BRAF mutations ($P = 0.338$), were not also substantially prognostic in this dataset (**Figure 4E–H**).

To assess the prognostic reproducibility of these bridge proteins and other expression-signature markers, we then classified patients in an independent dataset [22] as having good or bad prognoses, through a supervised classification system, using previous dataset [26] as the training dataset (**Figure 5**). Patients in the test data were classified, using their expression levels, based on correlation coefficients to mean expression levels of poor-prognosis-group patients in the training data, like previously performed [29]; we assigned patients into a poor-prognosis group if their correlation coefficients were high. We obtained thresholds of correlation coefficients to decide poor-prognosis patients with the highest statistical significance, through cross-validation procedures on the training data (See Materials and Methods). Noteworthy, patients in the test data can be significantly distinguished between good and poor prognoses when we used expression levels of bridge proteins as features for correlation coefficients; survival outcomes, i.e., CRC-specific survivals, of classified groups by the bridge proteins were significantly different when the Kaplan-Meier method with the log-rank test was used ($P = 2.70 \times 10^{-2}$) (**Figure 5A**). Other expression signatures, including ColoPrint ($P = 0.210$) and Wang's ($P = 0.558$) (**Figure 5B and C**), were not prognostic in the independent test dataset, suggesting that only bridge proteins were reproducibly prognostic. These results underline the potential and reliability of bridge proteins as prognostic markers.

Discussion

By investigating genes involved in the regulation of BA homeostasis, this study has identified numerous genes for prognostic biomarkers of CRC, with showing mechanistic relevance to CRC pathogenesis. Although various prognostic biomarkers have been proposed based on biological hypotheses [4], these biomarkers have shown limited clinical usefulness. The hypothesis, that BAs play pivotal roles in CRC, provides clues to

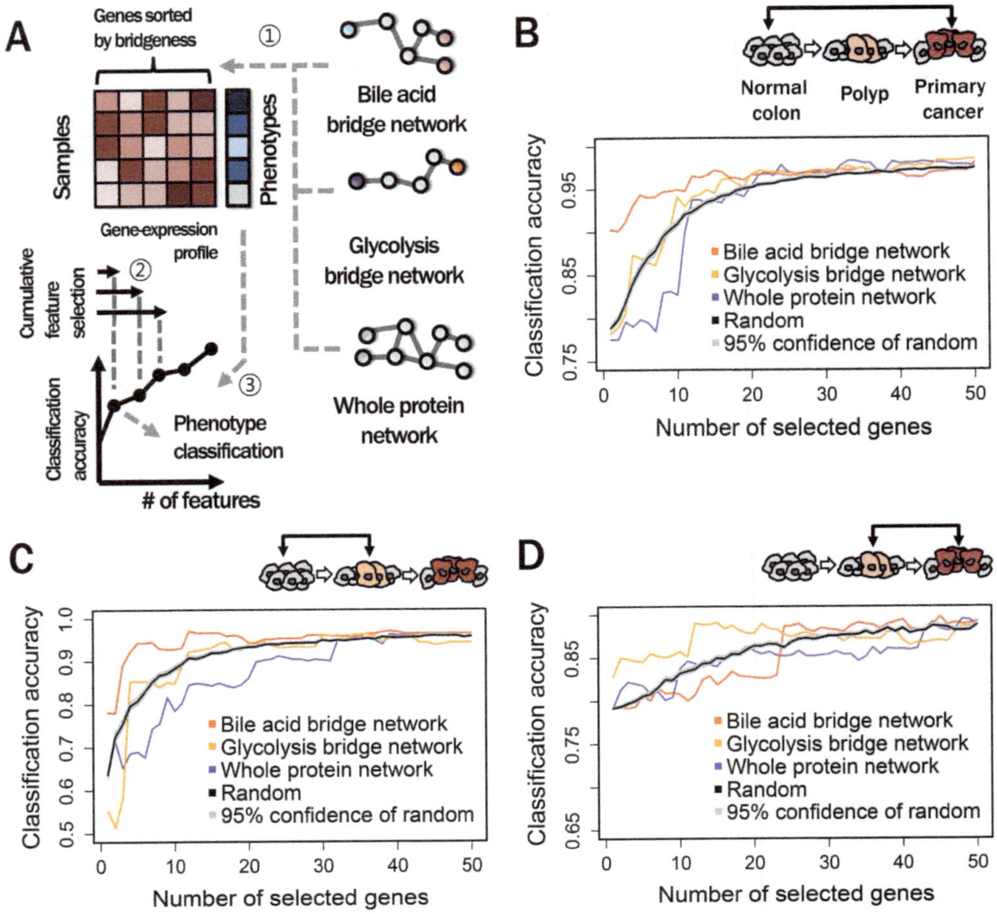

Figure 3. Multivariate analysis of bridge proteins from different networks. (**A**) Overall process of multivariate classifications using features from bridge proteins of different networks. After sorting bridge proteins by their bridgeness (①), features were extracted cumulatively from top-ranked bridge proteins (②). Samples were subsequently classified by cumulatively selected features and calculated classification accuracies (③). (**B**) Accuracies of classifications between normal colon and primary CRC tissues. For classifications, bridge proteins were obtained from (i) a bile acid bridge network (red), (ii) a glycolysis bridge network (yellow) and (iii) an whole protein network (purple). Classification accuracies were also calculated using randomly selected proteins (black) with 95% confidence intervals (gray) on the mean classification accuracies of repeated random selections (**C**) Accuracies of classifications between normal colon and polyp tissues. (**D**) Accuracies of classifications between polyp and primary CRC tissues.

understanding the pathogenesis of this disease. However, rather than focusing on BAs themselves, we focused on the genes involved in regulating BA metabolism by linking metabolic sensors and metabolic enzymes. Based on a devised metric, "bridgeness", numerous bridge proteins were selected from a reference, or bridge, network, and their prognostic abilities were analyzed. Bridge proteins could distinguish between normal and diseased tissues and are therefore relevant to the pathogenesis of CRC. These bridge proteins had greater and reproducible prognostic ability, as shown by statistical significance, than previously identified prognostic markers, suggesting that they are reliable prognostic markers in patients with CRC.

Interestingly, however, neither sensor nor enzyme proteins could significantly distinguish between normal colon tissue and CRC, a finding that may result from the housekeeping roles of these sensor and enzyme proteins for cell survival. Cells lack proteins with molecular functions similar to those of most of these sensor and enzyme proteins; thus, defects in their expression would have detrimental effects on cellular functions. Thus, evolutionarily, genetic anomalies in bridge proteins may have survival advantages over anomalies in sensor and enzyme proteins. Indeed, some bridge proteins, including caspase 8, apoptosis-related cysteine peptidase (CASP8), p53 and catenin (cadherin-associated protein) beta 1, 88 kDa (CTNNB1, also known as β-catenin), showed high mutational frequencies in CRC samples, whereas sensor and enzymes proteins for BA metabolism did not [30]. This evolutionary pressure, including during CRC tumorigenesis, would accelerate the acquisition of anomalies by bridge proteins.

In previous studies, notably, one bridge protein, STK11, was shown to have particular mechanistic potential to promote colorectal tumorigenesis [31–33]. STK11 has been associated with Peutz-Jeghers syndrome (PJS), a condition that enhances the formation of gastric adenomatous polyps and hepatocellular carcinoma [31]. In most PJS patients, one allele of STK11 is mutated, causing multiple gastric adenomatous polyps or hepatocellular carcinoma [32,33]. Similarly, STK11 may have the mechanistic potential to promote colorectal tumorigenesis. Other bridge proteins may also have prognostic value in CRC pathogenesis.

STK11 is also associated with energy metabolism, either alone or by interacting with AMPK, making it a potential bridge protein involved in the regulation of energy metabolism [34,35]. Among the other bridge proteins involved in energy metabolism are PPARGC1A, GSK3B, PPARA, peroxisome proliferator-activated

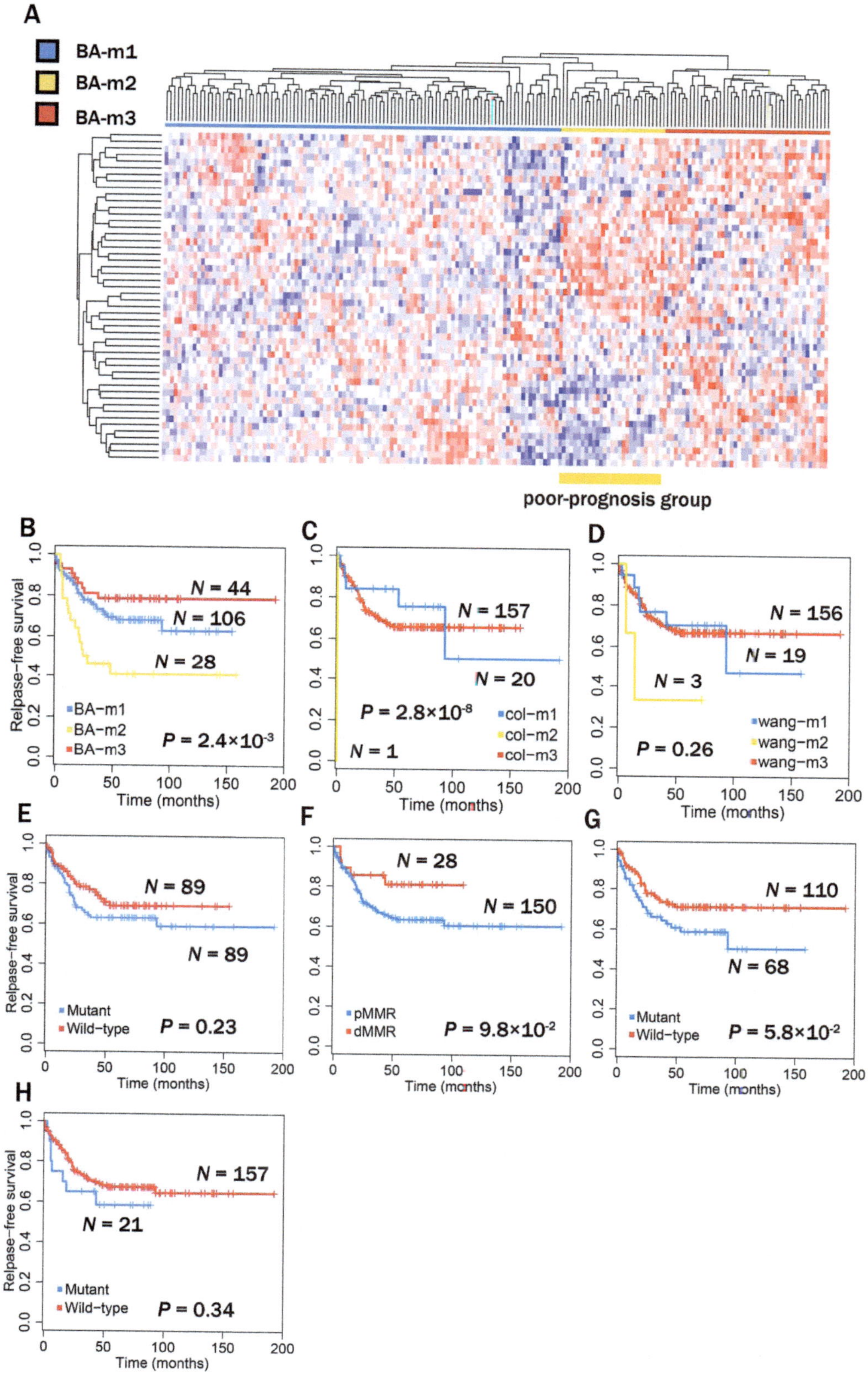

Figure 4. Identification of the prognostic ability of markers. Their prognostic ability was examined using a dataset of tissue samples from patients with CRC [26]. (**A**) Heatmap of CRC tumor samples with subgroups classified by the expression patterns of bridge proteins: BA-m1 (blue), BA-m2 (yellow) and BA-m3 (red). Prognostic ability was assessed by Kaplan-Meier survival analyses. The BA-m2 group showed the poorest prognosis. (**B**) Prognostic ability of our bridge proteins. (**C**) Prognostic ability of the ColoPrint gene set [28], with subgroups classified as col-m1 (blue), col-m2 (yellow) and col-m3 (red). (**D**) Prognostic ability of the Wang et al. signature gene set [27], with subgroups classified as wang-m1 (blue), wang-m2 (yellow) and wang-m3 (red). (**E**) Prognostic ability of p53 mutation status, mutant and wild-type. (**F**) Prognostic ability of mismatch repair gene (MMR)

status, deficient (dMMR) and proficient (pMMR). (**G**) Prognostic ability of KRAS mutation status, mutant and wild-type. (**H**) Prognostic ability of BRAF mutation status, mutant and wild-type.

receptor gamma (PPARG), solute carrier family 2 (facilitated glucose transporter) member 4 (SLC2A4, also known as GLUT4), glyceraldehyde-3-phosphate dehydrogenase (GAPDH), and lactate dehydrogenase A (LDHA), all important regulators of or enzymes involved in energy metabolism. Thus, their molecular functions may explain the activities of BAs that increase energy expenditure [36]. Assessments of the molecular functions of bridge proteins may provide novel insights on their as yet unidentified roles in BA homeostasis.

Despite bridge proteins showing prognostic potential, BA bridge networks show limited ability to identify other known CRC-susceptibility genes. For example, we found that a BA bridge network was unable to identify several well-known CRC-susceptibility genes, such as APC, KRAS, and BRAF. Inaccuracies originating from large-scale interactome data could impede in-depth analysis of bridge networks. Also, the interrelations of metabolic pathways, such as lipid, cholesterol, and glucose metabolism, would extend the ability to investigate all risk factors for CRC pathogenesis. This approach could also be applied to other diseases vulnerable to metabolic anomalies, including obesity, type-2 diabetes and Alzheimer's disease once metabolic sensors, enzymes and proper interactome data are generated for these diseases. The determination of proper and accurate bridge networks for metabolic pathways can allow the identification of disease-susceptibility genes and their clinical use as prognostic markers.

In summary, we found that bridge proteins, which are involved in the regulation of BA metabolism, have prognostic potential in patients with CRC. Despite their potential to promote CRC pathogenesis, bridge proteins had not been systematically investigated in previous studies. Based on a devised metric for "bridgeness", we computationally selected bridge proteins from a reference network and examined their prognostic potential in CRC. We also tested whether differences in their discriminative

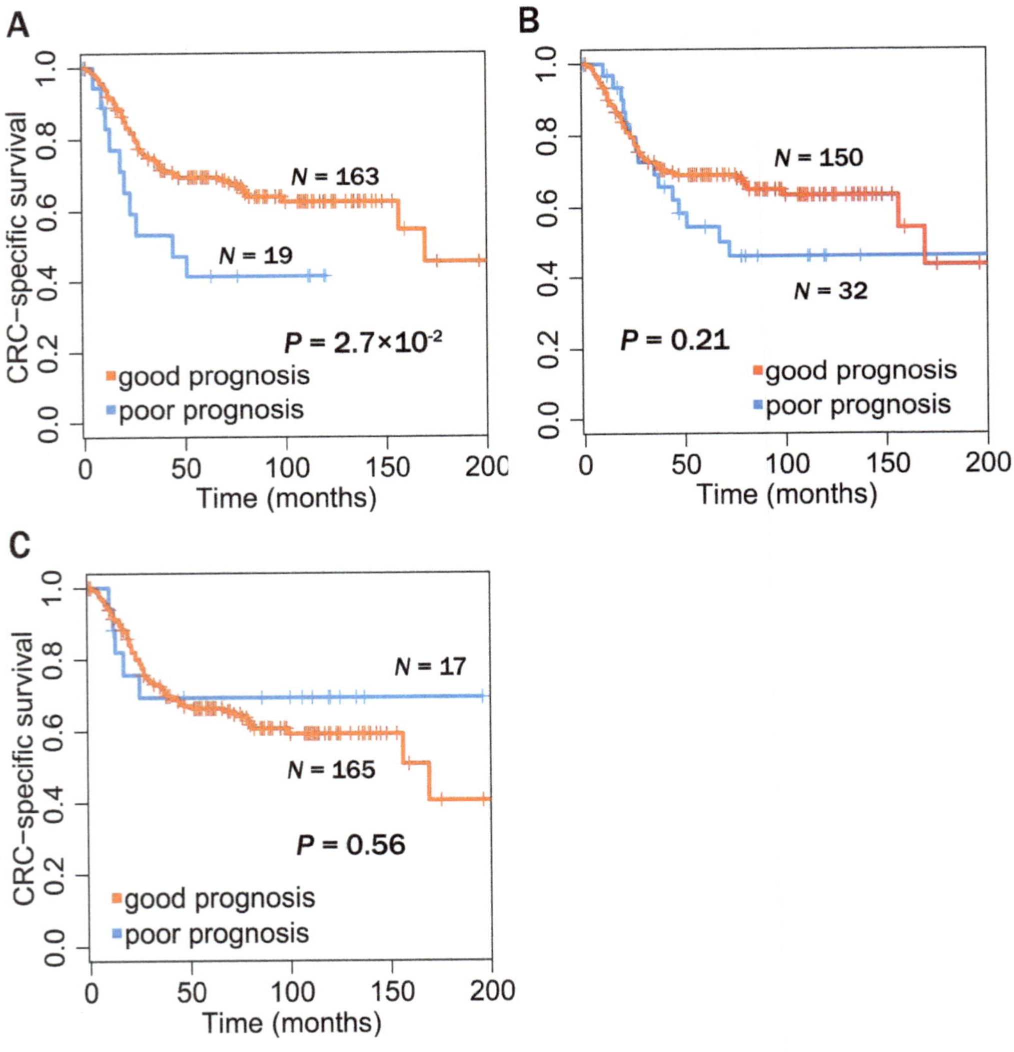

Figure 5. Identification of the prognostic reproducibility of markers. Their prognostic ability was examined in an independent test data [22] by supervised classifications and thus confirmed their prognostic reproducibility. (**A**) Prognostic ability of our bridge proteins (**B**) Prognostic ability determined by the ColoPrint gene set in reference [28] (**C**) Prognostic ability determined by the Wang et al. gene set in reference [27].

expression patterns in normal colon and CRC made them relevant to CRC pathogenesis. The findings indicate that bridge proteins involved in the regulation of BA metabolism may be reliable prognostic markers for CRC patients.

Materials and Methods

Bridge network construction

The reference network for BA metabolism consisted of metabolic sensors, metabolic enzymes and proteins interacting with both. The selected BA sensor was FXR and the BA enzymes were those designated in the EHMN human metabolic network database as enzymes involved in the "bile acid biosynthesis" pathway [19]. Possible interactions between the sensor and the enzymes were integrated using protein-protein interactions (PPI) described in the HPRD human protein information database [20] and protein-DNA interactions (PDI) from the commercial TF binding site database, TRANSFAC (Ver. 11.1) [21]. PPIs were regarded as bidirectional interactions and PDIs as unidirectional interactions from TFs to target genes. Next, we imposed constraints on the integrated network. On edges, we assigned distance values using a co-expression measure (i.e., the distance d_{ij} between genes i and j was defined as $d_{ij} = 1 - r^2_{ij}$ where r_{ij} is Pearson's correlation coefficient for the correlation in expression between genes i and j). Co-expression, defined as the functional relationship of a pair of proteins [37], was calculated using recently published FACS-sorted cell expression profiles from 52 patients with CRC [38], obtained from the public gene expression profile database, GEO (ID: GSE39397). On nodes, we imposed constraints regarding colonic gene expression. Using human whole-tissue gene expression data obtained from the public database, BioGPS [39], we determined the colonic expression of individual genes and compared the colonic and tissue-wide expression of each (total 176 samples with 84 tissue types; two samples for a colon tissue). If the average ratio of colonic to tissue-wide expression was lower than our criterion, that gene was removed. The criterion for gene removal was determined by comparing the p-value distribution of 50 bridge proteins with a background p-value distribution, as described in Results (**Figure S3**). In those comparisons, a 40th percentile cutoff produced the highest significance of the shifted p-value distribution.

Similarly, we constructed bridge networks relative to glycolysis and all proteins without specification for network comparisons. All the processes were identical to those used to construct the BA bridge network, except for the selection of metabolic sensors and enzymes. For glycolysis, we selected the metabolic sensors egl nine homolog 2 (C. elegans) (EGLN2, also known as PHD1), egl nine homolog 1 (C. elegans) (EGLN1, also known as PHD2), egl nine homolog 3 (C. elegans) (EGLN3, also known as PHD3) and hypoxia inducible factor 1 alpha subunit inhibitor (HIF1AN, also known as FIH). Their sensing of glycolysis metabolites was determined in vitro and in vivo [40]. Metabolic enzymes for glycolysis pathway were obtained from the "glycolysis and gluconeogenesis" pathway in the EHMN database [19]. For the whole protein network, we regarded metabolic sensors and enzymes as all the genes in the network in order to avoid specification by certain types of metabolism.

Bridgeness score

The bridgeness metric of a gene i with a set of sensors S and a set of enzymes T was calculated as:

$$B_{i,S,T} = \frac{1}{|S| \times |T|} \sum_{s \in S, t \in T, s \neq t} \frac{d(s,t)}{d_i(s,t)} = \frac{1}{|S| \times |T|} \sum_{s \in S, t \in T, s \neq t} \frac{d(s,t)}{d(s,i) + d(i,t)}$$

where $d(s,t)$ represents the distance of the shortest path between a sensor s and an enzyme t, and $d_i(s,t)$ represents the distance of the shortest path between node s and node t via node i. If gene i in the network is far from the shortest path between sensors and enzymes (i.e., $d_i(s,t) \gg d(s,t)$), then the addend tends to zero. Therefore, the bridgeness of gene i would be high if it is located near the shortest paths between sensors and enzymes, thus avoiding unrelated paths in cellular signaling networks. All calculations of network features and bridgeness were determined using R language and the *igraph* package [41].

Univariate and multivariate analysis

For univariate and multivariate analyses, we used a gene expression profile from CRC patients [22], which we obtained from the GEO database (ID: GSE41258). This dataset includes gene expression in 54 normal colons, 49 adenomatous polyps and 186 primary CRC tissue samples. Before using gene expression profiles to distinguish among tissue types, we performed gene-wise normalization on the profile using Z score transformation. In univariate analysis, the ability of each gene's expression to distinguish normal colon and primary CRC tissues was assessed by Student's t-test. We also calculated the statistical significance of the shifted p-value distribution of genes of interest against a background p-value distribution using the two-sample Kolmogorov-Smirnov one-sided test with the support of R package, *stats*. In multivariate analysis, we identified a bridge protein's discriminative ability, at the transcriptome level, using a logistic regression model with the support of java machine learning API, *Weka* [42]. Multivariate features were cumulatively selected from top-ranked bridge proteins of networks. The ability of each selection to classify samples as normal colon or primary CRC was evaluated using the five-fold cross-validation method with five repeats. The ability of each to distinguish between normal colon and polyp tissues, and between polyps and primary CRCs tissues, was assessed using the same features. We also simultaneously evaluated randomly selected proteins with an equal number of features. At each evaluation step, classification accuracy (i.e. accuracy $= \frac{TP+TN}{TP+FP+TN+FN}$) was measured and averaged after five repeats. In assessing features of randomly selected proteins, we calculated the mean classification accuracy after 100 repeats of selections and afterward calculated a 95% confidence interval of mean classification accuracies.

Survival analysis

First, the prognostic ability of bridge proteins was determined using related information from the Marisa et al. dataset [26] in the GEO database (ID: GSE39582). Information was available about gene expression; CRC recurrence-free survival event and time; treatment status; molecular marker status, including p53, KRAS, and BRAF mutations; and mismatch repair gene status. In this dataset, we used 178 tumor samples of patients to assess the prognostic ability; samples from treated patients or with missing information about survival outcomes or molecular status were excluded, avoiding unexpected effects of treatment on survival outcomes or unknown information. Identifying prognostic ability, we clustered patients, based on Euclidean distances between gene expressions of patients, by an unsupervised hierarchical clustering algorithm and measured the difference of survival outcomes among the patient clusters by the Kaplan-Meier method with the log-rank test. To compare prognostic abilities with other gene-expression signature markers, we used two gene sets, Wang's ($N = 21$) [27] and ColoPrint ($N = 15$) [28], and assessed their prognostic ability using their expression profiles from patients with CRC. In the two comparative gene sets [27,28], we only utilized genes detected in microarray data that we applied.

We also assessed the prognostic reproducibility of bridge proteins through a supervised classification system (**Figure S4**). In this classification system, the previous dataset [26] were used as a training data and the Sheffer et al. dataset [22] were used as an independent test data during supervised classifications. Total 182 tumor samples of patients from the Sheffer et al. data were used, after excluding samples that were not used in the original study [22]. This dataset contains information about gene expression and CRC-specific survival event and time. Performing supervised classification, we first determined a patient group with the poorest prognosis from the training data, after clustering patients by a hierarchical clustering and comparing survival probabilities among patient clusters. Referencing mean expression levels of patients in the poorest prognosis group (i.e. BA-m2 in **Figure 4A**) as a criterion, we classified patients of the test data into poor prognoses if their correlations of gene expressions with the reference expression levels are higher than a threshold, like existing study [29]. We calculated the correlations based on Pearson's correlation coefficients. A threshold of a correlation coefficient deciding prognosis was obtained through cross-validated procedures using the training data [26]. In this data set, we performed supervised classifications through five-fold cross-validations with various thresholds and selected the best threshold that can distinguish patients into a good or poor prognosis group with the most statistical significance. The statistical significance was measured by the Kaplan-Meier method with the log-rank test. We repeated cross-validations 100 times and averaged best thresholds in all repeats as a final threshold to use. Based on the final threshold, at last, we classified patients in an independent test data with learning a training data. We performed supervised classifications by other expression signatures in a similar way. All the statistical analyses, including Kaplan-Meier survival analysis, were performed by R packages.

Supporting Information

Figure S1 A final reference network for bile acid metabolism. This network is composed of a metabolic sensor (red), metabolic enzymes (blue) and interplay proteins (the outer layer of the largest circle). The Top-50 bridge proteins (black) are also shown. The edges representing the shortest paths between a sensor or an enzyme and a top-50 bridge protein are underlined (red edges).
(TIF)

Figure S2 p-value distributions of bridge proteins with varying numbers of selections. They stand for p-value distributions of the (**A**) top-10, (**B**) top-20, (**C**) top-30, (**D**) top-40, (**E**) top-50, (**F**) top-60, (**G**) top-70, (**H**) top-80, (**I**) top-90, and (**J**) top-100 bridge proteins. Statistical significance was highest for the top-50 bridge proteins when the shifted degrees of background (gray) and selected (blue) p-value distributions were measured using the one-sided Kolmogorov-Smirnov test.
(TIF)

Figure S3 p-value distribution of bridge proteins according to imposed constraints. (**A**) without node or edge constraints, (**B**) without node constraints, (**C–F**) with node constraints of (**C**) 10%, (**D**) 20%, (**E**) 30%, and (**F**) 40% removal criteria. Node removals within 40% were the most feasible for network construction.
(TIF)

Figure S4 An overview of a supervised classification system. The pipeline of supervised classification system was demonstrated. We used Marisa et al. dataset as a training data and Sheffer et al. dataset as a test data, after filtering out samples of patients in undesired conditions (1). Supervised classifications were based on correlations of gene expressions between the reference from the training data and samples from the test data. To select the threshold of correlation coefficients for deciding prognosis, we performed cross-validation procedure; we repeated five-fold cross-validation 100 times and averaged best threshold in all repeats (2). Based on the threshold obtained, we classified patients in the test data (3) and compared survival outcomes among classified patient groups, having a good or poor prognosis, through the Kaplan-Meier method with the log-rank test.
(TIF)

Table S1 Top-50 bridge protein information. We showed statistics of each bridge protein about discriminative power (T-score and T-test P) using datasets of Sheffer et al.
(DOCX)

Table S2 Evidence of bridge proteins involved in the regulation of bile acid homeostasis. Shown was previous literature that identified bridge proteins as being involved in the regulation of bile acid homeostasis. In the second column, we provided literature with definitive evidence that defects of some bridge proteins cause abnormal changes of bile acid levels. In the third and fourth columns, we provided literature with indirect evidence: studies in the third column showing that bridge proteins were regulated by or co-activated with a bile acid sensor; studies in the fourth column showing that bridge proteins regulated enzymes in bile acid metabolism.
(DOCX)

Table S3 Sensor and enzyme proteins in bile acid or glucose metabolism.
(DOCX)

Table S4 Enriched GO terms under corrected p-value<0.01.
(DOCX)

Table S5 Enriched KEGG non-metabolic pathways under FDR-adjusted hypergeometric p-value<0.01.
(DOCX)

Dataset S1 A source code and a dataset for extracting bridge proteins involved in bile acid metabolism. Performing a source code with a dataset will provide an output file showing top-50 bridge proteins we used.
(ZIP)

Dataset S2 A source code and a dataset for survival analyses in Figure 4 and 5. Performing a source code with a dataset will provide figures shown in our manuscripts.
(ZIP)

Text S1 Characteristics of bridgeness scores.
(DOCX)

Author Contributions

Conceived and designed the experiments: SL KL JWL DL. Performed the experiments: SL SY. Analyzed the data: SL KL SY JWL DL. Wrote the paper: SL KL.

References

1. Ferley J, SoerjomataramI I, Ervik M, Dikshit R, Eser S, et al. (2013) GLOBOCAN 2012 v1.0, Cancer Incidence and Mortality Worldwide: IARC CancerBase No. 11 [Internet]. Lyon, Fr Int Agency Res Cancer. Available from: http://globocan.iarc.fr, accessed on day/month/year.
2. Joensuu H (2008) Risk stratification of patients diagnosed with gastrointestinal stromal tumor. Hum Pathol 39: 1411–1419.
3. Chang HY, Sneddon JB, Alizadeh AA, Sood R, West RB, et al. (2004) Gene expression signature of fibroblast serum response predicts human cancer progression: similarities between tumors and wounds. PLoS Biol 2: E7.
4. Walther A, Johnstone E, Swanton C, Midgley R, Tomlinson I, et al. (2009) Genetic prognostic and predictive markers in colorectal cancer. Nat Rev Cancer 9: 489–499.
5. Willett WC, Stampfer MJ, Colditz GA, Rosner BA, Speizer FE (1990) Relation of meat, fat, and fiber intake to the risk of colon cancer in a prospective study among women. N Engl J Med 323: 1664–1672.
6. Bernstein H (2009) Bile acids as endogenous etiologic agents in gastrointestinal cancer. World J Gastroenterol 15: 3329.
7. Bernstein C, Holubec H, Bhattacharyya AK, Nguyen H, Payne CM, et al. (2011) Carcinogenicity of deoxycholate, a secondary bile acid. Arch Toxicol 85: 863–871.
8. Chey WD, Camilleri M, Chang L, Rikner L, Graffner H (2012) Response to Drs Trivedi and Ward. Am J Gastroenterol 107: 140–141.
9. Lage R, Diéguez C, Vidal-Puig A, López M (2008) AMPK: a metabolic gauge regulating whole-body energy homeostasis. Trends Mol Med 14: 539–549.
10. Guarani V, Potente M (2010) SIRT1 - a metabolic sensor that controls blood vessel growth. Curr Opin Pharmacol 10: 139–145.
11. Makishima M (1999) Identification of a Nuclear Receptor for Bile Acids. Science 284: 1362–1365.
12. Sinal CJ, Tohkin M, Miyata M, Ward JM, Lambert G, et al. (2000) Targeted disruption of the nuclear receptor FXR/BAR impairs bile acid and lipid homeostasis. Cell 102: 731–744.
13. Schwarz M, Russell DW, Dietschy JM, Turley SD (2001) Alternate pathways of bile acid synthesis in the cholesterol 7alpha-hydroxylase knockout mouse are not upregulated by either cholesterol or cholestyramine feeding. J Lipid Res 42: 1594–1603.
14. Wertheim BC, Smith JW, Fang C, Alberts DS, Lance P, et al. (2012) Risk modification of colorectal adenoma by CYP7A1 polymorphisms and the role of bile acid metabolism in carcinogenesis. Cancer Prev Res 5: 197–204.
15. Maran RRM, Thomas A, Roth M, Sheng Z, Esterly N, et al. (2009) Farnesoid X receptor deficiency in mice leads to increased intestinal epithelial cell proliferation and tumor development. J Pharmacol Exp Ther 328: 469–477.
16. Kim D-H, Lee JW (2011) Tumor suppressor p53 regulates bile acid homeostasis via small heterodimer partner. Proc Natl Acad Sci U S A 108: 12266–12270.
17. Goodwin B, Jones SA, Price RR, Watson MA, McKee DD, et al. (2000) A regulatory cascade of the nuclear receptors FXR, SHP-1, and LRH-1 represses bile acid biosynthesis. Mol Cell 6: 517–526.
18. Kim D-H, Kim J, Lee JW (2011) Requirement for MLL3 in p53 regulation of hepatic expression of small heterodimer partner and bile acid homeostasis. Mol Endocrinol 25: 2076–2083.
19. Hao T, Ma H-W, Zhao X-M, Goryanin I (2010) Compartmentalization of the Edinburgh Human Metabolic Network. BMC Bioinformatics 11: 393.
20. Keshava Prasad TS, Goel R, Kandasamy K, Keerthikumar S, Kumar S, et al. (2009) Human Protein Reference Database–2009 update. Nucleic Acids Res 37: D767–72.
21. Matys V, Kel-Margoulis O V, Fricke E, Liebich I, Land S, et al. (2006) TRANSFAC and its module TRANSCompel: transcriptional gene regulation in eukaryotes. Nucleic Acids Res 34: D108–10.
22. Sheffer M, Bacolod MD, Zuk O, Giardina SF, Pincas H, et al. (2009) Association of survival and disease progression with chromosomal instability: a genomic exploration of colorectal cancer. Proc Natl Acad Sci U S A 106: 7131–7136.
23. Vander Heiden MG, Cantley LC, Thompson CB (2009) Understanding the Warburg effect: the metabolic requirements of cell proliferation. Science 324: 1029–1033.
24. Koppenol WH, Bounds PL, Dang C V (2011) Otto Warburg's contributions to current concepts of cancer metabolism. Nat Rev Cancer 11: 325–337.
25. Davies RJ, Miller R, Coleman N (2005) Colorectal cancer screening: prospects for molecular stool analysis. Nat Rev Cancer 5: 199–209.
26. Marisa L, de Reyniès A, Duval A, Selves J, Gaub MP, et al. (2013) Gene expression classification of colon cancer into molecular subtypes: characterization, validation, and prognostic value. PLoS Med 10: e1001453.
27. Wang Y, Jatkoe T, Zhang Y, Mutch MG, Talantov D, et al. (2004) Gene expression profiles and molecular markers to predict recurrence of Dukes' B colon cancer. J Clin Oncol 22: 1564–1571.
28. Salazar R, Roepman P, Capella G, Moreno V, Simon I, et al. (2011) Gene expression signature to improve prognosis prediction of stage II and III colorectal cancer. J Clin Oncol 29: 17–24.
29. Van de Vijver MJ, He YD, van't Veer LJ, Dai H, Hart AAM, et al. (2002) A gene-expression signature as a predictor of survival in breast cancer. N Engl J Med 347: 1999–2009.
30. The Cancer Genome Atlas Network (2012) Comprehensive molecular characterization of human colon and rectal cancer. Nature 487: 330–337.
31. Giardiello FM, Brensinger JD, Tersmette AC, Goodman SN, Petersen GM, et al. (2000) Very High Risk of Cancer in Familial Peutz-Jeghers Syndrome. Gastroenterology 119: 1447–1453.
32 Nakau M, Miyoshi H, Seldin MF, Imamura M, Oshima M, et al. (2002) Hepatocellular carcinoma caused by loss of heterozygosity in Lkb1 gene knockout mice. Cancer Res 62: 4549–4553.
33. Thorgeirsson SS (2003) Hunting for tumor suppressor genes in liver cancer. Hepatology 37: 739–741.
34. Shackelford DB, Shaw RJ (2009) The LKB1-AMPK pathway: metabolism and growth control in tumour suppression. Nat Rev Cancer 9: 563–575.
35. Gurumurthy S, Xie SZ, Alagesan B, Kim J, Yusuf RZ, et al. (2010) The Lkb1 metabolic sensor maintains haematopoietic stem cell survival. Nature 468: 659–663.
36. Watanabe M, Houten SM, Mataki C, Christoffolete MA, Kim BW, et al. (2006) Bile acids induce energy expenditure by promoting intracellular thyroid hormone activation. Nature 439: 484–489.
37. Xulvi-Brunet R, Li H (2010) Co-expression networks: graph properties and topological comparisons. Bioinformatics 26: 205–214.
38. Calon A, Espinet E, Palomo-Ponce S, Tauriello DVF, Iglesias M, et al. (2012) Dependency of colorectal cancer on a TGF-β-driven program in stromal cells for metastasis initiation. Cancer Cell 22: 571–584.
39. Su AI, Wiltshire T, Batalov S, Lapp H, Ching K a, et al. (2004) A gene atlas of the mouse and human protein-encoding transcriptomes. Proc Natl Acad Sci U S A 101: 6062–6067.
40. Chen N, Rinner O, Czernik D, Nytko KJ, Zheng D, et al. (2011) The oxygen sensor PHD3 limits glycolysis under hypoxia via direct binding to pyruvate kinase. Cell Res 21: 983–986.
41. Csardi G, Nepusz T (2006) The igraph software package for complex network research. InterJournal, Complex Syst.
42. Hall M, Frank E, Holmes G, Phahringer B, Reuteman P, et al. (2009) The WEKA data mining software: an update. ACM SIGKDD Explor 11.

4

PF-4708671 Activates AMPK Independently of p70S6K1 Inhibition

Gilad W. Vainer[1]*, Ann Saada[2], Juliane Kania-Almog[1], Adir Amartely[3], Jacob Bar-Tana[4], Rachel Hertz[4]

1 Department of Pathology, Tel-Aviv Sourasky Medical Center, Tel-Aviv, Israel, **2** Monique and Jacques Roboh Department of Genetic Research, and Department of Genetics and Metabolic Diseases, Hadassah-Hebrew University Medical Center, Jerusalem, Israel, **3** Department of Pathology, Hadassah-Hebrew University Medical Center, Jerusalem, Israel, **4** Human Nutrition and Metabolism, Hebrew University Medical School, Jerusalem, Israel

Abstract

The P70 ribosomal protein S6 kinase 1 (P70S6K1) is activated by the mammalian target of rapamycin (mTORC1) and regulates proliferation, growth, and metabolism. PF-4708671 is a novel, cell-permeable, has been proposed to be a highly specific inhibitor of p70S6K1. It is used in micromolar concentration range to dissect signaling pathways downstream of mTORC1 and to study the function of p70S6K1. Here we show that PF-4708671 induces AMP-activated protein kinase (AMPK) phosphorylation and activation in immortalized mouse embryonic fibroblasts (MEF) independently of p70S6K1, due to specific inhibition of mitochondrial respiratory chain Complex I.

Editor: Petras Dzeja, Mayo Clinic, United States of America

Funding: This work was financed in part by the Tel-Aviv Sourasky Medical Center and Mr. Daniel Levin donation (GWV); the Israeli Ministry of Health grant #5938 (GWV); The Israeli Cancer Association (GWV); and Israeli Science Foundation (ISF) grant #1462 (AS). The funders had no role in study design, data collection and analysis, decision to publish, or preparation of the manuscript.

Competing Interests: The authors have declared that no competing interests exist.

* Email: giladwv@tlvmc.gov.il

Introduction

P70 ribosomal protein S6 kinases (p70S6K) 1 & 2 (S6K1, S6k2) are two isoforms of the AGC kinase (protein kinase A/protein kinase G/protein kinase C) family. The AGC kinase family consists of about 60 members that share a conserved catalytic kinase domain [1], making it difficult to find a specific inhibitor for each individual member of this family.

p70S6K isoforms are activated by the phosphorylation of Thr389 catalyzed by the rapamycin-sensitive mammalian TOR complex 1 (mTORC1) [2–5]. Activated p70S6K1 is considered a key kinase in body metabolism. Activated p70S6Ks phosphorylates rpS6 on five clustered residues, allowing the initiation of protein translation [6]. Also, mitochondria number and size is increased in p70S6K1-deficient mice, accompanied by enhanced beta-oxidation, increase in muscle AMP-activated protein kinase (AMPK), and increased life span [7,8].

PF-4708671 is a cell-permeate p70S6K1-specific inhibitor [9]. In vitro, PF-4708671 inhibits p70S6K1 at mid nanomolar concentrations (IC_{50} of 160 nM). In a panel of ~90 protein and lipid kinases of most closely related AGC family members, PF-4708671 was found to be highly selective for p70S6K1, being 400-fold less potent for inhibiting S6K2. Furthermore, PF-4708671 does not inhibit pyruvate dehydrogenase kinase, isozyme 1 (PDK1), which acts upstream of several AGC kinases, including p70S6K1. Among the AGC kinase family, mitogen- and stress-activated protein kinase 1 (MSK1) was found to be the next most sensitive to PF-470867, with an IC_{50} of 950 nM in vitro. However, PF-470867 did not appear to inhibit cellular MSK1 within the 1 to 10 micromolar concentration range, where it strongly inhibits p70S6K1 [9]. Due to its high specificity, PF-470867 is extensively used by studies that focus on verifying the role of p70S6K1 and mTOC1 in regulating metabolism.

Here we show that PF-4708671 activates AMPK in p70S6K 1 & 2 double knockout (DKO) immortalized mouse embryonic fibroblasts (MEF), implying AMPK activation independently of P70S6K1 inhibition. It does so directly by inhibiting mitochondrial respiratory chain Complex I. Since AMPK inhibits mTORC1 and p70S6K1 activities, its direct activation by PF-4708671 may amplify p70S6K inhibition by PF-4708671.

Methods

Cell culture

Immortalized mouse embryonic fibroblasts (MEF) derived from S6K double knockout or S6K wild type cells were a kind gift of Mario Pende (Inserm, Paris, France) [10] (figure S1). MEFs were cultured in DMEM (GIBCO) supplemented with 10% fetal calf serum and Pen-Strep (Biological industries, Beit HaEmek, Israel), Rapamycin and PF-4708671(Sigma-Aldrich) as indicated. A validation batch of original PF-4708671 was kindly provided by Dario Alessi (University of Dundee, Dundee, Scotland). The compound was dissolved in dimethyl sulphoxide (DMSO) and 10 mM stock solution aliquots were kept in −80°C.

Isolation of Mitochondria

MEF mitochondria were isolated by differential centrifugation: cells were homogenized in buffer A (320 mmol/L sucrose, 5 mmol/L Tris-HCl, 2 mmol/L EGTA, pH 7.4) with a Dounce homogenizer (Teflon glass) and centrifuged for 3 min at 2,000 g to remove nuclei and cell debris. The supernatant was centrifuged for 10 min at 12,000 g at 4°C, and the pellet was re-suspended in

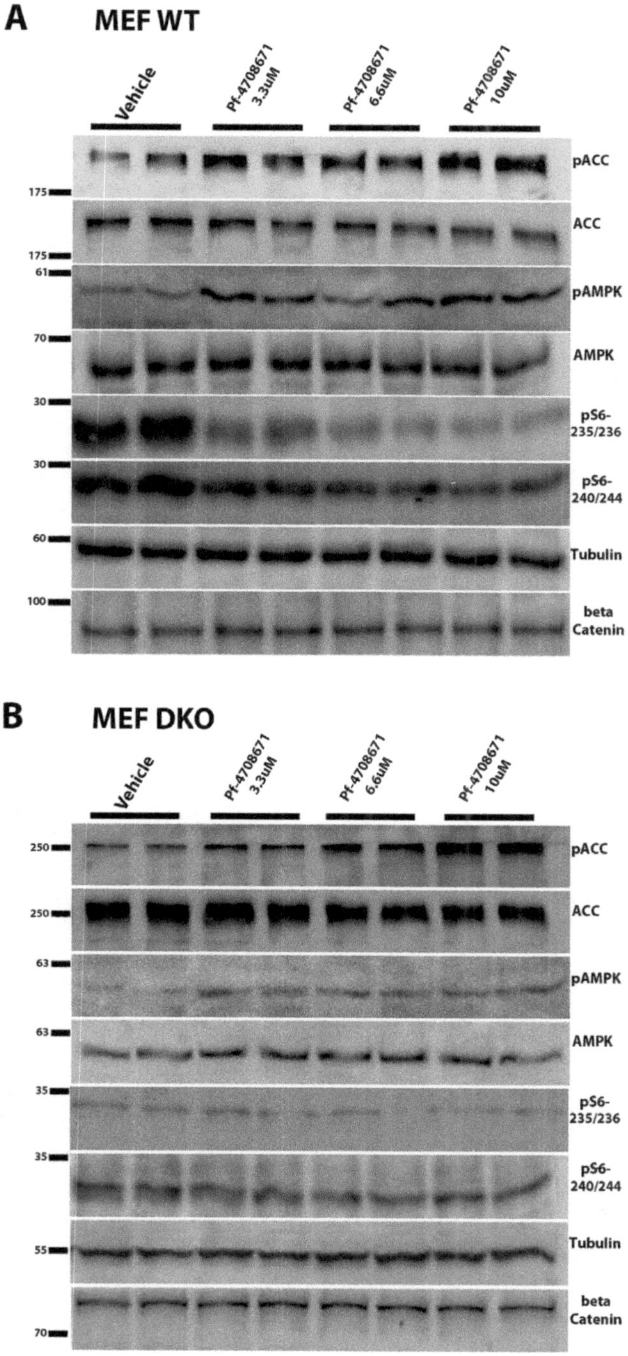

Figure 1. AMPK activation by PF-4708671. Wild type MEF (**A**) and p70S6K double knockout MEF (**B**) were treated with PF-4708671 as indicated. Phospho-ACC (Ser79), ACC, phospho-AMPK(Thr172), AMPK, phospho-S6 and (235/236 and 240/244 were determined by SDS-PAGE as described in Methods. Tubulin and beta-Catenin served as loading controls.

buffer A containing 0.02% digitonin (Sigma-Aldrich) and re-centrifuged. The mitochondrial pellet was washed again twice with buffer A and kept at –80°C until use.

Activity of mitochondrial electron transport components

Enzymatic activities of respiratory chain complexes were measured at 37_C by standard spectrophotometry, as previously described [11]. Briefly, Complex I was measured as rotenone-sensitive NADH-CoQ reductase, by monitoring the oxidation of NADH at 340 nm in the presence of coenzyme Q1. Complex II was measured at 600 nm by succinate-mediated phenazine methosulfate reduction of dichloroindophenol by succinate dehydrogenase (SDH). Complexes II+III were measured at 550 nm by succinate-mediated cytochrome c reduction. Complex IV (cytochrome c oxidase) was measured by the oxidation of reduced cytochrome c at 550 nm.

Citrate synthase (CS), a ubiquitous mitochondrial matrix enzyme serving as normalizer, was measured at 412 nm by the release of CoASH coupled to 50,50-dithiobis (2-nitrobenzoic) acid, in the presence of acetyl-CoA and oxaloacetate. Activities of mitochondrial respiratory chain complexes in the presence of added PF-4708671 are presented relative to respective activities in the presence of vehicle (DMSO). Protein concentration was determined by the Lowry method and calculated according to a bovine serum albumin (BSA) standard curve.

Seahorse oxygen consumption assay

L-15medium (Beit-Haemek biological industries; http://www.bioind.com/page_14372) containing 5% FCS (Beit-Haemek biological industries) was used as the standard medium in the XF and referred to as "assay medium". The Seahorse platform can calculate the extracellular acidification rate (ECAR) in a non-buffered medium. Thus, because L-15 is HEPES buffered extracellular acidification rate (ECAR) was not used.

MEF cells were seeded in XF 24-well cell culture microplates (Seahorse Bioscience) at 20×10^3 cells/well (0.32 cm2) in 500 µl complete growth medium (DMEM +10% FCS) and then incubated at 37°C/5% CO_2 for 15 h, followed by discarding the growth medium and washing the cells three times with 750 µl of assay medium. The cells were further incubated at 37°C with no CO_2 supplementation for 4 hours to allow temperature and pH equilibration. Prior to each measurement, the XF24 Analyzer mixed the assay media for 3 min followed by 2 minutes wait to allow for oxygen pressure equilibration. Then oxygen consumption rate (OCR) was measured for 3 min. This was repeated to establish the OCR baseline. PF-4708671 or Rapamycin were injected into each well to reach the final working concentration of 2.5/5 µM and 50 nM, respectively. DMSO prepared in assay medium was injected as control. After the compounds injection, OCR measurements were made. Responses were expressed as a percentage of the baseline rate prior to compound addition. Statistical analysis was done using Seahorse analysis software using OCR area under the curve ANOVA test, as recommended by the manufacturer.

Western blot

Cells were washed briefly in ice cold PBS, followed by adding lysis buffer (0.1M Tris pH6.8 containing 1% SDS). Lysate was incubated on ice for 30 minutes, and then centrifuged at 14,000 g at 4°C for 10 minutes. For Western blot analysis, 20 µg of protein extract was subjected to 8% or 10% SDS/PAGE and transferred to nitrocellulose membranes. After blocking in 5% non-fat milk, blots were incubated with AMPK (#2532), Phospho-AMPK (#2531), ACC (#3662), Phospho-ACC (#3661), P70S6K1 (#9202), Phospho-S6 ribosomal protein 240/244 (#5364),

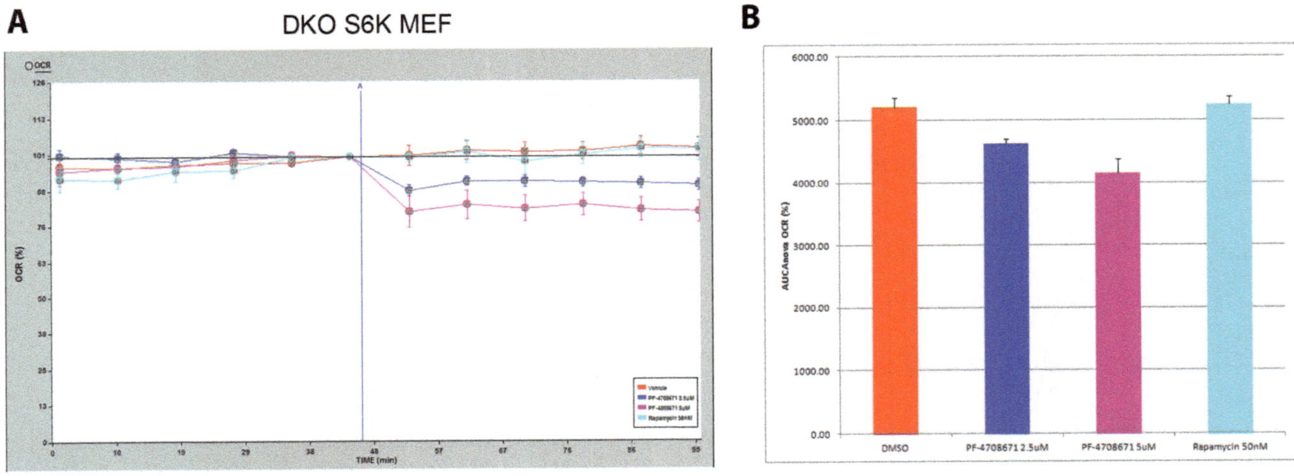

Figure 2. Inhibition of oxygen consumption by PF-4708671. (**A**) S6K double knockout MEF were treated with vehicle (red line); rapamycin 50 nM (turquoise); PF-4708671 2.5 μM (blue) and 5 μM (pink). Oxygen consumption rate (OCR) of was determined by the Seahorse platform as described in Methods. (**B**) ANOVA statistical analysis of the OCR area under the curve showed that after 2.5 uM or 5 uM PF-4708671 treatment the OCR was significantly decreased ($p=2.6*10^{-5}$ and $1.97*10^{-9}$, respectively; see Table S1).

Phospho-S6 ribosomal protein 235/236 (#4858), beta-actin (#4967) antibodies for overnight, at 4°C. All antibodies were from Cell Signaling Technology. Blots were washed in Tris buffered saline - 0.2% Tween20 (TBS-T) three times, and incubated for one hour at room temperature with anti-Rabbit or anti-Mouse secondary antibody diluted 1:350 in 5% no-fat milk (ImmPress peroxidase, Vector). Detection was performed using chemiluminescence reagent on microchemi digital platform (DNR bio-imaging systems). Protein concentration in extracts was determined using the BCA protein reagent (Pierce).

Results

In line with previous reports [9], phosphorylation of ribosomal protein S6 (rpS6) at positions 235/236 and 240/244 was dose-dependently inhibited in wild type MEF cultured in the presence of PF-4708671 (Figure 1A and S3 for densitometry data). Concomitantly with P70S6K1 inhibition, added PF-4708671 resulted in dose-dependent phosphorylation of AMPK(Thr172) and its ACC(Ser79) downstream substrate to an extent similar to that induced by oligomycin (Figure 1A). Phosphorylation of AMPK(Thr172) and ACC(Ser79) by added PF-4708671 was similarly observed in p70S6K 1/2 double knockout MEF (S6K DKO MEF) (Figure 1B and S3 for densitometry data), indicating that AMPK activation by PF-4708671 was p70S6K1 independent. Of note, PF-4708671 decreased the phosphorylation level of rpS6(235/236, 240/244) in S6K DKO MEF, namely, independently of p70S6K1. Furthermore, the effect of PF-4708671 is not MEF specific as other cell lines, as BT-474, a human breast cancer cell line, show elevation in ACC(Ser79) phosphorylation (Figure S2).

Concomitantly with AMPK activation, PF-4708671 inhibited dose dependently the oxygen consumption of DKO MEF (Figure 2; and of BT-474 see Figure S2), resulting in 10 and 20% decrease by PF-4708671 concentrations of 2.5 μM and 5 μM, respectively. PF-4708671 effect in suppressing oxygen consumption rate was immediate, steady, and maintained for over one hour.

Inhibition of oxygen consumption by PF-4708671 was further pursued by studying PF-4708671 effects in isolated mitochondria. Mitochondrial citrate synthase, Complex II, succinate dehydrogenase, Complex II+III, and complex IV activities were unaffected by PF-4708671 (Figure 3A). However, mitochondrial respiratory chain Complex I activity was inhibited in a concentration dependent manner, with an IC50 of 5.2 μM (Figure 3A–B). Complex I inhibition was highly statistically significant between DMSO and 7.5 μM PF-4708671 (t-Test $1.4*10^{-6}$; Table S1). Inhibition of mitochondrial Complex I by 5 μM PF-4708671 was comparable to that of 5 mM metformin. Mitochondrial Complex I activity was similarly inhibited by an original batch of PF-4708671 (data not shown), implying that inhibition was inherently due to PF-4708671, rather than an impurity in the commercial sample.

Discussion

The present study describes activation of AMPK by PF-4708671, a novel cell-permeable P70S6K1 inhibitor. AMPK activation by PF-4708671 is independent of p70S6K1, and may be ascribed to mitochondrial Complex I inhibition. Thus, AMPK activation and suppression of oxygen consumption by PF-4708671 were both observed in p70S6K null cells. The calculated IC_{50} of PF-4708671 in inhibiting mitochondrial Complex I was ~5 μM, namely, within the working concentration range of 3–10 μM used to effectively block p70S6K1 [9].

p70S6K1 is activated by phosphorylation of its Thr389 by mTORC1 [3]. However, mTORC1 is one of the downstream targets of AMPK, being inhibited by AMPK via multiple mechanisms [12,13]. Hence, inhibition of p70S6K1 by PF-4708671 may be a result of both, direct inhibition of its kinase activity complemented by inhibition of its upstream mTORC1 activator by PF-4708671-activated AMPK. In line with that,

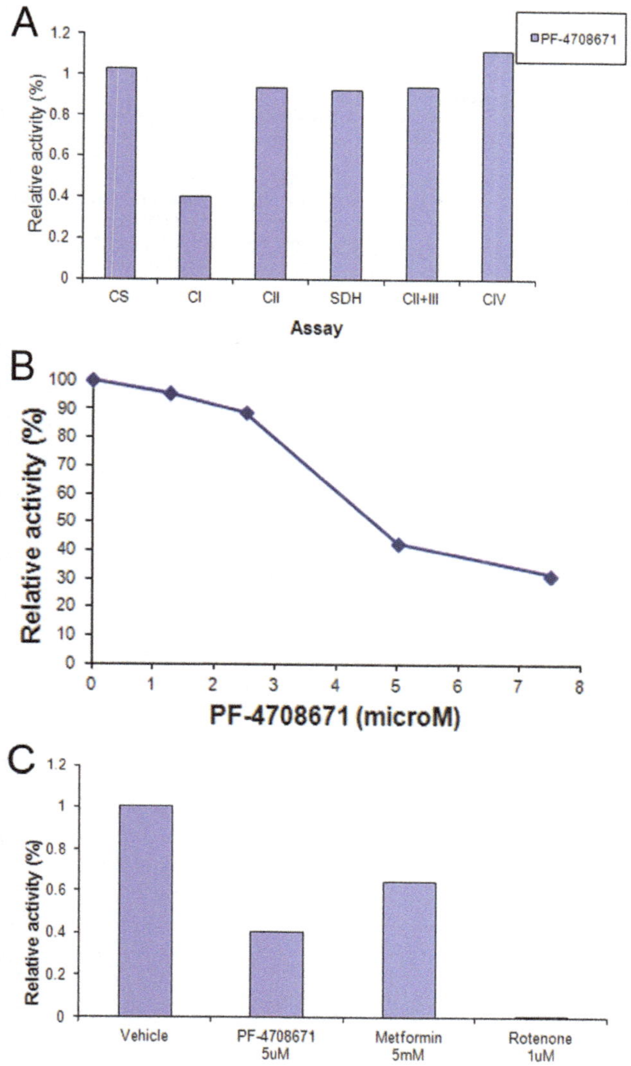

Figure 3. Inhibition of mitochondrial complex I by PF-4708671. MEF mitochondria were prepared as described in Methods and incubated in the presence of PF-4708671, metformin or rotenone as indicated. (**A**) Activities of citrate synthase and mitochondrial respiratory chain complexes were determined as described in Methods. Activities of mitochondrial respiratory chain complexes in the presence of added PF-4708671 are presented relative to respective activities in the presence of vehicle. (**B**) Relative activity of mitochondrial Complex I in the presence of increasing PF-4708671 concentrations as indicated. (**C**) Relative activity of mitochondrial Complex I in the presence of PF-4708671, metformin or rotenone as indicated. Activities of mitochondrial respiratory chain complexes are presented relative to (DMSO).

AMPK activation by PF-4708671 may partially account for PF-4708671 suppression of phospho-rpS6(235/235, 240/244), independently of P70S6K1 (Fig 1B). Indeed, in addition to p70S6K, rpS6 may directly be phosphorylated by p90S6K (RSK) [6,14] that may cross talk with AMPK, independently of p70S6K1.

In conclusion, direct activation of AMPK by PF-4708671, combined with inhibition of P70S6K1 activity, may open new prospects for PF-4708671 in modifying mTORC1 and p70S6K1 control of metabolism, protein synthesis, autophagy, and proliferation.

Supporting Information

Figure S1 DKO and WT MEFs comparison. p70S6K double knockout MEF show lower levels of phospho-rpS6. Wild type MEF and p70S6K double knockout MEF were grown in complete medium. p70S6K1 and phospho-S6 (235/236 and 240/244) were determined by SDS-PAGE as described in Methods. Tubulin served as loading control.
(TIF)

Figure S2 BT-474 show pACC elevation due to PF-4708671. BT-474 were treated with PF-4708671 as indicated. Phospho-ACC (Ser79), phospho- AKT (Ser473), AKT, and phospho-S6 (235/236 and 240/244) were determined by SDS-PAGE as described in Methods. Tubulin served as loading control.
(TIF)

Figure S3 Densitometry of Figure 1. AMPK activation by PF-4708671. Wild type MEF (**A**) and p70S6K double knockout MEF (**B**) were treated with PF-4708671 as indicated. Densitometry of Phospho-ACC (Ser79), ACC, phospho-AMPK(Thr172), AMPK, and phospho-S6 (235/236 and 240/244) were determined by chemiluminescence as described in Methods. ACC, AMPK or Tubulin served as loading controls, as indicated in the figures.
(TIF)

Figure S4 Inhibition of BT-474 oxygen consumption by PF-4708671. BT-474 were treated with vehicle (turquoise line) or 3 μM of PF-4708671(pink). Oxygen consumption rate (OCR) of was determined by the Seahorse platform as described in Methods. As shown before, PF-4708671 effect in suppressing oxygen consumption rate was immediate, steady, and maintained for over one hour.
(TIF)

Table S1 Seahorse platform and mitochondrial assays statistics. ANOVA statistical analysis of the OCR area under the curve showed that after 2 5 uM or 5 uM PF-4708671 treatment the OCR was significantly decreased (manufacturer output), and mitochondrial assay statistics.
(XLSX)

Author Contributions

Conceived and designed the experiments: GWV JBT RH. Performed the experiments: GWV AS JKA AA RH. Analyzed the data: GWV AS JKA AA. Contributed reagents/materials/analysis tools: GWV AS RH. Wrote the paper: GWV JBT RH.

References

1. Pearce LR, Komander D, Alessi DR (2010) The nuts and bolts of AGC protein kinases. Nat Rev Mol Cell Biol 11: 9–22.
2. Avruch J, Hara K, Lin Y, Liu M, Long X, et al. (2006) Insulin and amino-acid regulation of mTOR signaling and kinase activity through the Rheb GTPase. Oncogene 25: 6361–6372.
3. Weng Q, Kozlowski M, Belham C, Zhang A, Comb M, et al. (1998) Regulation of the p70 S6 kinase by phosphorylation in vivo. Analysis using site-specific anti-phosphopeptide antibodies. J Biol Chem 273: 16621–16629.
4. Ben-Sahra I, Howell JJ, Asara JM, Manning BD (2013) Stimulation of de novo pyrimidine synthesis by growth signaling through mTOR and S6K1. Science 339: 1323–1328.
5. Dowling RJ, Topisirovic I, Fonseca BD, Sonenberg N (2010) Dissecting the role of mTOR: lessons from mTOR inhibitors. Biochim Biophys Acta 1804: 433–439.
6. Meyuhas O (2008) Physiological roles of ribosomal protein S6: one of its kind. Int Rev Cell Mol Biol 268: 1–37.

7. Selman C, Tullet JM, Wieser D, Irvine E, Lingard SJ, et al. (2009) Ribosomal protein S6 kinase 1 signaling regulates mammalian life span. Science 326: 140–144.
8. Um SH, Frigerio F, Watanabe M, Picard F, Joaquin M, et al. (2004) Absence of S6K1 protects against age- and diet-induced obesity while enhancing insulin sensitivity. Nature 431: 200–205.
9. Pearce LR, Alton GR, Richter DT, Kath JC, Lingardo L, et al. (2010) Characterization of PF-4708671, a novel and highly specific inhibitor of p70 ribosomal S6 kinase (S6K1). Biochem J 431: 245–255.
10. Shima H, Pende M, Chen Y, Fumagalli S, Thomas G, et al. (1998) Disruption of the p70(s6k)/p85(s6k) gene reveals a small mouse phenotype and a new functional S6 kinase. EMBO J 17: 6649–6659.
11. Saada A, Bar-Meir M, Belaiche C, Miller C, Elpeleg O (2004) Evaluation of enzymatic assays and compounds affecting ATP production in mitochondrial respiratory chain complex I deficiency. Anal Biochem 335: 66–72.
12. Xu J, Ji J, Yan XH (2012) Cross-talk between AMPK and mTOR in regulating energy balance. Crit Rev Food Sci Nutr 52: 373–381.
13. Alers S, Loffler AS, Wesselborg S, Stork B (2012) Role of AMPK-mTOR-Ulk1/2 in the regulation of autophagy: cross talk, shortcuts, and feedbacks. Mol Cell Biol 32: 2–11.
14. Pende M, Um SH, Mieulet V, Sticker M, Goss VL, et al. (2004) S6K1(−/−)/S6K2(−/−) mice exhibit perinatal lethality and rapamycin-sensitive 5′-terminal oligopyrimidine mRNA translation and reveal a mitogen-activated protein kinase-dependent S6 kinase pathway. Mol Cell Biol 24: 3112–3124.

5

Copper-Deficiency in *Brassica napus* Induces Copper Remobilization, Molybdenum Accumulation and Modification of the Expression of Chloroplastic Proteins

Vincent Billard[1,2,3], Alain Ourry[1,2,3], Anne Maillard[1,2,3], Maria Garnica[4], Laurent Coquet[5], Thierry Jouenne[5], Florence Cruz[6], José-Maria Garcia-Mina[4], Jean-Claude Yvin[6], Philippe Etienne[1,2,3]*

1 Normandie Université, Caen, France, **2** UNICAEN, UMR 950 Ecophysiologie Végétale, Agronomie et nutritions N, C, S, Caen, France, **3** INRA, UMR 950 Ecophysiologie Végétale, Agronomie et nutritions N, C, S, Caen, France, **4** Timac Agro Spain, Poligono de Arazuri-Orcoyen, Orcoyen, Spain, **5** Plateforme de protéomique PISSARO, UMR6270 CNRS Faculté des Sciences de Rouen, Mont-Saint-Aignan, France, **6** Centre de Recherche International en Agroscience, CRIAS-TAI, Groupe Roullier, Dinard, France

Abstract

During the last 40 years, crop breeding has strongly increased yields but has had adverse effects on the content of micronutrients, such as Fe, Mg, Zn and Cu, in edible products despite their sufficient supply in most soils. This suggests that micronutrient remobilization to edible tissues has been negatively selected. As a consequence, the aim of this work was to quantify the remobilization of Cu in leaves of *Brassica napus* L. during Cu deficiency and to identify the main metabolic processes that were affected so that improvements can be achieved in the future. While Cu deficiency reduced oilseed rape growth by less than 19% compared to control plants, Cu content in old leaves decreased by 61.4%, thus demonstrating a remobilization process between leaves. Cu deficiency also triggered an increase in Cu transporter expression in roots (*COPT2*) and leaves (*HMA1*), and more surprisingly, the induction of the *MOT1* gene encoding a molybdenum transporter associated with a strong increase in molybdenum (Mo) uptake. Proteomic analysis of leaves revealed 33 proteins differentially regulated by Cu deficiency, among which more than half were located in chloroplasts. Eleven differentially expressed proteins are known to require Cu for their synthesis and/or activity. Enzymes that were located directly upstream or downstream of Cu-dependent enzymes were also differentially expressed. The overall results are then discussed in relation to remobilization of Cu, the interaction between Mo and Cu that occurs through the synthesis pathway of Mo cofactor, and finally their putative regulation within the Calvin cycle and the chloroplastic electron transport chain.

Editor: Paul A. Cobine, Auburn University, United States of America

Funding: VB's PhD grant was funded by the Regional Council of Basse-Normandie (http://www.region-basse-normandie.fr/) and Timac Agro International (http://www.timacagro.fr/). Regional Council of Basse-Normandie had no role in study design, data collection and analysis, decision to publish, or preparation of the manuscript. Timac Agro Internationnal had a role in study design and analysis.

Competing Interests: The authors have declared that no competing interests exist.

* Email: philippe.etienne@unicaen.fr

Introduction

While N, P, K and more recently S fertilizations are common practices in agriculture, the supply of micronutrients is less frequently considered. However, according to the World Health Organization [1], two billion people around the world suffer from micronutrient deficiencies (mostly Fe, Cu, Zn and Mg), causing 7.3% of disease burden. Among the micronutrients involved in human metabolism, Cu deficiency, for example, causes symptoms of diseases such as immune defects or anaemia [2]. Without stock or sufficient intake, undernourished children are the first affected [3], however Cu deficiency is also observed in developed countries [4,5]. In the human diet, Cu is mainly present in meat (especially liver and offal), fish and nuts [6]. However, these food supplies are not available to most populations suffering from micronutrients deficiencies. As a consequence, staple foods (cereals, tubers and oils) must be fortified in order to provide the minimum dose of essential micronutrients, such as Cu, for these populations [7].

Over the past 60 years, the micronutrient content (mostly Fe, Zn, Mg and Cu) has been reduced in edible products despite sufficient micronutrient availability in most cultivated soils, and this is the result of varietal selection that aimed to achieve higher yields [8,9]. Consequently, whether a genetic improvement of micronutrient uptake by roots would increase the edible plant content remains questionable. Alternatively, improving the transfer of such micronutrients in edible parts from remobilization of vegetative tissue could be an alternative for breeding if we assume that the recycling (i.e. remobilization) of such nutrients is a significant process. In addition, it has been shown that some soils can lack Cu resulting in dramatic agricultural effects [10]. Thus, Cu deficiency in plants affects human health either directly (decrease in yield) or indirectly (symptoms induced by a lack of micronutrients) [11,12].

As a transition metal [13], Cu is involved in numerous processes in plants. Most of Cu's functions in plants rely on enzymatically bound Cu (more than one hundred proteins identified so far [14]), mostly catalysing redox reactions. As a consequence, only 2% of plant Cu occurs in its free form. Chloroplasts contain about half of the total plant Cu, where it is especially associated with proteins involved in the electron transport chain such as plastocyanin [15]. Three Cu ions are also required for assembly of the active cytochrome c oxydase complex in the mitochondrial electron transport chain. Additionally, numerous proteins known to bind Cu or to be regulated by Cu are involved in the Calvin cycle or in the Tricarboxylic Acid Cycle (TCA), for example Fructose 1–6 Bis Phosphatase (FBPase), Glyceraldehyde 3 Phosphate DeHydrogenase (G3PDH) [16] and PhosphoGlycerate Kinase (PGK) [17]. Cu is also important for enzymes involved in cell detoxification, such as glutathione-S-transferase [18] or Cu-Zn superoxide dismutase [19], or in cell wall metabolism such as ascorbate and polyphenol oxidases [20].

From a genetic point of view, some studies have reported a high variation in Cu content between accessions of the same species such as spinach, pea [21] and cassava [22]. While it has been suggested that Cu management should be easily improved by breeding [4], mechanisms that could increase the Cu content in edible tissues, such as remobilization, have been poorly described in the literature.

An important agricultural crop, such as *Brassica napus*, usually requires high levels of nutrients and hence fertilization to reach an optimal yield with preserved quality. Indeed, winter oilseed rape is highly sensitive to sulphur (S) and nitrogen (N) deficiencies with negative consequences for yield and seed quality [23] and therefore, requires high doses of N and S fertilizers. These strong needs for fertilizers are partly due to the low (macro)-nutrient use efficiency (defined by the ratio of seed to plant nutrient contents) of oilseed rape. For example, it has been reported that its low N use efficiency results from a relatively inefficient endogenous N mobilization [24,25] that occurs mostly during leaf senescence [26]. In contrast, S remobilization from leaves to the seeds (linked to induction of tonoplastic sulphate transporters allowing the mobilization of sulphate previously stored in leaves [27]) is not triggered by leaf senescence but rather by S deficiency. Therefore, without S deficiency, the S content in fallen leaves remains high, up to 2% of dry weight. According to these studies, *Brassica napus* constitutes an important model of defective agro-environmental interactions in studies of nutrient remobilization from leaves. Moreover, it constitutes a relevant plant species model because its genetic proximity to *Arabidospis thaliana* provides easy access to molecular tools while maintaining an agronomic relevance.

To date, deciphering of the plant response to Cu deficiency has been performed through Fe and Zn interactions, identification and characterization of genes encoding transporters (such as *COPT* and *HMA* families), and monitoring target proteins and metabolites [28–31]. However, to our knowledge, the effect of Cu deficiency on growth and plant metabolism is currently poorly reported. Indeed, only one article [32] describes the use of a microarray approach to study modifications of the *Arabidopsis thaliana* transcriptome in response to Cu deficiency.

The first aim of this study was to quantify the effect of Cu deficiency on growth of *Brassica napus* and to assess the mobility of Cu between plant tissues (i.e. Cu remobilization). Moreover, a wider analysis of plant nutrients has been used to monitor the effect of Cu-deficiency on the uptake of some macro (Ca, K, Mg, N, P and S) and micronutrients (B, Fe, Mn, Mo, Na and Zn). Finally, a molecular approach combining proteomic and targeted transcriptomic studies has been used to identify the main metabolic pathways affected by Cu deficiency.

Materials and Methods

Growth conditions

Seeds of *Brassica napus* var. Boheme were surface-sterilized by exposure to 80% ethanol for 30 s followed by 20% sodium hypochlorite for 10 min. After 10 washes with demineralised water, seeds were germinated on perlite over demineralised water for 2 days in the dark and 1 week under natural light in a greenhouse. Just after first leaf emergence, seedlings were transferred to a 20 L tank containing the following nutrient solution: KNO_3 1.25 mM, $Ca(NO_3)_2$ 1.25 mM, KH_2PO_4 0.25 mM, $MgSO_4$ 0.5 mM, EDTA, 2 NaFe 0.2 mM, H_3BO_3 0.01 mM, $MnSO_4$ 5 µM, $ZnSO_4$ 3 µM, $(NH_4)_6Mo_7O_{24}$ 0.7 µM, $CoCl_2$ 0.1 µM, $NiCl_2$ 0.04 µM, SiO_2 0.1 mM, $CaCl_2$ 20 1.25 mM, KCl 0.25 mM. Control nutrient solution also contained NaOH 0.1 mM and $CuSO_4$ 0.7 µM while Na_2SO_4 1 µM was added to Cu-depleted nutrient solution. These nutrient solutions were renewed every two days. Plants were grown under greenhouse conditions with a thermoperiod of 20°C/17°C day/night and a photoperiod of 16 h. Natural light was supplemented with high pressure sodium lamps (Philips, MASTER GreenPower T400W) supplying an average photosynthetically active radiation of 280 µmol photons.$m^{-2}.s^{-1}$ at canopy height. After one week of growth, plants were separated into 2 sets: control plants receiving normal nutrient solution and Cu-deficient plants receiving the Cu-depleted nutrient solution (for details, see above) over 25 days. Four independent samples, each consisting of three plants, were harvested at the beginning of Cu depletion (t = 0) and after 25 days with (control) or without Cu (–Cu). Leaves and petioles present at the beginning of Cu depletion (referred as "old leaves" and "old petioles", respectively) were distinguished from leaves appearing during Cu depletion (referred as "young leaves" and "young petioles", respectively). At each date of harvest (t = 0 and t = 25 days), whole roots from control and Cu-depleted plants were collected. An aliquot of each tissue was weighed and dried in an oven (60°C) for dry weight (DW) determination and ground to fine powder for IRMS and ICP-OES analysis. Thereafter, the remaining fresh tissues were frozen in liquid nitrogen and stored at –80°C for transcriptomic and proteomic analyses.

Every five days throughout the culture period, non-destructive determination of chlorophyll content of young and old leaves was performed using a SPAD chlorophyll meter (SPAD-502 model, Minolta, Tokyo, Japan). The determination was carried using three replicates of ten measurements performed on independent leaves.

Analysis of nutrients in plant tissues

For the analysis of total N and S contents, an aliquot of around 4 mg DW of each plant organ sample was placed in tin capsules for total N and S analysis using an IRMS spectrometer (Isoprime, GV Instrument, Manchester, UK) linked to a C/N/S analyser (EA3000, Euro Vector, Milan, Italy). The total amount of N or S (N_{tot} or S_{tot}) in a tissue "i" at a given time "t" is calculated:

$$N_{tot}(\text{or } S_{tot}) = \%\ N_{i,t}(\text{or } S_{i,t}) \times DWi,t/100$$

Other nutrients (K, Ca, S, P, Mg, Fe, Na, Mn, B, Si, Zn and Cu) were analysed by Inductively Coupled Plasma Optical Emission Spectrometry (ICP-OES, Thermo Elemental Co. Iris Intrepid II XDL) with prior microwave acid sample digestion (8 mL of concentrated HNO_3 and 2 mL of H_2O_2 for 0.5 g DW), using a protocol previously described by Mora *et al.* [33].

RNA extraction

Total RNA was extracted from 200 mg of frozen samples ground to a powder with a pestle in a mortar containing liquid nitrogen. The resulting powder was suspended in 750 µl of extraction buffer [0.1 M TRIS, 0.1 M LiCl, 0.01 M EDTA, 1% SDS (w/v), pH 8] and 750 µl of hot phenol (80°C, pH 4). This mixture was vortexed for 30 s and, after addition of 750 µl of chloroform/isoamylalcohol (24:1, v/v), the homogenate was centrifuged (15 000 g, 5 min, 4°C). The supernatant was transferred into a 4 M LiCl solution and incubated overnight at 4°C. After centrifugation (15 000 g, 30 min, 4°C), the pellet was suspended in 100 µl of sterile water. RNA was then purified with an RNeasy mini kit according to the manufacturer's protocol (Qiagen, Courtaboeuf, France). Quantification of total RNA was performed by spectrophotometry at 260 nm (BioPhotometer, Eppendorf, Le Pecq, France) before Reverse Transcription (RT) and real time Quantitative Polymerase Chain Reaction (Q-PCR) analysis.

Reverse transcription (RT) and Q-PCR analysis

For RT, 1 µg of total RNA was converted to cDNA with an "iScript cDNA synthesis kit" according to the manufacturer's protocol (Bio-Rad, Marne-la-Coquette, France).

Q-PCR amplifications were performed using specific primers for each housekeeping gene (*EF1-α* and *18S rRNA*) and target genes (Table 1). Q-PCRs were performed with 4 µl of 200x diluted cDNA, 500 nM of primers, and 1x SYBR Green PCR Master Mix (Bio–Rad, Marne–la–Coquette, France) in a real-time thermocycler (CFX96 Real Time System, Bio–Rad, Marne–la–Coquette, France). A 2 step program, composed of 42 cycles of a denaturing step at 95°C for 15 s followed by an annealing and extending step at 60°C for 40 s, was used for all pairs of primers (Table 1) except for *COPT2*, for which we used a 3 step program. In this case the denaturing step was as described previously. Annealing was at 58.6°C for 10 s and was followed by an extending step at 72°C for 30 s. For each pair of primers, a threshold value and PCR efficiency were determined using a cDNA preparation diluted >10-fold. For all pairs of primers, PCR efficiency was around 100%. The specificity of PCR amplification was examined by monitoring the presence of the single peak in the melting curves after Q-PCRs and by sequencing the Q-PCR product to confirm that the correct amplicon was produced from each pair of primers (Eurofins, Ebersberg, Germany). The relative expression of the genes in each sample was compared with the control sample [corresponding to untreated plants at the same time of harvest] and was determined with the delta delta Ct (ΔΔCt) method using the following equation [34]: Relative expression $= 2^{-\Delta\Delta Ct}$, with $\Delta\Delta Ct = \Delta Ct_{treated} - \Delta Ct_{control}$ and with $\Delta Ct = Ct_{target\ gene} - [\sqrt{(Ct_{EF1-\alpha} \times Ct_{18S\ rRNA})}]$, where Ct refers to the threshold cycle determined for each gene in the exponential phase of PCR amplification and $[\sqrt{(Ct_{EF1-\alpha} \times Ct_{18S\ rRNA})}]$ corresponds to the geometric average of Ct of the reference genes. Using this analysis method, the relative expression of the different genes in the control sample of the experiment was equal to 1 [34], and on this basis the relative expression of other treatments was then compared with the control. For *COPT2*, because no transcript was detected in control plants (*i.e.* $Ct_{control}$ is undetermined), values are not expressed relatively to control but as $2^{-\Delta Ct}$.

Extraction and determination of total proteins

Two hundred mg of fresh matter from old leaf samples were ground to a fine powder in liquid nitrogen in the presence of 50 mg of poly(vinylpolypyrrolidone) (PVPP). The addition of PVPP is used to fix plant polyphenols that might interfere with the quantification of proteins or during separation of proteins by electrophoresis. The ground material was dissolved in 1.75 mL of TCA/acetone solution (10% TCA (w/v) prepared in acetone). After centrifugation (3 min, 16 000 g, 4°C), the protein pellet was purified according to the protocol adapted from Wang *et al.* [35]. The protein pellet obtained after precipitation with TCA/acetone (10% TCA (w/v) was suspended in 1.75 mL of 0.1 M ammonium acetate dissolved in 80% methanol. After homogenization and centrifugation (16 000 g, 3 min, 4°C), the pellet was washed with 1 75 mL of 80% acetone and centrifuged again (16 000 g, 3 min, 4°C). The supernatant was removed and the pellet was dried under vacuum (Speedvac Concentrator 5301, Eppendorf, France) for 5 min at 50°C and then suspended with 0.8 mL of phenol at pH 7.9 and in 0.8 mL of dense SDS buffer (30% sucrose, 2% SDS, 0.1 M Tris–HCl, pH 8.0, 0.5% 2–mercaptoethanol). After 5 min incubation at 4°C and centrifugation (16 000 g, 3 min, 4°C), the phenol phase was transferred to a new tube and supplemented with 1.75 mL of 0.1 M ammonium acetate and stored at –20°C overnight. Afterwards, ammonium acetate was used to precipitate proteins to enable their collection by centrifugation (16 000 g, 5 min, 4°C). The protein pellet was then washed with 1.75 mL of 100% methanol and again with 1.75 mL of 80% acetone. Residual acetone was removed by vacuum evaporation over a few minutes. The pellet was resuspended in 400 µL of R2D2 rehydration buffer [5 M urea, 2 M thiourea, 2% CHAPS, 2% N–decyl–N,N–dimethyl–3–ammonio–1–propanesulfonate, 20 mM dithiothreitol, 5 mM Tris (2–carboxy– ethyl) phosphine, 0.5% IPG buffer] (GE Healthcare, Saclay, France), pH 4 to 7 [36]. The total protein concentration was determined by the method previously described by Bradford [37] using bovine serum albumin as standard.

Table 1. Q-PCR primer sets.

Gene	Accession number	Forward	Reverse	Gene Function
EF1-α	DQ312264	5'-gcctggtatggttgtgacct-3'	5'-gaagttagcagcacccttgg-3'	
18S rRNA	GQ380689	5'-cggataaccgtagtaattctag-3'	5'-gtactcattccaattaccagac-3'	
COPT2	NM_114557	5'-tgcacatgaccttcttttgg-3'	5'-gtcatcggagggtttgttga-3'	Cu uptake
HMA1	NM_119890.6	5'-gtacagctgaccgaggaagc-3'	5'tgcccataaatgggttcaat-3'	Cu allocation to chloroplast
MOT1	NM_128127	5'-ctcgccaggatttggactta-3'	5'agatccccaacacgaacaag-3'	Mo uptake

EF1α and *18S rRNA* were housekeeping genes used for relative gene expressions by Q-PCR analysis.

Two-dimensional electrophoresis (2-DE) and image analysis

2-DE was performed according to the protocol detailed by Desclos *et al.* [25]. Gels were stained using the silver-staining procedure described by Blum *et al.* [38] and scanned with a ProXPRESS 2D proteomic imaging system (Perkin-Elmer, Courtaboeuf, France). Images of the 2-DE gels were analysed using the Progenesis SameSpots software v3.0 (Nonlinear Dynamics, Newcastle upon Tyne, UK) according to the manufacturer's protocol. Gels from three independent biological replicates were used. Spot detection, warping, and matching were performed automatically by the software and manually validated. Artefacts due to non-specific silver nitrate staining or spots that could not be confidently verified as true matches were disregarded rather than manually edited, and misalignments were corrected by manual warping when appropriate. Mw and *pI* were calculated using SameSpots software calibrated with commercial molecular mass standards (prestained precision protein standards; Bio-Rad, Marne-la-Coquette, France) run in a separate marker lane on 2-DE gel.

RuBisCO relative quantification

In order to observe maximum spots, silver staining was performed on 2-DE with a saturated RuBisCO signal. The amount of RuBisCO was then measured by Experion Pro260 Analysis Kit (Bio-Rad, Marne-la-Coquette, France). The Experion automated electrophoresis system applies a combination of microfluidic separation technology and sensitive fluorescent detection of proteins [39]. The protein extracts were treated with the Experion reagents and then separated on the Experion automated electrophoresis station according to the manufacturer's instruction. Peak areas corresponding to large and small RuBisCO subunits were then compared using software given by manufacturer.

Protein Identification by ESI LC-MS/MS

Spots of interest were excised from 2D gels and washed several times with water and dried for a few minutes. Trypsin digestion was performed overnight with a dedicated automated system (MultiPROBE II, Perkin-Elmer). The gel fragments were subsequently incubated twice for 15 min in a 0.1% CH_3CN solution in water to allow extraction of peptides from the gel pieces. Peptide extracts were then dried and dissolved in starting buffer for chromatographic elution, consisting of 3% CH_3CN and 0.1% HCOOH in water. Peptides were enriched and separated using lab-on-a-chip technology (Agilent, Massy, France) and fragmented using an on-line XCT mass spectrometer (Agilent). The ESI LC-MS/MS data were converted into DTA- format files that were further searched for proteins with MASCOT Daemon (Matrix Science [40]). For protein identification, two strategies were employed to mine the maximum data. Measured peptides were searched in the NCBI nr-protein sequence database, *viridiplantae* (green plants [41]), and in the *Brassica* EST database (*Brassica* Genome Gateway 2007, [42]). Proteins with two or more unique peptides matching the protein sequence with a score >53, as defined by MASCOT, were considered as a positive identification. The spectra of each peptide were verified manually. In cases where protein identification data were lacking, BLAST analysis was performed [43].

Data and statistical analysis

Regarding the growth, ICP-OES and IRMS analyses, all experiments were conducted with 4 independent biological replicates each corresponding to 3 plants. Proteomics and Q-PCR analyses were performed on 3 independent biological replicates each corresponding to 3 plants. Data are given as mean ± SE for n = 3 or 4. For Q-PCR, SE and Student's T-test were based on ΔΔCt (or ΔCt for *COPT2*, see above for more information), when the relation between values is still linear. According to Yuan *et al.* 2006 [44], a confidence interval of each mean value (with min and max values) can be calculated and has been presented in the results section as [mean of relative expression] (Min value-Max value) with Min value = $2^{-(\Delta\Delta Ct+SE)}$ and Max value = $2^{-(\Delta\Delta Ct-SE)}$ (Min value = $2^{-(\Delta Ct+SE)}$ and Max value = $2^{-(\Delta Ct-SE)}$ for *COPT2*). Conversion of the results of an exponential process into a linear comparison of amounts leads to an asymmetric distribution [34].

All data were analysed with Student's T-test ($p<0.05$) and marked by an asterisk (*) or cross (†) when significantly different.

Results

Cu depletion affects the growth and Cu content of *B. napus*

After 25 days of culture, control whole plant DW increased from 3.23±0.24 g to 39.67±1.26 g (table 2). The total DW of Cu-depleted plants was decreased by 18.4% relative to control plants (table 2, from 39.67 to 32.37 g DW.plant^{-1}). This was the result of a significant reduction in root DW (from 4.11±0.23 in control plants to 3.73±0.06 g DW.plant^{-1} in Cu-depleted plants) and secondly, of old petioles DW (from 8.07±0.60 to 6.31±0.49 g DW.plant^{-1}). No significant differences in growth rates were found for other organs. Control plants accumulated 251.5 µg of Cu while the Cu content in Cu-depleted plants only rose from 29.0±4.0 µg to 74.8±3.0 µg. The slight increase in total Cu in Cu-depleted plants was the result of a trace of Cu found in the mineralized water used for the nutrient solution (0.47±0.00 µM, data not shown). In roots of control plants, the amount of Cu increased from 14.7±3.0 µg to 155.2±13.2 µg but only reached 43.4±1.5 µg in Cu deficient plants. In aerial tissues that developed under Cu depletion, the Cu amount was reduced from 11.8±1.9 µg to 2.6±0.7 µg and from 76.2±4.8 µg to 16.2±2.3 µg in petioles and leaves, respectively. In petioles already present at the time of the Cu depletion, the amount of Cu rose from 2.1±0.2 µg to 8.0±0.7 µg during depletion while it reached 20.3±2.4 µg in control plants at the end of the experiment. The amount of Cu in old leaves of control plants did not vary significantly during the experiment, while in Cu-depleted plants, it was reduced from 12.2±1.1 µg to 4.7±1.0 µg indicating a remobilization of 61.4±4.2% of the Cu initially present in these leaves.

No significant difference was observed for RuBisCO by *Experion* method (Figure 1.A). During the whole experiment, the leaf chlorophyll content (SPAD values, Figure 1.B) were not significantly different between control and Cu-depleted plants. At the end of the experiment, this was confirmed in old leaves (Figure 1.C). Also, N content in old leaves didn't present any difference between the two conditions (Figure 1.D).

Cu depletion increases expression of Cu transporters in *Brassica napus*

The relative expression of genes involved in Cu transport (*HMA1, COPT2*) is given in Figure 1. Expression of the *HMA1* gene encoding a chloroplastic Cu transporter was monitored in leaves, whereas *COPT2*, which encodes a transporter involved in Cu uptake, was monitored in roots. The relative number of *HMA1* transcripts increased 2.68 fold (2.23-3.24) after 25d of Cu

Table 2. Biomass and Cu content in *Brassica napus* L. at t=0 and after 25 days of culture with (control) or without Cu (−Cu).

		t=0	Control (25d)	-Cu (25d)
Old leaves	DW (g)	2.13±0.17	4.78±0.79	4.83±0.93
	Cu amount (µg)	12.2±1.1	17.0±2.5	4.7±1.0*†
	% remobilized	NA	NA	60.36±4,15
Old petioles	DW (g)	0.63±0.04	8.07±0.60	6.31±0.49*†
	Cu amount (µg)	2.1±0.2	20.3±2.4*	8.0±0.7*†
Young leaves	DW (g)	NA	12.7±0.94	8.83±1.64
	Cu amount (µg)	NA	76.2±4.8	16.2±2.3 †
Young petioles	DW (g)	NA	10.02±0.78	8.66±0.55
	Cu amount (µg)	NA	11.8±1.9	2.6±0.7 †
Roots	DW (g)	0.48±0.03	4.11±0.23*	3.73±0.06*
	Cu amount (µg)	14.7±3.0	155.2±13.2*	43.4±1.5*†
Total	DW (g)	3.23±0.24	39.67±1.26*	32.37±1.71*†
	Cu amount (µg)	29.0±4.0	280.5±13.5*	74.8±3.0*†

* and † represent significant differences at $p<0.05$ compared to t=0 and control, respectively. NA: Not Applicable.

depletion (Figure 2.A). Transcripts of *COPT2* were detected only in roots from Cu-depleted plants suggesting that the expression of the *COPT2* transporter was up-regulated and/or de-repressed in *Brassica napus* roots in response to Cu deficiency (Figure 2.B).

Cu depletion affects uptake of some mineral nutrients

Figure 3 shows that Cu deprivation decreased Cu uptake by 81.8±2.0%, but also affected the uptake of Na (−41.0±3.8%), B (−31.6±2.1%) and Mg (−28.5±2.8%). while plant dry weight was reduced by 18.4±2.5% (Table 2). Cu deficiency had no effect on total plant uptake for nutrients such as N, Ca, K, S, P, B, Fe, Mn and Zn, despite the difference of biomass. In contrast, uptake of Mo was increased by 121.8% despite the reduced plant growth and was associated with a very strong up-regulation of the molybdenum transporter 1 gene (*MOT1*, Figure 2.C), having an important role in efficient Mo uptake [45,46]. Indeed, in Cu-depleted plants, the number of *MOT1* transcripts was found to be 159.61 fold (127–200) higher than in control plants.

Cu depletion induces proteomic modifications

Leaves already present at the beginning of Cu depletion were the most affected tissues, not only due to their decreased Cu content and hence the mobilization of Cu (by about 60%) to other tissues, but also according to their decrease of Mg (−42.1±11.0%), Na (−42.6±13.6%), Ca (−35.8±10.0%) and Fe (−27.9±13.5%) contents. Therefore, leaves were selected for a comparative proteomic approach in order to highlight the main metabolic modifications induced by Cu depletion. Representative 2D gels corresponding to protein extracts from leaves of control and Cu-depleted plants are shown in Figure 4. On each gel, 693 proteins were identified. Their comparative analysis revealed 48 proteins significantly up- or down-regulated by Cu depletion. Because different spots correspond to different isoforms of the same protein, only 33 different proteins were identified. Among them, 25 proteins were accumulated in Cu-depleted old leaves. Transketolase-like protein (#7 to 11, Figure 4 and Table 3) was the strongest protein induced with four isoforms that were accumulated more than fourfold compared to control. On the other hand, repression was significant in only 8 of the 33 identified

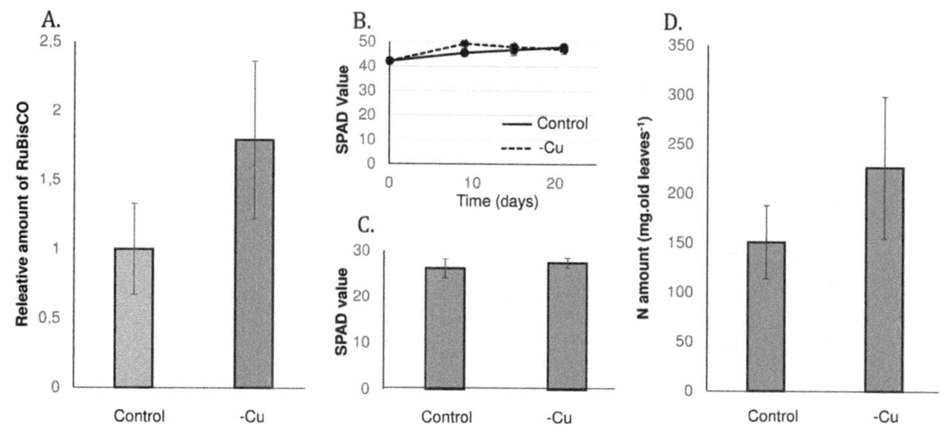

Figure 1. Parameters associated with senescence. (A.) Relative amount of RuBisCO in old leaves. Relative chlorophyll content (SPAD value) during (B.) the whole experiment at whole plant level and (C.) after 25 days of depletion in old leaves (D.) N content in old leaves.

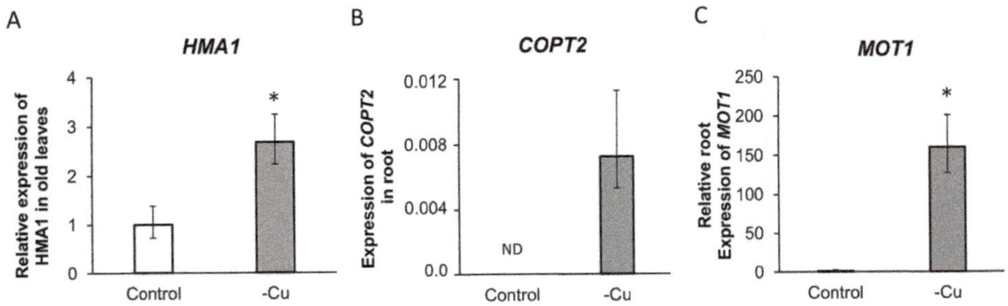

Figure 2. Relative accumulation of transcript of Cu transporters (A. *HMA1* in old leaves and B. *COPT2* in roots) and Molybdenum transporter 1 (C. *MOT1* in roots) in *Brassica napus* L after 25 days of culture with (control) or without Cu (−Cu). Control is represented by white bars and –Cu by grey bars. *MOT1* and *HMA1* values are expressed as $2^{-\Delta\Delta Ct}$ while *COPT2 value* is expressed as $2^{-\Delta Ct}$ (for details, see "Material and Methods"). * represents significant differences at $p<0.05$. ND: not detected.

proteins, with the Oxygen Evolving Enhancer (OEE, #33 to 37) having the greatest repression in our study (−2.3 to −3.93). Among all proteins (Table 3), five were involved in stress responses (disease/defence or defence/secondary metabolite), for example glutathione reductase (#18), and seventeen proteins were involved in energy processes, for example aldolase (#27) and ATP synthase (#17). Protein annotations in databases also indicated that chloroplasts were the main locations of the identified proteins with 17 references, including the most repressed, OEE (#33 to 37), and the most induced, Transketolase-like protein (#7 to 11).

Considering some of the characteristic Cu proteins reported in the literature [47], Plastocyanin was too small (around 10 kDa) to be found on our gels while Cu-Zn superoxide dismutase and Polyphenol oxidase were not differentially expressed and/or detected.

Discussion

Cu remobilization from old leaves is induced by Cu deficiency

In this study, the effectiveness of Cu deficiency can be attested by the low Cu concentration found in old and young leaves of Cu depleted plants (0.97±0.03 $\mu g.g^{-1}$ DW and 1.66±0.40 $\mu g.g^{-1}$ DW, respectively) which was lower than the usually admitted minimum concentration (5 $\mu g.g^{-1}$ DW) [48]. Using Cu deficiency we were able to demonstrate a decrease in the Cu content of leaves by 61.4±4.2% (Table 2). To our knowledge, alongside work on *Arabidopsis thaliana* [49], this is one of the few studies that shows that Cu can be recycled between leaves. However, the Cu remobilization could be underestimated in our conditions because some traces of Cu were found in the nutrient solution, as suggested by the slight increase in the overall Cu content in Cu-depleted plants (Table 2). A precise estimation of Cu remobilization would require the use of stable isotopes. This remobilization of Cu in leaves was concomitant with an increase in *COPT2* and *HMA1* expression. Despite a tendency of an increase of RuBisCO and N amounts, the RuBisCO, N amounts and chlorophyll contents in old leaves were not significantly differentially affected between control and Cu-depleted plants (Figure 1). Then we could assume that Cu remobilization was not associated with leaf senescence. This is also consistent with the down-regulation of senescence-associated proteases such as FtsH2 (#14, Table 3), a protease involved in the degradation of photosystem II [26,50]. Thus, this study suggests that Cu depletion induces a Cu mobilization independent of the distinguishing marks of senescence. This is different to what has been found for N for example, whose deficiency is known to increase leaf senescence in *Brassica napus* [25].

Effect of Cu deficiency on metabolic pathway in chloroplasts from old leaves of *B. napus*

Because the spots of our 2D–E gels correspond to the most abundant proteins in old leaves, numerous proteins (especially transporters and transcription factors) corresponding to transcripts

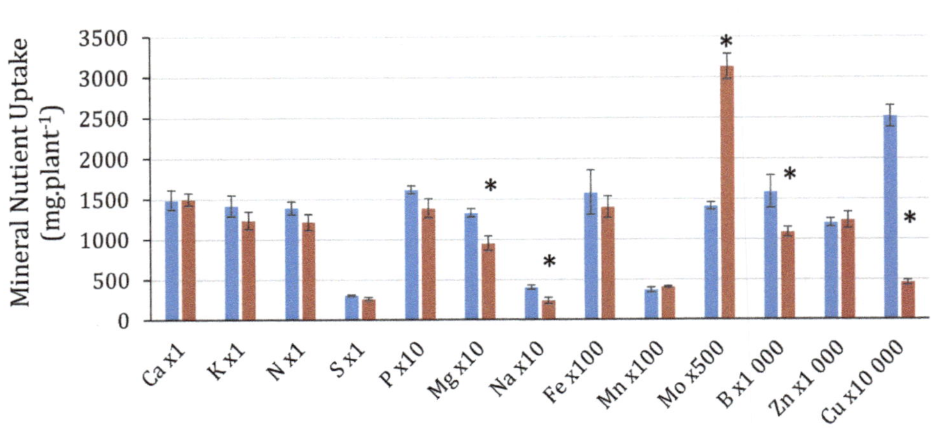

Figure 3. Mineral nutrient uptake in control plants (blue) and Cu-depleted plants over 25 Days (red). Due to large differences between nutrients, factors have been attributed to each nutrient to allow representation on the same graph. * indicates significant difference between control and Cu depleted plants at $p=0.05$.

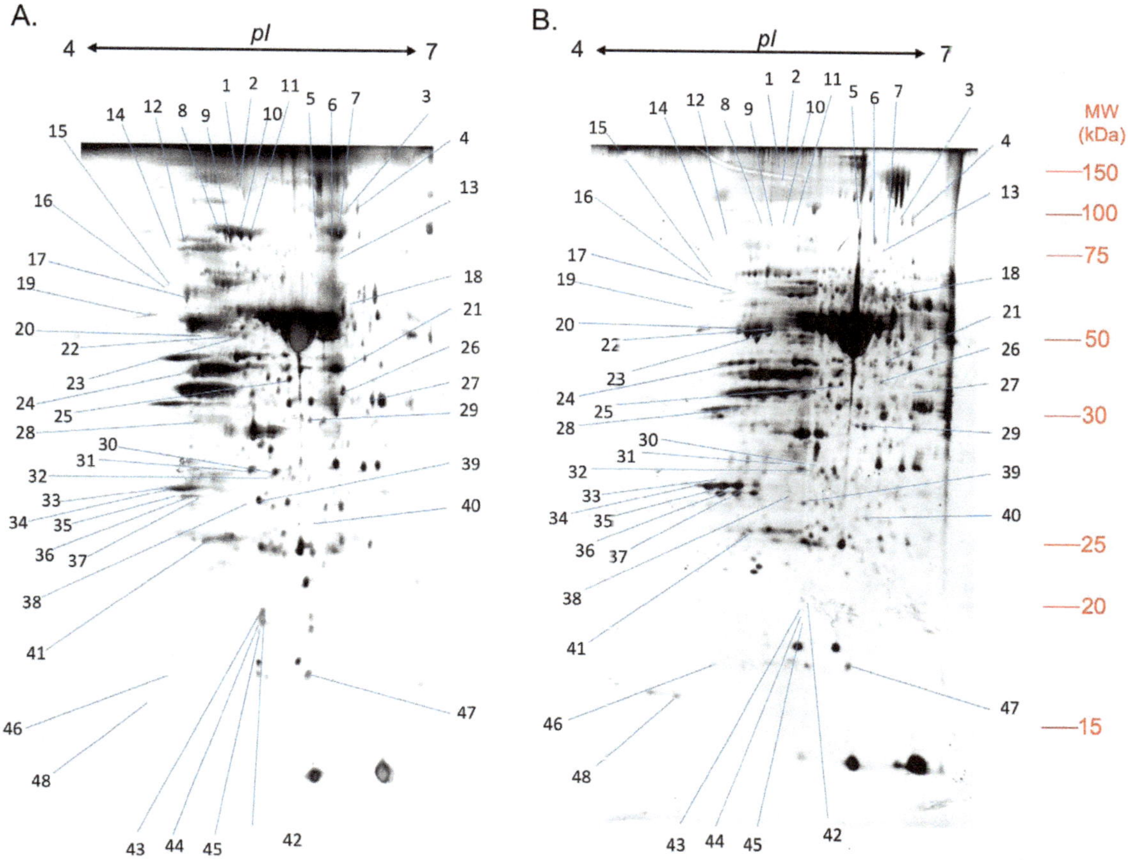

Figure 4. Representative Silver-stained 2-DE gels of proteins from old leaves of *Brassica napus* L. after 25 days of culture with (control, A) or without Cu (−Cu, B). A total of 150 µg of total proteins was loaded on an IEF strip forming an immobilized pH linear gradient from 4 to 7 (for details, see "Materials and Methods"). Numbers indicate sequenced and identified proteins presented in table 3. Molecular mass markers (MW in kDa) are listed on the right side of the gel.

identified in previous studies as regulated by Cu depletion could not be observed in the present work. Thus, this study complement previous transcriptomic data on Cu remobilisation [51,52]. This proteomic analysis of Cu-deficient old leaves showed that more than half of the up or down-regulated proteins were localised in chloroplasts (Table 3) indicating potential disturbance, despite an absence of chlorophyll content variation (Figure 1.B, C). Five of these proteins are involved in the Calvin cycle and 5 are involved in the thylakoid electron transport chain (Figure 5). From a general point of view, among the 33 proteins that were differentially expressed relatively to control plants (Table 3), 11 of them are known to rely on the presence of Cu for their synthesis and/or activity, and 9 were over expressed (#15, 16, 17, 19, 27, 29, 39 and 43). As a consequence, it could be suggested that the lack of Cu reduced their activity which may explain their up-regulation (or alternatively reduced their down-regulation). A similar regulation would also occur for enzymes that are located upstream or downstream of such copper deficiency up-regulated proteins. For example, this is the case (Figure 5) within the Calvin cycle for Transketolases (#6–11) or TPI (#41), which are located upstream and downstream of Aldolase (#27), a copper dependent protein that was up-regulated by Cu-deficiency.

Among Cu binding or regulated enzymes within the Calvin cycle (FBPase [53], GAPDH, Aldolase [16], RuBisCO [18], PGK [17]), only Aldolase (#27) and PGK (#21) were found to be over represented (Table 3 and Figure 5). However, whilst RuBisCO was not significantly affected (Figure 1), its main regulator RuBisCO activase (#24) was increased by about 3 folds (Table 3). It should be noted that PGK, Aldolase, Transketolase and TPI are also involved in glycolysis (in the cytosol or plastids [54]) and their regulation should therefore induce mitochondrial defects.

GAP is a crosspoint between numerous pathways, with B6 vitamin biosynthesis being one of them [55]. However, under an impaired Calvin cycle, more GAP would be available for this cycle (Figure 5). Vitamin B6 can also be synthesized through Pyridoxin biosynthesis protein (#32, Table 3) which is over represented in old leaves of Cu-depleted *B. napus*. The B6 dependent protein Glutamate-1-semialdehyde aminomutase (GSA, #25) [56] with a role in chlorophyll biosynthesis also accumulated (Figure 5).

Differentially expressed proteins located in thylakoids (Figure 5) could be subdivided into two groups: the first one including OEE (#33 to 37), NADPH subunit 48 (#20) and ATP synthase (#17) were down-regulated, which is consistent with previous work reporting a down-regulation of photosynthesis-related genes in response to Cu deficiency [51]. The second group corresponded to [Fe-S] cluster containing proteins such as Ferredoxin (#30 and 31) and Rieske domain protein (#47), which were over expressed in response to Cu depletion. We suggest that overall the electron transport chain was down-regulated, but an impairment of [Fe-S] cluster biosynthesis could be compensated by an up-regulation of *de novo* synthesis of [Fe-S] proteins. This stimulation of [Fe-S] cluster proteins can be linked to the accumulation of Rhodanese

Table 3. Abundance of identified proteins over (positive value) or under (negative value) accumulated in old leaves of Cu-depleted *B. napus* relative to control plants.

Spot no.	Relative Abundance	Exp. pI/ Mw	Theo. pI/Mw	PM	SC (%)	Protein name/Organism/NCBI accession no.	Functional classification/Sub-Cellular localization	Note
1	3.12	5.7/115	5.39/104.3	30	20	Myrosinase binding protein [*Brassica napus*]/gi|1711296	11. Disease/Defence	
2	2.11	6.3/99	6.59/102.2	117	40	Chaperone protein ClpC, chloroplastic-like [*Fragaria vesca* subsp. vesca]/gi|470122943	08. Intracellular traffic/Chloroplast	
3	1.75	6.64/90	6.12/84.3	58	24	Cobalamin-independent methionine synthase [*Arabidopsis lyrata* subsp. lyrata]/gi|297807807	01. Metabolism	**Zn** [73]
4	2.78	6.53/91	6.12/84.3	29	19	Cobalamin-independent methionine synthase [*Arabidopsis lyrata* subsp. lyrata]/gi|297807807	01. Metabolism	**Zn** [73]
5	4.16	6.26/80	6.16/81.2	13	13	NADH Ubiquinone oxydoreductase 75 kD subunit [*Arabidopsis thaliana*]/gi|30693102	02. Energy/Chloroplast-Mitochondria	[Fe-S] cluster
6	5.23	6.5/81	5.80/81.5	15	18	Transketolase-like protein [*Arabidopsis thaliana*]/gi|7329685	02. Energy/Chloroplast-Mitochondria	
7	2.22	6.41/80	6.12/79.9	13	10	Transketolase [*Arabidopsis thaliana*]/gi|30689983	02. Energy/Chloroplast-Mitochondria	
8	4.79	5.74/75	6.16/81.2	46	24	Transketolase-like protein [*Arabidopsis thaliana*]/gi|7329685	02. Energy/Chloroplast-Mitochondria	
9	4.17	5.78/75	6.16/81.2	61	27	Transketolase-like protein [*Arabidopsis thaliana*]/gi|7329685	02. Energy/Chloroplast-Mitochondria	
10	4.46	5.67/76	6.16/81.2	62	29	Transketolase-like protein [*Arabidopsis thaliana*]/gi|7329685	02. Energy/Chloroplast-Mitochondria	
11	4.23	5.54/74	6.16/81.2	44	26	Transketolase-like protein [*Arabidopsis thaliana*]/gi|7329685	02. Energy/Chloroplast-Mitochondria	
12	2.06	5.25/73	5.08/70.7	49	37	Heat shock cognate protein HSC70 [*Brassica napus*]/gi|2655420	08, intracellular traffic/Chloroplast	*Zn* [16]
13	−2.07	6.48/70	5.56/63.4	10	11	Phosphoglucomutase [*Arabidopsis thaliana*]/gi|15223226	02. Energy/Chloroplast	
14	−2.03	5.17/69	5.62/72.0	25	16	ATP-dependent zinc metalloprotease FTSH 2, chloroplastic-like [*Brachypodium distachyon*]/gi|357123383	06. Prot. Dest. & Storage/Chloroplast	***Zn*** [74]
15	1.54	6.14/64	5.22/49.4	11	7	Thylakoid rhodanese-like protein [*Arabidopsis thaliana*]/gi|18411523	02. Energy/Chloroplast	***Cu*** and ***Zn*** [75]
16	1.4	6.08/62	5.66/47.6	5	4	Thylakoid rhodanese-like protein [*Arabidopsis thaliana*]/gi|18411523	02. Energy/Chloroplast	***Cu*** and ***Zn*** [75]
17	2.63	5.25/62	5.97/47.0	85	42	ATPase subunit I [*Arabidopsis lyrata* subsp. lyrata]/gi|297837977	02. Energy/Chloroplast-Mitochondria	*Cu* and *Zn* [16]
18	−2.67	6.52/61	8.03/60.6	15	19	Glutathione reductase [*Brassica juncea*]/gi|13448672	20. Defence & Secondary metabolism/Vacuole	*Cu* [16]
19	2.09	4.92/55	5.23/50.6	7	5	Rhodanese-like domain-containing protein 4, chloroplastic-like [*Cicer arietinum*]/gi|502179307	01. Metabolism/Chloroplast	***Cu*** [75]
20	−2.27	5.45/50	6.49/51.0	19	16	NAD(P)H dehydrogenase subunit 48 [*Arabidopsis thaliana*] gi|18394307	02. Energy/Chloroplast	
21	1.52	6.44/47	5.91/50.1	25	21	Phosphoglycerate kinase 1 [*Arabidopsis thaliana*]/gi|15230595	02. Energy/Chloroplast-Mitochondria	*Cu* [17]
22	1.92	5.58/53	6.06/54.2	19	19	Mitochondrial processing peptidase alpha subunit, putative [*Arabidopsis thaliana*] gi|21594004	06. Prot. Dest. & Storage/ Mitochondria	
23	2.81	5.53/52	6.06/54.2	7	11	Mitochondrial processing peptidase alpha subunit, putative [*Arabidopsis thaliana*] gi|21594004	06. Prot. Dest. & Storage/ Mitochondria	
24	2.97	5.11/46	6.78/47.9	61	46	Chloroplast ribulose-1,5-bisphosphate carboxylase/oxygenase activase [*Brassica oleracea*]/gi|383470439	02. Energy/Chloroplast	

Table 3. Cont.

<table>
<tr><th>Spot no.</th><th>Relative Abundance</th><th>Exp. pI/Mw</th><th>Theo. pI/Mw</th><th>PM</th><th>SC (%)</th><th>Protein name/Organism/NCBI accession no.</th><th>Functional classification/Sub-Cellular localization</th><th>Note</th></tr>
<tr><td>25</td><td>1.7</td><td>6.03/44</td><td>6.43/50.2</td><td>44</td><td>32</td><td>Glutamate-1-semialdehyde 2,1-aminomutase 2 [*Arabidopsis thaliana*]/gi|15229018</td><td>01. Metabolism</td><td>B6 dependent [56]</td></tr>
<tr><td>26</td><td>1.42</td><td>6.44/42</td><td>5.91/15.1</td><td>16</td><td>45</td><td>Cinnamyl alcohol dehydrogenase [*Brassica napus*]/gi|6683959</td><td>01. Metabolism</td><td></td></tr>
<tr><td>27</td><td>1.46</td><td>6.64/40</td><td>6.22/38.4</td><td>19</td><td>37</td><td>Fructose-bisphosphate aldolase [*Arabidopsis thaliana*]/gi|15231715</td><td>02. Energy/Chloroplast-Mitochondria</td><td>*Cu* [16] and ***Zn*** [76]</td></tr>
<tr><td>28</td><td>1.47</td><td>5.24/37</td><td>5.14/32.0</td><td>9</td><td>35</td><td>40S ribosomal protein [*Brassica napus*]/gi|15214300</td><td>05. Protein synthesis/Ribosome</td><td></td></tr>
<tr><td>29</td><td>1.33</td><td>6.16/37</td><td>8.54/35.8</td><td>29</td><td>31</td><td>Mitochondrial NAD-dependent malate dehydrogenase [*Arabidopsis thaliana*]/gi|21592905</td><td>02. Energy/Mitochondria</td><td>*Cu* [16] and ***Zn*** [77]</td></tr>
<tr><td>30</td><td>1.72</td><td>5.7/32</td><td>8.32/40.3</td><td>49</td><td>34</td><td>Ferredoxin-NADP(+)-oxidoreductase 1 [*Arabidopsis thaliana*]/gi|15239282</td><td>02. Energy/Chloroplast</td><td>[Fe-S] cluster</td></tr>
<tr><td>31</td><td>1.56</td><td>5.8/32</td><td>8.66/40.1</td><td>34</td><td>29</td><td>Ferredoxin-NADP+ reductase [*Arabidopsis thaliana*]/gi|5730139</td><td>02. Energy/Chloroplast</td><td>[Fe-S] cluster</td></tr>
<tr><td>32</td><td>2.1</td><td>5.84/32</td><td>6.12/33.0</td><td>13</td><td>25</td><td>Pyridoxine biosynthesis protein [*Lotus japonicus*]/gi|72256517</td><td>11. Disease/Defence/Cytosol</td><td>B6 biosynthesis [55]</td></tr>
<tr><td>33</td><td>−3.93</td><td>4.88/31</td><td>5.92/35.0</td><td>5</td><td>15</td><td>Oxygen-evolving enhancer protein 1–2 [*Arabidopsis thaliana*]/gi|15230324</td><td>02. Energy/Chloroplast</td><td>***Mn*** [78]</td></tr>
<tr><td>34</td><td>−2.3</td><td>4.92/30</td><td>5.92/35.0</td><td>81</td><td>54</td><td>Oxygen-evolving enhancer protein 1–2 [*Arabidopsis thaliana*]/gi|15230324</td><td>02. Energy/Chloroplast</td><td>***Mn*** [78]</td></tr>
<tr><td>35</td><td>−2.56</td><td>4.94/30</td><td>5.93/35.1</td><td>36</td><td>48</td><td>Photosystem II subunit O-2 [*Arabidopsis lyrata* subsp. lyrata]/gi|297819782</td><td>02. Energy/Chloroplast</td><td>***Mn*** [78]</td></tr>
<tr><td>36</td><td>−2.95</td><td>5.03/30</td><td>5.93/35.1</td><td>67</td><td>52</td><td>Photosystem II subunit O-2 [*Arabidopsis lyrata* subsp. lyrata]/gi|297819782</td><td>02. Energy/Chloroplast</td><td>***Mn*** [78]</td></tr>
<tr><td>37</td><td>−2.8</td><td>5.32/30</td><td>5.92/35.0</td><td>84</td><td>55</td><td>Oxygen-evolving enhancer protein 1–2 [*Arabidopsis thaliana*]/gi|15230324</td><td>02. Energy/Chloroplast</td><td>***Mn*** [78]</td></tr>
<tr><td>38</td><td>1.69</td><td>5.61/29</td><td>8.65/33.7</td><td>13</td><td>11</td><td>Thioredoxin-like protein CDSP32 [*Arabidopsis thaliana*]/gi|15222954</td><td>11. Disease/Defence/Cytosol</td><td></td></tr>
<tr><td>39</td><td>3.81</td><td>5.73/29</td><td>5.54/29.5</td><td>23</td><td>17</td><td>Carbonic anhydrase 1 [*Arabidopsis thaliana*]/gi|30678347</td><td>01. Metabolism/Chloroplast</td><td>***Cu*** and ***Zn*** [79]</td></tr>
<tr><td>40</td><td>−1.81</td><td>6.17/27</td><td>6.28/27.6</td><td>29</td><td>34</td><td>Mitochondrial F1F0-ATP synthase subunit Fad [*Arabidopsis thaliana*]/gi|15227104</td><td>02. Energy/Mitochondria</td><td>Cu [18] and *Zn* [16]</td></tr>
<tr><td>41</td><td>1.55</td><td>5.28/26</td><td>7.67/33.3</td><td>10</td><td>16</td><td>Triosephosphate isomerase [*Arabidopsis thaliana*]/gi|15226479</td><td>02. Energy/Chloroplast-Mitochondria</td><td></td></tr>
<tr><td>42</td><td>2.41</td><td>5.73/20</td><td>5.37/19.9</td><td>9</td><td>26</td><td>Uncharacterized protein [*Arabidopsis thaliana*]/gi|18391006</td><td>13. Unclassified</td><td></td></tr>
<tr><td>43</td><td>2.79</td><td>5.7/19</td><td>5.47/35.7</td><td>41</td><td>38</td><td>Chloroplast beta-carbonic anhydrase [*Brassica napus*]/gi|297787439</td><td>01. Metabolism/Chloroplast</td><td>***Cu*** and ***Zn*** [79]</td></tr>
<tr><td>44</td><td>2.6</td><td>5.75/19</td><td>6.81/21.8</td><td>6</td><td>17</td><td>Germin-like protein [*Arabidopsis thaliana*]/gi|1755154</td><td>12. Unclear Classification</td><td></td></tr>
<tr><td>45</td><td>1.99</td><td>5.7/19</td><td>6.27/22.0</td><td>7</td><td>23</td><td>RecName: Full = Germin-like protein 1; Flags: Precursor [*Sinapis alba*]/gi|1169944</td><td>12. Unclear Classification</td><td></td></tr>
<tr><td>46</td><td>−2.6</td><td>4.99/17</td><td>9.12/24.7</td><td>18</td><td>31</td><td>Peroxiredoxin 2E [*Arabidopsis thaliana*]/gi|15231718</td><td>11. Disease/Defence/Cytosol</td><td></td></tr>
<tr><td>47</td><td>2.74</td><td>6.11/17</td><td>8.80/24.3</td><td>22</td><td>22</td><td>Rieske FeS protein [*Arabidopsis thaliana*]/gi|9843639</td><td>02. Energy/Chloroplast</td><td>[Fe-S] cluster</td></tr>
<tr><td>48</td><td>−1.76</td><td>4.58/16</td><td>5.24/17.9</td><td>6</td><td>15</td><td>Probable glycine cleavage system H protein 2 [*Arabidopsis thaliana*] gi|15223217</td><td>02. Energy/Chloroplast</td><td></td></tr>
</table>

Experimental and theoretical pI/Mw, the number of LC–MS/MS matched peptides (PM), the SCORE and the percentage of sequence coverage (SC) obtained are also indicated. For each protein, the assigned best–matched protein is listed with the organism in which it was identified and its GenBank protein accession number is indicated. Elements given in notes correspond to ligand (in italics) or regulators (in bold) of the corresponding protein described in previous studies (indicated in brackets).

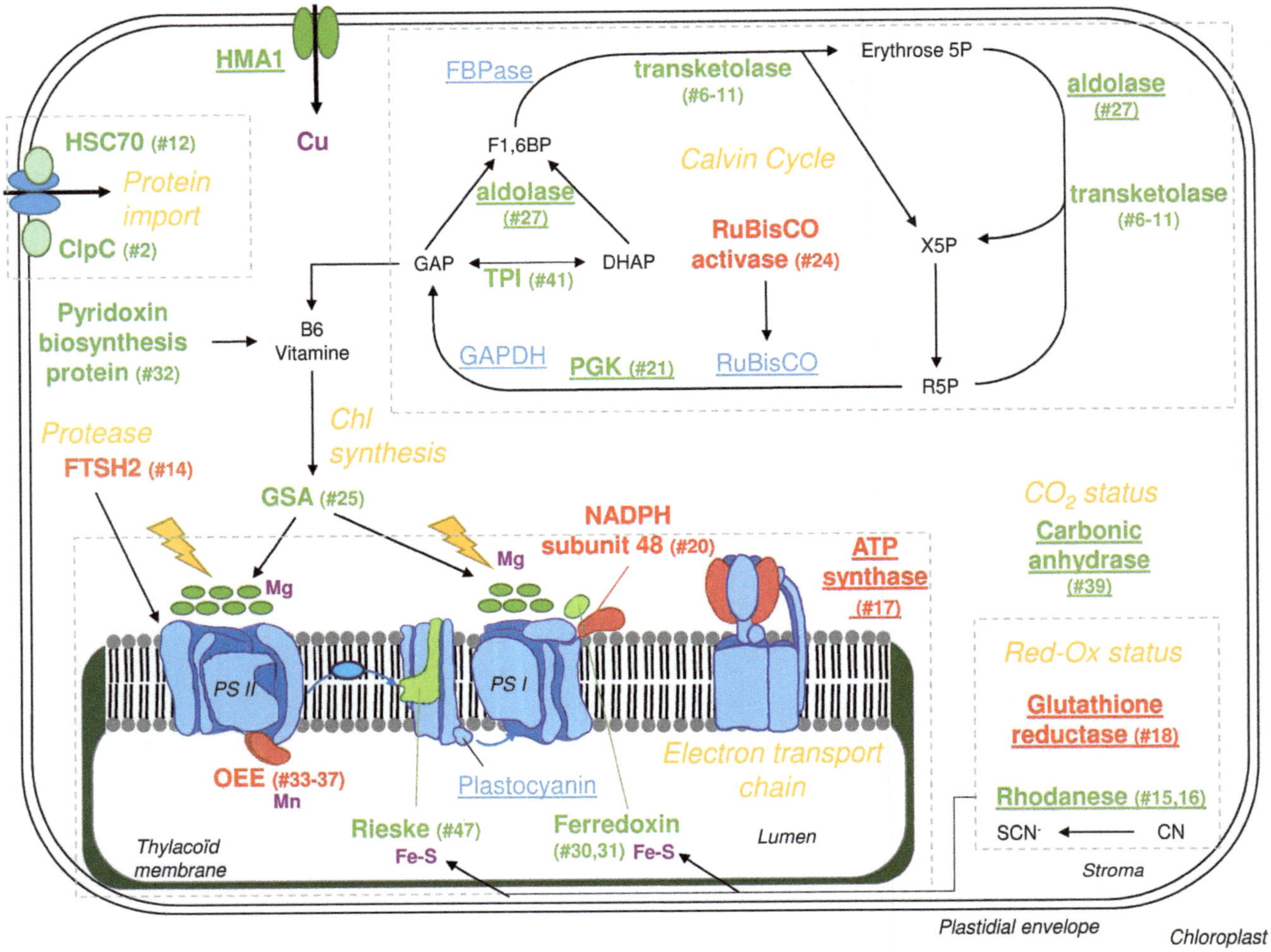

Figure 5. Effect of Cu deficiency on metabolic pathwayspathway in chloroplasts from old leaves of *B. napus*. Bold red indicates a decrease in the amount of protein while bold green indicates an increase. Spot number(s) corresponding to proteins identified in table 3 are indicated in parentheses. Intermediate metabolites are in black, proteins not identified in 2D-E analysis are in blue. Pathways are encircled by grey dashed lines and named in orange italics. Underlines indicate Cu binding and/or known Cu regulation. Other elements known to bind and/or regulate proteins are indicated in purple. PS I: Photosystem I; PS II Photosystem II; R5P: ribulose-5-phosphate; X5P: xylose-5-phosphate; F1,6BP: Fructose-1,6-bisphosphate; GAPDH: GlycerAldehyde Phosphate DeHydrogenase; PGK: PhosphoGlucoKinase; GAP: GlycerAldehyde-3-Phosphate; FBPase: Fructose-1,6-BisPhosphatase; TPI: Triose Phosphate Isomerase; GSA: Glutamate-1-Semialdehyde 2,1-Aminomutase; OEE: Oxygen Evolving Enhancer; Chl: Chlorophyll; CN: cyanate; SCN: Thiocyanate.

family members (#15 to 16), previously described as protecting [Fe-S] clusters [57] in addition to their role in cyanate detoxification.

Previous studies [52,58] highlighted the transcription factor SPL7 (SQUAMOSA-promoter binding link protein 7) which seems to coordinate the early response to Cu deficiency by targeting a Cu response element (CuRE) in the promoter of genes differentially expressed during Cu deficiency. Unfortunately, we couldn't performed an analysis on promoters of genes corresponding to the proteins presented in table 3 as *B. napus* genome sequence is not yet available. However, regulation seems to be complex as Cu deficiency induces a down-regulation at transcript level of carbonic anhydrase [52], but it was up-regulated at protein level in our study (#39). It should be noted that the Carbonic anhydrase regulation is independent from SPL7 [52] and thus its regulation should not be directly induced by Cu deficiency. Moreover, previous studies on Cu deficiency also reported a variation of microRNAs [51,52] or proteases ([59] as well as FTSH2 #14, Table 3 and Figure 5) that could impact on the translation and the proteolysis and thus lead to differences between transcriptomic and proteomic results.

Crosstalk between Cu, [Fe-S] and Mo

Surprisingly, among all quantified mineral nutrients, only Mo uptake was strongly increased in Cu-depleted plants (Figure 1) and this was further linked to strong up-regulation of *MOT1* gene in roots (Figure 2).

The relation between Mo and Cu homeostasis has been reported several times but remains poorly explained [60–63]. As an example, in lesser yam (*Dioscorea esculenta*) [63], molybdenum reached a higher concentration in Cu depleted plants than in plants submitted to any other nutrient deficiencies.

The crosstalk between Cu-Mo and Cu-[Fe-S] established from the literature and the present study are summarised in Figure 6 and can explain the up-regulation of Mo uptake (Figure 1) and expression of [Fe-S] proteins (Rieske #47 and Ferredoxin #30, 31, table 3) by Cu deficiency. Indeed, the chloroplast produces its own [Fe-S] clusters through the SUF (sulfur assimilation proteins)

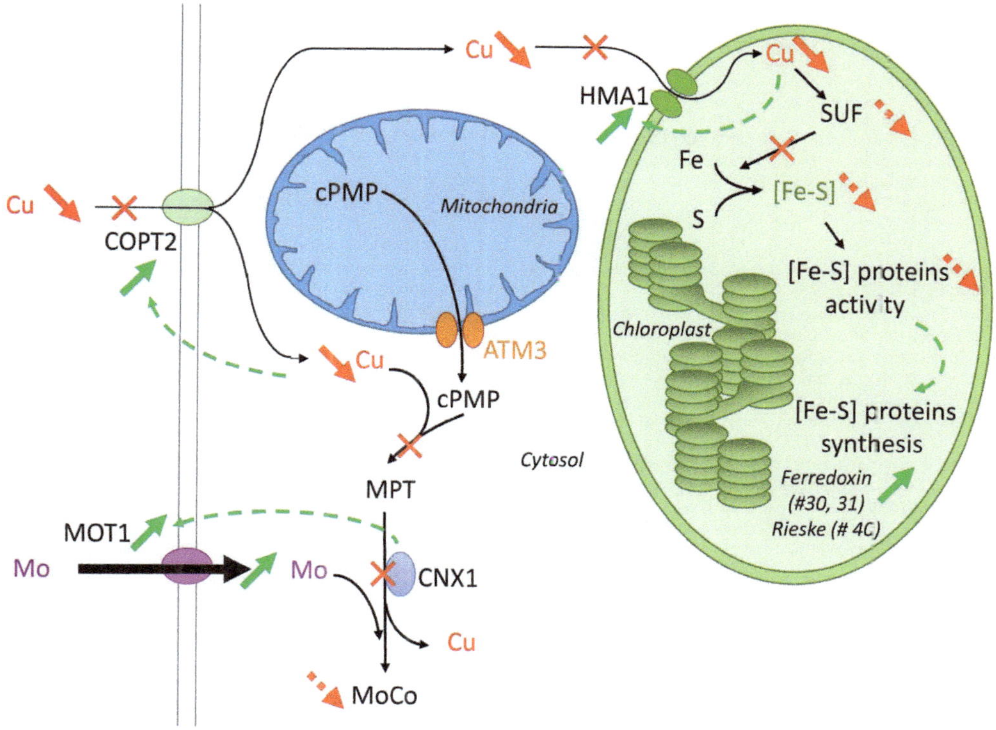

Figure 6. Crosstalk between Cu, [Fe-S] and Mo. Bold arrows indicate variation (green: increase; red: decrease). Full arrows represent observations from this study while dashed arrows have been deduced from the literature. The synthesis of chloroplastic [Fe-S] clusters is controlled by SUF (Sulphur Assimilation Protein), itself controlled by the level of Cu. On the other hand, in the mitochondria, the MoCo precursor cPMP, is exported through mitochondrial membrane by support of the ATM3 transporters. In the cytosol, the second and third steps of MoCo biosynthesis are highly dependent on Cu levels. MoCo is then incorporated into Mo proteins such as Nitrate reductase and into Mo-[Fe-S] proteins such as Xanthine dehydrogenase. The up-regulation of Copt2, HMA1 and Mot1 could be explained by feedback signals due to a low [Cu] in cell, in cytosol and an impairment of CNX1 activity respectively. cPMP: cyclic Pyranopterin MonoPhosphate; MPT: MolybdoPTerin; ATM: ATP-binding cassette Transporters of Mitochondria.

machinery [64], which may be inhibited by Cu deficiency, as previously described in *Bacillus subtilis* [65]. As previously discussed, our leaf proteomic analysis revealed two identified proteins that were over-expressed during Cu deficiency, Ferredoxin and Rieske protein which contain [Fe-S] clusters (#30, 31, 47, Table 3, Figure 6). We may hypothesized that their activity was lowered due to [Fe-S] cluster destabilization by Cu-depletion. This should lead to the observed up-regulation of both proteins to compensate for a possible loss of activity. To minimize the effect of Cu deficiency in chloroplast, the Cu transport across Chloroplast membrane is stimulated through the up regulation of *HMA1* (Figure 2.C, 5, 6).

MOT1 was previously identified as a S transporter (*SULTR5;2*) and it was identified very recently as a high affinity transporter of Mo [45]. Its localization remains unclear (mitochondrial membrane [46] or plasma membrane [45]). Its precise function has not been reported yet, but its importance for an efficient Mo uptake is acknowledged [66]. At cellular level, Mo is essentially used by the Molybdenum Cofactor (MoCo) which is essential for key enzymes such as nitrate reductase and aldehyde oxidase [66]. The first step of the MoCo biosynthesis occurs in the mitochondria and produces cPMP (Figure 6, [67]). cPMP is then exported across the mitochondria membrane by support of the ATM3 transporter [68]. In the cytosol, it must bind Cu in order to be metabolised into MPT representing a step that should be affected in Cu depleted plants. Then, the CNX1 multidomain protein releases Cu and inserts Mo from MPT to give a functional MoCo molecule [69]. It has been reported that CNX1 binds to the cytoskeleton [70] and could have a role as a Mo sensor, interacting with MOT1 transporter to regulates Mo concentration in the cell [71]. Overall, it would explain how the Cu deficiency indirectly up-regulates the Mo uptake through the increase of MOT1 expression. On the other side, a transcriptomic analysis of *Arabidopsis thaliana* WT and *mot1* mutants submitted to Mo deficiency [72] did not reveal a modification of the expression of genes involved in Cu metabolism.

The impairment of Mo metabolism could also be the cause of Ferredoxin and Rieske up-regulation. Indeed, a strong crosstalk exists between Fe and Mo metabolisms: some proteins involved in MoCo biosynethesis are Fe-S (but these Fe-S clusters assembly are independent from the SUF machinery) [66], and Ferredoxin transcripts were up-regulated during an Mo deficiency in *mot1* mutants [72].

Acknowledgments

We are most grateful to the PLATIN' (Plateau d'Isotopie de Normandie) core facility for all element analysis used in this study and the authors deeply appreciate the technical help of Josette Bonnefoy and Marie-Paule Bataillé. The authors thank Laurence Cantrill for improving the English in the manuscript. The authors deeply thank the reviewers for greatly improving the quality of this manuscript with their comments.

Author Contributions

Conceived and designed the experiments: VB AO PE JCY FC. Performed the experiments: VB AM MG JMGM TJ LC. Analyzed the data: VB AM

MG LC. Contributed reagents/materials/analysis tools: MG JMGM TJ LC. Contributed to the writing of the manuscript: VB AO PE.

References

1. Allen L, World Health Organization, Food and Agriculture Organization of the United Nations (2006) Guidelines on food fortification with micronutrients. Geneva; Rome: World Health Organization; Food and Agriculture Organization of the United Nations.
2. Collins JF, Klevay LM (2011) Copper. Adv Nutr 2: 520–522. doi:10.3945/an.111.001222.
3. Steinnes E (2009) Soils and geomedicine. Environ Geochem Health 31: 523–535. doi:10.1007/s10653-009-9257-2.
4. White PJ, Broadley MR (2005) Biofortifying crops with essential mineral elements. Trends Plant Sci 10: 586–593. doi:10.1016/j.tplants.2005.10.001.
5. Ruston D, Great Britain. Office for National Statistics. Social Survey Division, Medical Research Council. Human Nutrition Research, Great Britain. Food Standards Agency (2004) The national diet & nutrition survey: adults aged 19 to 64 years: volume 4: nutritional status (anthropometry and blood analytes), blood pressure and physical activity. London: TSO.
6. ANSES (2012) Anses Table Ciqual 2012 Composition nutritionnelle des aliments. Ciqual Fr Food Comp Table Version 2012. Available: http://www.ansespro.fr/TableCIQUAL/index.htm. Accessed 11 February 2014.
7. Bouis I (2012) Biofortification: Leveraging agriculture to reduce hidden hunger. Ed Shenggen Fan Rajul Pandya-Lorch: 83.
8. Fan M-S, Zhao F-J, Fairweather-Tait SJ, Poulton PR, Dunham SJ, et al. (2008) Evidence of decreasing mineral density in wheat grain over the last 160 years. J Trace Elem Med Biol Organ Soc Miner Trace Elem GMS 22: 315–324. doi:10.1016/j.jtemb.2008.07.002.
9. Garvin DF, Welch RM, Finley JW (2006) Historical shifts in the seed mineral micronutrient concentration of US hard red winter wheat germplasm. J Sci Food Agric 86: 2213–2220. doi:10.1002/jsfa.2601.
10. Peverill KI, Sparrow LA, Reuter DJ (1999) Soil Analysis: An Interpretation Manual. Csiro Publishing. 427 p.
11. Alloway BJ (2013) Bioavailability of Elements in Soil. In: Selinus O, editor. Essentials of Medical Geology. Springer Netherlands. 351–373. Available: http://link.springer.com/chapter/10.1007/978-94-007-4375-5_15. Accessed 11 February 2014.
12. Andersen P (2007) A Review of Micronutrient Problems in the Cultivated Soil of Nepal. Mt Res Dev 27: 331–335. doi:10.1659/mrd.0915.
13. Yruela I (2013) Transition metals in plant photosynthesis. Metallomics 5: 1090–1109. doi:10.1039/C3MT00086A.
14. Yruela I (2009) Copper in plants: acquisition, transport and interactions. Funct Plant Biol 36: 409. doi:10.1071/FP08288.
15. Hänsch R, Mendel RR (2009) Physiological functions of mineral micronutrients (Cu, Zn, Mn, Fe, Ni, Mo, B, Cl). Curr Opin Plant Biol 12: 259–266. doi:10.1016/j.pbi.2009.05.006.
16. Tan Y-F, O'Toole N, Taylor NL, Millar AH (2010) Divalent Metal Ions in Plant Mitochondria and Their Role in Interactions with Proteins and Oxidative Stress-Induced Damage to Respiratory Function. Plant Physiol 152: 747–761. doi:10.1104/pp.109.147942.
17. Bailey E, Stirpe F, Taylor CB (1968) Regulation of rat liver pyruvate kinase. The effect of preincubation, pH, copper ions, fructose 1,6-diphosphate and dietary changes on enzyme activity. Biochem J 108: 427–436.
18. Kung C-CS, Huang W-N, Huang Y-C, Yeh K-C (2006) Proteomic survey of copper-binding proteins in Arabidopsis roots by immobilized metal affinity chromatography and mass spectrometry. Proteomics 6: 2746–2758. doi:10.1002/pmic.200500108.
19. Tainer JA, Getzoff ED, Richardson JS, Richardson DC (1983) Structure and mechanism of copper, zinc superoxide dismutase. Nature 306: 284–287. doi:10.1038/306284a0.
20. Broadley M, Brown P, Cakmak I, Rengel Z, Zhao F (2012) Chapter 7 - Function of Nutrients: Micronutrients. In: Marschner P, editor. Marschner's Mineral Nutrition of Higher Plants (Third Edition). San Diego: Academic Press. 191–248. Available: http://www.sciencedirect.com/science/article/pii/B9780123849052000078. Accessed 14 April 2014.
21. Grusak MA, Cakmak I (2005) 12 Methods to improve the crop-delivery of minerals to humans and livestock. Plant Nutr Genomics: 265.
22. Chavez AL, Bedoya JM, Sánchez T, Iglesias C, Ceballos H, et al. (2000) Iron, carotene, and ascorbic acid in cassava roots and leaves. Food Nutr Bull 21: 410–413.
23. Dubousset L, Etienne P, Avice JC (2010) Is the remobilization of S and N reserves for seed filling of winter oilseed rape modulated by sulphate restrictions occurring at different growth stages? J Exp Bot 61: 4313–4324. doi:10.1093/jxb/erq233.
24. Etienne P, Desclos M, Le Gou L, Gombert J, Bonnefoy J, et al. (2007) N-protein mobilisation associated with the leaf senescence process in oilseed rape is concomitant with the disappearance of trypsin inhibitor activity. Funct Plant Biol 34: 895–906.
25. Desclos M, Dubousset L, Etienne P, Le Caherec F, Satoh H, et al. (2008) A proteomic profiling approach to reveal a novel role of Brassica napus drought 22 kD/water-soluble chlorophyll-binding protein in young leaves during nitrogen remobilization induced by stressful conditions. Plant Physiol 147: 1830–1844. doi:10.1104/pp.108.116905.
26. Avice J-C, Etienne P (2014) Leaf senescence and Nitrogen Remobilization Efficiency in oilseed rape (Brassica napus L.). J Exp Bot in press. doi;JEXBOT/2014/119172, IF: 5,54.
27. Abdallah M, Dubousset L, Meuriot F, Etienne P, Avice J-C, et al. (2010) Effect of mineral sulphur availability on nitrogen and sulphur uptake and remobilization during the vegetative growth of Brassica napus L. J Exp Bot 61: 2635–2646. doi:10.1093/jxb/erq096.
28. Waters BM, Armbrust LC (2013) Optimal copper supply is required for normal plant iron deficiency responses. Plant Signal Behav 8.
29. Garcia-Molina A, Andrés-Colás N, Perea-García A, Neumann U, Dodani SC, et al. (2013) The Arabidopsis COPT6 transport protein functions in copper distribution under copper-deficient conditions. Plant Cell Physiol 54: 1378–1390. doi:10.1093/pcp/pct088.
30. Deng F, Yamaji N, Xia J, Ma JF (2013) A member of the heavy metal P-type ATPase OsHMA5 is involved in xylem loading of copper in rice. Plant Physiol 163: 1353–1362. doi:10.1104/pp.113.226225.
31. Higashi Y, Takechi K, Takano H, Takio S (2013) Involvement of microRNA in copper deficiency-induced repression of chloroplastic CuZn-superoxide dismutase genes in the moss Physcomitrella patens. Plant Cell Physiol 54: 1345–1355. doi:10.1093/pcp/pct084.
32. Wintz H, Fox T, Wu Y-Y, Feng V, Chen W, et al. (2003) Expression Profiles of Arabidopsis thaliana in Mineral Deficiencies Reveal Novel Transporters Involved in Metal Homeostasis. J Biol Chem 278: 47644–47653. doi:10.1074/jbc.M309338200.
33. Mora V, Bacaicoa E, Zamarreño A-M, Aguirre E, Garnica M, et al. (2010) Action of humic acid on promotion of cucumber shoot growth involves nitrate-related changes associated with the root-to-shoot distribution of cytokinins, polyamines and mineral nutrients. J Plant Physiol 167: 633–642. doi:10.1016/j.jplph.2009.11.018.
34. Livak KJ, Schmittgen TD (2001) Analysis of Relative Gene Expression Data Using Real-Time Quantitative PCR and the 2−ΔΔCT Method. Methods 25: 402–408. doi:10.1006/meth.2001.1262.
35. Wang W, Scali M, Vignani R, Spadafora A, Sensi E, et al. (2003) Protein extraction for two-dimensional electrophoresis from olive leaf, a plant tissue containing high levels of interfering compounds. ELECTROPHORESIS 24: 2369–2375. doi:10.1002/elps.200305500.
36. Méchin V, Consoli L, Le Guilloux M, Damerval C (2003) An efficient solubilization buffer for plant proteins focused in immobilized pH gradients. Proteomics 3: 1299–1302. doi:10.1002/pmic.200300450.
37. Bradford MM (1976) A rapid and sensitive method for the quantitation of microgram quantities of protein utilizing the principle of protein-dye binding. Anal Biochem 72: 248–254.
38. Blum H, Beier H, Gross HJ (1987) Improved silver staining of plant proteins, RNA and DNA in polyacrylamide gels. ELECTROPHORESIS 8: 93–99. doi:10.1002/elps.1150080203.
39. Bradová J, Matějová E (2008) Comparison of the Results of SDS PAGE and Chip Electrophoresis of Wheat Storage Proteins. Chromatographia 67: 83–88. doi:10.1365/s10337-008-0545-2.
40. Matrix science (n.d.) Mascot search engine | Protein identification software for mass spec data. Available: http://www.matrixscience.com/. Accessed 11 February 2014.
41. RefSeq: NCBI Reference Sequence Database (n.d.). Available: http://www.ncbi.nlm.nih.gov/refseq/. Accessed 22 April 2014.
42. Brassica Genome Gateway (n.d.). Available: http://brassica.nbi.ac.uk/. Accessed 11 February 2014.
43. BLAST: Basic Local Alignment Search Tool (n.d.). Available: http://blast.ncbi.nlm.nih.gov/Blast.cgi?CMD=Web&PAGE_TYPE=BlastHome. Accessed 14 February 2014.
44. Yuan JS, Reed A, Chen F, Stewart CN (2006) Statistical analysis of real-time PCR data. BMC Bioinformatics 7: 85. doi:10.1186/1471-2105-7-85.
45. Tomatsu H, Takano J, Takahashi H, Watanabe-Takahashi A, Shibagaki N, et al. (2007) An Arabidopsis thaliana high-affinity molybdate transporter required for efficient uptake of molybdate from soil. Proc Natl Acad Sci 104: 18807–18812. doi:10.1073/pnas.0706373104.
46. Baxter I, Muthukumar B, Park HC, Buchner P, Lahner B, et al. (2008) Variation in molybdenum content across broadly distributed populations of Arabidopsis thaliana is controlled by a mitochondrial molybdenum transporter (MOT1). PLoS Genet 4: e1000004.
47. Ravet K, Danford FL, Dihle A, Pittarello M, Pilon M (2011) Spatiotemporal analysis of copper homeostasis in Populus trichocarpa reveals an integrated molecular remodeling for a preferential allocation of copper to plastocyanin in the chloroplasts of developing leaves. Plant Physiol 157: 1300–1312. doi:10.1104/pp.111.183350.
48. Marschner P (2012) Marschner's Mineral Nutrition of Higher Plants. Academic Press. 668 p.

49. Del Pozo T, Cambiazo V, González M (2010) Gene expression profiling analysis of copper homeostasis in Arabidopsis thaliana. Biochem Biophys Res Commun 393: 248–252. doi:10.1016/j.bbrc.2010.01.111.
50. Burda K, Kruk J, Schmid GH, Strzalka K (2003) Inhibition of oxygen evolution in Photosystem II by Cu(II) ions is associated with oxidation of cytochrome b559. Biochem J 371: 597–601. doi:10.1042/BJ20021265.
51. Bernal M, Casero D, Singh V, Wilson GT, Grande A, et al. (2012) Transcriptome sequencing identifies SPL7-regulated copper acquisition genes FRO4/FRO5 and the copper dependence of iron homeostasis in Arabidopsis. Plant Cell Online 24: 738–761.
52. Andrés-Colás N, Perea-García A, de Andrés SM, Garcia-Molina A, Dorcey E, et al. (2013) Comparison of global responses to mild deficiency and excess copper levels in Arabidopsis seedlings. Metallomics 5: 1234–1246.
53. Slovacek RE, Monahan BC (1983) Reductive activation of fructose-1,6-bisphosphatase and the peroxide effect on chloroplast photosynthesis. Arch Biochem Biophys 224: 310–318.
54. Munoz-Bertomeu J, Cascales-Minana B, Mulet JM, Baroja-Fernandez E, Pozueta-Romero J, et al. (2009) Plastidial Glyceraldehyde-3-Phosphate Dehydrogenase Deficiency Leads to Altered Root Development and Affects the Sugar and Amino Acid Balance in Arabidopsis. Plant Physiol 151: 541–558. doi:10.1104/pp.109.143701.
55. Mooney S, Hellmann H (2010) Vitamin B6: Killing two birds with one stone? Phytochemistry 71: 495–501. doi:10.1016/j.phytochem.2009.12.015.
56. Hennig M, Grimm B, Contestabile R, John RA, Jansonius JN (1997) Crystal structure of glutamate-1-semialdehyde aminomutase: an alpha2-dimeric vitamin B6-dependent enzyme with asymmetry in structure and active site reactivity. Proc Natl Acad Sci U S A 94: 4866–4871.
57. Cereda A, Carpen A, Picariello G, Tedeschi G, Pagani S (2009) The lack of rhodanese RhdA affects the sensitivity of Azotobacter vinelandii to oxidative events. Biochem J 418: 135. doi:10.1042/BJ20081218.
58. Yamasaki H, Hayashi M, Fukazawa M, Kobayashi Y, Shikanai T (2009) SQUAMOSA promoter binding protein–like7 is a central regulator for copper homeostasis in Arabidopsis. Plant Cell Online 21: 347–361.
59. Castruita M, Casero D, Karpowicz SJ, Kropat J, Vieler A, et al. (2011) Systems Biology Approach in Chlamydomonas Reveals Connections between Copper Nutrition and Multiple Metabolic Steps. Plant Cell Online 23: 1273–1292. doi:10.1105/tpc.111.084400.
60. Lidon FC, Henriques FS (1993) Effects of copper toxicity on growth and the uptake and translocation of metals in rice plants. J Plant Nutr 16: 1449–1464. doi:10.1080/01904169309364626.
61. Wallace A, Romney EM, Alexander GV, Kinnear J (1977) Phytotoxicity and some interactions of the essential trace metals iron, manganese, molybdenum, zinc, copper, and boron. Commun Soil Sci Plant Anal 8: 741–750. doi:10.1080/00103627709366768.
62. Basak A, Mandal LN, Haldar M (1982) Interaction of phosphorus and molybdenum in relation to uptake and utilization of molybdenum, phosphorus, zinc, copper and manganese by rice. Plant Soil 68: 261–269.
63. O'Sullivan JN, Ernest J (2007) Nutrient deficiencies in lesser yam (Dioscorea esculenta) characterized using constant–water table sand culture. J Plant Nutr Soil Sci 170: 273–282. doi:10.1002/jpln.200625049.
64. Pilon M, Abdel-Ghany SE, Van Hoewyk D, Ye H, Pilon-Smits EAH (2006) Biogenesis of iron-sulfur cluster proteins in plastids. Genet Eng (N Y) 27: 101–117.
65. Chillappagari S, Seubert A, Trip H, Kuipers OP, Marahiel MA, et al. (2010) Copper Stress Affects Iron Homeostasis by Destabilizing Iron-Sulfur Cluster Formation in Bacillus subtilis. J Bacteriol 192: 2512–2524. doi:10.1128/JB.00058-10.
66. Bittner F (2014) Molybdenum metabolism in plants and crosstalk to iron. Front Plant Sci 5. Available: http://www.ncbi.nlm.nih.gov/pmc/articles/PMC3916724/. Accessed 18 August 2014.
67. Mendel RR (2013) The Molybdenum Cofactor. J Biol Chem 288: 13165–13172. doi:10.1074/jbc.R113.455311.
68. Teschner J, Lachmann N, Schulze J, Geisler M, Selbach K, et al. (2010) A Novel Role for Arabidopsis Mitochondrial ABC Transporter ATM3 in Molybdenum Cofactor Biosynthesis. Plant Cell Online 22: 468–480. doi:10.1105/tpc.109.068478.
69. Kuper J, Llamas A, Hecht H-J, Mendel RR, Schwarz G (2004) Structure of the molybdopterin-bound Cnx1G domain links molybdenum and copper metabolism. Nature 430: 803–806. doi:10.1038/nature02681.
70. Schwarz G, Schulze J, Bittner F, Eilers T, Kuper J, et al. (2000) The Molybdenum Cofactor Biosynthetic Protein Cnx1 Complements Molybdate-Repairable Mutants, Transfers Molybdenum to the Metal Binding Pterin, and Is Associated with the Cytoskeleton. Plant Cell Online 12: 2455–2471. doi:10.1105/tpc.12.12.2455.
71. Tejada-Jiménez M, Galván A, Fernández E, Llamas Á (2009) Homeostasis of the micronutrients Ni, Mo and Cl with specific biochemical functions. Curr Opin Plant Biol 12: 358–363. doi:10.1016/j.pbi.2009.04.012.
72. Ide Y, Kusano M, Oikawa A, Fukushima A, Tomatsu H, et al. (2011) Effects of molybdenum deficiency and defects in molybdate transporter MOT1 on transcript accumulation and nitrogen/sulphur metabolism in Arabidopsis thaliana. J Exp Bot 62: 1483–1497. doi:10.1093/jxb/erq345.
73. Ferrer J-L, Ravanel S, Robert M, Dumas R (2004) Crystal Structures of Cobalamin-independent Methionine Synthase Complexed with Zinc, Homocysteine, and Methyltetrahydrofolate. J Biol Chem 279: 44235–44238. doi:10.1074/jbc.C400325200.
74. Komenda J, Knoppová J, Krynická V, Nixon PJ, Tichý M (2010) Role of FtsH2 in the repair of Photosystem II in mutants of the cyanobacterium Synechocystis PCC 6803 with impaired assembly or stability of the CaMn4 cluster. Biochim Biophys Acta BBA - Bioenerg 1797: 566–575. doi:10.1016/j.bbabio.2010.02.006.
75 Volini M, Van Sweringen B, Chen F-S (1978) Effects of metal-ion binding on rhodanese activity. Arch Biochem Biophys 191: 205–215. doi:10.1016/0003-9861(78)90083-8.
76. Berry A, Marshall KE (1993) Identification of zinc-binding ligands in the class II fructose-1,6-bisphosphate aldolase of Escherichia coli. FEBS Lett 318: 11–16.
77. Mahmoud YA, el Souod SM, Niehaus WG (1995) Purification and characterization of malate dehydrogenase from Cryptococcus neoformans. Arch Biochem Biophys 322: 69–75.
78. Ikeuchi M, Inoue Y (1986) Characterization of O2 evolution by a wheat photosystem II reaction center complex isolated by a simplified method: disjunction of secondary acceptor quinone and enhanced Ca2+ demand. Arch Biochem Biophys 247: 97–107.
79. Giulia M, Caricato R, Elena M, Erroi E, Schettino T (2012) Carbonic Anhydrase and Heavy Metals. In: Ekinci D, editor. Biochemistry. InTech. Available: http://www.intechopen.com/books/biochemistry/carbonic-anhydrase-and-heavy-metals. Accessed 19 February 2014.

6

Proteomic Analysis of Nasal Epithelial Cells from Cystic Fibrosis Patients

Ludovic Jeanson[1,2], Ida Chiara Guerrera[2,3], Jean-François Papon[4,5], Cerina Chhuon[2,3], Patricia Zadigue[4], Virginie Prulière-Escabasse[5], Serge Amselem[1], Estelle Escudier[1], André Coste[4,5], Aleksander Edelman[2,3]*

1 Service de Génétique et Embryologie Médicales, Unité Mixte de Recherche_Scientifique 933, Institut National de la Santé et de la Recherche Médicale, Université Pierre et Marie Curie – Paris 6, and Assistance Publique – Hôpitaux de Paris, Hôpital Armand Trousseau, Paris, France, **2** Unité_1151, Institut National de la Santé et de la Recherche Médicale, Université Paris Descartes, Paris, France, **3** Plateau Proteome Necker, Structure Fédérative de Recherche de Necker, Université Paris Descartes, Paris, France, **4** Unité Mixte de Recherche_Scientifique 855, Institut National de la Santé et de la Recherche Médicale, Université Paris 12, Faculté de Médecine, Créteil, France, **5** Service d'Otorhinolaryngologie et de chirurgie cervico-faciale, Assistance Publique – Hôpitaux de Paris, Hôpital inter-communal et Groupe Hospitalier Henri Mondor-Albert Chenevier, Créteil, France

Abstract

The pathophysiology of cystic fibrosis (CF) lung disease remains incompletely understood. New explanations for the pathogenesis of CF lung disease may be discovered by studying the patterns of protein expression in cultured human nasal epithelial cells (HNEC). To that aim, we compared the level of protein expressions in primary cultures of HNEC from nasal polyps secondary to CF (CFNP, n = 4), primary nasal polyps (NP, n = 8) and control mucosa (CTRL, n = 4) using isobaric tag for relative and absolute quantification (iTRAQ) labeling coupled with liquid chromatography (LC)-MS-MS. The analysis of the data revealed 42 deregulated protein expressions in CFNP compared to NP and CTRL, suggesting that these alterations are related to CF. Overall, AmiGo analysis highlighted six major pathways important for cell functions that seem to be impaired: metabolism, G protein process, inflammation and oxidative stress response, protein folding, proteolysis and structural proteins. Among them, glucose and fatty acid metabolic pathways could be impaired in CF with nine deregulated proteins. Our proteomic study provides a reproducible set of differentially expressed proteins in airway epithelial cells from CF patients and reveals many novel deregulated proteins that could lead to further studies aiming to clarify the involvement of such proteins in CF pathophysiology.

Editor: Andy T. Y. Lau, Shantou University Medical College, China

Funding: Funding was provided by Agence Nationale pour la Recherche -05-MRAR-02201; European Community: Life Sciences and Health Grant (LSHG-CT-2005-512044) NEUPROCF; French Foundation for Cystic Fibrosis: "Vaincre la Mucoviscidose"; Legs Poix from Chancellerie des Universités. The funders had no role in study design, data collection and analysis, decision to publish, or preparation of the manuscript.

Competing Interests: The authors have declared that no competing interests exist.

* Email: aleksander.edelman@inserm.fr

These authors contributed equally to this work.

Introduction

Chronic inflammation, which represents one of the most serious causes of morbidity/mortality in the world, is a severe complication common to both frequent and rare disorders. Among chronic inflammatory disorders, cystic fibrosis (CF) is the most common recessively inherited disease in the Western world. CF is caused by mutations in the gene encoding the cystic fibrosis transmembrane conductance regulator (CFTR) protein [1], which is expressed in airway epithelial cells and acts as a chloride channel activated by cyclic AMP. The most frequent mutation of CF alleles is F508del, a deletion of phenylalanine in the intra-cytosolic nucleotide binding domain 1 of CFTR.

Primary nasal polyposis, a common disorder of the upper airways [2] which affects almost 4% of the population, is an inflammatory disease of unknown etiology. Interestingly, CF is often associated with nasal polyps [3] that constitute an accessible model to study CF pathophysiology.

Global approaches may be useful to unmask abnormal cellular pathways in complex chronic inflammatory diseases. A few proteomic studies have been performed in CF with different sample sources: sputum [4,5], bronchio-alveolar lavage fluid [6] or serum [7]. Only three global proteomic studies have been performed in airway epithelial cells which are one of the main targets in CF: two in lung epithelial cell lines IB3-1 (i.e. immortalized bronchial epithelial cell line isolated from a patient homozygous for the F508del mutation) and C38 (i.e. IB3-1 "corrected" by transfection with wild type *CFTR*) [8,9] and the last one in cells obtained by nasal brushing (including blood, inflammatory and epithelial cells) from CF patients [10]. To our knowledge, no proteomic studies have been done on primary cultures of pure airway epithelial cells from CF patients (i.e. not contaminated with blood and/or inflammatory cells). This culture system allows epithelial cell differentiation at the air/liquid interface and is useful to study chronic inflammation in airways [11].

The effects of F508del CFTR are complex at the cellular level and are known to alter many different functions [11–13]. Among them, metabolism, G protein process, inflammation, oxidative stress response, protein folding, proteolysis and structural proteins are known to be abnormal but the proteins implicated in those malfunctions are largely unknown.

The aim of the present study was to compare the levels of protein expressions in primary cultures of human nasal epithelial cells (HNEC) from control mucosa (CTRL), primary nasal polyps (NP) and NP secondary to CF (CFNP). We chose differential quantitative global proteomic analysis based on isobaric tag for relative and absolute quantification (iTRAQ) labeling coupled with liquid chromatography (LC)-MS-MS that permitted reproducible protein quantification in the different patients evaluated in this study. The iTRAQ method is based on the covalent labeling of the amines from proteins with isobaric tags containing a different distribution of heavy isotopes. The fragmentation of the attached tag generates a low molecular mass reporter ion that can be used to relatively quantify the peptides and the proteins from which they originated.

The results provided new insights into the comprehension of abnormal airway epithelial cell functions that might exist in CF pathogenesis.

Materials and Methods

Subjects

In order to evaluate the differential protein expression related to CF four patients with CFNP were randomly matched to four non-CF patients affected by NP (CFNP *vs.* NP), and four other non-CF patients with NP were matched to four CTRL patients without nasal polyposis (NP *vs.* CTRL) (Figure 1).

All NP samples were collected from patients requiring surgery for their nasal polyposis. Patients were requested to stop all treatments at least one month before surgery except for CF patients who were prepared for surgery by systemic antibiotics adapted to nasal bacteriology. Four control mucosa from the inferior turbinate were obtained from individuals who underwent turbinectomy for snoring. All the patients did not mention any allergic background and were negatively tested with skin prick tests. All non-CF patients with NP presented typical primary nasal polyposis and did not suffer from recurrent airway infections. In order to avoid, as much as possible, heterozygous carriers of the F508del mutation, nasal polyps occurring in a context of purulent chronic sinusitis were not included [14]. All four CF patients were homozygous for the F508del mutation. The CTRL and NP patients were not genotyped for CF mutations. Characteristics of the patients are given in Table 1. This protocol was approved by the ethics committee of our institution (Comité de Protection des Personnes-Ile-de-France IX CHU Henri Mondor). The participants provided their verbal consent to participate in the study and their non opposistion to the use of their data. Parental consent was obtained for minor subjects. The nasal mucosa collected for the experiments were either nasal polyps or turbinates obtained during ethmoidectomy or turbinectomy, respectively. All these samples were surgical waste, usually discarded during surgery. Our procedure is based on French law n°2004-800 of 6 August 2004 on Bioethics ("loi relative à la bioéthique") which was codified in the French Public Health Code (L. 1235-2 and L. 1245-2). The surgeon recorded in the patient's file the non-opposition for using the discarded nasal mucosa.

Material

Reagents for mass spectrometry and protein biochemistry experiments. dithiothreitol (DTT), pronase, ethylenediamine tetra-acetic acid (EDTA), collagen IV, urea, thio-urea, orthovanadate, glycerol, TRIS, iodoacetamide, NaCl, detergents (CHAPS, sodium deoxychotate, Triton X100), NH4CO3 and the Glu-fibrinopeptide were from Sigma-Aldrich (Saint-Quentin Fallavier, France). Anti-proteases were from Roche (Neuilly sur Seine, France). Bromophenol blue, SDS, acrylamide, silver staining kit were from Biorad (Marnes la Coquette, France). Trypsin for mass spectrometry was of highest grade purchased from Promega (Charbonnieres-les-bains, France). Acetonitril (ACN) and Tri-fluoro-acetic acid (TFA) were from Carlo Erba (Val de Reuil, France), Alpha-Cyano-4-Hydroxycinnamic Acid

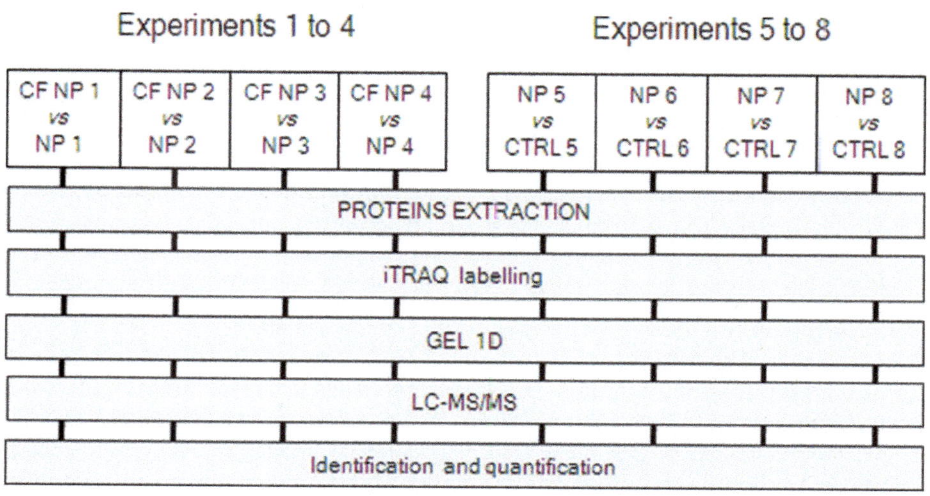

Figure 1. Workflow for the study. Two series of independent experiments were compared: (i) experiments 1 to 4 for proteins derived from HNEC cultures obtained from 4 CF NP patients *vs.* 4 NP patients (non CF); (ii) experiments 5 to 8 for proteins derived from HNEC cultures obtained from 4 NP patients (non CF) *vs.* 4 controls patients. Each couple of samples was matched randomly. For quantification of proteins, iTRAQ labeling coupled to LC-MS/MS was performed (see methods for details).

Table 1. Description of patients.

Patient	Sex	Age (years)	Associated disease	Cigarette smoke
CTRL				
CTRL 5	M	32	None	No
CTRL 6	F	24	None	No
CTRL 7	M	28	None	No
CTRL 8	M	20	None	Yes
CFNP*				
CFNP 1	M	21	None	No
CFNP 2	F	15	None	No
CFNP 3	M	22	None	No
CFNP 4	F	25	None	No
CFNP 5	F	19	None	No
CFNP 6	M	21	None	No
CFNP 7	M	17	None	No
NP				
NP 1	M	25	Samster's disease	No
NP 2	F	29	Aspirin intolerance	Yes
NP 3	M	35	Asthma	No
NP 4	F	34	Asthma	No
NP 5	M	33	Samster's disease	No
NP 6	F	23	Asthma	No
NP 7	M	37	Asthma	No
NP 8	F	41	Samster's disease	No

*All CFNP patients are homozygous for the F508del *CFTR* mutation.

(HCCA) from Laser Biolabs (Sophia-Antipolis, France). iTRAQ reagents were from Applied Biosystems (Courtaboeuf, France)

Columns. C18 trap column (C18 pepmap100, 5 µm particles, 100 Å pores, 300 µm i.d., 5 mm length), and C18 analytical column (C18 pepmap100, 3 µm particles, 15 cm length, 75 µm inner diameter., 100 Å pores) were from Acclaim (Vandoeuvre, France). The Ultimate3000 series HPLC and the Probot fraction collector were purchased from Dionex (Voisins le Bretonneux, France).

Primary cultures of HNEC

Nasal samples were transported to the laboratory in order to perform primary cultures of HNEC as previously described [11]. Mucosa samples were rinsed in phosphate-buffered saline (PBS) with dithiothreitol (DTT, 5 mM) and antibiotics (100 U/ml of penicillin, 100 mg/ml of streptomycin, 2.5 µg/ml of amphotericin B and 100 mg/ml of gentamicin) and then placed overnight at 4°C in a PBS-antibiotics solution containing 0.1% pronase. The samples were incubated in DMEM/F12 with 5% fetal calf serum (FCS) before centrifugation (800 g, 7 min). Cell pellets were then suspended in 0.25% trypsin-EDTA solution for 3 min and incubated in DMEM/F12-antibiotics with 10% FCS. Finally, HNEC were plated on permeable polycarbonate supports at 1×10^6 cells/cm^2. All inserts had a diameter of 12-mm and were coated with type IV collagen. HNEC were incubated at 37°C in 5% CO_2. For the first 24 h, HNEC were incubated with 1 ml of DMEM/F12-antibiotics with 2% Ultroser G outside the insert and DMEM/F12-antibiotics with 10% FCS inside the insert. After 24 h, medium was removed inside the inserts in order to place the cells at an air-liquid interface, and medium outside the inserts was then changed daily. Transepithelial resistance and potential difference were measured every three days using a microvoltmeter (World Precision Instruments, Astonbury, UK). The primary culture of HNEC, differentiated at the air-liquid interface, provides functional cells closely resembling normal airway epithelium within one week that maintain their differentiation for at least four weeks [15]. We previously controlled that all cultured cells are epithelial cells, with viability around 95% during the first four weeks. For all experiments the cells were studied at day 7 after seeding, corresponding to the establishment of cell differentiation.

Protein preparation for proteomic experiments

Experiments for protein quantification were performed in primary cultures of HNEC from CFNP, NP and CTRL. Cells in Transwell filters were washed, scraped off the filters in ice-cold PBS, and centrifuged (800 g, 10 min, 4°C). The pellet was suspended in 100 µl/culture plate of lysis buffer, constituted of 0.5% w/v SDS, 50 mM Tris-HCl pH 7, before incubation at 95°C for 5 min. The total cellular proteins obtained were supplemented by 250 µl/culture plate of dissolution buffer, constituted of 8 M urea, 2.5 M thio-urea, 4% w/v CHAPS, 50 mM DTT, anti-protease 1X, 1 mM orthovanadate, and mixed one h at room temperature. Labeling of proteins with iTRAQ reagents, protein fractionation and digestion were performed according to previously described protocols [16].

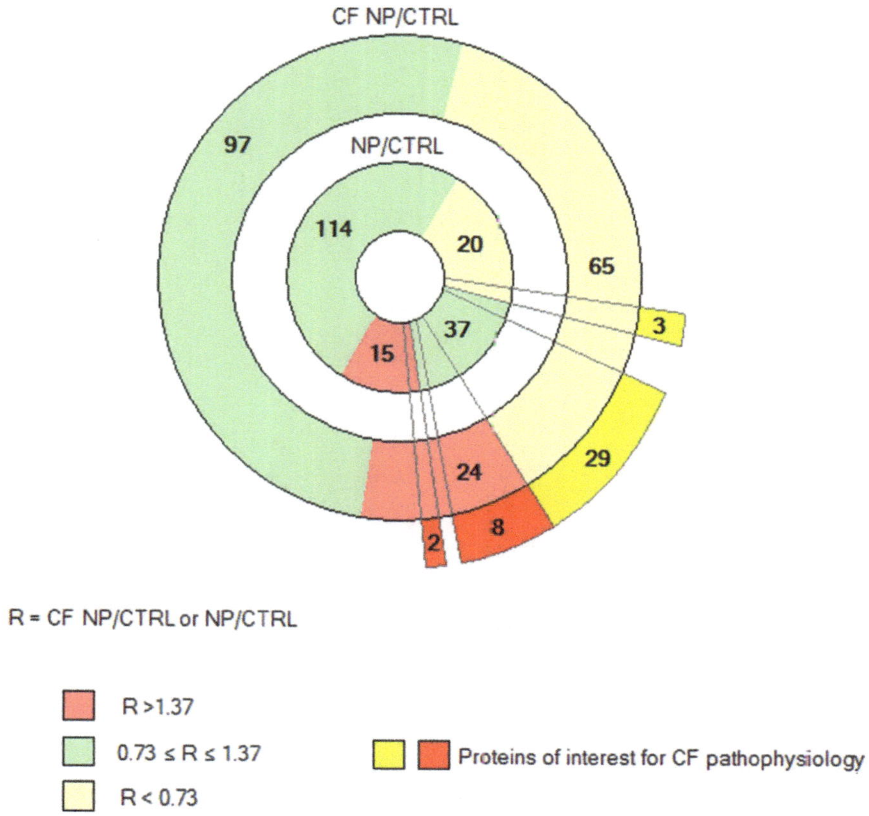

Figure 2. Distribution of the 186 common quantified proteins in CFNP (outer donut) and NP (inner donut) *vs* CTRL. Downregulated proteins are highlighted in yellow, upregulated proteins are highlighted in red and stable proteins are highlighted in green according to the ratio thresholds (R = CF NP/CTRL or NP/CTRL). 15 proteins are upregulated in NP compared to CTRL (inner donut) and 21 proteins are upregulated in CFNP compared to CTRL (outer donut). Among them, 11 proteins are upregulated in both CFNP and NP. 20 proteins are downregulated in NP compared to CTRL (inner donut) and 48 proteins are downregulated in CFNP compared to CTRL (outer donut). Among them, 16 proteins are downregulated in both CFNP and NP. Finally, among the proteins deregulated in common between CFNP and NP, two are more regulated in CFNP and three are more downregulated in CFNP. The 42 proteins of interest for CF pathophysiology are represented in intense red and intense yellow outside the outer donut.

iTRAQ labeling

After protein preparation, 25 µg of each sample were precipitated overnight at −20°C using 100% acetone and dissolved in 20 µl of denaturing buffer containing 1 µl denaturant and 2 µl reducing reagent (TCEP) as provided in the iTRAQ kit, followed by vortexing and incubation at 60°C for 1 h. One µl of cysteine blocking solution (iodoacetamide) was then added to each sample, followed by incubation at room temperature for 30 min. iTRAQ labeling was carried out by adding iTRAQ reagents 114, 115, 116 or 117 to either the NP, CF NP or CTRL samples followed by incubation at room temperature for 2 h. For the NP/CFNP experiments, iTRAQ reagent 114 was used for NP and the 116 was used for CFNP. For the NP/CTRL experiments: iTRAQ reagent 115 was used for NP and the 117 was used for CTRL. Subsequently, two samples/experiment (NP/CF NP or NP/CTRL) were mixed by vortexing, pooled together and dried in a Speedvac evaporator (Thermo Fisher Scientific, Courtaboeuf, France).

Trypsin digestion

Samples were digested according to Bensalem et al [17] with minor modifications. Briefly, prior to mass spectrometry analysis, the dry iTRAQ sample pellet was dissolved in Laemmli buffer, constituted of 10% glycerol, 0.05 M Tris-HCl (pH 6.8), 0.1% SDS, 50 mM DTT, and 0.01% (w/v) bromophenol blue, then supplemented by 1/5 H_2O dilution. Dissolved proteins were separated by electrophoresis (SDS-PAGE). Briefly, protein samples were loaded onto 10% w/v polyacrylamide gels and run at 25 mA for 2 to 3 h at room temperature. After electrophoresis, protein visualization was carried out by silver staining compatible with MS/MS analysis.

12 protein bands were excised from the SDS-PAGE gel and prepared as follows: each gel piece was destained for 5 min with 200 µl of destaining buffer (silver staining kit). Before drying the gel pieces in a Speedvac evaporator, they were washed with deionised water (Millipore, Guyancourt, France) and shrunk in the presence of Acetonitril for 15 min. The gel pieces were dried at room temperature for 20 min and followed by digestion with 0.8 µg (4 U) of trypsin in 8 µl of 25 mM NH_4/CO_3, overnight at 37°C. The peptide mixture was extracted from the gel plug with 10 µl of a 50% ACN, 0.1% tri-fluoro-acetic acid solution and dried in a Speedvac evaporator.

Nano-LC-MALDI-TOF-TOF

Nanochromatography protocol. Digested peptides were suspended in water with 0.1% TFA and 10% acetonitrile, separated with an Ultimate3000 series HPLC. Briefly: 10 µL of eluate were injected and trapped using solvent A (0.1%

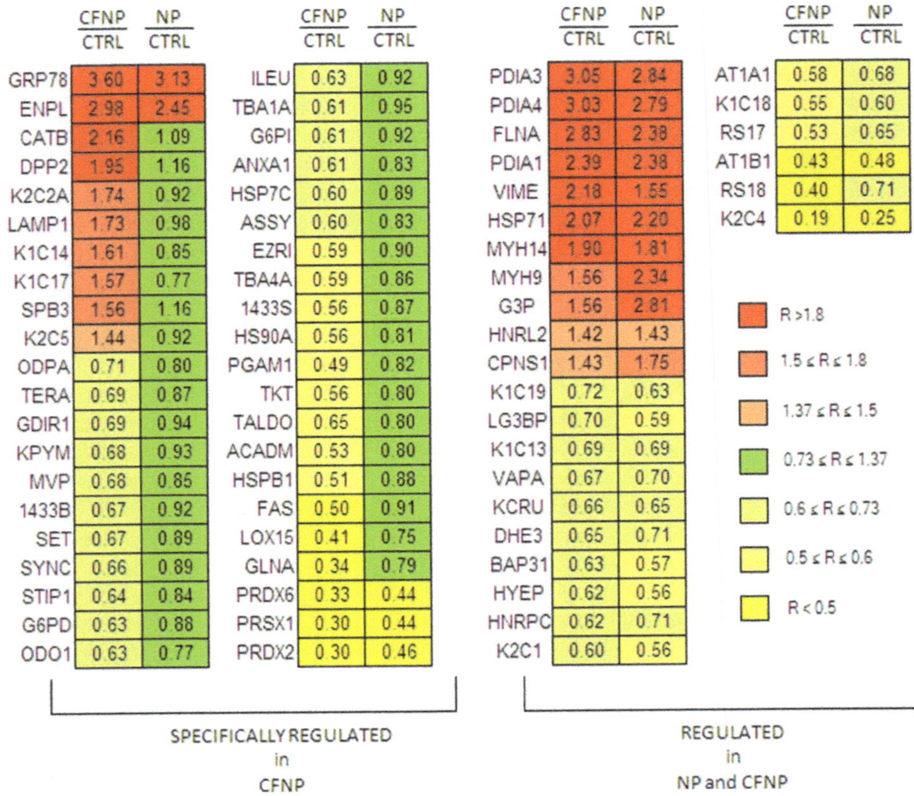

Figure 3. Comparison of differentially expressed proteins in CFNP compared to NP and CTRL. On the left, we report the ratios of 42 proteins significantly modulated in CFNP/CTRL *vs* NP/CTRL patients. On the right, we report proteins that are modulated in both CFNP/CTRL and NP/CTRL patients. To obtain the CFNP/CTRL ratios, relative protein expressions ratios from CF NP/NP were normalized to the average of the ratios obtained in NP *vs* CTRL (CF NP/CTRL = CFNP/NP × NP/CTRL).

Trifuoroacetic acid, 2% ACN) at a 30 µl/min loading flow for 3 min in a C18 trap column. The microvalve then oriented the elution to back flow (300 nl/min) the peptides toward the analytical column with a gradient rising from 7% solvent B (80% acetonitrile, 20% solvent A) at microvalve switch to 16% in 7 min, from 16% to 32% in 20 min and from 32% to 50% in 7 min. Fractions were spotted on-line on a MALDI target using a Probot fraction collector. Spotted fractions were mixed 1:4 with 2 mg/ml of HCCA in 70% ACN 0.1% TFA and Glu fibrinopeptide at 30 fmoles/spot. 96 fractions were collected and analysed using a 4800 MALDI TOF/TOF analyser (ABI, Lyon - Rhône-Alpes, France).

MS mode. Spectra acquisition and processing were performed using the 4000 series explorer software (ABI) version 3.5.28193 build 1011 in positive reflectron mode at fixed LASER fluency with low mass gate and delayed extraction. External plate calibration was performed using 4 calibration points spotted throughout the plate, additional internal calibration was performed using the Glu-fibrinopeptide (M/Z = 1570.677) allowing a mass accuracy of few ppm. For each fraction, steps of 50 spectra in the range of 700 to 4000 Da were acquired at a 200 Hz LASER shot frequency. 500 spectra per sample were summed and processed to obtain monoisotopic values from isotopes clusters with a raw spectra s/n ratio of 20.

MS/MS mode. In each MS spectrum, the 8 most abundant peaks were selected for fragmentation starting with the least abundant. A cut off was applied at a minimum s/n of 20. Neighboring precursors within resolution of 200 were excluded. 1000 MS/MS spectra per precursor were summed by increments of 50. Processing included baseline subtraction and Stavitsky Golay smoothing with 15 points across peak and a polynomial order of 4. Peaklists reflect monoisotopic values from isotope clusters with a s/n ratio of minimum 22.

Peak list and quantification analysis. Generated MS/MS peaklists were subsequently submitted to an in-house Mascot version 2.2 search engine (Matrix science). The Swiss-Prot 55.5 release Swiss-Prot 20080610 (389046 sequences; 139778124 residues) release database was used with the Homo sapiens species. Parent and fragment mass tolerances were respectively set to 50 ppm and 0.3 Da. Variable modification, oxidation (M), iTRAQ 4plex (Y) and fixed modification, iTRAQ 4plex (N-term), iTRAQ 4plex (K), carbamidomethylation (C) were allowed. Trypsin digestion and possibility of one missed cleavage were selected. A filter was applied to the search in order to reduce false positives and matching redundancies of the same peptide in several hits. All matches above 1% risks of random matching were eliminated ($p<0.01$). The probability score calculated by the software was used as a primary criterion for correct identification. For Mascot scores slightly above the threshold of p value, measurement errors were carefully watched to detect possible false positives that usually give random mass measurement errors, while relevant identifications give constant measurement errors. Molecular weights from experimental and Mascot results were also compared. Whenever the result was ambiguous, the spectra either in MS or MS/MS were manually checked for proper matching and the result was accordingly rejected or accepted.

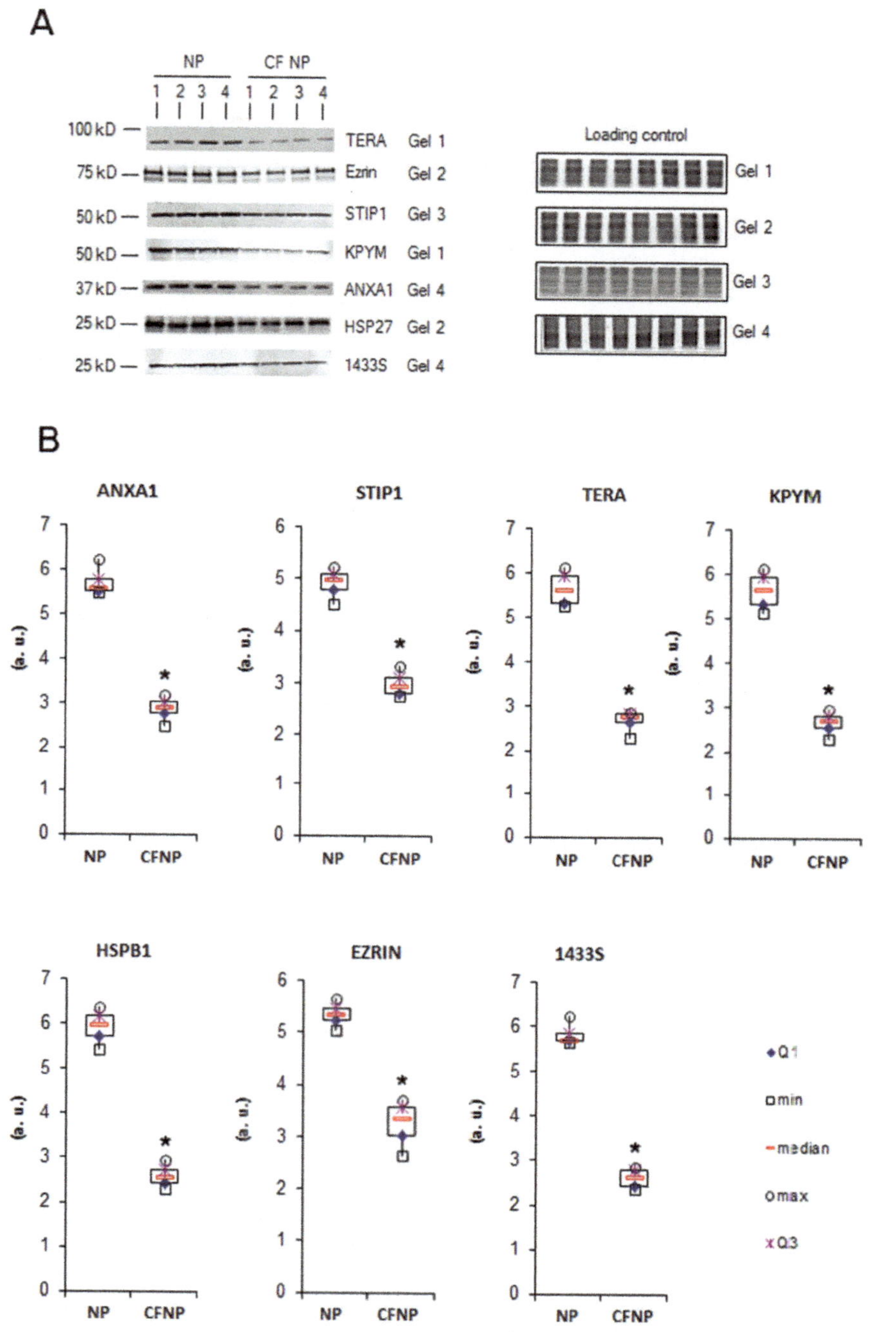

Figure 4. Western blot analysis of seven differentially expressed proteins. The Western blots from the 4 NP patients and the 4 CFNP patients used for proteomic experiments are shown with their respective loading controls. **A**: Each protein is analyzed in three technical replicates. For normalization, the acrylamide gels were stained with Coomassie blue (total migrated protein detection) just after electro-transfer and scanned using the Odyssey LI-COR. Using the Odyssey software, protein intensity values were normalized with their corresponding loading control intensity values. **B**: Box plots of the intensity calculated with the Odyssey software, *p<0.01, Mann and Whitney test. (a.u.: arbitrary units). We report Maximum (max), minimum (min), first quartile (Q1), third quartile (Q3).

Statistical analysis for proteomic experiments

For the quantification, the protein ratio type was the 'weighted' geometric mean, only 'bold face ratios' were considered (significantly different from 1 at a 95% confidence level), outlier removal was 'automatic' (Dixon's method and Rosner's method), the minimum number of peptides was two and at least two peptides were required to be the top ranking peptide matches ('bold and red type faces'). Selective criteria were adopted to obtain a consistently quantified set of proteins: (i) the intensity ratio must be calculated with at least two unique quantified peptides and (ii) the protein must be quantified in at least 3/4 experiments. Differences between expression of proteins in CFNP vs NP were evaluated by Mann-Whitney test p-value<0.05 (mostly confirmed by Student test with Aspin-Welch correction, p-value<0.05) (Table S1). Proteins were therefore considered up/down-regulated if the average of the ratio of the protein expression were >1.37 and $<$

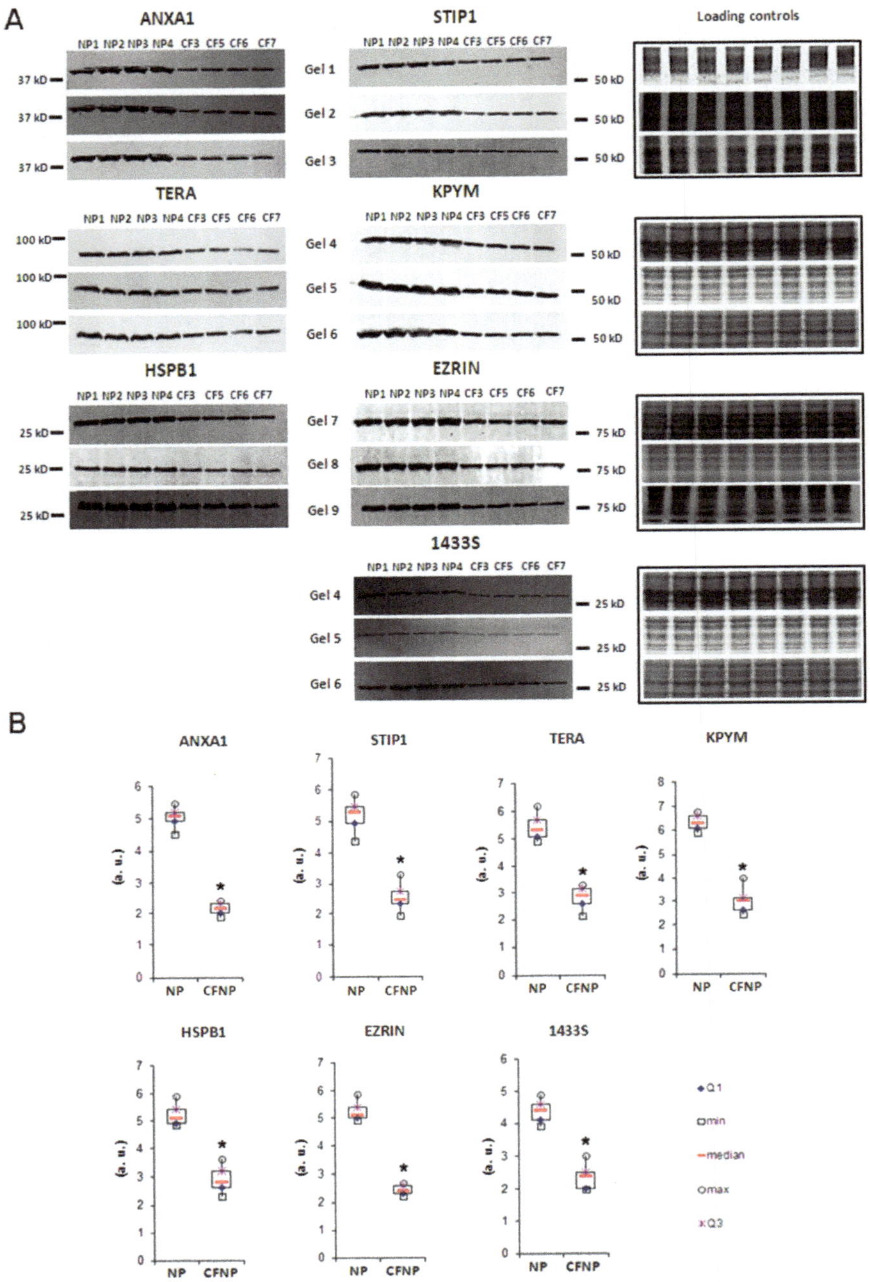

Figure 5. Western blot analysis of ANXA1, STIP1, TERA, KPYM, HSPB1, EZRIN and 1433S in primary cultures of human nasal epithelial cells (HNEC) from NP (n = 4) and CFNP (n = 4). A: Each protein was analyzed in three technical replicates. For normalization, the acrylamide gels were stained with Coomassie blue (total migrated protein detection) just after electro-transfer and scanned using the Odyssey LI-COR. Using the Odyssey software, protein intensity values were normalized with their corresponding loading control intensity values. **B**: Box plots of the intensity calculated with the Odyssey software, *$p<0.01$, Mann and Whitney test. (a.u.: arbitrary units). We report Maximum (max), minimum (min), first quartile (Q1), third quartile (Q3).

0.73. To determine this threshold, we calculated the standard deviations of the protein ratios for each experiment, we normalized the Gaussian distributions and we calculated the average threshold to select the proteins upregulated (last 20% on the right of the Gaussian) and the more downregulated under-expressed (first 20% on the left of the Gaussian). These thresholds are in agreement with other studies using iTRAQ labeling [18]. Finally, the reliably quantified protein data set for both series of experiments were analysed using the AmiGO toolbox kit (http://amigo.geneontology.org/cgi-bin/amigo/term_enrichment and Cytoscape software).

SDS PAGE and Western blot

All Western blot analyses were performed according to previously described protocols [11]. Analysis of protein expression in HNEC was performed using antibodies directed against human STIP1, TERA, EZRI, ANXA1, 1433S, KPYM and HSPB1. All the used primary and secondary antibodies are listed in the

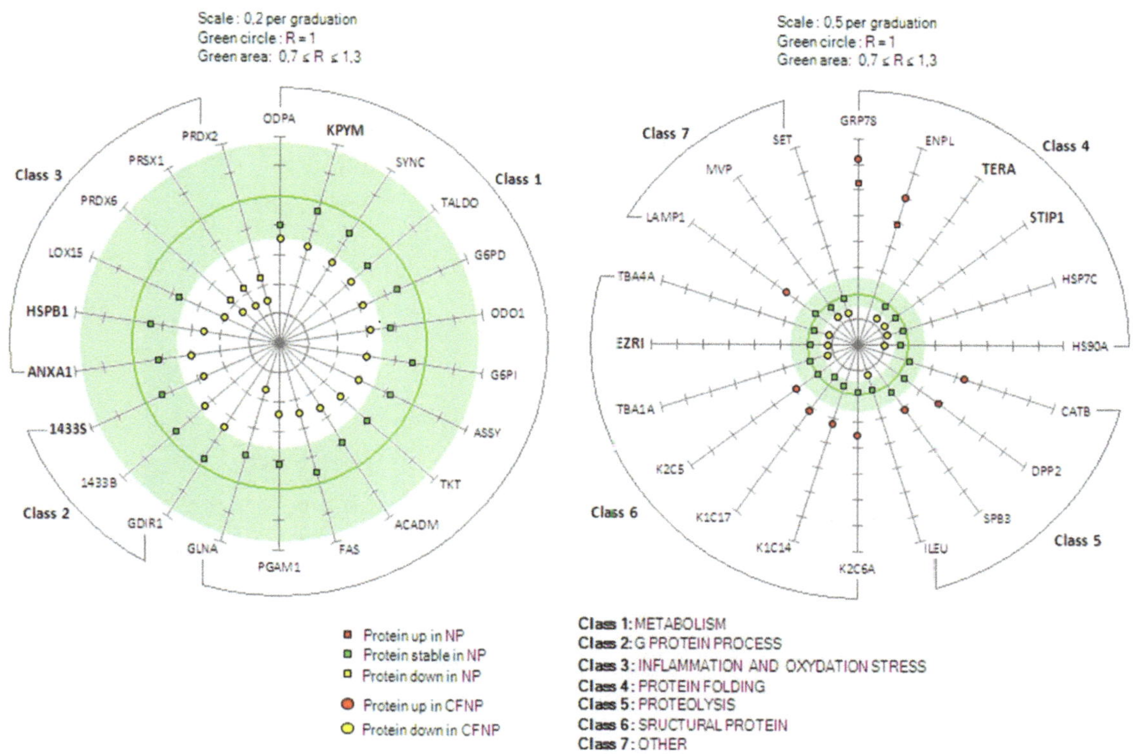

Figure 6. Expression pattern of the 42 differential proteins of interest in CF. The 42 differential proteins are displayed in seven different classes related to their functions (according to the AmiGO toolbox kit). Left graph: metabolism (class 1), G protein process (class 2), inflammation and oxidative stress (class 3); right graph): protein folding (class 4), proteolysis (class 5), structural protein (class 6), other (class 7). The green circle represent the absolute stable ratio value (R = 1) and the green area represent the ratio thresholds (0.73≤R≤1.37) (R = CFNP/CTRL or NP/CTRL). Scale graduations for R are 0.2 and 0.5 for left and right graph, respectively. The CFNP/CTRL ratio values of the differential proteins are represented by yellow dots (downregulated in CFNP) and red dots (upregulated in CFNP); the NP/CTRL ratio values are represented by yellow squares (downregulated in NP), red squares (upregulated in NP), and green squares (stable in NP). For the proteins in bold, the modulation was confirmed by Western blot. Protein names are from Swissprot, ratio values and p values are available in supplemental material (Table S1).

supplemental data with all the details including stock concentrations, used dilutions, manufacturer details and specificity (Table S2). For Western blots realized from NP 1, 2, 3, 4 and CFNP 3, 5, 6, 7, experiments were performed on four patients per category and in triplicate. Mann and Whitney test was used for comparison of data series and P values≤0.01 were considered statistically significant. Data are expressed with Box plots including median, maximum and minimum values and first and third quartiles.

SDS PAGE. Proteins were prepared as follows: first, HNEC from CF NP, NP and CTRL in Transwell filters were washed twice, scraped off the filters in ice-cold PBS, and centrifuged at 800 g for 10 min at 4°C. Second, the pellet was re-suspended in 500 µl of RIPA lysis buffer constituted of 150 mM NaCl, 50 mM Tris-HCl (pH 7.6), 1% v/v Triton X-100, 0.1% w/v SDS, 1% w/v sodium deoxychotate and antiprotease 1X, and kept on ice for 30 min under agitation. Third, the cell lysates were then centrifuged (20 000 g, 15 min) at 4°C. Finally, samples of the supernatants (20 µg protein in 1 volume Laemmli buffer containing 10% glycerol, 0.05 M Tris-HCl (pH 6.8), 0.1% SDS, 50 mM DTT, and 0.01% (w/v) bromophenol blue) were loaded onto 10% w/v polyacrylamide gels and run at 25 mA for 2 to 3 h at room temperature and electrically transferred at 200 mA for 90 min to nitrocellulose membranes.

Western blot analysis. The proteins of interest were detected using appropriate first and secondary antibodies all used at 1:15000 dilution. The infra-red signal was obtained by scanning with the Odyssey LI-COR (Techsciences, Cergy Saint Christophe, France). Quantification of protein levels was obtained using the Odyssey LI-COR software. For normalization, the acrylamide gels were stained with Coomassie blue (total migrated protein detection) just after electro-transfer [19] and scanned using the Odyssey LI-COR.

Results

Comparison of protein expression patterns in CFNP, NP and CTRL

186 proteins were quantified across eight patients in both sets of experiments (CFNP *vs.* NP and NP *vs.* CTRL). The distribution of modulated proteins was different in CF and NP: CFNP presented a higher proportion of deregulated proteins (24 upregulated and 65 downregulated) than NP patients (15 upregulated and 20 downregulated). In order to identify the proteins that were specifically deregulated in CFNP patients, we selected proteins that are significantly modulated in CFNP patients compared to NP patients. 8 proteins were upregulated in CFNP, but stable in NP patients, 29 were downregulated in CFNP but stable in NP, 3 proteins were significantly more downregulated and 2 proteins more upregulated in CFNP than in NP (Figure 2). In addition, 11 proteins were equally upregulated and 16 proteins were equally downregulated in both CFNP and NP. The expression ratios of the modulated proteins are presented in Figure 3 (Table S1 for full details). Seven of these 42 differential proteins (TERA, EZRIN, STIP1, KPYM, ANXA1, HSP27 and 1433S) were analyzed by

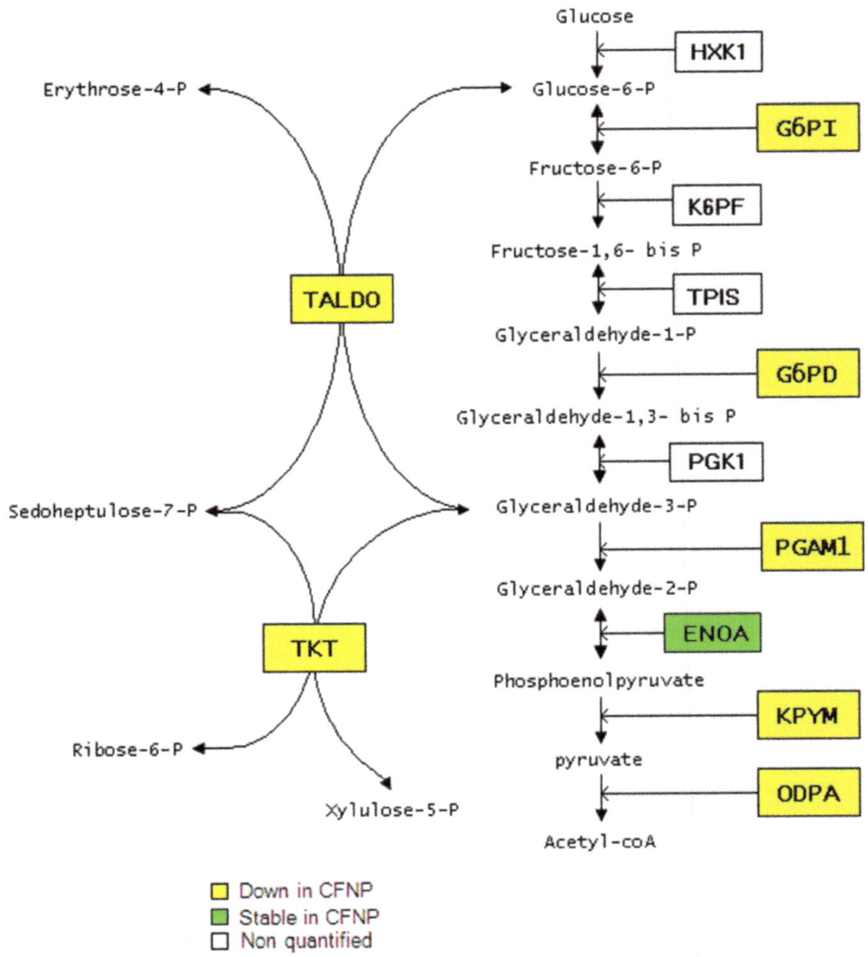

Figure 7. Diagram of the glycolysis/pentose pathway indicating the 7 downregulated proteins in CFNP.

Western blot on patients CFNP1, 2, 3, 4 and NP1, 2, 3, 4 (Figure 4). To extend our validation we repeated the Western blot analysis, on three new additional CFNP patients (CFNP5, 6, 7). Furthermore, technical reproducibility of Western blot analysis, was also assessed by performing three technical replicates for the seven above-mentioned proteins derived from NP1, 2, 3, 4 and CFNP3, 5, 6, 7 (Figure 5A). The statistical analysis of Western blot results confirmed fully the differential expressions observed in the proteomic quantification data (Figure 5B).

Functional distribution of the differential proteins in CFNP/CTRL

To further investigate the altered molecular mechanisms in CF epithelial cells, the differential proteins were analyzed using the AmiGO toolkit which allows highlighting the most represented classes of proteins in a given set. For the retained 42 proteins specifically modulated in CFNP, the AmiGO analysis revealed six major classes of proteins important for cell functions: metabolism, G protein process, inflammation and oxidative stress response, protein folding, proteolysis and structural proteins (Figure 6).

Metabolism (class 1). Thirteen proteins implicated in metabolism were downregulated specifically in CFNP. Among them, seven proteins implicated in glycolysis related metabolism (KPYM, G6PD, ODPA, G6PI, PGAM1, TALDO and TKT) were found downregulated in CFNP and stable in NP suggesting metabolism abnormalities related to CF. Four of these proteins are directly implicated in the canonical glycolytic pathway, pyruvate kinase (KPYM), glucose phosphate isomerase (G6PI), glucose phosphate dehydrogenase 1 (G6PD) and pyruvate dehydrogenase (ODPA). Three other proteins, phosphoglycerate mutase 1 (PGAM1), transketolase (TKT) and transaldolase (TALDO), are implicated in the interconnection between glycolysis and the pentose pathway (Figure 7). In addition, four proteins implicated in amino-acid metabolism, arginosuccinate synthase (ASSY), glutamine synthetase (GLNA), asparaginyl-tRNA synthetase (SYNC), oxoglutarate dehydrogenase (ODO1), and two proteins implicated in fatty acid metabolism, medium-chain specific acyl-CoA dehydrogenase (ACADM) and fatty acid synthase, (FAS) were also significantly downregulated in CFNP compared to NP and CTRL.

G protein process (class 2). Three proteins implicated in these pathways were significantly downregulated in CFNP compared to NP and CTRL, Rho GDP-dissociation inhibitor 1 (GDIR1), 14-3-3 proteins sigma and beta (1433S, 1433B), suggesting possible abnormalities in the ligand/receptor responsiveness in CF.

Inflammation and oxidative stress response (class 3). Six proteins implicated in enzymatic antioxidant response and inflammation were found to be downregulated in CFNP compared to NP and CTRL. Among them, three inflammation-related proteins, arachidonate 15-lipoxygenase (LOX15), annexin

A1 (ANXA1) and heat shock protein beta 1 (HSPB1) were stable in NP compared to control suggesting a CF specificity. Three proteins related to oxidative stress, the peroxiredoxins 1, 2 and 6, (PRDX1, 2 and 6), were downregulated in CFNP and NP compared to CTRL. However, these downregulations in CFNP were significantly lower compared to NP suggesting that these anomalies are present in both nasal polyposis but they are more severe in CF.

Protein folding (class 4). Four protein folding-related proteins were found significantly downregulated in CFNP whilst stable in NP. Interestingly, these proteins, heat shock protein 90-alpha and cognate 71 (HS90A, HSP7C), transitional endoplasmic reticulum ATPase (TERA), stress-induced-phosphoprotein 1 (STIP1), are known to participate in proper folding and degradation of CFTR [20–22]. In addition, two proteins related to protein folding in endoplasmic reticulum were upregulated in CFNP with significant differences with NP, glucose-regulated protein 78 (GRP78) and endoplasmin (ENPL). These proteins are known to be upregulated in the unfolded protein response (UPR) [23].

Proteolysis (class 5). Three proteins implicated in proteolysis were upregulated, serpin B3 (SPB3), cathepsin B (CATB) and dipeptidyl peptidase 2 (DPP2) and one downregulated, serpin B1 (ILEU), in CFNP and stable in NP suggesting proteolysis abnormalities related to CF.

Structural proteins (class 6). Focusing on the cytoskeleton, we observed four proteins upregulated, keratins 6A, 14, 17 and 5 (K2C6A, K1C14, K1C17, K2C5) and three downregulated, ezrin (EZRI), tubulins alpha 1A and 4A (TBA1A and TBA4A) in CFNP and stable in NP compared to CTRL suggesting a cytoskeleton reshuffle in CF.

Other (class 7). The HLA class II-associated protein I (SET) was downregulated in CF and is implicated in MAP kinase signaling since it regulates ERK1/2 phosphorylation [24]. We found an upregulation of the lysosome-associated membrane protein 1 (LAMP1), a well-known lysosomal membrane marker [25] suggesting an increased degradation activity in CF. Finally, the major vault protein (MVP) was downregulated in CF and is implicated in the bacterial resistance of lung epithelial cells [26].

Discussion

In this study we found 42 proteins significantly and specifically modulated in CFNP patients. For seven of these proteins the results were validated by Western blot in the eight patients analysed by proteomics and in three additional CF patients. Our data are also supported by previously published studies by us [11,13,27] and by others [9,10] showing deregulation of ANXA1 (in CFTR−/− mouse), of GRP78 for UPR activation (in both CFNP and NP), of PRDX6, PRDX2, CATB, HSPB1, PDIA3, ANXA1, VIME and EZRI (in CF cells). Importantly, we found many novel differentially expressed proteins in CF NP that could be of importance in CF pathophysiology.

Several proteins involved in CFTR synthesis/degradation were found to be dowregulated in CF (HSP7C, HS90A, STIP1, TERA and LAMP1), supporting previous observations of alteration of folding and degradation pathways in CF [20,21]. HSP7C, HS90A and STIP1 form a protein complex that assists correct CFTR protein folding and participates in the degradation of misfolded F508del CFTR by promoting ubiquitination [20,21,28]. In addition, TERA plays an essential role in endoplasmic reticulum-associated degradation [29] and is known to be involved in the ER-associated degradation of CFTR [22]. Of note, TERA, HSP7C and HS90A are present in the network of the CFTR physical interaction proteins (Figure 8). These downregulations observed in CF are surprising since F508del CFTR is mostly retained in ER for degradation and remain to be investigated. In addition, we observed an upregulation of a lysosomal membrane marker, LAMP1 [25], perhaps related to a raised turnover of the rescued F508del CFTR by the lysosomal pathway [30]. Altogether, these data may suggest a diminished folding-control/ER-degradation of F508del CFTR associated with a rise in lysosomal-mediated turnover, suggesting that F508del CFTR may be preferentially degraded by the lysosomal pathway in our model.

Several proteins involved in inflammation and oxidative stress were found to be deregulated in CF (PRDX1, 2 and 3, HSPB1, LOX15, ANXA1 and MVP), supporting previous studies [13,31,32]. PRDX1, 2 and 6 are implicated in enzymatic antioxidant defenses and we have previously shown a downregulation of PRDX6 under oxidative challenge in CFTR−/− mice lungs [13]. In another study, PRDX6 has also been found to be downregulated in HNEC of CF patients [10]. Another protein linked to oxidative stress, HSPB1 is able to raise intracellular glutathione levels [33] and activate ROS detoxification enzymes such as the peroxiredoxins mentioned above [34]. The decreased expression of these proteins in CFNP may result in lower protection against oxidative stress suggesting a susceptibility to oxidative damage in CF. LOX15 produces 15-hydroxyeicosatetraenoic acid (15-HPETE), an inhibitor of inflammatory mediators by promoting TNF mRNA decay [35]. ANXA1 is also known to have anti-inflammatory properties [36]. The downregulation of these two proteins participates in inflammation in CF by deregulating the eicosanoid-related pathway. Interestingly, MVP is known to participate in the resistance of airway cells to bacteria and may participate in the well-known lower resistance of CF patients to Pseudomonas aeruginosa infection [26].

Eight proteins belonging to the glucose and fatty acid metabolic pathways were found to be deregulated in CF (KPYM, G6PI, G6PD, ODPA, PGAM1, TALDO, TKT and ACADM). KPYM, G6PI, G6PD, ODPA, PGAM1, TALDO and TKT are directly implicated in the glycolysis/pentose pathway leading to NADPH and acyl-CoA production (Figure 7). Interestingly, a recent metabolomic study showed a diminished amount of glycolysis-related metabolites in CF cells (*i.e.* glucose, glucose-6-phosphate, fructose-6-phosphate) and concluded that glycolysis could be suppressed in CF [12]. However, a recent study showed upregulations of KPYM in IB3-1 cells compared to C38 cells, concluding that glycolysis could be increased in CF [8]. This discrepancy could be related to the differences between primary cells and transfected cellular lineages. ACADM is the medium chain of acyl-CoA dehydrogenase that is a key enzyme of fatty acid beta-oxidation leading to acyl-CoA production. The protein expression and activity of all these metabolic enzymes have never been directly investigated in CF. Altogether, our data show that the two major pathways leading to the production of NADPH and acetyl-CoA seem to be altered in CF and could further aggravate oxidative stress conditions.

Deregulation of proteases (SPB3 and CATB) and of a protease inhibitor (ILEU) may participate in lung fibrosis frequently observed in CF. SPB3, ILEU and CATB are important inflammation modulators. SPB3 and CATB (upregulated in this study) have been shown to promote lung, intestinal and liver fibrosis, respectively, in mouse models [37,38]. Furthermore, ILEU (downregulated in this study) is a leukocyte elastase inhibitor that prevents lung fibrosis in mice [39]. These anomalies, detected by us in upper airway epithelial cells, may represent at the cellular level the molecular anomalies that could lead to lung fibrosis. In addition, the upregulation of cytoplasmic structural proteins in CF

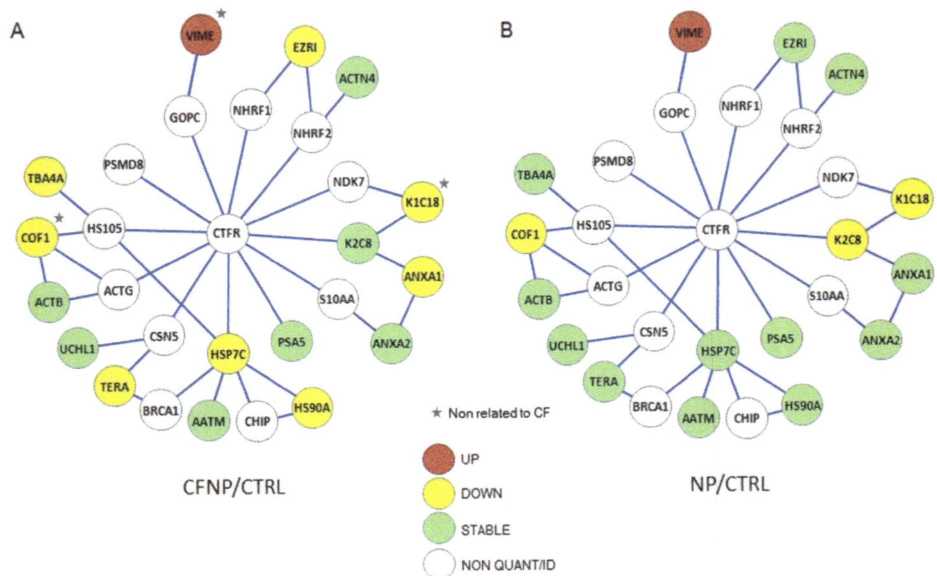

Figure 8. Quantified proteins present in the CFTR physical network. Proteomics quantitative analysis allowed us to reproducibly quantify 186 proteins. Of those, 16 could be mapped on a hand curated CFTR physical network based on the Biogrid interaction database (http://thebiogrid.org).This network is composed of 198 proteins selected according to experimental criteria for direct protein-protein interactions (surface plasmon resonance (SPR), pull-down and FRET experiments). To this bibliographic network, we have added proteins identified in our laboratory using SPR and DNA ligation assays. We built the CFTR protein interaction network by taking into account only proteins that directly interact with one another, at first and second degrees. In the graph, we represent the 16 proteins with a color code corresponding to their expression status in CF (**A**) and NP (**B**) compared to CTRL: upregulated (red), downregulated (yellow), stable (green). The network is completed by a minimal number of connecting proteins of the CFTR network (white) that could not be identified or quantified by MS/MS. Proteins that are up or downregulated in both CFNP and NP are marked by a grey star.

(K2C5, K1C14, K2C6A and K1C17) and the downregulation of TBA1A, TBA4A and EZRI, known to interact with CFTR by anchoring it to the tubulin cytoskeleton [40,41], suggest a possible intracellular remodeling of airway epithelial cells that could be related to trafficking and/or stabilization processes. All together, these results suggest a possible remodeling at the intra and extra cellular levels.

G protein related pathways were found to be altered in CF, suggesting numerous dysfunctions in CF cells. Three proteins related to G protein signaling (GDIR1, 1433B and 1433S) and one protein related to MAP kinase signaling (SET) (all downregulated) are multifunctional proteins that regulate most of the major cellular functions including proliferation, apoptosis, inflammatory response, epithelial cell polarization and secretory processes [42]. Those downregulations could be related to some or many alterations described above.

Finally, based on the Biogrid interaction database (http://thebiogrid.org), we built the CFTR protein interaction network by taking into account only proteins that directly interact with one another, at first and second degrees. This network is composed of 198 proteins selected according to experimental criteria for direct protein-protein interactions (surface plasmon resonance (SPR), pull-down and FRET experiments). To this bibliographic network we have added proteins identified in our laboratory using SPR and DNA ligation assays [43]. It is of interest to note that 16 of the 186 quantified proteins in the present study could be mapped on our handcurated CFTR physical network (Figure 8). Remarkably, 9 proteins out of these 16 were deregulated in CF patients (8 down and 1 up), while only 4 of the 16 proteins were downregulated (3) or upregulated (1) in NP patients (Figure 8). Those differentially expressed proteins are cytoskeleton or chaperon proteins that are all implicated in the folding, stability or degradation of CFTR. We suggest that some of the deregulated proteins in CF patients may be related to the perturbation of the CFTR/F508delCFTR physical network.

Overall, our proteomic study provides a reproducible set of differentially expressed proteins in airway epithelial cells from CF patients. While some of our findings correlate with previous studies, we have identified many novel differentially expressed proteins that have not been previously studied in CF pathophysiology.

Supporting Information

Table S1 List of quantified proteins: Ratio values, Mann-Whitney test (p-value<0.05) and Student test with Aspin-Welch correction (p-value<0.05).
(PDF)

Table S2 Antibodies used in this study: Target, stock concentration, dilution used, manufacturer, reference, host and clonality.
(XLSX)

Acknowledgments

We thank François Guillonneau and all the personnel of the 3P5 Proteomics facility, University Paris Descartes, Nicolas Cagnard of the Bioinformatic Platform University Paris Descartes and Dr. Charlotte Sumida for correcting the manuscript.

Author Contributions

Conceived and designed the experiments: AE EE LJ ICG. Performed the experiments: LJ ICG CC. Analyzed the data: ICG LJ. Contributed reagents/materials/analysis tools: AC JFP VPE PZ. Contributed to the writing of the manuscript: LJ ICG AE EE SA AC.

References

1. Rommens JM, Iannuzzi MC, Kerem B, Drumm ML, Melmer G, et al. (1989) Identification of the cystic fibrosis gene: chromosome walking and jumping. Science 245: 1059–1065.
2. Fokkens W, Lund V, Bachert C, Clement P, Helllings P, et al. (2005) EAACI position paper on rhinosinusitis and nasal polyps executive summary. Allergy 60: 583–601. doi:10.1111/j.1398-9995.2005.00830.x.
3. Beer H, Southern KW, Swift AC (2013) Topical nasal steroids for treating nasal polyposis in people with cystic fibrosis. Cochrane Database Syst Rev 4: CD008253. doi:10.1002/14651858.CD008253.pub3.
4. Sloane AJ, Lindner RA, Prasad SS, Sebastian LT, Pedersen SK, et al. (2005) Proteomic analysis of sputum from adults and children with cystic fibrosis and from control subjects. Am J Respir Crit Care Med 172: 1416–1426. doi:10.1164/rccm.200409-1215OC.
5. Pedersen SK, Sloane AJ, Prasad SS, Sebastian LT, Lindner RA, et al. (2005) An immunoproteomic approach for identification of clinical biomarkers for monitoring disease: application to cystic fibrosis. Mol Cell Proteomics MCP 4: 1052–1060. doi:10.1074/mcp.M400175-MCP200.
6. McMorran BJ, Patat SAO, Carlin JB, Grimwood K, Jones A, et al. (2007) Novel neutrophil-derived proteins in bronchoalveolar lavage fluid indicate an exaggerated inflammatory response in pediatric cystic fibrosis patients. Clin Chem 53: 1782–1791. doi:10.1373/clinchem.2007.087650.
7. Charro N, Hood BL, Faria D, Pacheco P, Azevedo P, et al. (2011) Serum proteomics signature of cystic fibrosis patients: a complementary 2-DE and LC-MS/MS approach. J Proteomics 74: 110–126. doi:10.1016/j.jprot.2010.10.001.
8. Ciavardelli D, D'Orazio M, Pieroni L, Consalvo A, Rossi C, et al. (2013) Proteomic and ionomic profiling reveals significant alterations of protein expression and calcium homeostasis in cystic fibrosis cells. Mol Biosyst 9: 1117–1126. doi:10.1039/c3mb25594h.
9. Pollard HB, Eidelman O, Jozwik C, Huang W, Srivastava M, et al. (2006) De novo biosynthetic profiling of high abundance proteins in cystic fibrosis lung epithelial cells. Mol Cell Proteomics MCP 5: 1628–1637. doi:10.1074/mcp.M600091-MCP200.
10. Roxo-Rosa M, da Costa G, Luider TM, Scholte BJ, Coelho AV, et al. (2006) Proteomic analysis of nasal cells from cystic fibrosis patients and non-cystic fibrosis control individuals: search for novel biomarkers of cystic fibrosis lung disease. Proteomics 6: 2314–2325. doi:10.1002/pmic.200500273.
11. Jeanson L, Kelly M, Coste A, Guerrera IC, Fritsch J, et al. (2012) Oxidative stress induces unfolding protein response and inflammation in nasal polyposis. Allergy 67: 403–412. doi:10.1111/j.1398-9995.2011.02769.x.
12. Wetmore DR, Joseloff E, Pilewski J, Lee DP, Lawton KA, et al. (2010) Metabolomic profiling reveals biochemical pathways and biomarkers associated with pathogenesis in cystic fibrosis cells. J Biol Chem 285: 30516–30522. doi:10.1074/jbc.M110.140806.
13. Trudel S, Kelly M, Fritsch J, Nguyen-Khoa T, Thérond P, et al. (2009) Peroxiredoxin 6 fails to limit phospholipid peroxidation in lung from Cftr-knockout mice subjected to oxidative challenge. PloS One 4: e6075. doi:10.1371/journal.pone.0006075.
14. Coste A, Girodon E, Louis S, Prulière-Escabasse V, Goossens M, et al. (2004) Atypical sinusitis in adults must lead to looking for cystic fibrosis and primary ciliary dyskinesia. The Laryngoscope 114: 839–843. doi:10.1097/00005537-200405000-00009.
15. Papon J-F, Coste A, Gendron M-C, Cordonnier C, Wingerstmann L, et al. (2002) HLA-DR and ICAM-1 expression and modulation in epithelial cells from nasal polyps. The Laryngoscope 112: 2067–2075. doi:10.1097/00005537-200211000-00030.
16. Guerrera IC, Quetier I, Fetouchi R, Moreau F, Vauloup-Fellous C, et al. (2014) Regulation of interleukin-6 in head and neck squamous cell carcinoma is related to papillomavirus infection. J Proteome Res 13: 1002–1011. doi:10.1021/pr401009f.
17. Bensalem N, Masscheleyn S, Mozo J, Vallée B, Brouillard F, et al. (2007) High sensitivity identification of membrane proteins by MALDI TOF-MASS spectrometry using polystyrene beads. J Proteome Res 6: 1595–1602. doi:10.1021/pr0606272.
18. Unwin RD, Griffiths JR, Whetton AD (2010) Simultaneous analysis of relative protein expression levels across multiple samples using iTRAQ isobaric tags with 2D nano LC-MS/MS. Nat Protoc 5: 1574–1582. doi:10.1038/nprot.2010.123.
19. Welinder C, Ekblad L (2011) Coomassie staining as loading control in Western blot analysis. J Proteome Res 10: 1416–1419. doi:10.1021/pr1011476.
20. Rubenstein RC, Zeitlin PL (2000) Sodium 4-phenylbutyrate downregulates Hsc70: implications for intracellular trafficking of DeltaF508-CFTR. Am J - Physiol Cell Physiol 278: C259–267.
21. Loo MA, Jensen TJ, Cui L, Hou Y, Chang XB, et al. (1998) Perturbation of Hsp90 interaction with nascent CFTR prevents its maturation and accelerates its degradation by the proteasome. EMBO J 17: 6879–6887. doi:10.1093/emboj/17.23.6879.
22. Boyault C, Gilquin B, Zhang Y, Rybin V, Garman E, et al. (2006) HDAC6-p97/VCP controlled polyubiquitin chain turnover. EMBO J 25: 3357–3366. doi:10.1038/sj.emboj.7601210.
23. Ma Y, Hendershot LM (2004) ER chaperone functions during normal and stress conditions. J Chem Neuroanat 28 51–65. doi:10.1016/j.jchemneu.2003.08.007.
24. Yu L-G, Packman LC, Weldon M, Hamlett J, Rhodes JM (2004) Protein phosphatase 2A, a negative regulator of the ERK signaling pathway, is activated by tyrosine phosphorylation of putative HLA class II-associated protein I (PHAPI)/pp32 in response to the antiproliferative lectin, jacalin. J Biol Chem 279: 41377–41383. doi:10.1074/jbc.M400017200.
25. Sajjan US, Yang JH, Hershenson MB, LiPuma JJ (2006) Intracellular trafficking and replication of Burkholderia cenocepacia in human cystic fibrosis airway epithelial cells. Cell Microbiol 8: 1456–1466. doi:10.1111/j.1462-5822.2006.00724.x.
26. Kowalski MP, Dubouix-Bourandy A, Bajmoczi M, Golan DE, Zaidi T, et al. (2007) Host resistance to lung infection mediated by major vault protein in epithelial cells. Science 317: 130–132. doi:10.1126/science.1142311.
27. Bensalem N, Ventura AP, Vallée B, Lipecka J, Tondelier D, et al. (2005) Down-regulation of the anti-inflammatory protein annexin A1 in cystic fibrosis knock-out mice and patients. Mol Cell Proteomics MCP 4: 1591–1601. doi:10.1074/mcp.M500019-MCP200.
28. Onuoha SC, Coulstock ET, Grossmann JG, Jackson SE (2008) Structural studies on the co-chaperone Hop and its complexes with Hsp90. J Mol Biol 379: 732–744. doi:10.1016/j.jmb.2008.02.013.
29. Lim PJ, Danner R, Liang J, Doong H, Harman C, et al. (2009) Ubiquilin and p97/VCP bind erasin, forming a complex involved in ERAD. J Cell Biol 187: 201–217. doi:10.1083/jcb.200903024.
30. Cushing PR, Fellows A, Villone D, Boisguérin P, Madden DR (2008) The relative binding affinities of PDZ partners for CFTR: a biochemical basis for efficient endocytic recycling. Biochemistry (Mosc) 47: 10084–10098. doi:10.1021/bi8003928.
31. Kelly-Aubert M, Trudel S, Fritsch J, Nguyen-Khoa T, Baudouin-Legros M, et al. (2011) GSH monoethyl ester rescues mitochondrial defects in cystic fibrosis models. Hum Mol Genet 20: 2745–2759. doi:10.1093/hmg/ddr173.
32. Chen J, Kinter M, Shank S, Cotton C, Kelley TJ, et al. (2008) Dysfunction of Nrf-2 in CF epithelia leads to excess intracellular H_2O_2 and inflammatory cytokine production. PloS One 3: e3367. doi:10.1371/journal.pone.0003367.
33. Mehlen P, Kretz-Remy C, Préville X, Arrigo AP (1996) Human hsp27, Drosophila hsp27 and human alphaB-crystallin expression-mediated increase in glutathione is essential for the protective activity of these proteins against TNFalpha-induced cell death. EMBO J 15: 2695–2706.
34. Préville X, Salvemini F, Giraud S, Chaufour S, Paul C, et al. (1999) Mammalian small stress proteins protect against oxidative stress through their ability to increase glucose-6-phosphate dehydrogenase activity and by maintaining optimal cellular detoxifying machinery. Exp Cell Res 247: 61–78. doi:10.1006/excr.1998.4347.
35. Ferrante JV, Ferrante A (2005) Novel role of lipoxygenases in the inflammatory response: promotion of TNF mRNA decay by 15-hydroperoxyeicosatetraenoic acid in a monocytic cell line. J Immunol Baltim Md 1950 174: 3169–3172.
36. Jorge YC, Mataruco MM, Araújo LP, Rossi AFT, de Oliveira JG, et al. (2013) Expression of annexin-A1 and galectin-1 anti-inflammatory proteins and mRNA in chronic gastritis and gastric cancer. Mediators Inflamm 2013: 152860. doi:10.1155/2013/152860.
37. Lunardi F, Villano G, Perissinotto E, Agostini C, Rea F, et al. (2011) Overexpression of SERPIN B3 promotes epithelial proliferation and lung fibrosis in mice. Lab Investig J Tech Methods Pathol 91: 945–954. doi:10.1038/labinvest.2011.1.
38. Moles A, Tarrats N, Fernández-Checa JC, Marí M (2012) Cathepsin B overexpression due to acid sphingomyelinase ablation promotes liver fibrosis in Niemann-Pick disease. J Biol Chem 287: 1178–1188. doi:10.1074/jbc.M111.272393.
39. Takemasa A, Ishii Y, Fukuda T (2012) A neutrophil elastase inhibitor prevents bleomycin-induced pulmonary fibrosis in mice. Eur Respir J 40: 1475–1482. doi:10.1183/09031936.00127011.
40. Playford MP, Nurminen E, Pentikäinen OT, Milgram SL, Hartwig JH, et al. (2010) Cystic fibrosis transmembrane conductance regulator interacts with multiple immunoglobulin domains of filamin A. J Biol Chem 285: 17156–17165. doi:10.1074/jbc.M109.080523.
41. Castellani S, Guerra L, Favia M, Di Gioia S, Casavola V, et al. (2012) NHERF1 and CFTR restore tight junction organisation and function in cystic fibrosis airway epithelial cells: role of ezrin and the RhoA/ROCK pathway. Lab Investig J Tech Methods Pathol 92: 1527–1540. doi:10.1038/labinvest.2012.123.
42. Samanta AK, Chakraborty SN, Wang Y, Kantarjian H, Sun X, et al. (2009) Jak2 inhibition deactivates Lyn kinase through the SET-PP2A-SHP1 pathway, causing apoptosis in drug-resistant cells from chronic myelogenous leukemia patients. Oncogene 28: 1669–1681. doi:10.1038/onc.2009.7.
43. Colas J, Faure G, Saussereau E, Trudel S, Rabeh WM, et al. (2012) Disruption of cytokeratin-8 interaction with F508del-CFTR corrects its functional defect. Hum Mol Genet 21: 623–634. doi:10.1093/hmg/ddr496.

7

Systems Responses to Progressive Water Stress in Durum Wheat

Dimah Z. Habash[1*¤], Marcela Baudo[1], Matthew Hindle[2], Stephen J. Powers[2], Michael Defoin-Platel[2], Rowan Mitchell[2], Mansoor Saqi[2], Chris Rawlings[2], Kawther Latiri[†3], Jose L. Araus[4], Ahmad Abdulkader[5], Roberto Tuberosa[6], David W. Lawlor[1], Miloudi M. Nachit[7]

1 Plant Biology and Crop Science, Rothamsted Research, Harpenden, United Kingdom, **2** Computational and Systems Biology, Rothamsted Research, Harpenden, United Kingdom, **3** Laboratoire D'agronomie, National Agricultural Research Institute of Tunisia, Ariana, Tunisia, **4** Dept. of Vegetal Biology, Faculty of Biology, Barcelona, Spain, **5** Biotechnology Department, General Commission for Scientific Agricultural Research, Damascus, Syria, **6** Dept. of Agroenvironmental Science and Technology, University of Bologna, Bologna, Italy, **7** Biodiversity and Integrated Gene Management Program, International Center for Agricultural Research in the Dry Areas, Rabat, Morocco

Abstract

Durum wheat is susceptible to terminal drought which can greatly decrease grain yield. Breeding to improve crop yield is hampered by inadequate knowledge of how the physiological and metabolic changes caused by drought are related to gene expression. To gain better insight into mechanisms defining resistance to water stress we studied the physiological and transcriptome responses of three durum breeding lines varying for yield stability under drought. Parents of a mapping population (Lahn x Cham1) and a recombinant inbred line (RIL2219) showed lowered flag leaf relative water content, water potential and photosynthesis when subjected to controlled water stress time transient experiments over a six-day period. RIL2219 lost less water and showed constitutively higher stomatal conductance, photosynthesis, transpiration, abscisic acid content and enhanced osmotic adjustment at equivalent leaf water compared to parents, thus defining a physiological strategy for high yield stability under water stress. Parallel analysis of the flag leaf transcriptome under stress uncovered global trends of early changes in regulatory pathways, reconfiguration of primary and secondary metabolism and lowered expression of transcripts in photosynthesis in all three lines. Differences in the number of genes, magnitude and profile of their expression response were also established amongst the lines with a high number belonging to regulatory pathways. In addition, we documented a large number of genes showing constitutive differences in leaf transcript expression between the genotypes at control non-stress conditions. Principal Coordinates Analysis uncovered a high level of structure in the transcriptome response to water stress in each wheat line suggesting genome-wide co-ordination of transcription. Utilising a systems-based approach of analysing the integrated wheat's response to water stress, in terms of biological robustness theory, the findings suggest that each durum line transcriptome responded to water stress in a genome-specific manner which contributes to an overall different strategy of resistance to water stress.

Editor: Pilar Hernandez, Institute for Sustainable Agriculture (IAS-CSIC), Spain

Funding: DH, RT, JA, AA, KL, and MN gratefully acknowledge support from the European Union FPVI programme (INCO-CT-509136) which funded material, overheads and salaries. MH was supported by a Lawes Agricultural Trust studentship. DH, CR, SP, MD-P, RM, MB, and MS acknowledge strategic funding support from the Biotechnology and Biological Sciences Research Council (BBSRC) of the UK, and CR was also supported by BBSRC grant BB/F006039/. MN received support from the International Center for Agricultural Research in the Dry Areas. The funders had no role in study design, data collection and analysis, decision to publish, or preparation of the manuscript.

Competing Interests: The authors have declared that no competing interests exist.

* Email: Dimah.Habash@securewheat.com

¤ Current address: Securewheat Research, Securewheat, St Albans, United Kingdom

† Deceased.

Introduction

Durum wheat is a major staple crop in the Mediterranean basin as a source of semolina for the production of pasta, couscous, burghul and other local products. It is cultivated primarily under rain-fed conditions across a wide range of environments varying in the profile and quantity of rainfall with drought as a major abiotic stress [1,2]. Regional to global-scale climate models predict, with high confidence, that the Mediterranean region will face soil drying in the future as temperatures rise and this may increase incidents of agricultural drought [3,4,5]. This will have serious negative implications for crop productivity and food security and thus crop yield and its component traits are targets for wheat breeders in their efforts to design cultivars for drought-prone environments as a means to ameliorate the climatic problems [6]. However, crop yield and drought resistance are complex genetic traits exacerbated by the fact that drought is often encountered with other abiotic and biotic constraints and these factors have contributed to the slow progress, so far, in breeding for relative drought resistance.

To address these challenges, various national and international efforts have been proposed to enhance wheat production by integrating technologies and approaches in plant science to underpin breeding efforts [7]. Thus, improving our understanding of the crop's responses to drought stress has emerged as a major target. Genetic studies on durum wheat have identified quantitative trait loci (QTL) for yield and yield component traits as a first important step in understanding and deconstructing the complex genetic factors defining drought resistance [8,9,6]. In parallel, studies of plants exposed to water stress under controlled environmental conditions enable the mechanistic understanding of the complex responses, adaptations, acclimation and recovery from stresses at the molecular, biochemical and physiological levels [10,11,12]. Genomic tools and 'omic' technologies have recently been employed to screen for the genes, proteins and metabolites responsive to water stress, and have identified key molecular elements [13,14]. Physical stress signals activate a series of molecular cascades and signalling components, both abscisic acid (ABA) and non-ABA-dependent, responsive to abiotic stresses in plants [15]. Common features have been identified in Arabidopsis and grass species [16] which have enabled the initial modelling of gene network responses under water stress [17]. These studies form a mechanistic basis for our knowledge of how physical signals are transduced to biochemical processes which ultimately translate into adjustments of metabolism and physiology to stress. Whilst genetic manipulation of key candidate genes, arising from such studies, has resulted in some form of resistance to water stress in individual transgenic plants, transferring such benefits from model species to improve crop performance under drought has been markedly less successful [18]. There is an urgent need to improve our understanding of drought effects on crops by combining comparisons of cultivars of different field performance with detailed analysis of their resistance mechanisms under well-defined plant water deficits. In addition, it is increasingly argued that systems-based approaches are necessary to integrate knowledge across genetics, genomics, breeding and plant physiology to tackle the complex genetic traits defining drought resistance [19,20,21]. Focusing system studies on crop species *per se* is also advocated considering the importance of genetic backgrounds and species-specific differences in adaptations to stress and for the immediate development of drought resistant crop cultivars, as demonstrated in bread *versus* durum wheat by [22].

Durum wheat grown in the Mediterranean is susceptible to post-anthesis terminal drought, a phase of growth which directly affects yield. Thus wheat breeders have targeted this type of drought by selecting for cultivars with a fast growth rate that rapidly accumulate biomass before flowering as a water stress avoidance strategy [2]. In addition, they have also developed wheat germplasm of promising yield potential by applying a double gradient selection strategy under both scarce and optimal water environments [8]. These efforts have been relatively successful in terms of yield improvement but the physiological and molecular basis of these drought resistant cultivars remains complex and unclear. Our aim in this study was to dissect the responses of drought resistant durum breeding lines to water stress in an attempt to identify molecular and physiological properties defining stress resistance and thus to build knowledge to accelerate the breeding effort. We chose wheat lines from the breeding programme at the International Centre for Agricultural Research in the Dry Areas (ICARDA) with good yield under irrigated and drought prone environments established in multi-year trials in the field (unpublished observations). Our study targets the post-anthesis developmental phase which corresponds to the period when plants are susceptible to terminal drought under field conditions in the Mediterranean basin. We applied rigorous criteria for the development of the water stress experiments, both in design and measurements, following [23]. High-throughput gene expression technology was applied in parallel with physiological measurements to profile durum lines at a range of water deficits (water stress transients) designed to dissect the systems responses to stress. Mathematical and bioinformatics methods were used to deconstruct the complex datasets and results are presented in formats amenable for further mining and exploration. Analysis was undertaken on datasets at equivalent leaf relative water content (RWC) to establish evidence of differential plant genome stress responses amongst the three wheat lines. Results demonstrate a global transcriptome response to water stress in all three lines that has a systems feature of high levels of coordination and uncovered within that key genome-specific differences. These results are discussed using a systems-based framework of biological robustness theory and how it could impact the search for new breeding strategies for drought resistance in durum wheat.

Materials and Methods

Plant material

Two durum wheat [*Triticum turgidum* L. ssp. *durum* (Desf.) Husn.] parents plus one recombinant inbred line (RIL) from the Cham1 ×Lahn mapping population developed at ICARDA were selected for this study because of their combined drought resistance and yield potential established in multiple year × site field trials in the Mediterranean region. Lahn was characterised by high yield potential under favourable rain environments, Cham1 exhibited relatively higher yield under drought, and RIL2219 was established as being more drought resistant than the best parent in addition to showing good yield stability (unpublished observations). Data collected in multiple field trials (30 over four years) showed RIL2219 closer to Cham1 in its overall phenology so that it reached heading 1.2 and 6 days earlier, on average, than Cham1 and Lahn respectively; it's grain fill duration was 1.69 and 2.69 days later than Cham1 and Lahn respectively; and it was similar to cham1 for days to maturity with both reaching it 5 days earlier than Lahn (unpublished observations).

Plant growth and experimental design

Seeds were sown in peat-free soil-based compost enriched with slow release fertilizer (Osmocote, Scotts UK Professional, Ipswich, Suffolk) and seedlings were vernalized for 3 weeks to synchronise growth and flowering. Three plants per line were transplanted into 3.5 L pots and grown under controlled chamber conditions of 25°C day/20°C night, 14 hours photoperiod, 800 $\mu mol^{-2} s^{-1}$ photon flux density and 65% relative humidity. Pots were watered uniformly each day until the start of the experiment. Plants were arranged using a randomised block design with 3 biological replicates per combination of line, treatment (control or stressed) and time point of the stress. At anthesis, plants were subjected to four independent time series water deficit experiments (termed water stress transients) by withholding water from pots for a period of up to 6 days with parallel controls being well watered, all measurements were taken at 10 am on a daily basis following the initiation of stress. Flag leaves were sampled from 3 experiments for molecular and biochemical studies and from one representative experiment for photosynthetic studies, one week after flowering (Zadoks scale 70). For each experiment, a flag leaf was sampled for RNA extraction and ABA content, another was taken for photosynthetic and osmotic adjustment measurements and a third for leaf relative water content (%RWC) and water potential.

Measurements of plant water status

The water status of the plants was monitored by flag leaf % RWC (fresh mass – dry mass)/(turgid mass – dry mass) × 100), following [24] and leaf water potential was measured with a pressure chamber, after [25]. Difference between pot mass in the morning and afternoon gave the total plant plus soil water loss.

Physiological and biochemical measurements

Flag leaf transpiration rate, stomatal conductance to water, and CO_2 assimilation rate were measured at 800 µmol m^2 s^{-1} PFD using portable infrared gas analysers (LICOR LI-64 LI-COR Biosciences, Nebraska, USA and SIRAS PP Systems, Norfolk, UK). Plant height, tiller number, development stage and leaf area were measured to monitor growth. Osmotic potential and adjustment was measured using a vapour pressure Osmometer (5100C, Wescor, USA) on samples of sap from frozen, thawed, macerated and centrifuged leaves. ABA was quantified by GC-MS, essentially as described in [26], except that freeze-dried leaf samples were homogenised in 80% methanol-water using a Polytron homogeniser, after which [3-methyl-$^2H_3$]ABA (20–25 ng) was added as internal standard.

RNA isolation and array hybridisation

Individual flag leaves, one leaf per biological sample, were sampled for RNA extraction and transcript expression studies. For all three wheat lines, 3 biological replicates per line were taken at days 1, 2, 3, 4, 5 (all 3 lines) and 6 (RIL2219 only) of the stress plus day 0 and 1 from the well watered controls to give a total of 66 samples. Leaves were frozen in liquid N_2 and total RNA was extracted using TRIZOL Reagent (Invitrogen, UK) following the manufacturer's recommendations. Total RNA was DNase treated with TURBO DNase enzyme (Ambion) and cleaned using RNeasy columns (Qiagen) following the manufacturer's instructions. High quality RNA was submitted to transcriptome profiling at Bristol University (http://www.bristol.ac.uk/biology/research/transcriptomics/). Expression analysis was carried out on Affymetrix GeneChip Wheat genome arrays using Affymetrix GeneChip Eukaryotic One cycle target labelling reagents and protocols (701028 Rev 4). Five µg of total RNA was first reverse transcribed using a T7-Oligo(dT) promotor primer in a first strand cDNA synthesis reaction, followed by RNase H mediated 2nd strand cDNA synthesis. The double stranded cDNA was purified and served as a template in a subsequent in vitro transcription (IVT) reaction carried out in the presence of T7 RNA polymerase and a biotinylated nucleotide analog/ribonucleotide mix to produce complimentary biotinylated RNA (cRNA). The biotinylated cRNA targets were cleaned up, fragmented and hybridised to Wheat Genome arrays for 16 hours in a GeneChip Affymetrix Hybridisation oven 640. Hybridised probe arrays were then washed and stained with a streptavidin-conjugated fluorescent staining solution followed by antibody amplification of the fluorescent signal in a GeneChip fluidics 450 station using the appropriate fluidics scripts for the array format. Following washing and staining, arrays were scanned using the GeneChip scanner 3000. Cell intensity data was automatically computed from the image data using GeneChip Operating Software (GCOS) under default settings. The quality and integrity of both total RNA and labelled and fragmented cRNA was assessed on an Agilent bioanalyser RNA Nano assay.

Data processing and modelling

Microarray data are available in the ArrayExpress database (www.ebi.ac.uk/arrayexpress) under accession number E-MTAB-2410. Microarray data initial analysis was carried out using GenespringGX 8.0 (Agilent Technologies, Inc., Santa Clara, California, USA). The set of array raw data files (.cel files) were pre-processed using the gcRMA algorithm [27] to standardise the mean and variance of expression. Probesets were filtered to remove those with mean signal values less than 10, which were deemed not expressed. All probeset signal data (hereafter referred to as transcript or probe expression) from the filtered set were then exported for ANOVA and multivariate statistical analysis using GenStat (2010, thirteenth edition, VSN International Ltd, Hemel Hempstead, UK). To account for the time differences in wheat line responses to the stress and to present the transcript expression results in terms of equivalent leaf RWC, data were interpolated according to suggestions for handling results of time series experiments [28]. Firstly, the logistic relationship of RWC over time (days) was used to facilitate interpolation of transcript expression on the $\log_2$ scale with respect to leaf RWC at proposed levels of 90, 82, 74, 66, 58 and 50; these levels of RWC were chosen to reflect a range of water deficit, whilst maintaining 6 observed occasions. RWC was then modelled on time (days of stress), fitting a logistic curve to the RWC data collected from each wheat line, using nonlinear regression, $RWC(time) = a+b/[1+\exp((time - c)/d)]$, for estimable parameters a, b, c, and d. For each transcript by replicate (statistical block) by wheat line combination, the change in transcript expression data over the time course was interpolated using Lagrange polynomials [29,30] which enabled the transcript expression at the time points corresponding to the proposed RWC levels to be calculated. The results were further analysed by ANOVA and $\log_2$ transformation which considered the main effect of, and interaction between, the line and RWC factors, and also accounted for the block effect. The transcripts were filtered into groups given significant F-test ($p<0.01$, to ensure well-defined groups) results, after correcting p-values for multiple testing and false-discovery [31] and least significant difference (LSD) values ($p = 0.05$) were used to compare means. Principal Coordinates Analysis (PCo) was applied to the RWC-interpolated data for the 19062 probes derived from 54 samples (three replicates of each of six RWC levels- 0–5 days- by three wheat lines) [32]. The probes proposed to be responsible for the separation observed in a given dimension were found by correlating the coordinates of the points in that dimension to transcript expression values, using an approximate F-test and extracting those transcripts with the greatest (most statistically significant) F-values. Physiological and biochemical measures were modelled with respect to RWC using linear or non-linear regression, testing whether single, parallel or separate lines (or curves) were required (F-tests) for the wheat lines. Linear regression was used for leaf water potential, water loss and osmotic adjustment. Gompertz curves, of the form $y = C - C[\exp(\exp(-k(RWC - m))]$, were used for rate of CO_2 assimilation, transpiration rate and stomatal conductance, with parameters C being the asymptote, m the inflection point and k the exponential rate separate, or common, for each wheat line. ABA content showed an asymmetric rise to a peak, therefore a rational functions model was used, $\log(ABA) = a+[(b+cRWC)/(1+ dRWC+fRWC^2)]$, with estimable parameters a, b, c, d and f, taking ABA on the log (to base e) scale to account for greater variability in this measure.

To estimate the contribution of allelic differences in gene sequences to probe hybridisation, for each perfect match probe the significance of the ratio changes across lines was calculated in proportion to the corresponding mean probe-set abundance. Robust Multiarray Averaging (RMA) background correction and constant normalisation was applied to the probe intensities [33,27]. The probe intensities were transformed using the $\log_2$ of the ratio of the probe to the mean of the corresponding probe-set intensity. For probe-set intensities the same probe normalisation and background corrections were applied, together with perfect match correction and summarization using median polish [33,27]. The significance of the presence of allelic difference for lines was determined from two-way ANOVA of line and time factors, using the p-value for the significance of the difference across lines.

Annotation and MapMan pathways analysis

The wheat Affymetrix chip was designed with 11 probe sets per gene and the majority of the analysis in this study focused on the average expression value of the 11 probes in a probe set. In our analysis, we did not filter out any probes or identify unigenes, thus we present results for each probe set signal intensity that represent changes that could be due to unigenes, splice variants, homeoalleles or several members of a gene family. Probes were assigned to orthologous genes and metabolic bins, or pathways, using the MapMan programme following the standard protocol [34,35], bin # 35 was unassigned annotation. When referring to probe intensity or transcript expression of genes from a particular metabolic pathway, from now onwards in the text, this refers to an annotated orthologous gene. qPCR was done on eight genes and results showed similar changes to Affymetrix probe expression as a function of the stress transient in all lines (unpublished observations).

Results

Physiological responses of wheat lines to water stress

Cessation of watering and loss of water from the soil for 6 days progressively decreased flag leaf RWC and water potential for all wheat lines (Figure 1 A): the initial RWC was identical for the three cultivars. It was not possible to accurately measure RWC for Lahn on day 6 as it was severely wilted. RWC decreased more slowly in RIL2219 than in either parent, as demonstrated by its larger RWC after day 3, related to the slower rate of total water loss from its pots both as a function of time and of RWC (Figure 1A, 1F). The relation between RWC and water potential of the flag leaves was linear, and identical, in the three cultivars (Figure 1B) and therefore RWC was used as the reference for tissue water status. As the rates of water loss and decrease in RWC differed significantly between the cultivars, all subsequent analyses were made with reference to RWC. RIL2219 had higher photosynthetic rate (Figure 1 C), instantaneous transpiration rate (Figure1 D) and a larger stomatal conductance (Figure 1 E) in the early phases of water depletion (from ca. 100 to 80% RWC). However, these decreased substantially in all cultivars (and marginally more so in RIL2219) as RWC fell from ca. 80 to 60% and where we also observed that RIL2219 and Cham1 maintained a higher osmotic adjustment for equivalent RWC when compared to Lahn (Figure 1 G). Furthermore, the ABA content of turgid leaves was larger in RIL2219 than Cham1 or Lahn notably at the start of the experiments in leaves at equivalent RWC, but increased similarly as RWC decreased, rising greatly and reaching a maximum at about 65% (Figure 1 H) after which there was a general decline. There were no statistically significant ($p<0.05$, F-test) difference in flowering time, leaf area or plant height between the lines in any of the experiments (unpublished observations).

Dissection of transcriptome responses to progressive water stress

Flag leaf RNA samples from three wheat lines subjected to the transient of water stress were submitted for the study of global transcript expression using the wheat Affymetrix gene expression chip. Initial filtering showed that 19062 probes (32% of total probe sets) were modulated in value in response to water stress, genotype, or in combination, in all lines during the stress transient. In the first three days of the stress transient after watering was stopped, probes showed moderate changes in expression and this increased sharply as the stress progressed and leaf RWC declined with some delay shown by RIL 2219 (Figure 2). To analyse probe expression responses as a function of equivalent stress amongst the three wheat lines, this delay was taken into consideration by interpolation of expression values as a function of leaf RWC. ANOVA was then applied to 19062 probes across the three lines and RWC transient, and the results are available for each probe as mean expression values and LSDs for comparisons (File S1). This analysis identified four independent groups showing statistically significant changes with respect to *Line* only (932 probes in File S2), *Stress* only (2579 probes in File S3), *Line+Stress* independently (7644 probes in File S4), *Line* × *Stress* interaction (7699 probes in File S5), and a fifth group of 208 probes showing *no statistically significant effect* of either factor (File S6). A detailed analysis of probes annotated to metabolic pathways significantly ($p<0.05$, LSD) altered from 90-50% leaf RWC for each of the above ANOVA groups was done using the MapMan programme. Some probes annotated to the same gene model and subsequently allocated to more than one of the latter ANOVA groups may be a result of hybridisation to closely related sequences such as gene families, homeoalleles or paralogs.

Equivalent transcript responses in all lines

ANOVA identified probes in *Stress* (File S3), *Line+Stress* (File S4) and *Line* × *Stress* (File S5) groups that were statistically altered in expression in response to the water stress transient when compared to controls. Systematic analysis of each of these datasets using MapMan showed similar general trends of metabolic pathway responses to water stress and these are summarized in the following sections with focus on the *Line+Stress* (File S4) group as a representative. The 7644 probes in this latter group showed initial differences in expression between the wheat lines at the onset of the stress but this did not influence the profile of response further into the stress transient as RWC decreased; the lines showed similar and parallel stress responses. Probes annotated to specific metabolic pathways (bins) significantly altered from 90-50% leaf RWC are documented in individual spreadsheets in File S4 and Table 1 to enable further mining. Analysis revealed that all represented metabolic pathway bins showed some probes with significant ($p<0.05$, LSD) over-expression or under-expression of varying orders of magnitude at each leaf RWC (Figure 3). The profiles of the changes in expression of metabolic pathways are complex and varied as leaf RWC decreases, and we focus in this paper on general trends and particularly on changes in regulatory pathways, photosynthetic, primary and secondary metabolism (see following sections).

Early changes. The experimental design uncovered pathways specifically altered early in the stress response with an average of 26% of annotated probes showing changed expression when leaf RWC decreased from 90 to 82% (File S4 spreadsheet summary of early changes). Of these, 247 probes annotated across most of the metabolic bins showed a reduction in expression (File S4 early changes underexpressed spreadsheet) and an even larger number of probes (624) increased in expression with bias towards RNA and protein processing, signalling and transport (File S4 early changes overexpressed spreadsheet). Results also showed that 73% of all annotated probes showed changes in expression as the

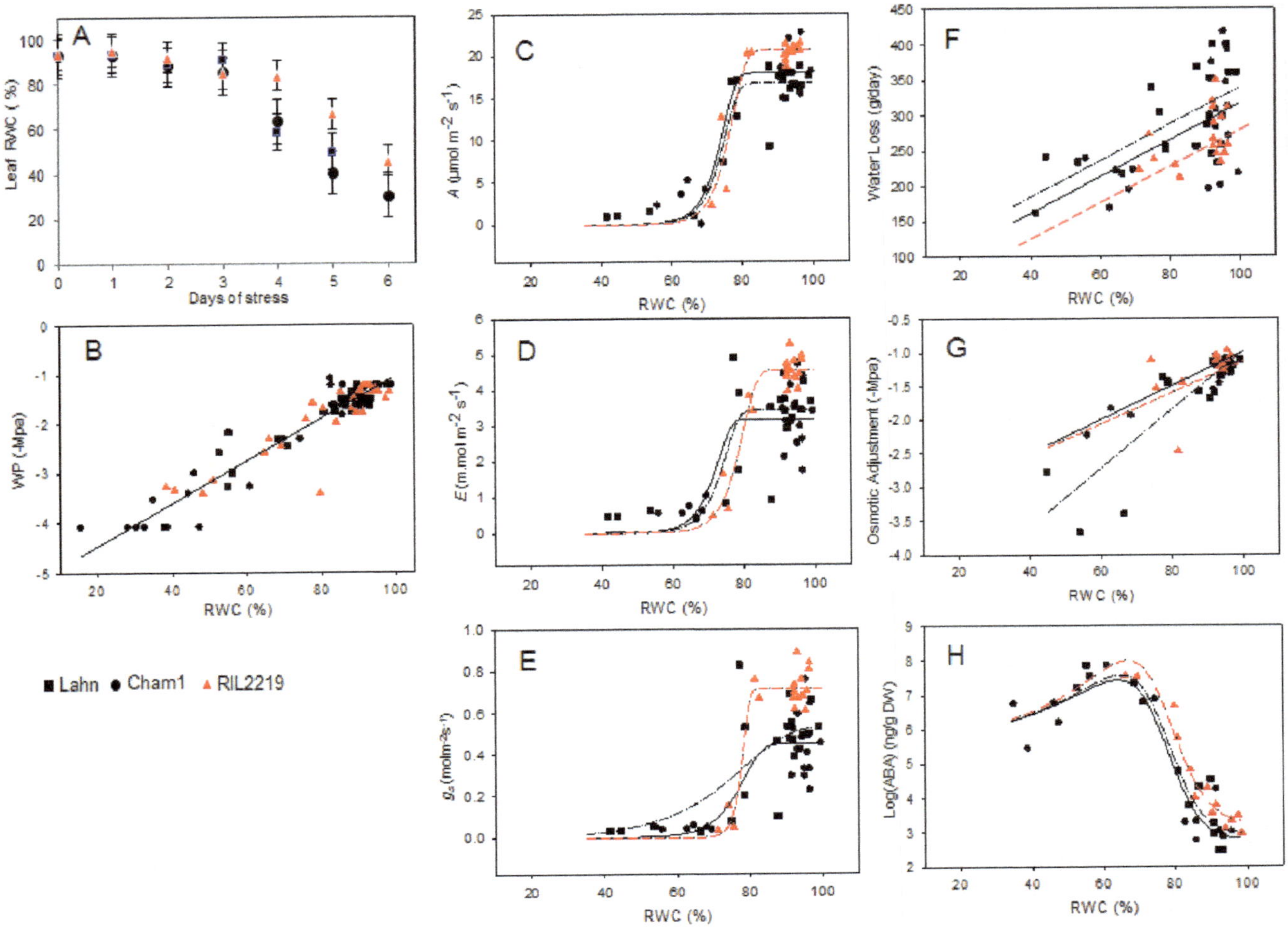

Figure 1. Physiological and biochemical parameters of leaves during water deficit. Plant water status was measured by flag leaf %RWC as a function of days of stress (A), leaf water potential (B) and total plant water loss (F) during a stress transient of 0–6 days for three wheat lines Lahn (■), Cham1 (●) and RIL2219 (▲). Leaf photosynthetic and biochemical parameters were CO_2 assimilation (C), transpiration (D), stomatal conductance (E), osmotic adjustment (G) and ABA content (H). Linear regression analysis gave a single line relationship for leaf water potential, parallel lines ($p< 0.05$, F-test) for water loss, and separate lines ($p<0.05$, F-test) for osmotic adjustment. Using the Gompertz curve, separate *C* and *k* parameters ($p< 0.05$, F-tests) were required for CO_2 assimilation, whereas for transpiration and stomatal conductance only separate *C* parameters ($p<0.05$, F-test) were required; (C–H) red line, black dashed line and black solid line are for RIL2219, Lahn and Cham1 respectively. For ABA, separate c parameters were significant ($p<0.05$, F-test) in the rational functions model.

leaf RWC declined to 74% therefore defining a transition to the next stage of stress response.

Regulatory pathways. The largest number of probes affected by the stress transient were annotated to regulatory pathways with several members showing changes at the very early stage of the stress (82%leaf RWC) (Figure 4). Across the time course of the stress transient, probes annotated to DNA repair and synthesis (chromatin structure: histone, retrotransposon/transposase.hat-like transposase) were identified as significantly altered compared to un-stressed controls with the majority showing under-expression (File S4 bin 28). RNA processing (ribonucleases, RNA helicase, siRNA methyltransferase, splicing), regulation of transcription (transcription factor families or TFs) and RNA binding were particularly affected (File S4 bin 27). Probes annotated to several TFs showed modulation during the whole stress transient (Figure 5), and members belonging to AP2-EREBP, ARF, bHLH, bZIP, C2C2, MADS, Histone ATase, NAC, SET families and MYB showed changes at the earliest stage of stress at 82% leaf RWC with a MYB4 transcript (ta.26049.1.s1_a_at) giving one of the highest measured increases in expression (Figure 5). At 74% leaf RWC, 78% of TF transcripts showed changes in expression profiles and by the final stress point of 50% leaf RWC, 13 showed small changes specific only to that stress point and belonged to ARF, ALFIN, HB MADS, WRKY and histone HDA5 families (Figure 5, File S4 bin 27).

A large number of transcripts in protein modification specifically protein assembly and cofactor ligation, protein degradation (cysteine protease, aspartate protease, autophagy, serine protease, metalloprotease, AAA type, ubiquitin, ubiquitin E1/E2/E3/ ubiquitin ubiquitin/ubiquitin ubiquitin protease/ubiquitin proteasome), protein folding, protein glycosylation, post-translational modification and protein targeting were also altered during the stress transient (Figure 4, and File S4 bin 29) with amino acid activation and protein synthesis transcripts showing predominant down-regulation (Figure 3). Transcripts for signalling components, especially a large number of calcium- and G-protein-related and various receptor kinases (leucine rich repeats classes II\ III\VI\-VIII-1\X\XI\VIX, DUF 26, legume-lectin, *Catharanthus roseus*-like RLK1, wheat LRK-10 like, proline extension like, wall associated kinase, S-locus glycoprotein like) and others included

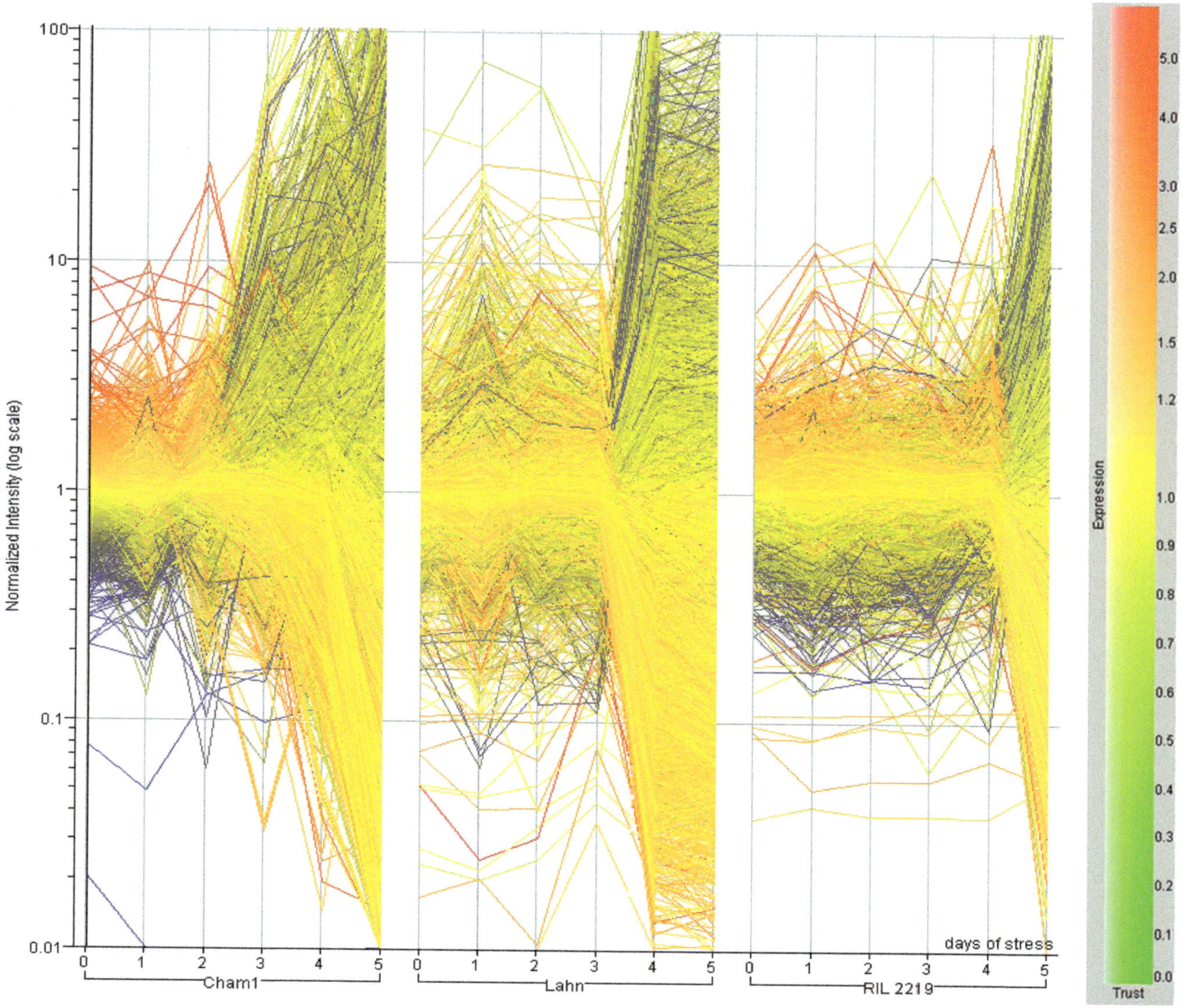

Figure 2. Mean normalised expression of Affymetrix probesets from leaves exposed to water stress. Filtered whole expression dataset of 19062 probes for the three wheat line leaves during the transient of six days of stress analysed using GenespringGX 8.0. Lines are coloured according to their normalised expression for genotype Cham1 at day 0.

members belonging to 14-3-3 proteins, sugar and nutrient physiology, light, lipids, MAP kinases, lysine motif and phosphinositides were particularly altered during the stress (Figure 4, File S4 bin 30). These were accompanied by changes in expression of transcripts annotated to redox control; ascorbate and glutathione, dismutases and catalases, glutaredoxins, peroxiredoxin and thioredoxin with most changes measured at 66% leaf RWC (Figure 4, File S4 bin 21).

Hormones and stress-specific responses. Probes belonging to each of the major categories of hormone metabolism were modulated in expression as a result of water stress with the majority showing over-expression (Figure 4, File S4 bin 17). They showed close to average change (25%) at the early stress point of 82% RWC except transcripts of abscisic acid (ABA) where 60% of probes showed early phase changes. Nine out of ten probes annotated to ABA metabolism showed over-expression across the transient with the largest change annotated to signal transduction DNA binding transcription activity ABA-responsive element-binding factor (3ABF3/DPBF5) (File S4 bin 17). At 74% leaf RWC most probes annotated to auxin, ethylene, jasmonic acid, salicylic acid and gibberellin metabolism showed modulation in expression and a probe annotated to gibberellin 2-beta-dioxygenase (ATGA2OX1) gave the largest increase amongst all the hormone classes measured at the end of the stress transient (Figure 4, File S4 bin 17; Table 1). Water stress also modulated the expression of a group of probes designated to biotic and abiotic stress-specific responses with 29% of all probes changed at 82% leaf RWC (File S4 bin 20 stress). This included universal stress proteins, genes that respond to cold, heat, touch, salt/drought stress (dehydration-responsive protein-related, drought-responsive family protein, early-responsive to dehydration protein-related/ERD protein-related, DREPP plasma membrane polypeptide family protein, EARLY-RESPONSIVE TO DEHYDRATION 3,4 (ERD3,4)) and a large number annotated to heat shock proteins (HSP: 17, 18, 22, 21, 23, 70, 81, 83, 88, 91, 98, 101, DNAJ, GFA2 and MTHSC70-1).

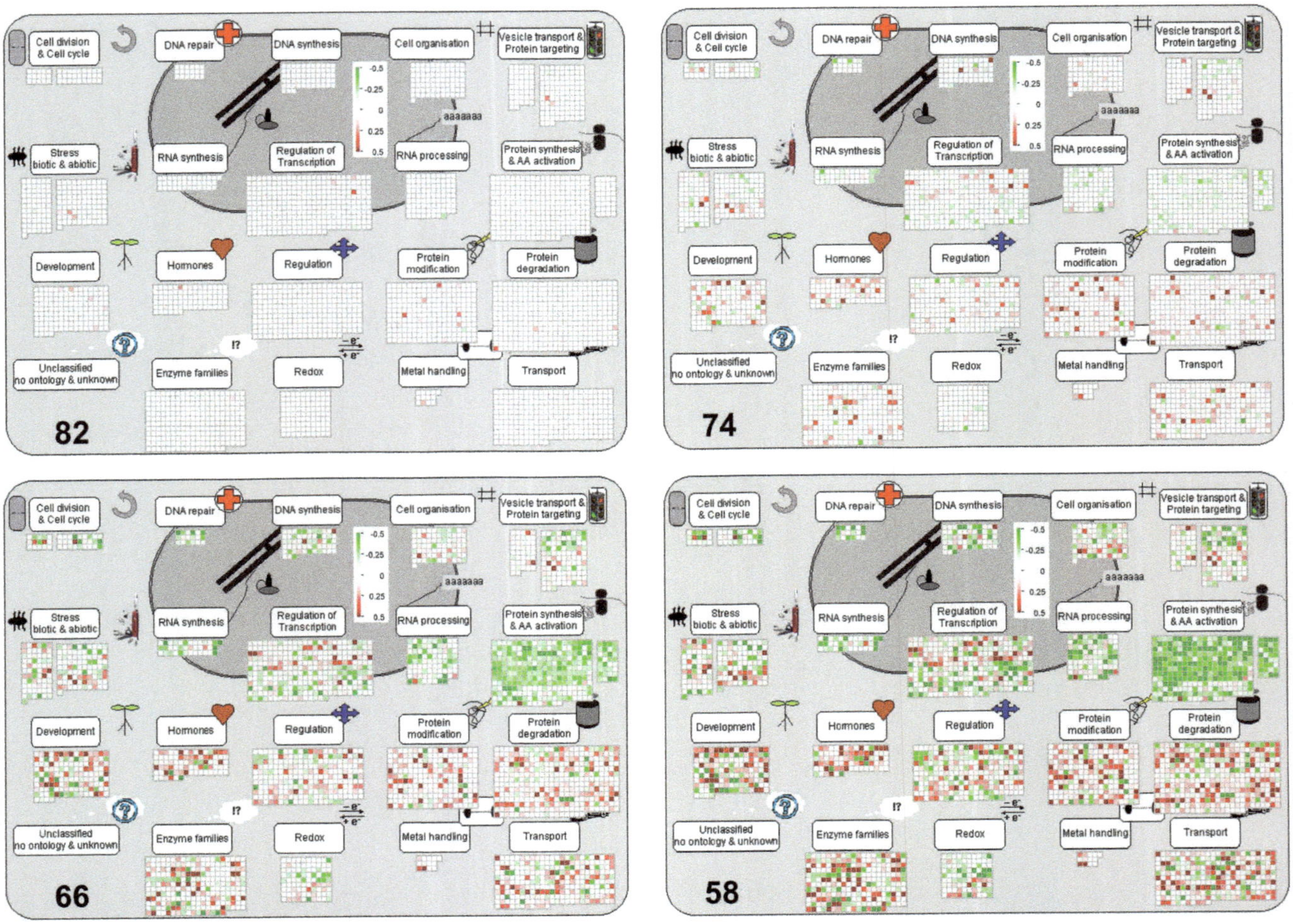

Figure 3. Overview of cell function transcript expression during the water stress transient. Transcript profile changes, at decreasing leaf %RWC (82,74,66,58), were similar across the stress transient for the three wheat lines in probes taken from MapMan bins 15,17,20,21,26,27,28,29,30,31,33,34 in the *Line+Stress* ANOVA group dataset (File S4). Results were visualised in MapMan and the colour scale for transcripts that were increased or decreased in abundance were denoted as red and green, respectively.

Photosynthesis and energy. The majority of probes annotated to photosynthesis were down-regulated during the stress transient with transcripts annotated to LHCII, PS1 and ATP synthase changing early at 82% leaf RWC (Figure 6, File S4 bin 1). Most other transcripts showed changes at 74% leaf RWC, they were annotated to light reactions (LHCII, PS2 and PS1 polypeptide subunits), Rubisco small and large subunit, plastoquinone dehydrogenase complex, ATP synthase beta chain, ferredoxin 3 (ATFD3), serine hydroxymethyltransferase (SHM4), cyclic electron flow (NDHH), phosphoglycerate kinase 1 (PGK1), triosephosphate isomerase (TIM) and phosphoribulokinase (PRK). A few were up-regulated and belonged to phosphoribulokinase (PGK), ATP synthase (VHA-A), subunit of ATP synthase (ATPA), cyclic electron flow (PIFI, NAD2.2) and PS2 polypeptide subunits. This predominant down-regulation of expression of photosynthetic transcripts was accompanied by a reduction in the expression of several probes annotated to tetrapyrrole biosynthesis (Figure 6, File S4 bin 19). Water stress also modulated the expression of transcripts of several enzymes annotated to glycolysis such as pyruvate kinase, phosphofructokinases, phosphopyruvate hydratase (LOS2), and gave increases specifically in phosphoenolpyruvate carboxylase (ATPPC3), fructose-2,6-bisphosphate 2-phosphatase (F2KP) and fructose-bisphosphate aldolase; with phosphofructokinase (PFK2) transcript showing the highest increase measured through the whole transient. Meanwhile phosphoenolpyruvate carboxylase 1 (ATPPC1), pyrophosphate-fructose-6-P phosphotransferase, triose-phosphate isomerase (ATCTIMC) and glyceraldehyde-3-phosphate dehydrogenase (GAPC1, 2) declined (File S4 bin 4). Changes in transcripts of the tricarboxylic acid cycle occurred mainly at 74% leaf RWC, with large changes in carbonic anhydrase (File S4 bin 8) and specific increases in probes annotated to gluconeogenesis at 82%leaf RWC: malate synthase, citrate synthase, malate dehydrogenase PMDH1 and pyruvate dikinase (PPDK) (File S4 bin 6). Various transcripts of mitochondrial electron transport also showed changes, such as ATP synthesis complex I and IV, cyt C and F1-ATPase, with specific increases in alternative oxidase (File S4 bin 9).

Sugar, amino acid, lipid and secondary metabolism. Water stress increased the expression of probes annotated to sucrose degradation (sucrose synthase SUS3, 4; fructokinase pfkB-type; hexokinase HXK1), some sucrose invertases (cell wall, neutral and vacuolar) and starch degradation (beta-amylase BMY8/BAM3 and alpha amylase ATAMY1). This trend was accompanied by large decreases in the expression of probes annotated to alpha-glucan phosphorylase 2, cell wall invertases such as beta-fructofuranosidase (ATFRUCT3,5), starch synthesis (synthase, branching, transporter and D-fructose-1,6-bisphosphate

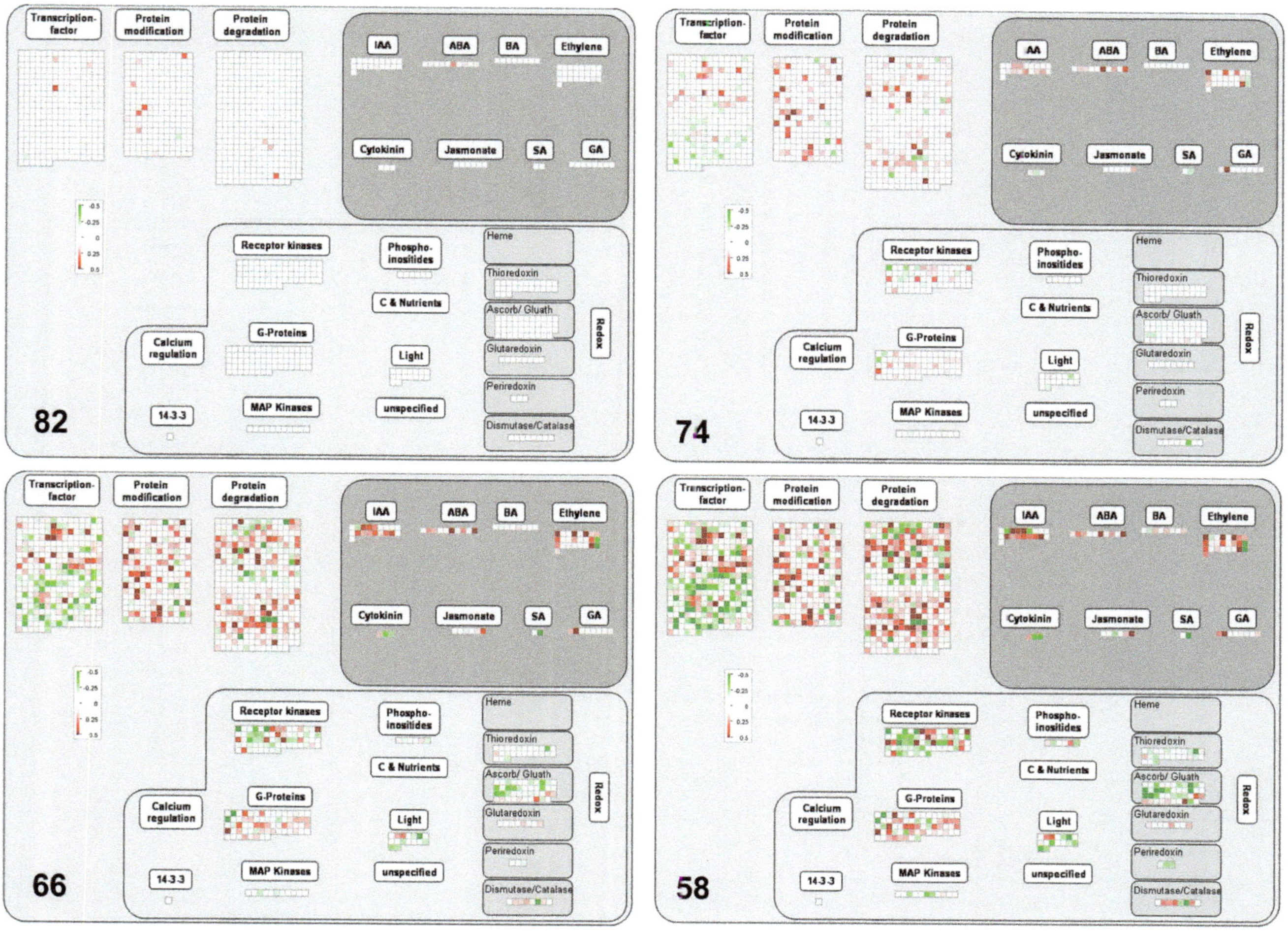

Figure 4. Expression of genes involved in regulatory pathways during water stress. Response of transcripts annotated to MapMan bins 17, 21, 27, 29 and 30 at decreasing leaf %RWC during the water stress transient. ANOVA Dataset and visualisation as for Figure 3.

1-phosphohydrolase) (Figure 6, Figure S1, File S4 bin2) and by changes in minor sugar metabolism (File S4 bin3). Various probes annotated to cell wall metabolism, including degradation and synthesis, showed large levels of change especially early in the stress at small water deficit of 82% RWC (Figure 6, File S4 bin10). Water stress produced major modulations in expression of transcripts of amino acid metabolism in both synthesis and degradation: there were decreases in the synthesis of arginine, histidine, threonine, lysine, leucine, cysteine, methionine, aspartate, asparagine, branched chain amino acids, serine, tryptophan, precursors for aromatic amino acids. There were however increases in transcripts for the synthesis of cysteine, methionine, glutamate decarboxylase 1, 4-aminobutyrate transaminase (GABA-T), aspartate transaminase, proline synthase (P5CS2), some branched-chain aminotransferase, acetolactate synthase small subunit and various aromatic amino acids. These changes were accompanied by decreases in transcripts for the degradation of histidine, GCN5-related N-acetyltransferase, threonine, methionine, lysine, asparagine, tyrosine and tryptophan and increases in those involved in the degradation of arginine, valine, branched-chain group, methionine, lysine, tyrosine and tryptophan (Figure 6, File S4 bin 13).

Water stress caused down-regulation of transcripts of nitrate reductase (NIA2) and increases in those for glutamine synthetase (cytosolic Gln1;4), glutamate dehydrogenase (GDH2) and glutamate synthase (GLT1) (Figure 6, File S4 bin12). Changes in the expression of probes annotated to lipid metabolism were also measured for biosynthetic enzymes of fatty acid desaturation, fatty acid synthesis and elongation, steroids, squalene and sphingolipid metabolism, lipid degradation (beta oxidation, lipases, lysophospholipases) lipid transfer proteins and phospholipid synthesis (Figure 6 and File S4 bin11). Probes were also altered in nucleotide metabolism with overall down-regulation in purine and ATP synthesis, whilst some coding for pyrimidine and CTP synthase were up-regulated (Figure 6, File S4 bin 23). Probes annotated to several secondary metabolic pathways were affected by the stress with predominant increases in expression of phenylpropanoids, lignin biosynthesis, flavanoids, betaine, and wax metabolism (Figure 7, File S4 bin16).

Cell organisation, transport and development. Probes annotated to cell cycle, cell division, cell organisation, development and vesicle transport were modulated by water stress (Figure 3, File S4 bin 31, 33) with probes annotated to late embryogenesis abundance showing over-expression at the earlier phases of the stress (File S4 bin 33). A large number of transcripts annotated to transport showed changes in expression such as those belonging to ABC transporters and multidrug resistance proteins; transporters for : amino acids, ammonium, calcium, sugars, sucrose, metals, cyclic nucleotide or calcium regulated channels, peptides and oligopeptides, phosphate, potassium, sulphate and

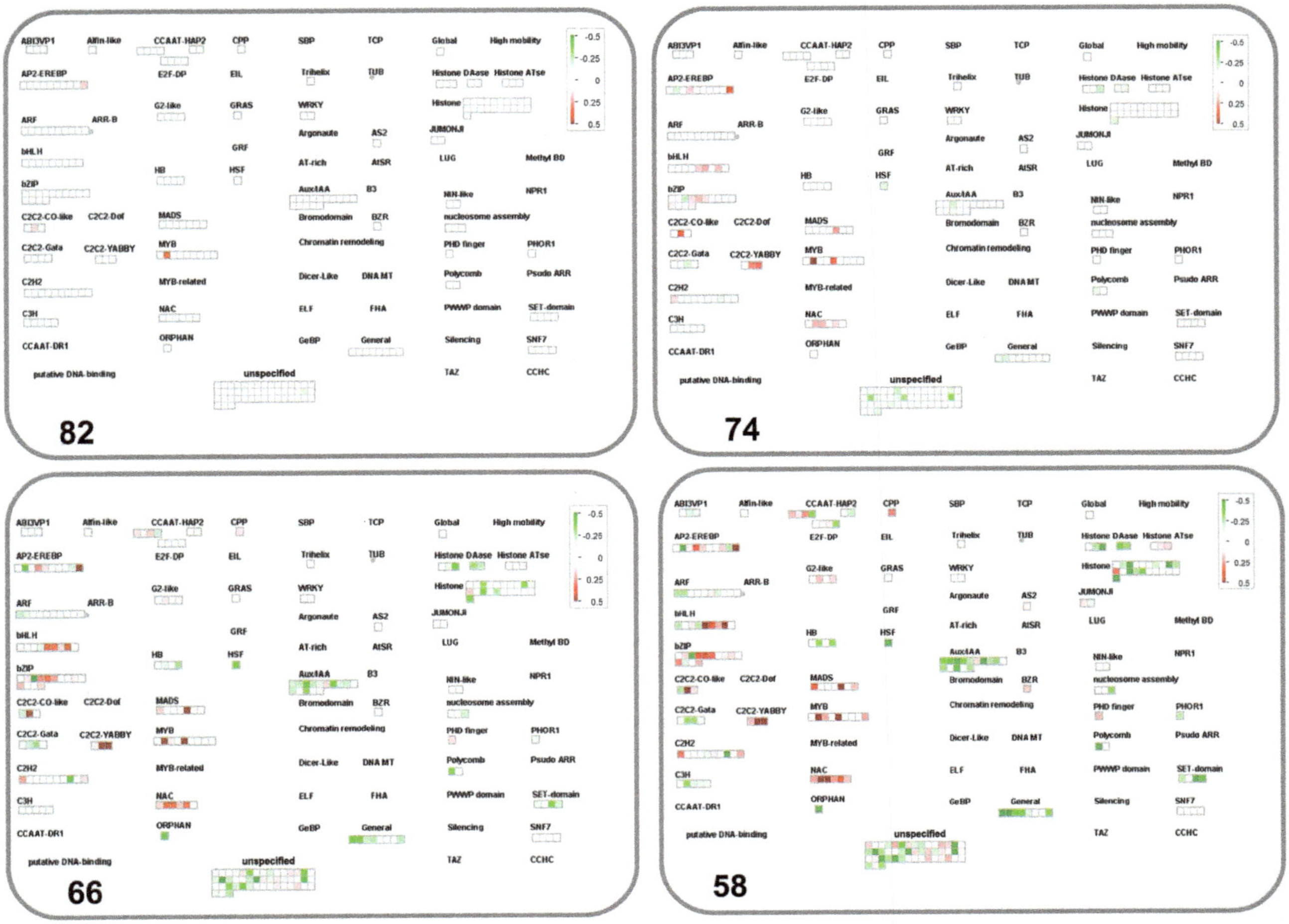

Figure 5. Changes in RNA regulation with focus on transcription during water stress. Expression of probes annotated to transcription from MapMan bin 27 and 28 at decreasing leaf %RWC during the water stress transient. ANOVA Dataset and visualisation as for Figure 3.

unspecified ions; H+ transporting pyrophosphatase; Major Intrinsic Proteins (PIP, TIP); metabolite transporters at the envelope and mitochondrial membrane; NDP-sugars at the ER; p- and v-ATPases, p- and v-ATPases.H+-transporting two-sector ATPase (File S4 bin 34).

Another group of probes, assigned to miscellaneous enzyme families involved in various metabolic processes, were modulated in response to the stress and included several members of cytochrome P450 which showed relatively large changes in expression (File S4 bin 26). Annotated probes from all metabolic bins sorted by the largest increase in expression measured at the last time point of the stress are summarised in Table 1. Probes contributing to most differences across the whole stress response were identified using PCo, see below.

Genotype-dependent responses

The ANOVA group *Line* × *Stress* comprised 7699 probes documenting an interaction between the two independent variables wheat line (two parents and RIL 2219) and stress (transient); probes that showed statistically significant responses to the stress and also measured differences between the lines in the way they responded to the water deficit. The global nature of the stress response, in all three lines, is demonstrated by the many pathways affected and by the large number of affected probes annotated to regulatory pathways presented in File S5 and Figure 8. These were essentially similar to those documented by the other ANOVA groups *Stress* (File S3) and *Line+Stress* (File S4) in showing down-regulation of photosynthesis, reconfiguration of primary and secondary metabolism and a predominant transcript modulation of regulatory pathways for RNA and protein modification. Nevertheless, each of the 7699 probes showed a statistically significant difference in expression between the lines in terms of absolute value, starting value and/or pattern of change across the transient. These are documented for each probe and partitioned into MapMan metabolic bins highlighting probes showing changes early in the stress transient for each line (File S5 early changes sorted for each line spreadsheets, Table 2). Inspection of the early phase of the stress at 82% leaf RWC showed the greatest changes in gene expression had occurred in Lahn, with 10% less changing in Cham1 and even 60% less in RIL 2219 with the latter being the most conserved. Graphical plots of the two parent's transcripts showed distinctly different profile responses to water stress, at the same leaf %RWC, across several metabolic pathways, whilst RIL 2219 combined features derived from either parent and also showed some distinct responses either in the trend or in the magnitude of the transcript expression (Figure 8). Moreover, this plot uncovers a more responsive transcriptome in Lahn, with at least one probe per bin showing a higher expression value than either Cham1 or

Table 1. Probes with the largest over-expression or under-expression values achieved at the final stress point. Data represents the ratio of change in expression at the final stress level of 50% relative to that at 90 leaf %RWC taken from the ANOVA group dataset stress+line (File S4). Transcript profile changes, as a function of leaf %RWC, were similar across the stress transient for the three wheat lines and probes were annotated to MapMan pathways and bins.

MapMan Bin#	MapMan Pathway	Affymetrix Probe	MapMan Annotation	Expression Change 50:90
1	Photosynthesis	ta.3252.1.s1_at	PS2 polypeptide subunits	1.34
2	Sugar metabolism	ta.11114.1.a1_at	ATBETAFRUCT4/VAC-INV (VACUOLAR INVERTASE)	0.93
3	Minor sugars	ta.20649.2.s1_a_at	ATTPS6 (Arabidopsis thaliana trehalose phosphatase/synthase 6)	0.83
4	Glycolysis	taaffx.80151.1.s1_at	PFK2 phosphofructokinase family protein	1.08
5	Fermentation	taaffx.734.1.s1_x_at	ALDH7B4 (ALDEHYDE DEHYDROGENASE 7B4)	0.93
6	Gluconeogenesis	ta.23970.1.a1_at	MASY malate synthase	1.59
7	OPP	ta.9372.1.s1_at	transaldolase	0.57
8	Tricarboxylic acid cycle	ta.2022.2.s1_at	aconitate hydratase, cytoplasmic, putative	0.35
9	Mitochondrial electron transport	taaffx.121853.1.s1_at	AOX1C (alternative oxidase 1C)	0.70
10	Cell wall	taaffx.7177.1.s1_s_at	glycosyl hydrolase family 3 protein	1.91
11	Lipid metabolism	ta.9528.1.a1_at	oxidoreductase, acting on the CH-CH group of donors	0.91
12	N metabolism	ta.1870.1.s1_a_at	GDH2 (GLUTAMATE DEHYDROGENASE 2)	0.78
13	Amino acid metabolism	taaffx.95414.1.s1_at	ATBCAT-3/BCAT3 (BRANCHED-CHAIN AMINOTRANSFERASE 3)	1.50
14	S assimilation	ta.11132.1.s1_at	adenylylsulfate kinase	0.14
15	Metal handling	ta.17991.1.s1_x_at	binding, chelation and storage	0.49
16	Secondary metabolism	ta.11147.1.a1_at	FAH1 (FERULATE-5-HYDROXYLASE 1)	1.11
17	Hormones	taaffx.12175.1.s1_at	ATGA2OX1 (GIBBERELLIN 2-OXIDASE 1)	1.60
18	Cofactors and vitamins	ta.13745.1.s1_a_at	LIP1 (LIPOIC ACID SYNTHASE 1	0.10
19	Tetrapyrrole biosynthesis	ta.11708.1.s1_at	HEMD; uroporphyrinogen-III synthase	0.08
20	Abiotic stress	ta.1316.2.s1_x_at	universal stress proteins (USP)family protein	1.26
20	Stress (biotic and abiotic)	ta.13785.1.s1_at	(CHIB1) acidic endochitinase	1.66
21	Redox	ta.5610.1.s1_at	AHB1 (ARABIDOPSIS HEMOGLOBIN 1)	1.04
22	Polyamine synthesis	ta.3005.1.s1_a_at	ATAIH/EMB1873 (AGMATINE IMINOHYDROLASE)	0.13
23	Nucleotide metabolism	taaffx.97704.1.a1_at	EMB2742 (EMBRYO DEFECTIVE 2742)	0.73
24	Degradation of xenobiotics	ta.8571.1.s1_a_at	lactoylglutathione lyase family protein/glyoxalase I family protein	1.83
25	C1 metabolism	ta.20549.1.s1_x_at	FDH (FORMATE DEHYDROGENASE	0.32
26	Misc enzyme families	ta.11025.1.a1_at	FAD-binding domain-containing protein	1.41
27	RNA processing	ta.26049.1.s1_a_at	MYB4 (myb domain protein 4); transcription factor	1.39
28	DNA synthesis	taaffx.128682.1.s1_at	ENDO4 (ENDONUCLEASE 4)	1.45
29	Protein modification	ta.3507.1.s1_at	SCPL16 (serine carboxypeptidase-like 16)	1.64
30	Signalling	ta.4334.1.s1_at	CPK6 (CALCIUM-DEPENDENT PROTEIN KINASE 6)	1.55
31	Cell cycle, division, vesicle transport and organisation	ta.214.1.s1_at	TUB6 (BETA-6 TUBULIN)	1.37
33	Development	ta.13396.1.s1_at	late embryogenesis abundant	1.87
34	Transport	ta.9295.1.a1_at	PGP21 (P-GLYCOPROTEIN 21); ATPase, coupled to transmembrane	1.42
1	Photosynthesis	ta.22984.2.s1_x_at	LHB1B2 (Photosystem II light harvesting complex gene 1.5); chlorophyll binding	−1.28
2	Sugar metabolism	ta.1197.1.s1_x_at	ATPHS2/PHS2 (ALPHA-GLUCAN PHOSPHORYLASE 2)	−1.04
3	Minor sugars	ta.12952.2.s1_at	pfkB-type carbohydrate kinase family protein	−0.65
4	Glycolysis	ta.9096.2.s1_x_at	ATPPC1 (PHOSPHOENOLPYRUVATE CARBOXYLASE 1)	−0.25
5	Fermentation	ta.28355.2.s1_x_at	ADH2 (ALCOHOL DEHYDROGENASE 2)	−0.52
6	Gluconeogenesis	ta.37.1.s1_at	PMDH2 (PEROXISOMAL NAD-MALATE DEHYDROGENASE 2)	−0.36
7	OPP	ta.29329.1.s1_at	phosphogluconate dehydrogenase	−0.56
8	Tricarboxylic acid cycle	taaffx.123182.1.s1_at	carbonic anhydrase family protein	−1.35

Table 1. Cont.

MapMan Bin#	MapMan Pathway	Affymetrix Probe	MapMan Annotation	Expression Change 50:90
9	Mitochondrial electron transport	ta.1147.3.s1_at	NDH-M (SUBUNIT NDH-M OF NAD(P)H:PLASTOQUINONE DEHYDROGENASE COMPLEX)	−0.56
10	Cell wall	ta.25823.1.a1_at	MUM4 | MUM4 (MUCILAGE-MODIFIED 4)	−1.12
11	Lipid metabolism	ta.4873.1.s1_at	FAD8 (FATTY ACID DESATURASE 8)	−1.32
12	N metabolism	taaffx.12494.1.s1_at	nitrogen regulation family protein	−0.58
13	Amino acid metabolism	ta.23170.1.a1_s_at	DHDPS2 (DIHYDRODIPICOLINATE SYNTHASE)	−1.06
14	S assimilation	ta.19092.1.s1_x_at	SIR (sulfite reductase)	−0.11
15	Metal handling	taaffx.86907.1.s1_at	heavy-metal-associated domain-containing protein / copper chaperone (CCH)-related	−0.13
16	Secondary metabolism	taaffx.12557.1.a1_at	terpenoid metabolism	−1.08
17	Hormones	ta.3526.1.s1_at	NFD5 (NUCLEAR FUSION DEFECTIVE 5)	−0.99
18	Cofactors and vitamins	ta.8003.1.s1_at	UbiE/COQ5 methyltransferase family protein	−0.70
19	Tetrapyrrole biosynthesis	ta.9574.1.s1_at	GUN5 (GENOMES UNCOUPLED 5)	−1.02
20	Abiotic stress	ta.3155.1.s1_at	GFA2 (GAMETOPHYTIC FACTOR 2); heat shock protein binding	−0.80
20	Stress (biotic and abiotic)	ta.27335.1.s1_at	ATMLO1/MLO1 (MILDEW RESISTANCE LOCUS O 1)	−0.84
21	Redox	taaffx.12477.1.s1_at	thioredoxin family protein.	−1.20
22	Polyamine synthesis	ta.24743.2.s1_a_at	SPDS1 (SPERMIDINE SYNTHASE 1)	−0.15
23	Nucleotide metabolism	ta.3179.1.s1_x_at	thymidylate kinase family protein	−1.20
24	Degradation of xenobiotics	ta.7031.3.s1_a_at	GLX2-2 (GLYOXALASE 2-2)	−0.10
25	C1 metabolism	ta.4843.1.a1_at	dihydropterin pyrophosphokinase	−0.73
26	Misc enzyme families	ta.385.2.s1_at	ATPRMT11/PRMT11 (ARABIDOPSIS ARGININE METHYLTRANSFERASE 11)	−1.20
27	RNA processing	ta.26907.1.s1_at	ribonuclease T2 family protein	−1.20
28	DNA synthesis	ta.13372.1.s1_at	hydrolase, alpha/beta fold family protein	−1.06
29	Protein modification	ta.19412.1.s1_at	ribosomal protein L20 family protein	−1.13
30	Signalling	ta.7378.4.s1_at	ATARCA (Arabidopsis thaliana Homolog of the Tobacco ArcA); nucleotide binding	−1.19
31	Cell cycle, division, vesicle transport and organisation	ta.14519.1.s1_x_at	regulators of chromosome condensation (RCC1family).	−0.89
33	Development	ta.7378.4.s1_at	ATARCA (Arabidopsis thaliana Homolog of the Tobacco ArcA); nucleotide binding	−1.19
34	Transport	taaffx.8804.2.s1_at	PIP2;5/PIP2D (plasma membrane intrinsic protein 2;5); water channel	−1.21

RIL2219; the only exceptions being for bins 10 (cell wall), 15 (metal handling), 16 (secondary metabolism) and 17 (Hormones).

Identifying probes responsible for most differences between lines. To explore patterns and sources of variation responsible for differences between wheat lines from those due to the stress transient, the whole RWC-interpolated dataset of 19062 probes was submitted to PCo analysis. The analysis identified three major PCos and revealed clear zoning of each line across the stress transient and also between the lines with little evidence of randomness, demonstrating the structure, global nature and tight regulation of the response to water stress (Figure 9). PCo1 accounted for 47.84% of the total variation in the sample-to-sample similarities and separated the samples with regards to increasing stress for all three lines (Figure 9 A, B). The plot suggests that each line transcriptome transitions smoothly across the stress in the direction of the arrows and the circles highlight possible system functional states Figure 9 A. PCo2, explaining 15.01% of the variance, separated the data as regards line-to-line differences, in particular separating Lahn from Cham1 and RIL 2219, these latter two lines showing more similar coordinates on this axis (Figure 9 A, C). PCo3, accounting for 9.95% of the variance, also pulled out line-to-line differences, but this time extracting differences for Cham1 from the other two lines (Figure 9 B, C). PCo2 and PCo3 therefore uncovered the global ways in which each of the two parents Cham1 (PCo2) and Lahn (PCo3) are similar to, or different from, the offspring, RIL 2219 which in turn combines transcriptome patterns derived from both parents in terms of its response to decreased RWC. The plot of PCo1 *vs.* PCo2 shows that RIL 2219 and Cham1 are closer, or more similar, throughout the transient, and that all lines become more similar as the stress progresses to the severe point at 50 leaf % RWC (Figure 9 A). Furthermore, the plot of PCo1 *vs.* PCo3 (Figure 9 B) showed that the response of Cham1 differed from the other two lines as RWC fell from 82 to 58%. Plotting PCo2 *vs.* PCo3 (Figure 9C) revealed that PCo3 can be used to consider relative within-line stability as the stress increases, three distinct

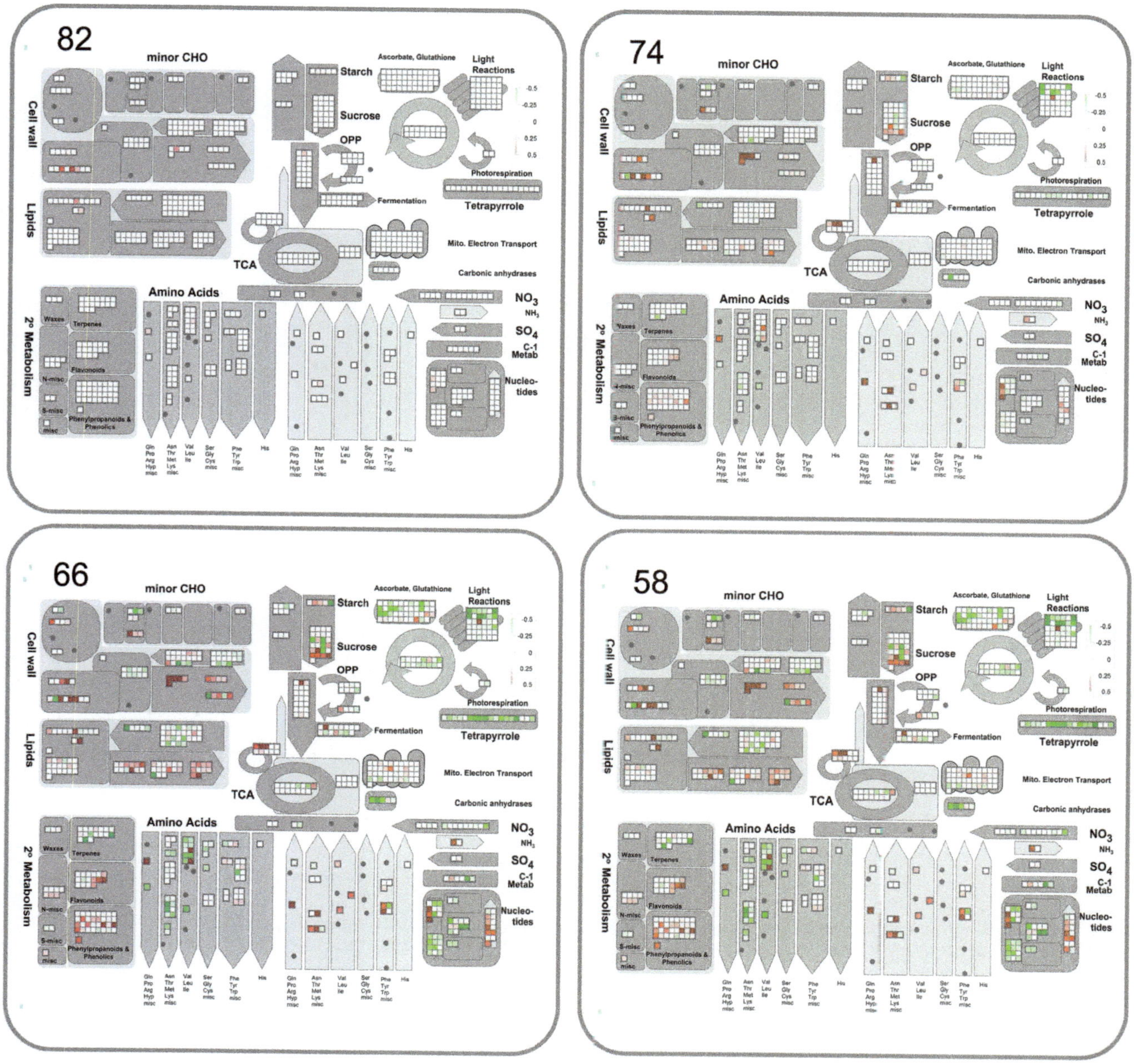

Figure 6. Overview of transcription of genes annotated to primary metabolism during water stress. Expression of probes from MapMan bins 1, 2, 3, 4, 5, 6, 7, 8, 9, 10, 11, 12, 13, 14, 16, 19, 21, 23 and 25, 30 at decreasing leaf %RWC during the water stress transient. ANOVA Dataset and visualisation as for Figure 3.

line-specific circle clouds of points are shown, the most disparate set being for Lahn (more susceptible parent) and the most clustered set being for Cham1 (more resistant parent) apart from its point at 50% RWC, which is a considerable distance away from the others. The set of points for the RIL 2219 is more similar to that of Cham1 than to that of Lahn, viewed from the PCo2 axis. Therefore, it is clear that PCo2 and PCo3 pull out differences between lines at all stress points including the relatively stress-free condition of 90% RWC. Importantly, the spread of coordinates of samples at 90% leaf RWC for all three lines demonstrates the intrinsic differences that exist between the transcriptomes of each line before the onset of water deficit.

The analysis also enabled the identification of the major probes responsible for differences between lines and those due to the stress transient. The 5% of probes most significantly correlated (F-tests) with the PCo coordinates PCo1, PCo2 and PCo3 (File S7) were imported into the MapMan programme for assignment to bins and results showed differences in the distribution of these probes across the metabolic pathways (Figure 10). Thus, PCo1 identified 1–3 probes in annotated bins 1 (photosynthesis), 10 (cell wall), 13 (amino acid metabolism), 23 (nucleotide synthesis), 26 (misc. enzyme families), 30 (signalling), 31 (cell organisation), 33 (development), 34 (transport), and several belonging to regulatory bins 27 (RNA processing), 29 (protein modification) and bin 34 (transport). The probe with the greatest F-value (Ta.3659.1.-

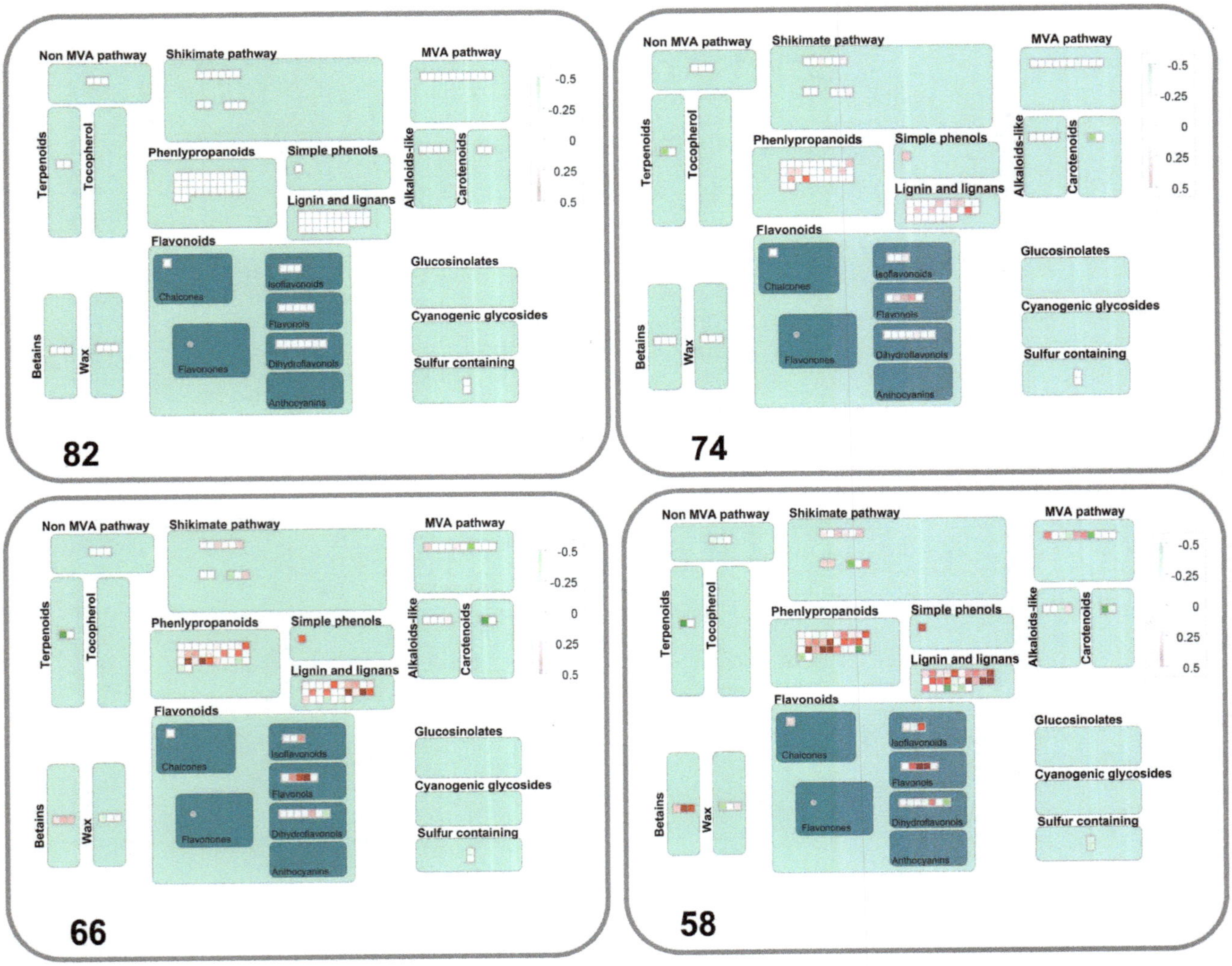

Figure 7. Changes in transcription of genes annotated to secondary metabolism during water stress. Expression of probes annotated to MapMan bin 16 at decreasing leaf %RWC during the water stress transient. ANOVA Dataset and visualisation as for Figure 3.

S1_a_at) gave no annotation using MapMan, so we explored this further using PLEXdb to obtain a putative annotation to heat shock factor-binding protein 1, using rice alignment at Gramene (LOC_Os09g20830.2). PCo2 identified probes in annotated bins 11 (lipid metabolism), 28 (DNA), 29 (protein modification) and 34 (transport); the probe with the highest F-value (taaffx.3401.1.s1_at) was annotated to vacuolar sorting protein 39, having small GTPase regulator activity and association with the TGF beta receptor. Finally, PCo3 identified probes in bins 2 (major sugars), 10 (cell wall), 11 (lipid metabolism), 16 (secondary metabolism), 17 (hormones), 21(redox), 27 (RNA regulation), 29 (protein modification), 30(signalling) and 34 (transport) with the most influential probe ta.16038.1.s1_at annotated to hormone metabolism- ABA inducible, encoding one of five HVA22 homologs in *Arabidopsis*; an ABA- and stress-inducible gene.

Another source of variation that can influence the expression values for probes between the three lines is allelic variation at the gene sequence level. This may affect the binding of a particular probe from a given probe set and the extent of its occurrence in our study was estimated by calculating the ratio of a probe signal to that averaged over the eleven probes in the corresponding probe set. A preliminary analysis was done for probe sets annotated to transcription factors, bin27, for the three ANOVA groups, *line*, *line+stress* and *line x stress* where such differential hybridisation may be an important consideration when comparing expression differences between the lines. The results showed that for the *line only* ANOVA group, 15 probes out of 396 probes (36 probe sets each containing 11 individual probes) showed some evidence of differential hybridisation between the lines but with no stress response (File S8). The analysis also showed that for the *line+stress* ANOVA group, 140 individual probes out of a total of 4862 showed statistically different hybridisation when compared to the average for probe sets; however, this component did not affect the response to water stress since all lines behaved similarly. The group for which such allelic hybridisation may contribute to the differential stress response amongst the three lines is the *line x stress* ANOVA group and results showed that 108 individual probes out of a total of 4807 (File S8) may have such an interaction; further verification in new studies would be required to confirm this.

Probes showing no response to stress

The ANOVA group, *Line*, identified 932 probes that were significantly different in transcript expression between the three

Table 2. Top two probes showing largest changes in expression during the early phase in the stress for each wheat line. Data represents the ratio of change in expression (over and under-expression values) at 82% relative to that at 90 leaf %RWC taken from the ANOVA group dataset *line x stress* (File S5).

MapMan Bin#	MapMan Annotation	Affymetrix Probe	82:90
			Lahn
26	misc.cytochrome P450	ta.3813.1.a1_at	**0.61**
27	RNA.regulation of transcription.Aux/IAA family	ta.10395.1.s1_a_at	**0.47**
34	transport.peptides and oligopeptides	taaffx.65026.1.a1_at	**−0.31**
29	protein.synthesis.ribosomal protein.prokaryotic.mitochondrion.50S subunit.L2	ta.28514.2.s1_at	**−0.46**
			Cham1
16	secondary metabolism.phenylpropanoids.lignin biosynthesis.CAD	ta.2659.1.s1_at	**0.38**
16	secondary metabolism.simple phenols	ta.4455.1.a1_at	**0.35**
34	transport.metal	ta.9290.1.s1_at	**−0.11**
20	stress.biotic	taaffx.108556.1.s1_at	**−0.11**
			RIL2219
17	hormone metabolism.ABA.synthesis-degradation.9-cis-epoxycarotenoid dioxygenase	taaffx.76007.1.s1_at	**0.46**
29	protein.degradation.ubiquitin.E3.SCF.FBOX	taaffx.12903.1.a1_at	**0.26**
8	TCA/org. transformation.carbonic anhydrases	taaffx.37789.1.a1_at	**−0.11**
20	stress.abiotic.heat. Heat shock protein HSP21	ta.202.1.s1_at	**−0.36**

lines but did not show changes in response to the water stress transient (File S2). We also identified 208 probes (File S6) that showed no statistical difference either between lines or in response to the stress but which were selected during the initial filtering step using Genespring GX. These two ANOVA groups could be used to provide potential control genes for future water stress transcript expression studies in durum wheat.

Discussion

Durum wheat gene expression responds to water stress in a global and regulated manner

The stress transient enabled the capture of the complexity of the plant's response to water stress as a first step in dissecting inherent features of the biological system under perturbation. The

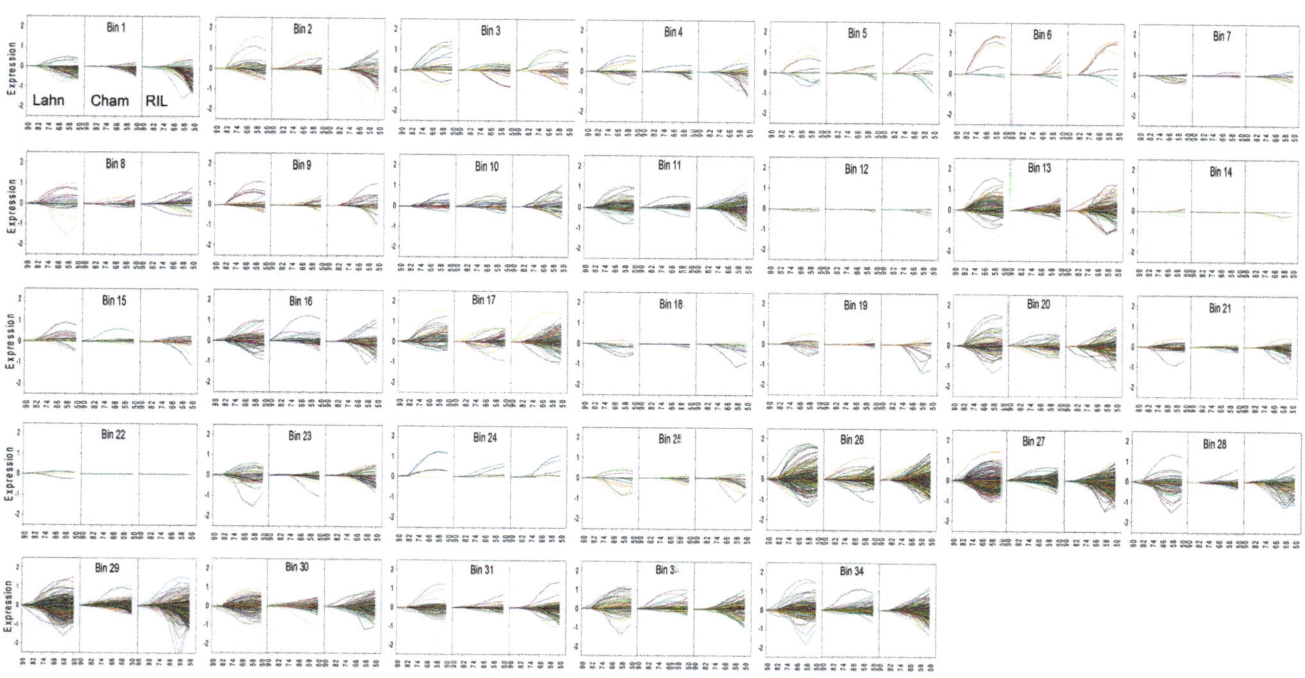

Figure 8. Differences in the global transcription profiles of the three lines under stress. Expression of probes from the three lines, Lahn, Cham1 and RIL2219, as a function of leaf %RWC during the water stress transient, visualised in MapMan for all metabolic bins. For every bin, the same colour probe is plotted for each wheat line to enable direct comparison. Transcript profiles belong to the *Line* × *Stress* ANOVA group dataset (File S5).

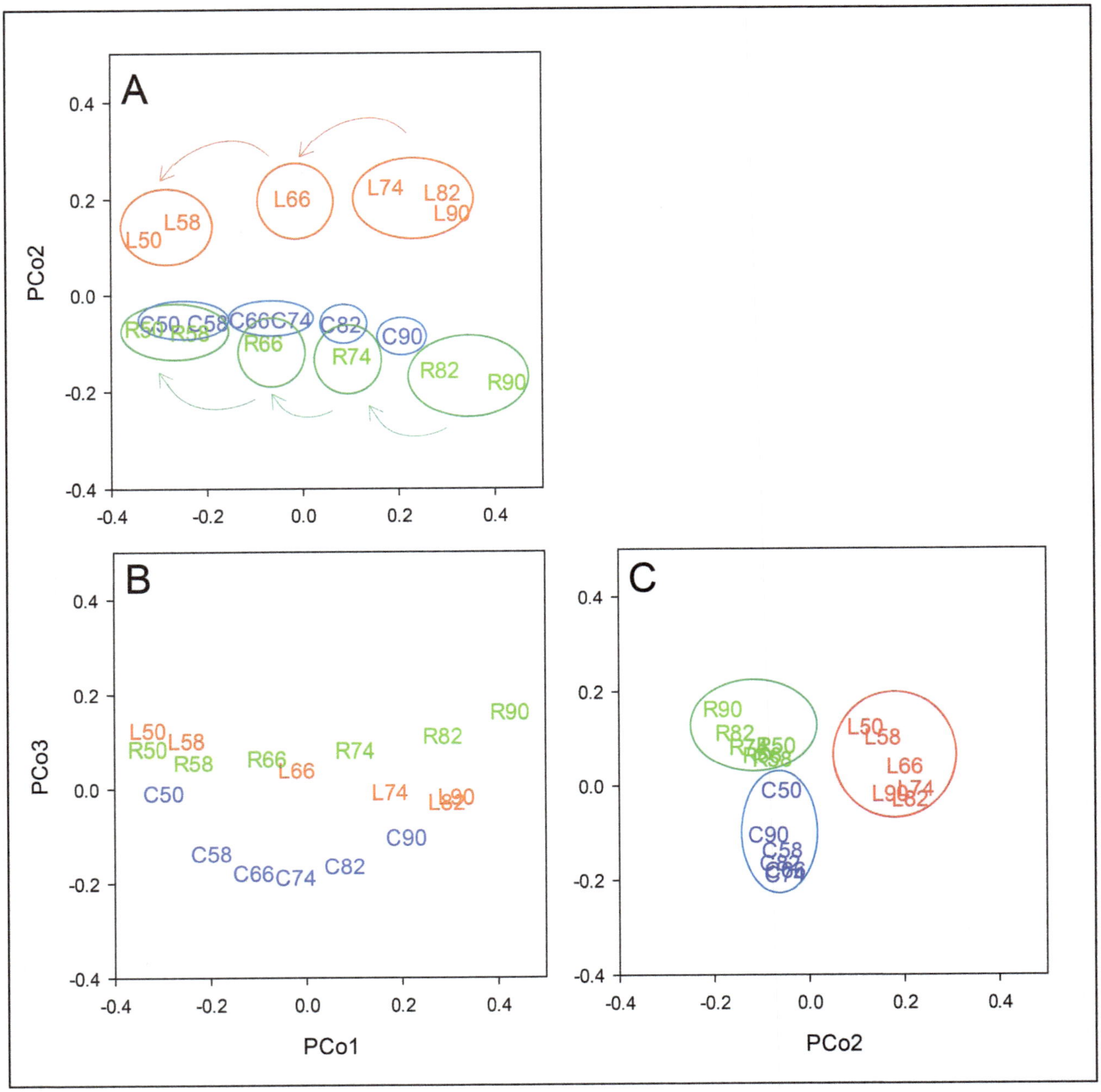

Figure 9. PCo plots for the expression of all 19062 probe dataset. The first three PCos account for 72.8% of the variation represented in the (reduced) similarity matrix for the 18 wheat line by RWC combinations. Visualisation of the combinations in the three dimensions was achieved by plotting PCo1 *vs.* PCo2 (A), PCo1 *vs.* PCo3 (B) and PCo2 *vs.* PCo3 (C). Letters L, C and R represent the lines Lahn, Cham1 and RIL 2219 respectively at 90-50 leaf %RWC values. Arrows on plot A show the direction of the stress transient of decreasing leaf RWC and the free-drawn circles around the points highlight potential functional states for each line, inferred from distance between points.

mathematical models identified all statistically significant responses, without filtering, which uncovered all changes in transcript expression, not just focussing on a particular arbitrary magnitude of change, or exploring only specific pathways; unlike most published studies. The results showed that the durum wheat transcriptome responded in a global manner to progressive water stress as demonstrated by the modulation of a large number of probes annotated to various metabolic pathways in all three lines; this has been observed in previous datasets on various species [36,37,38,39,22,40,41]. Our study also showed evidence of a highly regulated response in the proportionally large number of probes coding for regulatory genes (DNA, RNA and protein modification, transcription factors, hormones and signalling) specifically altering expression at the early phases of the stress. This supports and demonstrates the fundamental components of regulatory networks of sensing, transducing and responding that operate at the genome level when exposed to abiotic stress [16,42]. These are essential features of system robustness that enable the maintenance of function under stress through immediate changes in some metabolic pathways, documented in File S3, File S4, File S5, and also via the launch of downstream pathways such as protective secondary metabolism. Particular attention has been

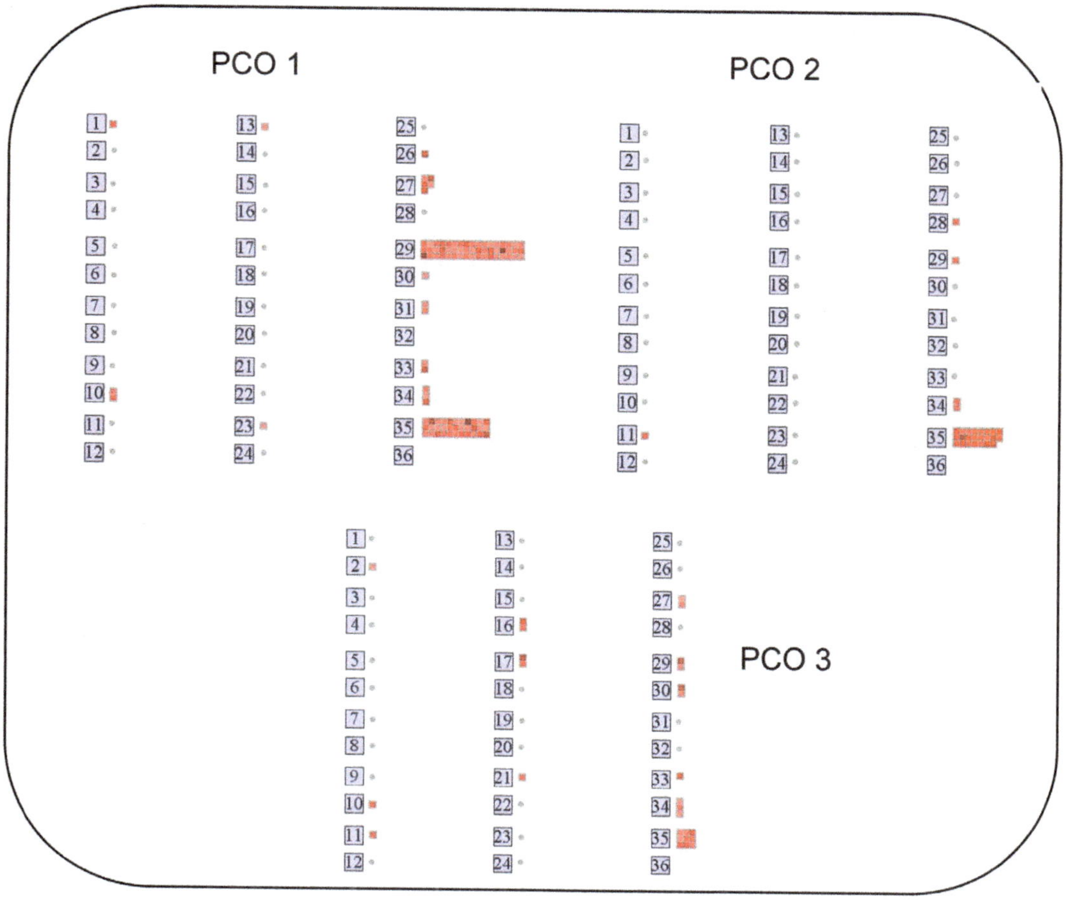

Figure 10. MapMan visualisation of the top PCo probes responsible for differences. 5% of probes with greatest F-values following PCO (data from File S7) showed differences in distribution across all metabolic bins for PCo1, PCo2, and PCo3. For presentation in Mapman, FPCO maximum values were set at 2000, 700 and 450 for PCo1, PCo2, and PCo3 respectively.

given to transcription factors as major control points in the plant system's response to water stress [43] and here we present the first overview of the water stress response profile of several durum wheat family members changing expression over the range of RWC. We identified particular groups of probes annotated to AP2-EREBP, ARF, bHLH, bZIP, C2C2, MADS, Histone ATase, SET and MYB4 families that were altered in leaf transcription early on in the stress transient when no other physiological changes were measurable. This reinforces the importance of transcription factors in the regulation and coordination of transcriptional and metabolic responses which form essential components of regulatory networks. Another interesting result is that of the modulations in transcripts for heat shock proteins throughout the stress transient. Molecular chaperones play a crucial role in the reestablishment of cellular homeostasis under abiotic stress [44] and alterations to heat shock proteins, caused by shared metabolic responses, are common to abiotic stresses such as water deficit and heat and thus share fundamental mechanisms as discussed by [17].

The changes in expression of a large number of probes annotated to protein modification and signalling in addition to hormones further supports this integrated global view of the wheat genome stress response as the system aims to balance key processes such as protein synthesis, cellular maintenance, metabolism, growth and development. Thus, these multifaceted changes in gene expression of the regulatory pathways are best interpreted as essential features of robustness as the system adjusts to perturbations and moves from one functional state to another, as envisaged in Figure 9A. Therefore, the tendency to consider particularly large increases in gene expression as all candidates for 'improving' plant growth under drought is flawed and such individual changes should be understood in the context of robustness with all its elements of degeneracy, feedback, homeostasis and environmental tracking operating within the biological system [45,46]. To establish how this highly regulated response is co-ordinated at the gene expression level across the stress transient, mathematical and bioinformatics models can be used to uncover new gene networks for water stress response in cereals and other species [47,42]. Recently, co-expression network algorithms have been successfully used to explore gene networks, network architecture and control hubs, and to sketch out new hypotheses for further testing [48,49]. Future studies would also enable a comparison of our datasets with similar ones from barley, bread wheat, maize and rice, in order to determine which transcriptome responses are conserved as a common framework amongst cereals and which are species- and environment-specific.

Water stress downregulates photosynthesis and alters transcripts for sugar metabolism

Durum wheat leaves exhibited a decline in photosynthetic parameters, osmotic adjustment and increases in ABA content as the stress progressed through the transient of declining leaf RWC, as expected [50]. This was accompanied by a lowering of photosynthetic, photorespiratory and tetrapyrrole biosynthetic transcripts which reflects a down-regulation of light capture and carbon fixation. Such patterns have been seen in previous studies in dicots, rice, barley and wheat and support a common framework for the plant's photosynthetic responses to water stress [48]. We also found that the lowered expression of photosynthetic transcripts was coupled by increases in expression for sucrose and starch breakdown, and a decrease for starch synthesis. This implies a requirement for soluble sugar availability for respiration, solute homeostasis and/or signalling throughout the plant [51]. The perturbations observed in key regulators of carbon metabolism and glycolysis such as phosphofructokinase, phosphoenolpyrovate carboxylase and pyruvate kinase with a concomitant increase in expression of enzymes of gluconeogenesis and sucrose invertases may also contribute to an increase in sucrose hydrolysis [52] in agreement with analysis of metabolites under water deficit [51]. These results provide some evidence for leaf osmotic adjustment as a strategy to maintain cell turgor pressure [51,53], it might also reflect the need for respiratory substrates as respiration is maintained when photosynthesis slows and ceases.

Water stress reconfigures transcription of primary and secondary metabolic pathways

The complex changes measured in transcripts coding for amino acid and lipid metabolism, GABA-T, and various aminotransferases, demonstrate the reconfiguration of primary metabolism under stress as the system moves between different metabolic 'states of function'. The overall downregulation of transcripts for synthesis of most amino acid groups and an increase in the expression of transcripts involved in amino acid breakdown occurred downstream in the stress response as the plant adjusts metabolism from synthesis towards remobilisation and protective metabolic pathways, in addition to providing substrate for respiration as photosynthesis declines. This is demonstrated by specific changes in transcripts for key enzymes involved in amino acid and inorganic nitrogen assimilation. For example, the down-regulation of nitrate reductase implies a decreased assimilation of inorganic nitrogen and this was accompanied by over-expression of transcript annotated to cytosolic glutamine synthetase, glutamate dehydrogenase and glutamate synthase which in turn would release glutamate and 2 oxoglutarate for recycling, thus accelerating remobilisation and senescence [54], and possibly increasing the production of proline via glutamate. This enables increased remobilisation of nitrogen assimilates as metabolism is adjusting the decline in growth with primary synthesis under stress conditions.

We also measured changes in transcripts coding for secondary metabolism such as phenylpropanoids, lignin biosynthesis, flavonoids, betaine, and wax metabolism, thus providing evidence for a protective adjustment role in the stress response. The increases we observed in the expression of an orthologue for proline biosynthesis, P5CS2 might be considered as providing proline for osmotic adjustment and/or reflecting an imbalance in amino acid regulation. Other adaptive responses such as changes in expression of transcripts for cell wall biosynthesis, measured early on in the stress at small changes in RWC, demonstrate the dynamism of mature cell walls in response to decreased turgor and supports recent studies [55,56,57,11]. We also observed that a range of responsive transcripts to both biotic and abiotic stress were also modulated and this supports the existence of common global gene networks for environmental stress responses; interestingly [35] also noted similar changes in biotic pathogen-related gene expression under water stress in wheat.

Wheat lines show constitutive differences in gene expression

The time series approach allowed the dissection of the complex transcriptional response into the wheat specific component and those specific to changes in RWC, as well as identifying probes showing an interaction between the two. A large number of probes showed inherent or constitutive differences in transcript expression between the wheat lines before the stress had started; all probes in File S2, File S4 and some probes in File S5. This is clearly seen in the PCO plots in Figure 9 where the values at leaf 90%RWC are dispersed across all axes for all three lines. Large systemic differences in gene expression between germplasm within a species, including between siblings and parent, have been shown to be strikingly common in rice [58], emmer wheat [41] and Arabidopsis [59]. It is unclear however, at this stage of study, how these differences contribute towards the ability of a line to resist stress and these results should be coupled to expression QTL studies dissecting GxE to enable further evaluation. Studying the transcriptome of wheat lines under both controlled and field conditions would then enable us to understand the importance of these global differences between lines in terms of the variation in the final integrated responses to the stress in the complex field environment and we are currently evaluating such datasets from field trials. Integrating genomic high-throughput datasets with fine mapped QTLs in future studies would consequently bridge the link between changes at the molecular level with downstream yield-related traits underpinning final grain yield [21].

Differential gene expression responses amongst wheat lines under stress

The three wheat lines responded to water stress in a global framework of common changes of gene expression in regulatory, primary, secondary, and protective metabolic pathways. In addition, we documented thousands of probes that showed significant differences in the transcript response for each wheat line during the stress transient (File S5). These differences were analysed at equivalent leaf RWC reflecting real differences in gene responses between the susceptible and resistant lines which may prove to be the most interesting in terms of targets for genotype-specific responses under water stress. The probe responsible for inducing most variation across the whole dataset in terms of line-to-line stability, derived from PCo3, was annotated to an ABA inducible HVA22E or a late embryogenesis abundant protein which is implicated in a protective stress response by regulating vesicular traffic in stressed cells [60]. This, and other probes identified by PCo, can now be tested in new mechanistic studies to establish whether they are new regulatory control points for drought resistance in durum wheat. An interesting result from our study was that Lahn, the drought sensitive parental line, had a higher number of modulated probes in response to the stress, especially at the initial phase of the transient. This was true at equivalent leaf %RWC, across most pathways, and in comparison to the other two lines, especially RIL2219 which had the lowest number of modulated probes at that stage. An examination of the literature highlighted few comparative analyses in rice [39], a maize drought resistant RIL line with fewer modulated genes under stress [61] whilst drought in maize landraces showed a mixed picture [62]. Therefore, it may also be important to consider the metrics of gene expression changes in addition to the nature of the genes and the networks involved.

In addition to differential gene expression documented between the three lines, we have estimated the contribution of allelic differential hybridisation to the transcripome response. We have identified a small proportion of such individual probes in bins annotated to transcription factor probe sets, most of which did not influence the response of probes to water stress (ANOVA group *line only* and *Line+stress*). Nevertheless, a handful were identified as promising in the ANOVA group *Line x stress* and one in particular, probe # 602715 belonging to the probe set ta.28513.1.s1_s_at and annotated as an Auxin Response Factor family, deserves close verification since it was also identified by PCo analysis as a major candidate for within-line stability under stress (File S7 and S8). These represent possible candidates for marker development and should be further examined in future studies to verify their potential for durum wheat marker assisted breeding.

One of the most striking findings is the co-ordinated and highly structured transcriptome response for each wheat line across the water stress transient as demonstrated by the lack of scatter and high level of structure in Fig. 9C. Each line started at a different point along the axes for the three PCos demonstrating inherent differences in transcript expression across most metabolic pathways and this distinctness is maintained during the transient with some indication of reaching closer values at the final stress point. This provides strong evidence for a distinct cultivar-specific global transcriptome response in each wheat line so that the whole genome is responding in a highly coordinated manner. Very few studies have undertaken such a global analysis and for those that have utilised PCo, a similar interpretation can also be derived [22]. These, together with constitutive differences are important findings and should be factored into current simplistic efforts to identify a singular mechanism for drought resistance and genetic modification of a particular selected metabolic pathway. We argue that a shift in thinking is required to integrate high throughput datasets to advance our understanding of plant responses under stress and that robustness theory may offer such a framework.

Physiological traits differ before and during the stress

The most drought resistant RIL 2219 line had higher photosynthetic and transpiration rates, greater stomatal conductance and ABA content prior to the initiation of stress and was also able to maintain greater RWC and leaf water potential for a longer period than either parent. It is interesting to note that parent Cham1 has relatively larger stomatal conductance when compared to other breeding varieties [63]. The greater rate of leaf transpiration may seem counterintuitive for a drought resistant line, but if coupled with lower leaf area and/or bigger capacity to capture water (for example from a more efficient root system) then it can afford to transpire slightly more and maintain its stomata open for longer to allow entry of CO_2 for fixation. We argue that this could provide an important advantage in the sustained accumulation of assimilate, especially during the fast vegetative growth period where water may be available, and hence increased grain yield if integrated over the whole growth season. In addition, the higher leaf transpiration rate for RIL2219 may also reduce the canopy temperature, this being another added benefit for durum wheat during terminal stress when drought is often compounded with heat stress [64,65,66]. We propose that these combined factors are responsible for RIL2219's inherent greater capacity for carbon assimilation and thus the ability to maintain higher yield stability than either parent under drought in the field, as supported by datasets from a large number of field trials under various drought gradients (unpublished observations). Evidence exists that wheat breeders have inadvertently selected for higher stomatal conductance in their quest for greater yield potential under optimal conditions, and this has also proved to be beneficial under water-limiting Mediterranean environments [67,64,68,69,70,65,71,72,73,74]. There was also evidence for greater osmotic adjustment ability in the two drought resistant lines Cham1 and RIL 2219 than in Lahn. Taken together, improving osmotic adjustment, and increasing carbon fixation and stomatal conductance may define one physiological strategy for success in breeding high yield stability wheat for Mediterranean drought prone environments. Further systems studies are now necessary to link the transcriptome, proteome, metabolome and phenome of wheat to establish how the different levels of function interact to produce a final yield under stress.

Conclusions

We found evidence of a common global framework of physiological, biochemical and gene expression responses in all three wheat lines under water stress. The stress transient identified the predominant modulation of the transcriptome of regulatory pathways early in the stress response, downregulation of photosynthesis, reconfiguration of primary metabolism and the launch of protective downstream pathways at various leaf RWC of relevance to terminal drought experienced by crops under field conditions. We also uncovered genotype-specific transcriptome and physiological differences before and during the stress amongst the three wheat lines which suggests that each wheat line functions within a genome-specific structure that form the basis of different stress-resistance strategies. We interpret the results by ascribing systems–based concepts of robustness as a fundamental framework of the plant's response to perturbation under stress. By targeting our studies to durum wheat lines from the breeding programme at ICARDA, our results should be useful to breeders searching for novel variation and new concepts to test and exploit in wheat cultivar improvement.

Supporting Information

Figure S1 Expression of genes in sugar metabolism during stress. Expression of probes annotated to MapMan bin 2 as a function of decreasing leaf %RWC. ANOVA Dataset and visualisation as for Figure 3.
(TIF)

File S1 ANOVA results of probe expression values of three wheat lines in response to water stress. Whole dataset of 19062 probes interpolated values, pre-ANOVA, for individual replicates at each leaf %RWC level and for the three lines Lahn, Cham1 and RIL2219 in spreadsheet 'Interpolated data at RWC levels'. ANOVA, log (base2), split the whole dataset into 5 independent statistical groups represented by individual spreadsheets: probes showing differences in expression between the lines but not during the stress in spreadsheet 'line only 932 probes'; probes showing differences in expression as a result of the stress transient but not between lines in spreadsheet 'stress only 2579 probes'; probes with differences in expression between the lines at the onset of the stress with the profile changing in the same way during the stress transient in spreadsheet 'line+stress 7644 probes'; probes with differences in expression due to line and stress combined, showing an interaction in spreadsheet 'line x stress 7699 probes'; and probes with no differences due to stress or line after initial filtering in spreadsheet 'no effect 208 probes'. Statistical results are given as least significant difference (LSD) at 5% and 1% levels of significance for each of the ANOVA groups.
(XLSX)

File S2 MapMan analysis of ANOVA group *line only* 932 probes. Probes showing statistically different expression values between the three lines Lahn, Cham1 and RIL2219 but with no effect of the water stress. Data were calculated as expression value beyond the relevant LSD (5%) given in File S1 when comparing means for each probe between the three lines Lahn, Cham1 and RIL2219, or otherwise zero when there was no significant difference. Probes are presented by Affymetrix id and MapMan annotation to bin name, pathway and description of gene model. (XLSX)

File S3 MapMan analysis of ANOVA group *stress only* 2579 probes. Probes showing no differences in expression between the three lines Lahn, Cham1 and RIL2219 but changing in response to the stress. Data were calculated as expression value beyond the relevant LSD (5%) given in File S1 when comparing means for stressed (82, 74, 66, 58 and 50% leaf RWC) to control (90% leaf RWC) conditions for each probe, or otherwise zero when there was no significant difference. Probes are presented by Affymetrix id and MapMan annotation to bin name, pathway and description of gene model. Analysis of results are presented in individual spreadsheets for all MapMan bins and sorted for different leaf %RWC. In addition, probes with the greatest levels of over-expression or under-expression measured at the earliest stress point of 82% are highlighted in spreadsheets 'early changes over-expression' and 'early changes under-expression', respectively.
(XLSX)

File S4 MapMan analysis of ANOVA group *line+stress* 7644 probes. Probes showing consistent differences in expression between the three lines Lahn, Cham1 and RIL2219 in response to the stress transient –*i.e.* they show a parallel profile of change in transcript response across the stress transient. Data were calculated as expression value beyond the relevant LSD (5%) of data in File S1 when comparing means for stressed (82, 74, 66, 58 and 50% leaf RWC) to control (90% leaf RWC) conditions for each probe, or otherwise zero when there was no significant difference. Data were also calculated as the differences in expression between the three lines. Probes are presented by Affymetrix id and MapMan annotation to bin name, pathway and description of gene model. Analysis of results are presented in individual spreadsheets for all MapMan bins and sorted for different leaf %RWC. In addition, probes with the greatest levels of over-expression or under-expression measured at the earliest stress point of 82% are highlighted in spreadsheets 'early changes over-expression' and 'early changes under-expression', respectively. Percentage changes in probes at 82:90 and 74:90% leaf RWC stress points are also included in spreadsheet 'summary of early changes' and calculated as a percentage of probes changed at 82:90 or 74:90 as a proportion of all probes showing changes in each specific bin across the whole transient.
(XLSX)

File S5 MapMan analysis of ANOVA group *line x stress* 7699 probes. Probes showing differences in expression between the three lines Lahn, Cham1 and RIL2219, *and* in response to the stress; they show an interaction. Data were calculated as expression value beyond the relevant LSD (5%) given in File S1 when comparing means for stressed (82, 74, 66, 58 and 50% leaf RWC) to control (90% leaf RWC) conditions for each probe from each of the three lines Lahn, Cham1 and RIL2219, or otherwise zero when there was no significant difference. Probes are presented by Affymetrix id and MapMan annotation to bin name, pathway and description of gene model. Analysis of results are presented in individual spreadsheets for all MapMan bins and sorted for different leaf %RWC for the three lines. In addition, probes from all the bins which showed expression changes at the earliest stress point of 82% are highlighted for the three lines in spreadsheet 'early changes' and sorted for the three lines in subsequent spreadsheets. A summary of two probes showing the largest changes measured at the early stage of 82%RWC for each line is presented in spreadsheet 'summary top changes early'.
(XLSX)

File S6 MapMan analysis of ANOVA '*no effect* 208 probes' group. Probes, post initial filtering, showing no difference in expression between the three lines Lahn, Cham1 and RIL2219, or in response to the stress from data in File S1. Probes are presented by Affymetrix id and MapMan annotation to bin name, pathway and description of gene model.
(XLSX)

File S7 MapMan annotation of PCo results. Mapman analysis of probes responsible for maximum separation observed, as defined by F-tests for the correlation between probes and principal coordinates (PCos), following PCo applied to the RWC-interpolated data for the 19062 probes derived from 54 samples (three replicates of each of six RWC levels by three wheat lines). Individual spreadsheets show annotation for the probes relevant for the first three principal coordinates retained from the analysis (PCo1, 2, 3). Probes are presented by Affymetrix id and MapMan annotation to bin name, pathway and description of gene model.
(XLSX)

File S8 The contribution of allelic differences in gene sequences to probe hybridisation for probes annotated to transcription factors. ANOVA for each perfect match probe and the significance of the ratio changes across lines was calculated in proportion to the corresponding mean probe-set abundance. The name of the probe and P value are given for the corresponding probe set taken from *line only*, *line+stress* and *Line x stress* ANOVA groups datasets from File S3,4,5 respectively focusing on MapMan bin 27.
(XLSX)

Acknowledgments

We acknowledge Peter Hedden for help with ABA measurements and ICARDA Durum Wheat Breeding Program for providing the lines from the Cham1xLahn mapping population. This work is dedicated to Kawther Latiri (deceased).

Author Contributions

Conceived and designed the experiments: DH SP KL AA RT DL JA MN. Performed the experiments: MB DL DH. Analyzed the data: MH SP MD-P RM CR MS DH. Contributed reagents/materials/analysis tools: DH KL AA RT JA MN. Wrote the manuscript: DH. Developed the genetic material: MN.

References

1. Nachit MM, Nachit G, Ketata H, Gauch HG, Zobel RW (1992) Use of AMMI and linear regression models to analyze genotype-byenvironment interaction in durum wheat. Theor. Appl. Genet 83: 597–601.
2. Araus JL, Slafer GA, Reynolds MP, Royo C (2002) Plant breeding and drought in C3 cereals: what should we breed for? Ann. Bot. 89, Special No, 925–940.
3. Stocker TF, Qin D, Plattner GK, Alexander LV, Allen SK, et al. (2013) Technical Summary. In: Stocker TF, Qin D, Plattner GK, Tignor M, Allen SK, Boschung J, Nauels A, Xia Y, Bex V and Midgley PM, editors. Climate Change 2013: The Physical Science Basis. Contribution of Working Group I to the Fifth Assessment Report of the Intergovernmental Panel on Climate Change Cambridge University Press, Cambridge, United Kingdom and New York, NY, USA.

4. Bates BC, Kundzewicz ZW, Wu S, Palutikof JP, Eds. (2008) Climate change and water. Technical Paper of the Intergovernmental Panel on Climate Change. IPCC Secretariat, Geneva.
5. Hoerling M, Eischeid J, Perlwitz J, Quan X, Zhang T, et al. (2012) On the Increased Frequency of Mediterranean Drought. J Clim 25: 2146–2161.
6. Habash DZ, Kehel K, Nachit MM (2009) Genomic approaches for designing durum wheat ready for climate change with a focus on drought. J. Exp. Bot. 60: 2805–2815.
7. WHEAT (2011) Global alliance for improving food security and the livelihoods of the resource-poor in the developing world. CIMMYT publication, Mexico.
8. Nachit MM, Elouafi I (2004) Durum adaptation in the Mediterranean dryland: breeding, stress physiology, and molecular markers. In: Rao SC, Ryan J, editors. Challenges and strategies for dryland agriculture. CSSA Special Publication 32. Madison, Wisconsin, USA: Crop Science Society of America Inc., American Society of Agronomy Inc, 203–218.
9. Maccaferri M, Sanguineti MC, Corneti S, Araus JL, Salem MB, et al. (2008) Quantitative Trait Loci for grain yield and adaptation of durum wheat (Triticum durum Desf.) across a wide range of water availability. Genetics 178: 489–511.
10. Moinuddin, Fischer RA, Sayre KD, Reynolds MP (2005). Osmotic Adjustment in Wheat in Relation to Grain Yield under Water Deficit Environments. Agron. J. 97: 1062–1071.
11. Alvarez S, Marsh EL, Schroeder SG, Schachtman DP (2008) Metabolomic and proteomic changes in the xylem sap of maize under drought. Plant Cell Environ. 31: 325–340.
12. Harb A, Krishnan A, Ambavaram MMR, Pereira A (2010) Molecular and Physiological Analysis of Drought Stress in Arabidopsis Reveals Early Responses Leading to Acclimation in Plant Growth. Plant Physiol. 154: 1254–1271.
13. Ahuja I, de Vos R, Bones AM, Hall RD (2010) Plant molecular stress responses face climate change. Trends Plant Sci. 15: 664–674.
14. Mochida K, Shinozaki K (2010) Genomics and Bioinformatics Resources for Crop Improvement. Plant Cell Physiol. 51: 497–523.
15. Yamaguchi-Shinozaki K, Shinozaki K (1994) A novel cis-acting element in an Arabidopsis gene is involved in responsiveness to drought, low temperature, or high-salt stress. Plant Cell 6: 251–264.
16. Nakashima K, Ito Y, Yamaguchi-Shinozaki K (2009) Transcriptional Regulatory Networks in Response to Abiotic Stresses in Arabidopsis and Grasses1. Plant Physiol. 149: 88–95.
17. Shinozaki K, Yamaguchi-Shinozaki K (2007) Gene networks involved in drought stress response and tolerance. J. Exp. Bot. 58: 221–227.
18. Reguera M, Peleg Z, Blumwald E (2011) Targeting metabolic pathways for genetic engineering abiotic stress-tolerance in crops. Biochim. Biophys. Acta. 2012: 186–94.
19. Tuberosa R, Salvi S (2006) Genomics-based approaches to improve drought tolerance of crops. Trends Plant Sci. 11: 405–12.
20. Cattivelli L, Rizza F, Badeck F, Mazzucotelli E, Mastrangelo A, et al. (2008) Drought tolerance improvement in crop plants: An integrated view from breeding to genomics. Field Crops Res. 105: 1–14.
21. Fleury D, Jefferies S, Kuchel H, Langridge P (2010) Genetic and genomic tools to improve drought tolerance in wheat. J. Exp. Bot. 61: 3211–3222.
22. Aprile A, Mastrangelo A, De Leonardis A, Galiba G, Roncaglia E, et al. (2009) Transcriptional profiling in response to terminal drought stress reveals differential responses along the wheat genome. BMC Genomics, 10: 279.
23. Lawlor DW (2013) Genetic engineering to improve plant performance under drought: physiological evaluation of achievements, limitations, and possibilities. J Exp Bot. 64: 83–108.
24. Barr HD, Weatherley PE (1962) A re-examination of the relative turgidity technique for estimating water deficit in leaves. Aust. J. Biol. Sci. 15: 413–428.
25. Scholander PF, Hammel HT, Hemmingsen EA, Bradstreet ED (1964) Hydrostatic pressure and osmotic potential in leaves of mangroves and some other plants. Proc. Natl. Acad. Sci. U.S.A. 52: 119–25.
26. Fernandez-Garcia N, Hernandez M, Casado-Vela J, Bru R, Elortza F, et al. (2011) Changes to the proteome and targeted metabolites of xylem sap in *Brassica oleracea* in response to salt stress. Plant Cell Environ. 34: 821–836.
27. Wu ZJ, Irizarry RA, Gentleman R, Martinez-Murillo F, Spencer F (2004) A model-based background adjustment for oligonucleotide expression arrays. J Am Stat Assoc 99: 909–917.
28. Bar-Joseph Z (2004) Analyzing time series gene expression data. Bioinformatics. 20: 2493–2503.
29. Mastroianni G, Milovanovic G (2008) Interpolation Processes: Basic Theory and Applications. Springer Monographs in Mathematics, Springer-Verlag, Berlin, Germany.
30. Gasca M, Sauer T (2000) On the history of multivariate polynomial interpolation, J Comput Appl Math 122: 23–25.
31. Benjamini Y, Hochberg Y (1995) Controlling the false discovery rate: a practical and powerful approach to multiple testing. J R Stat Soc Series B 57: 289–300.
32. Krzanowski WJ (2000) Principles of multivariate analysis: a user's perspective. Oxford University Press, Oxford, UK.
33. Gautier L, Cope L, Bolstad BM, Irizarry RA (2004) affy–analysis of Affymetrix GeneChip data at the probe level. Bioinformatics 20: 307–315.
34. Thimm O, Bläsing O, Gibon Y, Nagel A, et al. (2004) MAPMAN: a user-driven tool to display genomics data sets onto diagrams of metabolic pathways and other biological processes. Plant J. 37: 914–39.
35. Usadel B, Nagel N, Thimm O, Redestig H, Blaesing OE, et al. (2005) Extension of the Visualization Tool MapMan to Allow Statistical Analysis of Arrays, Display of Coresponding Genes, and Comparison with Known Responses. Plant Physiol 138: 1195–1204.
36. Xue GP, McIntyre CL, Chapman S, Bower NI, Way H, et al. (2006) Differential gene expression of wheat progeny with contrasting levels of transpiration efficiency. Plant Mol. Biol. 61: 863–81.
37. Kilian J, Whitehead D, Horak J, Wanke D, Weinl S, et al. (2007) The AtGenExpress global stress expression data set: protocols, evaluation and model data analysis of UV-B light, drought and cold stress responses. Plant J. 50: 347–63.
38. Talamè V, Ozturk NZ, Bohnert HJ, Tuberosa R (2007) Barley transcript profiles under dehydration shock and drought stress treatments: a comparative analysis. J. Exp. Bot. 58: 229–240.
39. Degenkolbe T, Do PT, Zuther E, Repsilber D, Walther D, et al. (2009) Expression profiling of rice cultivars differing in their tolerance to long-term drought stress. Plant Mol. Biol. 69: 133–53.
40. Ergen NZ, Thimmapuram J, Bohnert HJ, Budak H (2009). Transcriptome pathways unique to dehydration tolerant relatives of modern wheat. Funct. Integr. Genomics 9: 377–396.
41. Krugman T, Chagué V, Peleg Z, Balzergue S, Just J, et al. (2010) Multilevel regulation and signalling processes associated with adaptation to terminal drought in wild emmer wheat. Funct Integr Genomics 10: 167–186.
42. Cramer G R, Urano K, Delrot S, Pezzotti M, Shinozaki K (2011) Effects of abiotic stress on plants: a systems biology perspective. BMC Plant Biol. 11: 163.
43. Hirayama T, Shinozaki K (2010) Research on plant abiotic stress responses in the post- genome era: past, present and future. Plant J. 61: 1041–1052.
44. Wang W, Vinocur B, Shoseyov O, Altman A (2004) Role of plant heat-shock proteins and molecular chaperones in the abiotic stress response. Trends Plant Sci. 9: 244–252.
45 Kitano H (2007) Towards a theory of biological robustness. Mol. Syst. Biol. 3: 137.
46. Whitacre JM (2012) Biological robustness: paradigms, mechanisms, and systems principles. Front Genet. 11; 3: 67.
47. Claverie JM (1999) Computational methods for the identification of differential and coordinated gene expression. Hum. Mol. Genet. 8(10): 1821–32.
48. Pinheiro C, Chaves MM (2011). Photosynthesis and drought: can we make metabolic connections from available data? J. Exp. Bot. 62: 869–882.
49. Swarbreck SM, Defoin-Platel M, Hindle M, Saqi M, Habash DZ (2010) New perspectives on glutamine synthetase in grasses. J. Exp. Bot. 62: 1511–1522.
50. Lawlor DW, Tezara W (2009) Causes of decreased photosynthetic rate and metabolic capacity in water-deficient leaf cells: a critical evaluation of mechanisms and integration of processes. Ann. Bot. 103: 561–579.
51. Xue GP, McIntyre CL, Glassop D, Shorter R (2008) Use of expression analysis to dissect alterations in carbohydrate metabolism in wheat leaves during drought stress. Plant Mol. Biol. 67: 197–214.
52. van Dongen JT, Gupta KJ, Ramírez-Aguilar SJ, Araújo WL, Nunes-Nesi A, et al. (2011) Regulation of respiration in plants: a role for alternative metabolic pathways. J. Plant Physiol. 168: 1434–43.
53. Rosa M, Prado C, Podazza G, Interdonato R, González JA, et al. (2009) Soluble sugars–Metabolism, sensing and abiotic stress. Plant Signal Behav. 4: 388–393.
54. Miflin BJ, Habash DZ (2002) The role of glutamine synthetase and glutamate dehydrogenase in nitrogen assimilation and possibilities for improvement in the nitrogen utilization of crops. J Exp Bot. 53: 979–87.
55. Konno H, Yamasaki Y, Sugimoto M, Takeda K (2008) Differential changes in cell wall matrix polysaccharides and glycoside-hydrolyzing enzymes in developing wheat seedlings differing in drought tolerance. Plant Physiol. 165: 745–54.
56. Moura JC, Bonine CA, de Oliveira Fernandes Viana J, Dornelas MC, Mazzafera P (2010) Abiotic and biotic stresses and changes in the lignin content and composition in plants. J Integr Plant Biol. 52: 360–76.
57. Sasidharan R, Voesenek LACJ, Pierik R (2011) Cell wall modifying proteins mediate plant acclimatization to biotic and abiotic Stresses, CRC Crit Rev Plant Sci 30: 6, 548–562.
58. Moumeni A, Satoh K, Kondoh H, Asano T, Hosaka A, et al. (2011) Comparative analysis of root transcriptome profiles of two pairs of drought-tolerant and susceptible rice near-isogenic lines under different drought stress. BMC Plant Biol. 11: 174.
59. Gan X, Stegle O, Behr J, Steffen JG, Drewe P, et al. (2011) Multiple reference genomes and transcriptomes for Arabidopsis thaliana. Nature 477: 419–423.
60. Guo WJ, Ho TH (2008) An abscisic acid-induced protein, HVA22, inhibits gibberellin-mediated programmed cell death in cereal aleurone cells. Plant Physiol. 147: 1710–22.
61. Zheng J, Fu J, Gou M, Huai J, Liu Y, et al. (2010) Genome-wide transcriptome analysis of two maize inbred lines under drought stress. Plant Mol. Biol. 72: 407–21.
62. Hayano-Kanashiro C, Calderón-Vázquez C, Ibarra-Laclette E, Herrera-Estrella L, Simpson J (2009) Analysis of Gene Expression and Physiological Responses in Three Mexican Maize Landraces under Drought Stress and Recovery Irrigation. PLoS ONE 4: e7531.
63. Monneveux P, Rekika D, Acevedo E, Merah O (2006) Effect of drought on leaf gas exchange, carbon isotope discrimination, transpiration efficiency and productivity in field grown durum wheat genotypes. Plant Sci. 170: 867–872.
64. Fischer RA, Rees D, Sayre KD, Lu ZM, Condon AG, et al. (1998) Wheat Yield Progress Associated with Higher Stomatal Conductance and Photosynthetic Rate, and Cooler Canopies. Crop Sci. 38: 1467–1475.

65. Araus JL, Casadesus J, Bort J (2001) Recent Tools for the Screening of Physiological Traits Determining Yield. In Reynolds, MP, Ortiz-Monasterio JI, McNab A (eds.). Application of Physiology in Wheat Breeding. Mexico, D.F.: CIMMYT.
66. Reynolds MP, Nagarajan S, Razzaque MA, Ageeb OAA (2001) Heat tolerance. In Reynolds, MP, Ortiz-Monasterio JI, McNab A (eds.). Application of Physiology in Wheat Breeding. Mexico, D.F.: CIMMYT. 88–100.
67. Blum A (1996) Yield potential and drought resistance: are they mutually exclusive? In:. Reynolds MP, Rajaram S, McNab A (eds.). Increasing Yield Potential in Wheat: Breaking the Barriers. Mexico, D.F.: CIMMYT. 76–89.
68. Delgado MI, Reynolds MP, Larque-saavedra A, Nava T (1994) Mexico, Genetic diversity for photosynthesis in wheat under heat-stressed environments and its relationship to productivity. Wheat Special Report No. 30, CIMMYT, Mexico.
69. Brennan JP, Condon AG, Van Ginkel M, Reynolds MP (2007) An economic assessment of the use of physiological selection for stomatal aperture-related traits in the CIMMYT wheat breeding programme. J Agric Sci. 145: 187–194.
70. Condon AG, Richards RA, Rebetzke GJ, Farquhar GD (2004) Breeding for high water-use efficiency. J. Exp. Bot. 55: 2447–2460.
71. Giunta F, Motzo R, Pruneddu G (2008) Has long-term selection for yield in durum wheat also induced changes in leaf and canopy traits? Field Crops Res. 106: 68–76.
72. Bahar B, Yildirim M, Barutcular C (2009) Relationships between stomatal conductance and yield components in spring durum wheat under Mediterranean conditions. Not Bot Horti Agrobo 37: 45–48.
73. Richards RA, Rebetzke GJ, Watt M, Condon AG, Spielmeyer W, et al. (2010) Breeding for improved water productivity in temperate cereals: phenotyping, quantitative trait loci, markers and the selection environment. Functional Plant Biology, 2010, 37, 85–97.
74. Khamssi NN, Najaphy A (2012) Comparison of photosynthetic components of wheat genotypes under rain-fed and irrigated conditions. Photochem. Photobiol. 88: 76–80.

8

Genome-Based Comparative Analyses of Antarctic and Temperate Species of *Paenibacillus*

Melissa Dsouza[1]*, Michael W. Taylor[1], Susan J. Turner[1,2], Jackie Aislabie[3]

1 Centre for Microbial Innovation, School of Biological Sciences, University of Auckland, Auckland, New Zealand, **2** BioDiscovery New Zealand Limited, Parnell, Auckland, New Zealand, **3** Landcare Research, Hamilton, New Zealand

Abstract

Antarctic soils represent a unique environment characterised by extremes of temperature, salinity, elevated UV radiation, low nutrient and low water content. Despite the harshness of this environment, members of 15 bacterial phyla have been identified in soils of the Ross Sea Region (RSR). However, the survival mechanisms and ecological roles of these phyla are largely unknown. The aim of this study was to investigate whether strains of *Paenibacillus darwinianus* owe their resilience to substantial genomic changes. For this, genome-based comparative analyses were performed on three *P. darwinianus* strains, isolated from gamma-irradiated RSR soils, together with nine temperate, soil-dwelling *Paenibacillus* spp. The genome of each strain was sequenced to over 1,000-fold coverage, then assembled into contigs totalling approximately 3 Mbp per genome. Based on the occurrence of essential, single-copy genes, genome completeness was estimated at approximately 88%. Genome analysis revealed between 3,043–3,091 protein-coding sequences (CDSs), primarily associated with two-component systems, sigma factors, transporters, sporulation and genes induced by cold-shock, oxidative and osmotic stresses. These comparative analyses provide an insight into the metabolic potential of *P. darwinianus*, revealing potential adaptive mechanisms for survival in Antarctic soils. However, a large proportion of these mechanisms were also identified in temperate *Paenibacillus* spp., suggesting that these mechanisms are beneficial for growth and survival in a range of soil environments. These analyses have also revealed that the *P. darwinianus* genomes contain significantly fewer CDSs and have a lower paralogous content. Notwithstanding the incompleteness of the assemblies, the large differences in genome sizes, determined by the number of genes in paralogous clusters and the CDS content, are indicative of genome content scaling. Finally, these sequences are a resource for further investigations into the expression of physiological attributes that enable survival under extreme conditions and selection processes that affect prokaryotic genome evolution.

Editor: Patrick C. Y. Woo, The University of Hong Kong, Hong Kong

Funding: This research was supported by Core funding for Crown Research Institutes from the Ministry of Business, Innovation and Employment's Science and Innovation Group, New Zealand (www.mbie.govt.nz). The funders had no role in study design, data collection and analysis, decision to publish, or preparation of the manuscript.

Competing Interests: I have read the journal's policy and the authors of this manuscript have the following competing interests: one or more of the authors are employed by a commercial company - BioDiscovery New Zealand Limited.

* Email: melissa33in@gmail.com

Introduction

The phylum *Firmicutes* represents one of the most abundant and ubiquitous bacterial groups in the environment. Members of this phylum have been identified in a wide variety of habitats that vary in physical and biochemical characteristics, including the vertebrate gut [1], activated sludge [2], soil [3,4], sediment [5], ocean waters [6], and lakewater [7]. Within soil, *Firmicutes* typically form a minor, yet consistent, component of microbial communities [8–11]. In addition, members of *Firmicutes* can be locally abundant, as observed in grassland soils of The Netherlands [12], and significantly more abundant in compacted forest [13] and arid soils [14] as compared to unimpacted, control soils. Within *Firmicutes*, members of the genera *Bacillus*, *Clostridium*, and *Paenibacillus* are commonly identified in soil microbial communities [15].

The genus *Paenibacillus* contains a monophyletic lineage of endospore-forming bacteria represented by over 100 described species [16]. Members of this genus have largely been isolated from terrestrial environments including cold soils of the Antarctic Peninsula [17], the Transantarctic Mountains, the Kafni glacier, Himalayas [18], Alaska [19] and from temperate soil environments, particularly those rich in humus and plant material [20,21]. Their ability to successfully colonise these environments can be attributed to common physiological traits including formation of stress-resistant endospores, secretion of extracellular enzymes and anti-microbial compounds (that suppress growth of competing microorganisms), and the ability to hydrolyse a variety of carbohydrates including cellulose, starch, and xylan. They are also noted for their ability to form complex colonial patterns illustrating complex multicellular-like behaviours and for the production of phytohormones, antifungal compounds and nutrients including nitrate that promote plant health [16]. Whole genome sequencing can provide an insight into the molecular mechanisms of these physiological attributes and, more generally, elucidate the metabolic potential and ecological role of these species. To date, 67 *Paenibacillus* genomes have been sequenced, of which two, those of *P. larvae* [22,23] and *P. vortex* [24], are accompanied by publications [25]. *P. larvae* was investigated for its ability to cause American foulbrood, a disease of honey bees, and *P. vortex* for its social organisation and complex pattern-forming behaviours.

Here, we report genome analyses of three *P. darwinianus* strains, isolated from gamma-irradiated soils of the Ross Sea Region (RSR), Antarctica [26]. Soils of the RSR represent a unique environment characterised by physical extremes of low temperature, elevated ultraviolet radiation and geochemical extremes of high salinity, low water and low nutrient availability [27]. Despite the harshness of this environment, many bacterial species prevail, with 16S rRNA gene pyrosequencing- and clone library-based studies identifying members of 15 bacterial phyla, namely *Acidobacteria*, *Actinobacteria*, *Armatimonadetes*, *Bacteroidetes*, *Chloroflexi*, *Cyanobacteria*, *Deinococcus-Thermus*, *Firmicutes*, *Gemmatimonadetes*, *Nitrospira*, *Planctomycetes*, *Proteobacteria*, *Spirochaetes*, *Verrucomicrobia* and Candidate 'TM7' [28–35]. Of these, *Firmicutes* were identified in most soil microbial communities, but their relative abundance was variable, and they typically formed a minor component of the community. However, they can be abundant, as observed in ornithogenic soils of Cape Hallett [36] and in mineral soils of the Wright Valley [31]. In addition, a pyrosequencing-based diversity study of airborne bacteria over Miers Valley identified members of the family *Paenibacillaceae* (of phylum *Firmicutes*) amongst the most abundant OTUs (operational taxonomic units) [37].

Members of the genus *Paenibacillus* are commonly cultured and identified in polar and permafrost soil communities. Despite this, only six novel *Paenibacillus* spp., namely *P. antarcticus* [38], *P. cineris*, *P. cookii* [39], *P. darwinianus* [26], *P. macquariensis* [40], and *P. wynnii* [17] have been isolated from the Antarctic environment, and little is known about the potential functions, or the metabolic contributions, of these bacteria in the Antarctic environment. A key question is whether strains of *P. darwinianus* owe their resilience to substantial genomic changes compared to *Paenibacillus* spp. isolated from temperate soil environments. Therefore, the objective of this study was to compare genomes of three strains of *P. darwinianus* with nine temperate, soil-dwelling *Paenibacillus* spp., focusing on traits that may contribute to survival and growth in soil, including signal transduction pathways, sigma factors, sporulation, motility and their ability to cope with oxidative and osmotic stress. This study has demonstrated that, while the genomic content investigated is largely similar across all 12 *Paenibacillus* isolates, there is some evidence for genome content scaling.

Results and Discussion

Genome overview

The three *P. darwinianus* genomes are composed of the chromosome (approx. 3 Mbp), each containing a varying number of DNA scaffolds ranging from 107 to 111. The genome completeness was approx. 88%, assessed by the occurrence of essential, single-copy genes. Due to this incompleteness, we have regarded the apparent absence or low copy number of a given gene with caution. General genome features of the *P. darwinianus* strains compared with nine temperate *Paenibacillus* spp. are listed in Table 1. Dot plot diagrams comparing levels of homology between the three *P. darwinianus* genomes were generated on the IMG website. These plots revealed high levels of synteny between the three strains (Figure S1). The genome G+C content for the three strains was approx. 56%, in accordance with other members of *Paenibacillus*, yet significantly different to that of the nine temperate species ($P<0.01$). The genomes contained 3,101–3,123 open reading frames with an average coding density of 87.4%. These genomes comprised approx. 3,000 protein coding sequences (CDSs). For each genome, approx. 77% of the total CDSs were assigned to clusters of orthologous groups (COG) categories. Notably, the highest proportion of genes was assigned to COG categories including amino acid transport and metabolism [E], carbohydrate transport and metabolism [G], transcription [K], and translation, ribosomal structure and biogenesis [J] (Figure 1). Transmembrane proteins constituted approx. 25% of the total CDSs identified. Putative horizontally transferred genes constituted 1.87–2.29% of the total genes observed. Significantly fewer copies of the 16S rRNA gene were identified in *P. darwinianus* genomes as compared to the temperate *Paenibacillus* spp. ($P< 0.01$). Temperature range data indicated that *P. darwinianus* strains are psychrotolerant with optimal, minimal and maximal growth occurring at 18, 10 and 37°C, respectively [26]. The temperate *Paenibacillus* spp. exhibit growth over a wide range of temperatures (10–45°C) with optimal growth occurring between 25 and 30°C [16,20,41,42].

General genome comparisons

General comparisons between genomes of the three *P. darwinianus* strains and nine temperate *Paenibacillus* spp. were carried out using CMG-Biotools [43]. Firstly, the amino acid composition of protein sequences from the 12 *Paenibacillus* spp. was determined. This analysis revealed broad similarities in genome-wide amino acid usage profiles for all 12 *Paenibacillus* spp. with Ala, Leu, Gly, and Val the most frequently used amino acids across all genomes (Figure 2). Predicted proteome comparisons and a pan- and core-genome plot analysis were also performed on all 12 *Paenibacillus* genomes using CMG-Biotools [43]. Proteomes were predicted using Prodigal [44] and then BLAST algorithm (Basic Local Alignment Search Tool)-based proteome comparisons were performed to identify whether proteins are shared between predicted proteomes [45]. The main part of the matrix (shaded green) (Figure 3) consists of pairwise proteome comparisons and the bottom row (shaded red) represents a self-comparison where a hit within the proteome to a protein other than the query is identified as an internal homolog or a paralog. The BLAST matrix illustrates that the conservation between genomes is higher within species than within a genus. *P. darwinianus* strains exhibited a high level of conservation of gene families (89.2–90.3%). This observation was also supported by the pan- and core-genome analysis, as the *P. darwinianus* core- and pan-genome comprised 2,735 and 3,194 gene families, respectively. A large proportion of genes in the *P. darwinianus* core-genome were assigned to COG categories including carbohydrate transport and metabolism [G], amino acid transport and metabolism [E], and transcription [K]. In contrast, the temperate *Paenibacillus* core- and pan-genome comprised 1,139 and 25,493 gene families, respectively. Genes in the temperate *Paenibacillus* core-genome were assigned to COG categories including translation, ribosomal structure and biogenesis [J], amino acid transport and metabolism [E], and transcription [K]. Overall, the temperate and Antarctic core- and pan-genome comprised 998 and 26,612 gene families, respectively. Finally, up to 6% of the CDSs were in paralogous clusters for temperate *Paenibacillus* spp., whereas approx. 2.4% of the CDSs were in paralogous clusters for all *P. darwinianus* strains (Figure 3).

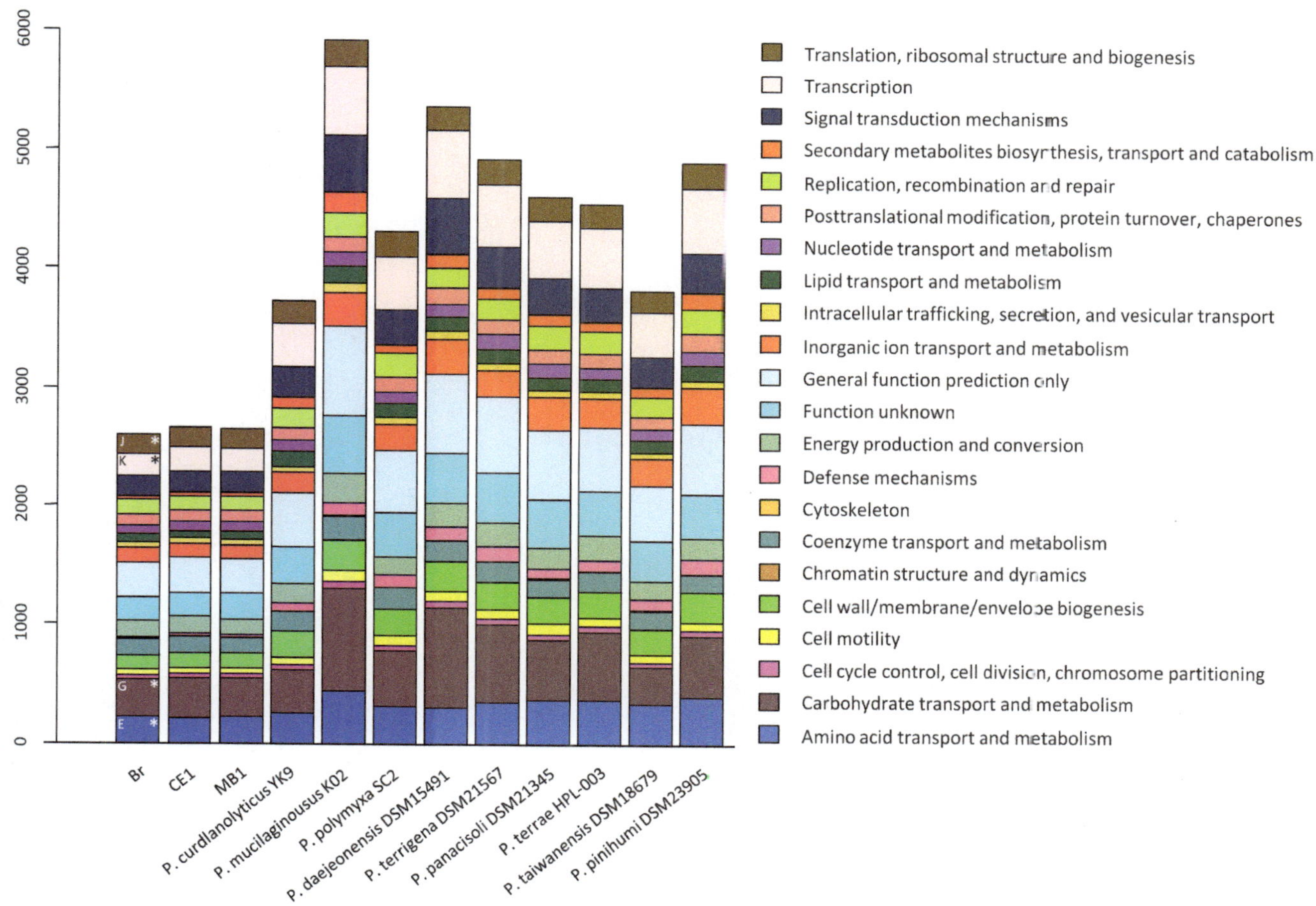

Figure 1. Comparison of gene content in temperate *Paenibacillus* spp. and *P. darwinianus* strains by COG categories.

Genomic features linked to survival in stressful conditions

Two-component signal transduction systems. Two-component signal transduction systems (TCS) represent a primary means by which a bacterial cell senses and responds to a variety of stresses and to the changing environment. *P. darwinianus* genomes contained 78–82 genes associated with TCS including approx. 40 genes for membrane-bound histidine kinases and 41 for response regulators (Table S1). A large number of genes for TCS were also identified in the psychrophilic archaeon *Methanococcoides burtonii* [46] and bacterium *Desulfotalea psychrophila* [47], further illustrating the importance of signal transduction systems for growth and survival in cold environments. In *P. darwinianus* genomes, approx. 2.6% of the total CDSs can be attributed to TCS. A similar proportion of genes (approx. 3% of the total CDSs) were attributed to TCS in genomes of the temperate *Paenibacillus* species. Genes for TCS, DesK-DesR for low temperature [48], CheA-CheY for chemotaxis [49], PhoR-PhoP for phosphate regulation [50], ResE-ResD for oxygen limitation [51], DegS-DegU for exoprotease production, competence development, biofilm, flagellum and complex colony formation [52], were identified in all 12 *Paenibacillus* genomes (Table S2). Notably, genes for TCS, LytS-LytR (for the regulation of autolysis), were identified in all *P. darwinianus* genomes and in the genome of just one temperate *Paenibacillus* species, *P. mucilaginosus*. However, the *lrgAB* operon, induced by the *lytSR* operon and responsible for blocking the activity of murein hydrolases (enzymes that have the ability to degrade bacterial cell wall), was not identified in *P. darwinianus* genomes [53]. In the psychrophilic bacterium, *Flavobacterium psychrophilum*, expression of the sensor kinase LytS was cold-induced, with expression significantly upregulated at 8°C compared to its expression at 20°C [54]. In the Antarctic soil environment, regulated autolysis may allow for the recycling of cellular components, thus providing bacterial communities with nutrients.

With the exception of ornithogenic soils formed under penguin rookeries, RSR soils typically contain low concentrations of organic carbon, ranging from 0.01 to 0.96 mg C g^{-1} soil [55]. Therefore the ability to detect labile C sources such as dicarboxylic acids can be crucial for survival in the Antarctic soil environment. Organic compounds including C_4-dicarboxylates, oxalate and succinate have been identified in aerosol particles over coastal East Antarctica [56], Showa Station [57] and Finnish Station Aboa [58] in Queen Maud Land. Genes for TCS, DctS-DctR (for the detection of aerobic C_4-dicarboxylates, namely succinate, fumarate, malate, and oxaloacetate) were identified in the *P. darwinianus* genomes and in the genomes of just two temperate *Paenibacillus* spp., *P. mucilaginosus* and *P. terrae*. Genes encoding additional proteins required for the function of sensor kinase, DctS including DctA and DctB, were also identified in all *P. darwinianus* genomes [59].

Sigma factors. Sigma factors are dissociable units of RNA polymerase that activate the conditional expression of a specific set of genes in response to a particular stress or stimulus, thus

Table 1. General genome features of *P. darwinianus* genomes (this study) vs. nine temperate, soil-dwelling *Paenibacillus* spp. [100].

	1	2	3	4	5	6	7	8	9	10	11	12
Genome data												
Genome size (bp)	3,021,550	3,056,161	3,051,566	5,452,778	8,770,140	6,241,931	7,464,058	6,361,561	6,326,414	6,083,395	5,247,653	6,760,575
DNA coding region (%)	2,645,264 (87.55)	2,672,123 (87.43)	2,663,886 (87.30)	4,835,968 (88.69)	7,182,175 (81.89)	5,308,414 (85.04)	6,619,323 (88.68)	5,583,508 (87.77)	5,389,440 (85.19)	5,167,317 (84.94)	4,508,951 (85.92)	5,851,762 (86.56)
G+C content (%)	56.23	56.14	56.08	51.94	58.24	44.62	53.22	46.08	48.22	46.77	44.79	48.53
Scaffold count	111	108	107	40	1	2	36	40	43	1	20	45
Total RNA genes	58	53	72	133	224	196	103	118	127	117	100	105
tRNA genes	23	21	34	101	189	14	61	71	87	89	68	64
16S rRNA genes	4	4	5	10	13	14	7	10	7	9	8	8
Other RNA genes	29	24	26	-	-	-	29	27	26	-	16	26
Total number of genes	3,101	3,123	3,163	4,957	7,476	6,228	6,575	5,932	5,718	5,642	4,779	6,169
Total protein CDSs (%)	3,043 (98.13)	3,070 (98.30)	3,091 (97.72)	4,824 (97.32)	7,252 (97)	6,032 (96.85)	6,472 (98.43)	5,814 (98.01)	5,591 (97.78)	5,525 (97.93)	4,679 (97.91)	6,604 (98.30)
With function prediction (%)	2,487 (80.20)	2,540 (81.33)	2,535 (80.15)	3,488 (70.37)	4,389 (58.71)	4,463 (71.66)	5,199 (79.07)	4,777 (80.53)	4,389 (76.76)	3,768 (66.78)	3,692 (77.52)	4,754 (77.06)
Without function prediction (%)	556 (17.93)	530 (16.97)	556 (17.58)	1,336 (26.95)	2,863 (38.30)	1,569 (25.19)	1,273 (19.36)	1,037 (17.48)	1,202 (21.02)	1,757 (31.14)	987 (20.65)	1,310 (21.24)
With COGs (%)	2,389 (77.04)	2,431 (77.84)	2,426 (76.70)	3,390 (68.39)	5,362 (71.72)	3,906 (62.72)	4,887 (74.33)	4,468 (75.32)	4,151 (72.60)	4,117 (72.97)	3,461 (72.42)	4,443 (72.02)
Coding for transmembrane proteins (%)	71 (2.29)	59 (1.89)	59 (1.87)	398 (8.03)	-	257 (4.13)	302 (4.59)	240 (4.05)	99 (1.73)	-	101 (2.11)	283 (4.59)
Genome Completeness (%)	87	88	88	-	-	-	-	-	-	-	-	-
Metadata												
Isolation source	Britannia, Darwin Mountains	Cape Evans, Ross Sea Region	Minna Bluff, Ross Sea Region	Kobe city, Japan	-	China	Daejeon, South Korea	Chiba, Japan	Pocheon Province, South Korea	Gara Mountains,	Wu-Feng, Taiwan	Daejeon, South Korea
Habitat	Gamma-irradiated soil	Gamma-irradiated soil	Gamma-irradiated soil	Soil	Soil	Rhizosphere soil	Soil	Soil	Soil	Soil	Farmland soil	Rhizosphere soil
Temperature range (°C)	10–37	10–37	10–37	30**	10–45*	30**	30**	4–32#	15–45*	10–40*	10–45$	15–37%

1, *P. darwinianus* Br; 2, *P. darwinianus* CE1; 3, *P. darwinianus* MB1; 4, *P. curdlanolyticus* YK9; 5, *P. mucilaginous* K02; 6, *P. polymyxa* SC2; 7, *P. daejeonensis* DSM 15491; 8, *P. terrigena* DSM 21567; 9, *P. panacisoli* DSM 21345; 10, *P. terrae* HPL-003; 11, *P. taiwanensis* DSM 18679; 12, *P. pinihumi* DSM 23905.

*, Data from [16];

**, Optimal temperature reported [16];

#, Data from [41];

$, Data from [42];

%, Data from [20].

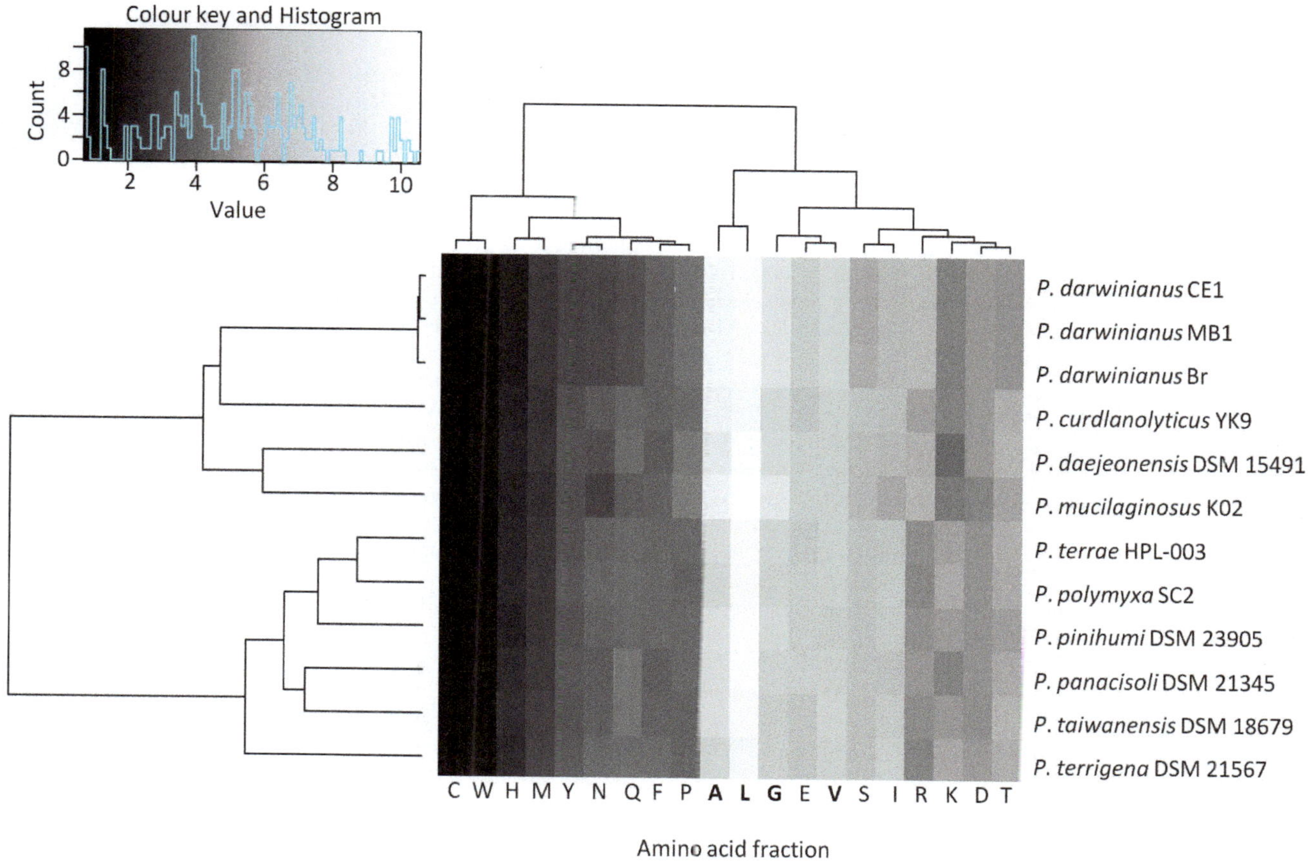

Figure 2. Amino acid usage heatmap of temperate *Paenibacillus* spp. and *P. darwinianus* strains.

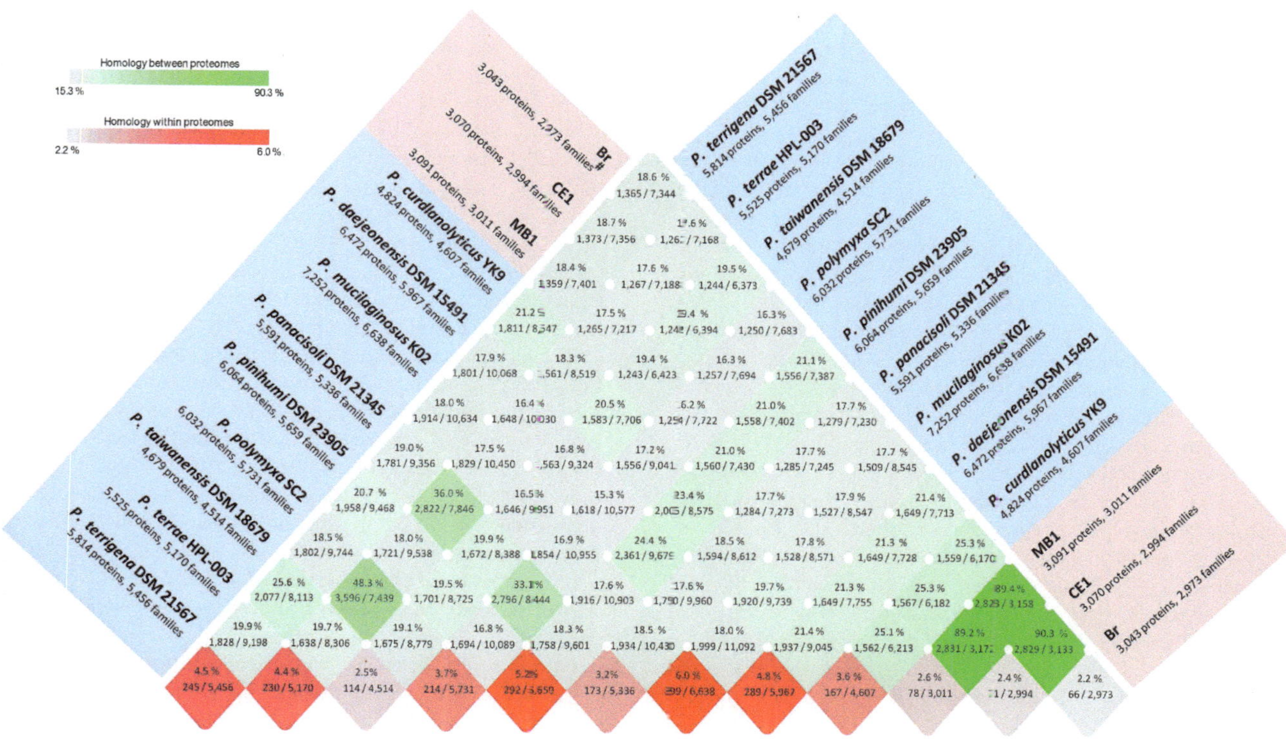

Figure 3. A BLAST matrix of an all against all protein comparison of 12 *Paenibacillus* genomes.

implementing compensatory physiological changes. *P. darwinianus* genomes contained genes associated with a diverse set of sigma factors similar to the multiple copies observed in psychrophilic bacteria *Planococcus halocryophilus* [60] and *Psychromonas ingrahamii* [61] (Table S3). These included the primary housekeeping factor, σ^A and alternative sigma factors including general stress response factor, σ^B, chemotaxis and flagellar motility associated factor, σ^D, cell envelope stress associated factor, σ^{ECF}, cold shock response factor, σ^L, and sporulation-specific factors, σ^E σ^F, σ^G, and σ^H [62,63]. Among these factors, σ^B is crucial as it controls the expression of an estimated 150 or more genes in response to a wide range of stress and starvation conditions [64]. Common regulators of σ^B, including RsbW and RsbV, were also identified in all *P. darwinianus* strains.

Oxidative and osmotic stress resistance. Reactive oxygen species (ROS) including superoxides (O_2^-), hydrogen peroxide (H_2O_2), hydroxyl radical molecules (·OH) and lipid peroxides, are both produced and accumulated as a result of aerobic metabolism [65]. Additionally, Antarctic soil bacteria are exposed to low temperatures that increase ROS stability and improve oxygen solubility [66]. Consequently, combating free radical damage is essential for survival in the Antarctic soil environment. Genomes of all three *P. darwinianus* strains contained two copies of the catalase gene, two of the superoxide dismutase gene *sodA* and one of the DNA oxidative damage protectant gene *dps*. Additionally, up to six copies of the peroxiredoxin genes, *ahpCF*, *bcp* and *tpx* and nine copies of the thioredoxin genes, *trxA* and *trxB* were identified in all *P. darwinianus* strains (Table 2, Table S10). Other bacteria that also contain multiple copies of genes for ROS detoxification include *Colwellia psychrerythraea* [67], *D. psychrophila* [47], and *P. halocryophilus* [60]. These genes were also identified in genomes of all temperate *Paenibacillus* spp. In *Bacillus subtilis*, the oxidative stress response is regulated by proteins, PerR and OhrR, activated by peroxides and by proteins, σ^B and Spx, that are in turn activated under diverse stress conditions [68,69]. Genes for regulatory proteins PerR, OhrR and σ^B were identified in all *P. darwinianus* genomes. However, Spx was only identified in *P. darwinianus* strain CE1. In genomes of the temperate *Paenibacillus* spp., these regulatory proteins were largely absent.

Salinity is a prominent feature of Antarctic soils, with some soils such as those in central Wright Valley containing water-soluble salts as high as 10 g cm^{-2} soil [70]. Accumulation of osmoprotectants such as glycine betaine and proline is an effective strategy to combat hyper-osmotic stress. Genes involved in glycine betaine uptake (*opuA*) comprising three components: OpuAA, the ATPase component, OpuAB, the permease component and OpuAC, the periplasmic component, were identified in all *P. darwinianus* strains and in most temperate *Paenibacillus* spp. [71]. Additionally, genes for high-affinity proline-specific uptake by the sodium:solute symporter, OpuE were identified in all *P. darwinianus* genomes and in the genome of just one temperate species, *P. terrae*. Proline biosynthesis enzymes, γ-glutamyl kinase (*proB*), γ-glutamyl phosphate reductase (*proA*) and Pyrroline-5-carboxylate reductase (*proC*) were also identified in all *Paenibacillus* genomes [72] (Table 2, Table S10). Multiple copies of genes for the uptake of these osmoprotectants were also identified in genomes of several cold-adapted bacteria including C. *psychrerythraea* [67], *P. halocryophilus* [60], *Pseudoalteromonas haloplanktis* [73], and *P. ingrahamii* [61]. Hypo-osmotic stress, a consequence of frequent freeze-thaw cycles, is also an important feature of the Antarctic soil environment. All *P. darwinianus* strains contained genes for mechanosensitive ion channels (MScL), responsible for the release of cytoplasmic solutes [74]. Genes for this transporter were also identified in genomes of all temperate *Paenibacillus* spp. (Table S4).

Transporter genes. Antarctic *Paenibacillus* genomes contained 411–422 membrane transport protein-related genes that constitute approx. 14% of the total CDSs (Table S4). Genomes of the temperate *Paenibacillus* spp. contained a similar proportion of membrane transport-related genes. Genes encoding ATP-binding cassette (ABC)-type transporters associated with amino acid, antimicrobial peptide, sugar, nitrate and proline/glycine betaine transport, comprised approx. 60% of the membrane transporter genes (Table S5). It is notable that significantly fewer copies of the ABC-type multidrug transport system were identified in *P. darwinianus* strains as compared to the temperate species ($P < 0.01$). While temperate *Paenibacillus* spp. contained several copies of genes encoding ABC-type oligopeptide, ABC-type dipeptide/oligopeptide/nickel, and ABC-type polysaccharide/polyol phosphate transport systems, no copies of genes associated with these transport systems were identified in the *P. darwinianus* genomes. However, a similar proportion of CDSs were associated with carbohydrate metabolism across all *Paenibacillus* genomes (Table S6). Therefore, the limited catabolic activity demonstrated by *P. darwinianus* strain Br^T in Biolog's phenotype microarray-based assay [26], may be due to its inability to transport polysaccharides and peptides, suggestive of an adaptation to the nutrient-limited Antarctic soil environment.

Cold shock response. During summer, diurnal temperature fluctuations are common in Antarctic soils. During this period, air temperatures are often below 0°C, however surface soils can be heated up to 10–15°C at midday depending on the position of the sun and cloud cover, and drop to below 0°C overnight [75,76]. Low temperatures have a major impact on the structure and function of cellular constituents including the membrane, the ribosome and nucleic acids by decreasing membrane fluidity, reducing ribosome function, unwinding the DNA double helix and by stabilizing secondary structures of nucleic acids, thus reducing mRNA transcription and translation [77,78]. Genome analysis of *P. darwinianus* strains revealed the presence of 4–5 copies of genes for cold-shock proteins similar to the multiple copies found in C. *psychrerythraea* [67], *Psychrobacter arcticus* [79] and *Shewanella oneidensis* [80]. In addition, several genes for cold-shock induced proteins, comparable to those identified in the transcriptome of cold-shocked *B. subtilis* [81] were also identified in all *P. darwinianus* genomes (Table 3, Table S11). A large proportion of these genes were also identified in temperate *Paenibacillus* spp.

Sporulation. Spores exhibit a high degree of resistance to various stresses including low temperatures, frequent freeze-thaw cycles, UV and gamma radiation, extreme desiccation, and low availability of nutrients that are all common features of the Antarctic soil environment. *P. darwinianus* genomes encode an extensive set of 63–78 genes (2.3% of the total CDSs) responsible for various facets of sporulation including DNA replication and translocation, formation of the sporulation septum, engulfment, spore morphogenesis and germination (Table S7). In comparison, genomes of temperate *Paenibacillus* spp. contained 94–112 genes (1.7% of the total CDSs) associated with sporulation (Table S8).

Motility and chemotaxis. In soil environments, active movement by bacteria towards regions that contain higher concentrations of beneficial compounds including water and nutrients may be crucial for survival. Over 40 genes are required for flagellar assembly and movement, including structural subunits for the synthesis of the basal body, the hook and the filament, regulatory proteins σ^D, FlgM and CodY, motor force generators MotA, MotB and chemosensory proteins [82,83]. Chemosensory

Table 2. Oxidative and osmotic stress response associated genes in genomes of temperate *Paenibacillus* spp. and *P. darwinianus* strains.

Product name	Gene Symbol	Enzymes	COGs	Pfams	1	2	3	4	5	6	7	8	9	10	11	12
Oxidative stress response																
Catalase	*katE*	EC:1.11.1.6	COG0753	pfam00199, pfam06628	1	1	1	3	3	1	4	2	1	1	-	1
Mn-containing catalase	-	-	COG3546	pfam05067	1	1	1	1	3	3	3	1	2	-	2	2
Superoxide dismutase	*sodA*	EC:1.15.1.1	COG0605	pfam00081 pfam02777	2	2	2	2	1	2	2	2	2	2	3	2
DNA-binding ferritin-like protein	*dps*	-	COG0783	pfam00210	1	1	1	1	4	1	1	1	1	1	1	2
Peroxiredoxin	*ahpCF*	-	COG0450	pfam00578	1	1	1	-	-	1	1	1	1	1	2	2
Peroxiredoxin	*tpx*	EC:1.11.1.15	COG2077	pfam08534	1	2	1	-	-	1	1	1	1	1	1	1
Peroxiredoxin	*bcp*	EC:1.11.1.15	COG1225	pfam00578, pfam08534	3	3	4	-	-	1	4	3	1	1	3	8
Thioredoxin/Thioredoxin domain-containing protein	*trxA*	-	COG0526, COG3118, COG0694	pfam00085	5	5	5	3	6	6	6	7	6	2	5	4
Thioredoxin reductase	*trxB*	EC:1.8.1.9	COG0492	pfam13738	5	5	5	2	-	3	10	6	5	5	5	8
Fe2+/Zn2+ uptake regulation proteins	*perR, fur*	-	COG0735	pfam01475	1	1	1	1	1	-	-	-	-	-	-	-
OhrR-like transcriptional regulator	*ohrR, marR*	-	COG1846	pfam01047	1	1	1	-	-	1	-	-	-	1	-	-
Transcriptional regulator Spx	*spx, arsC*	-	COG1393	pfam03960	-	1	-	-	-	-	1	1	1	-	1	1
Osmotic stress response																
ABC-type proline/glycine betaine transport system	*opuA*	EC:3.6.3.32	COG4175, COG4176, COG2113	pfam00005, pfam00571, pfam00528, pfam04069	3	3	3	-	-	2	3	3	3	2	-	3
Sodium:solute symporter family	*opuE*	-	COG4147, COG0591	-	2	2	2	-	-	-	-	-	-	1	-	-
γ-glutamyl phosphate reductase	*proA*	EC:1.2.1.41	-	pfam39821	1	1	1	1	1	1	1	1	1	1	1	1
γ -glutamyl kinase	*proB*	EC:2.7.2.11	COG0263	pfam00696, pfam01472	1	1	1	1	1	1	1	1	1	-	1	1
Pyrroline-5-carboxylate reductase	*proC*	EC:1.5.1.2	COG0345	pfam03807, pfam14748	2	2	2	2	1	2	2	1	2	1	2	1

1, *P. darwinianus* Br; 2, *P. darwinianus* CE1; 3, *P. darwinianus* MB1; 4, *P. curdlanolyticus* YK9; 5, *P. mucilaginous* K02; 6, *P. polymyxa* SC2; 7, *P. daejeonensis* DSM 15491; 8, *P. terrigena* DSM 21567; 9, *P. panacisoli* DSM 21345; 10, *P. terrae* HPL-003; 11, *P. taiwanensis* DSM 18679; 12, *P. pinihumi* DSM 23905.
Numbers in each column represent copy numbers per genome. Locus tags for each copy number are listed in Table S10.

Table 3. Cold-shock induced genes identified in genomes of temperate *Paenibacillus* spp. and *P. darwinianus* strains.

Product name	Gene symbol	COGs	Pfams	1	2	3	4	5	6	7	8	9	10	11	12
Adaptation to cold-shock															
Cold shock proteins	*cspB, cspC*	COG1278	pfam00313	5	5	4	5	1	3	3	3	3	3	3	4
Metabolism of lipids															
Fatty acid desaturase	*desA*	COG3239	pfam00487	1	1	3	1	1	-	1	1	1	1	1	2
Metabolism of amino acids															
Leucine dehydrogenase	*bcd*	COG0334	pfam00208, pfam02812	3	3	3	-	-	-	2	1	-	-	1	1
2-oxoisovalerate dehydrogenase (E3 subunit, dihydrolipoamide dehydrogenase)	*lpd*	COG1249	pfam00070, pfam02852, pfam07992	3	4	3	2	-	-	4	4	3	-	3	3
2-oxoisovalerate dehydrogenase (E1 alpha subunit)	*bkdAA*	COG1071	pfam00676	1	1	1	1	1	1	1	1	1	1	1	1
2-oxoisovalerate dehydrogenase (E1 beta subunit)	*bkdAB*	COG0022	pfam02779, pfam02780	1	1	1	1	1	1	1	1	1	1	1	1
2-oxoisovalerate dehydrogenase (E2 subunit, lipoamide acyltransferase)	*bkdB*	COG0508	pfam00198, pfam00364, pfam02817	3	3	3	-	-	-	2	3	3	1	3	2
Metabolism of nucleotides															
Adenylosuccinate synthetase	*purA*	COG0104	pfam00709	-	-	1	2	1	1	2	2	1	1	2	2
Metabolism of carbohydrates															
Fructose-1,6-bisphosphate aldolase	*fbaA*	COG0191	pfam01116	1	1	1	1	2	1	2	1	1	1	1	2
ABC-type Transporter															
ABC-type Mn/Zn transport systems, binding protein	*mntA*	COG0803	pfam01297	1	1	1	1	2	2	2	2	3	2	2	4
ABC-type Mn/Zn transport systems, ATPase component	*mntB*	COG1121	pfam00005	1	1	1	1	2	2	2	2	3	2	3	5
ABC-type Mn/Zn transport systems, permease component	*mntC*	COG1108	pfam00950	1	1	1	1	2	3	2	3	4	3	4	5
RNA synthesis - elongation															
DNA-directed RNA polymerase, delta subunit	*rpoE*	COG3343	-	1	1	1	1	1	1	1	1	1	1	1	1
Protein synthesis (ribosomal proteins)															
30S ribosomal protein S15	*rpsO*	COG0184	pfam00312	1	1	1	1	1	1	1	1	1	1	1	1
50S ribosomal protein L27	*rpmA*	COG0211	pfam01016	1	1	1	1	-	1	1	1	1	1	1	1
50S ribosomal protein L31	*rpmE*	COG0254	pfam01197	1	1	1	1	1	2	1	1	1	2	1	2
50S ribosomal protein L32	*rpmF*	COG0333	pfam01783	1	1	1	1	1	1	1	1	1	1	1	1
Unknown function															
Cold-inducible protein YdjO	*ydjO*	-	pfam14169	1	1	-	-	-	-	1	1	2	-	2	-

1, *P. darwinianus* Br; 2, *P. darwinianus* CE1; 3, *P. darwinianus* MB1; 4, *P. curdlanolyticus* YK9; 5, *P. mucilaginous* K02; 6, *P. polymyxa* SC2; 7, *P. daejeonensis* DSM 15491; 8, *P. terrigena* DSM 21567; 9, *P. panacisoli* DSM 21345; 10, *P. terrae* HPL-003; 11, *P. taiwanensis* DSM 18679; 12, *P. pinihumi* DSM 23905.
Numbers in each column represent copy numbers per genome. Locus tags for each copy number are listed in Table S11.

proteins comprise four groups, firstly a signal recognition and transduction group containing methyl-accepting chemotaxis proteins and glutamine deamidase, CheD, secondly an excitation group containing histidine kinase, CheA, coupling protein, CheW, and response regulator, CheY, an adaptation group containing methyl transferase, CheR and methyl esterase, CheB and finally a signal removal group containing phosphatase, CheC [84]. While approx. 50 genes (1.6% of the total CDSs) encoding components of the flagellum and chemosensory pathways were identified in *P. darwinianus* genomes, temperate *Paenibacillus* genomes contained 63–98 genes (1.3% of the total CDSs) associated with flagella biosynthesis and chemosensory pathways (Table S9). In addition, cells of *P. darwinianus* Br are known to possess a monotrichous flagellum. However, no motility was observed in cells of strain Br, indicating that further investigation is necessary to identify the stimulus needed for movement [26]. A similar observation was made for cells of *P. ingrahamii* that showed no motility despite the presence of a large cluster of flagellar genes [61]. It is hypothesized that the lack of observed motility may be due to a lack of appropriate stimulus or a defect in one of the essential flagellar proteins or in the expression or assembly processes.

Conclusions

Previous studies have shown that, in bacteria, an increase in genome size is often linked with an increase in metabolic complexity, allowing bacteria to produce new enzymes that exploit environmental conditions [85]. However, an increase in complexity is linked with a quadratic increase in regulatory proteins associated with transcription and two-component signal transduction systems [86,87]. In environments such as soil, efficient regulation of enzyme expression, enabling exploitation of scarce yet diverse, complex nutrients can offer a selective advantage, thus lowering the penalty of slow growth, common amongst dominant bacteria in soil environments [88]. Conversely, in the Antarctic soil environment, organic residues are scarce yet labile, with C and N being mineralisable within a relatively short period of time (90 d) under optimal conditions [89]. In the harsh Antarctic soil environment, maintenance of metabolic versatility comes at a higher cost and, more importantly, reproductive efficiency (promoted by smaller genomes) is crucial for survival and growth. Comparative genomic analyses with nine soil-dwelling, temperate *Paenibacillus* spp. have revealed that *P. darwinianus* genomes contained significantly fewer CDSs as compared to the temperate species. A significantly smaller proportion of genes was identified in paralogous clusters in the *P. darwinianus* genome as compared to the temperate *Paenibacillus* genomes ($P<0.01$). In addition, of the total CDSs identified in the *P. darwinianus* genomes, significantly fewer CDSs were assigned to COG category, transcription [K] ($P<0.01$). Finally, *P. darwinianus* strain Br^T showed limited catabolic activity [26], indicative of lowered metabolic complexity. It should be noted that although smaller genomes offer a metabolic advantage during reproduction, striking a balance between minimum cellular-doubling time and the ability to respond to, or exploit, changing environmental conditions is also crucial.

The *P. darwinianus* genomes contained several features that were also identified in genomes of cold-adapted bacteria and archaea. These included genes for signal transduction pathways, sigma factors, membrane transporters, motility and sporulation associated genes and mechanisms to deal with cold shock, oxidative and osmotic stresses, thus suggesting their importance in cold adaptation and survival. However, comparative analyses revealed that a large proportion of these features were also present in genomes of temperate species. This suggests that these physiological traits, while not unique to Antarctic soils, are beneficial for growth and survival in a range of soil environments. Transcriptomic- and proteomic-based studies comparing the expression profiles of *P. darwinianus* strains and temperate *Paenibacillus* spp., at relatively low versus high temperatures may elucidate the exact mechanisms for cold-adaptation in these strains.

In conclusion, fewer CDSs, lower paralogous content and the limited catabolic activity suggest the occurrence of genome content scaling, offering *P. darwinianus* a growth advantage in the Antarctic soil environment. Although this study has provided an insight into the metabolic potential of this species in RSR soils, it draws attention to our limited knowledge about the expression of physiological traits that enable survival under extreme conditions and, more importantly, to the lack of complete prokaryotic genomes from Antarctic soil environments. Complete genomes will not only reveal genes essential for survival in the harsh Antarctic soil environment but also assist with the characterisation of selection processes that affect prokaryotic genomes in this environment.

Materials and Methods

Habitat and isolation strategy

Soil samples were collected from three sites: Cape Evans (77°38′S, 166°24′E) on Ross Island, Minna Bluff (78°30′S, 169°E) on the mainland coast and the Britannia Drift, Lake Wellman Region (79°55′16.2″S, 156°55′30.7″E) in south-eastern Darwin Mountains [90,91]. Field permits for sample collection were issued by New Zealand Ministry of Foreign Affairs and Trade. These samples were irradiated with ^{60}Co γ-rays for 80 h at a dose of 288 Gy/h and plated onto PYGV (Peptone Yeast extract Glucose Vitamin) gellan gum-based solid medium at 15°C for up to two months [92]. Following incubation and purification, three isolates Br^T, CE1 and MB1 were identified as strains of *P. darwinianus* by 16S rRNA gene sequencing. *P. darwinianus* Br^T was deposited at the International Collection of Microorganisms from Plants (ICMP), Landcare Research, New Zealand (ICMP no. 19912) and at DSMZ, Germany (DSM 27245). *P. darwinianus* strains CE1 (ICMP no. 20538) and MB1 (ICMP no. 20539) were also deposited at ICMP.

Temperature range data were obtained for strains Br^T, CE1 and MB1 by growth on PYGV (Peptone Yeast Glucose Vitamin) medium solidified by gellan gum [92]. All PYGV plates were incubated for 56 days at 0, 5, 10, 15, 18, 20, 25, 37, 40 and 45°C.

DNA extraction and sequencing

The three Antarctic *Paenibacillus* strains were routinely cultured on PYGV gellan gum plates at 15–18°C. High molecular weight DNA was extracted by a modified CTAB (hexadecyltrimethylammonium) and protein lysis method [93]. Briefly, cells were scraped off PYGV gellan gum plates and re-suspended in 740 µl TE buffer, 20 µl lysozyme (100 mg/ml) and incubated for 10 min at room temperature. Then, 40 µl of 10% SDS and 8 µl of Proteinase K (10 mg/ml) were added and the reaction incubated overnight at 37°C. Following incubation, 100 µl of 5 M NaCl and CTAB/NaCl solutions were added to each reaction and incubated at 65°C for 10 min. Subsequently, 0.5 ml chloroform:isoamyl alcohol (24:1) was added, and the reaction was centrifuged at 16,000 g for 15 min. The aqueous phase was transferred to a clean eppendorf tube containing phenol:chloroform:isoamyl alcohol (25:24:1, pH 8) and centrifuged at 16,000 g for 15 min. The aqueous phase was transferred to a clean eppendorf tube containing 0.6 vol isopropanol. For DNA precipitation, reactions were incubated at room temperature for 60 min, then centrifuged at 16,000 g for 30 min. The DNA pellet was washed with 70% ethanol and re-suspended in TE buffer containing RNAse (99 μl TE buffer +1 μl RNAse (10 mg/ml)) and incubated at 37°C for 20 min. DNA extracts were quantified by Quant-iT PicoGreen dsDNA assay kit (Life Technologies) and their purity (A_{260}/A_{280}) was assessed on a NanoDrop ND-1000 Spectrophotometer (Biolab). The quality of each DNA extract was tested by electrophoresis on a 1% agarose gel. Following extraction, high molecular weight DNA was sent to Macrogen (Seoul, South Korea) for sequencing on the Illumina HiSeq 2000 platform using 100 bp paired end libraries. With a sequencing output of 35 Gb, estimated coverage was over 1,000× per genome.

De novo assembly, annotation and comparative analyses

FASTQ files obtained for each genome were trimmed by the FASTQ Trimmer tool of the FASTX-toolkit v0.0.13 [94] and quality filtered by Sickle (https://github.com/ucdavis-bioinformatics/sickle). High-quality reads (Q>30) were assembled into contigs by Velvet v1.2.10 [95]. All assemblies were further improved by tools of the PAGIT (post assembly genome improvement toolkit) pipeline. These included IMAGE (iterative mapping and assembly for gap elimination) for gap elimination and iCORN (iterative correction of reference nucleotides) for sequencing error correction [96]. Improved contigs were assembled into scaffolds by SSPACE basic version 1.0 (stand-alone scaffolder of pre-assembled contigs using paired-read data) [97]. Gene prediction and annotation was performed by the automated JGI (Joint Genome Institute) Pipeline [25] and the NCBI Prokaryotic Genome Annotation Pipeline (PGAP) [98]. Genome completeness was assessed as previously described [99].

For comparative analyses, nine temperate, soil-dwelling *Paenibacillus* species, namely *P. curdlanolyticus* YK9, *P. daejeonensis* DSM 15491, *P. mucilaginosus* K02, *P. panacisoli* DSM 21345, *P. pinihumi* DSM 23905, *P. polymyxa* SC2, *P. taiwanensis* DSM 18679, *P. terrae* HPL-003, and *P. terrigena* DSM 21567 were selected [100]. All analyses were performed using JGI-IMG/ER (Integrated Microbial Genomes/Expert Review) [25] and CMG biotools [43].

Genome Submission

This whole genome shotgun project has been deposited at DDBJ/EMBL/GenBank under the following accession numbers: JFHT00000000 for strain Br, JFHU00000000 for strain CE1, and JFHV00000000 for MB1.

Supporting Information

Figure S1 Dot plot comparisons of the three *P. darwinianus* strains.
(PPTX)

Table S1 List of Pfam domains associated with Two-Component Systems in genomes of temperate *Paenibacillus* spp. and *P. darwinianus* strains.
(XLSX)

Table S2 List of genes associated with Two-Component Systems in genomes of temperate *Paenibacillus* spp. and *P. darwinianus* strains.
(XLSX)

Table S3 List of sigma factors identified in genomes of the *P. darwinianus* strains.
(XLSX)

Table S4 List of genes associated with transporter proteins in genomes of temperate *Paenibacillus* spp. and *P. darwinianus* strains.
(XLSX)

Table S5 List of genes associated with ABC-type transport systems in genomes of temperate *Paenibacillus* spp. and *P. darwinianus* strains.
(XLSX)

Table S6 List of COGs associated with carbohydrate metabolism in genomes of temperate *Paenibacillus* spp. and *P. darwinianus* strains.
(XLSX)

Table S7 Sporulation associated genes identified in genomes of the *P. darwinianus* strains.
(XLSX)

Table S8 Sporulation associated genes identified in genomes of temperate *Paenibacillus* spp.
(XLSX)

Table S9 List of flagellar assembly and chemotaxis associated genes in genomes of temperate *Paenibacillus* spp. and *P. darwinianus* strains.
(XLSX)

Table S10 List of genes associated with oxidative and osmotic stress response in genomes of temperate *Paenibacillus* spp. and *P. darwinianus* strains.
(XLSX)

Table S11 List of cold-shock induced genes identified in genomes of temperate *Paenibacillus* spp. and *P. darwinianus* strains.
(XLSX)

Acknowledgments

We thank Antarctica New Zealand for providing logistic support during sample collection.

Author Contributions

Conceived and designed the experiments: MD MWT SJT JA. Performed the experiments: MD. Analyzed the data: MD. Contributed reagents/materials/analysis tools: MD MWT SJT JA. Wrote the paper: MD.

References

1. Turnbaugh PJ, Hamady M, Yatsunenko T, Cantarel BL, Duncan A, et al. (2008) A core gut microbiome in obese and lean twins. Nature 457: 480–484.
2. Zhang T, Shao M-F, Ye L (2012) 454 Pyrosequencing reveals bacterial diversity of activated sludge from 14 sewage treatment plants. The ISME Journal 6: 1137–1147.
3. Hirsch PR, Mauchline TH, Clark IM (2013) Culture-independent molecular approaches to microbial ecology in soil and the rhizosphere. Molecular Microbial Ecology of the Rhizosphere. Hoboken, NJ, USA: John Wiley & Sons, Inc. pp. 45–55.
4. Guan X, Wang J, Zhao H, Wang J, Luo X, et al. (2013) Soil bacterial communities shaped by geochemical factors and land use in a less-explored area, Tibetan Plateau. BMC Genomics 14: 820.
5. Schauer R, Bienhold C, Ramette A, Harder J (2009) Bacterial diversity and biogeography in deep-sea surface sediments of the South Atlantic Ocean. The ISME Journal 4: 159–170.

6. Campbell BJ, Yu L, Heidelberg JF, Kirchman DL (2011) Activity of abundant and rare bacteria in a coastal ocean. Proceedings of the National Academy of Sciences USA 108: 12776–12781.
7. Jiang H, Dong H, Zhang G, Yu B, Chapman LR, et al. (2006) Microbial diversity in water and sediment of Lake Chaka, an Athalassohaline Lake in Northwestern China. Applied and Environmental Microbiology 72: 3832–3845.
8. Roesch LFW, Fulthorpe RR, Riva A, Casella G, Hadwin AKM, et al. (2007) Pyrosequencing enumerates and contrasts soil microbial diversity. The ISME Journal 1: 283–290.
9. Lauber CL, Hamady M, Knight R, Fierer N (2009) Pyrosequencing-based assessment of soil pH as a predictor of soil bacterial community structure at the continental scale. Applied and Environmental Microbiology 75: 5111–5120.
10. Elshahed MS, Youssef NH, Spain AM, Sheik C, Najar FZ, et al. (2008) Novelty and uniqueness patterns of rare members of the soil biosphere. Applied and Environmental Microbiology 74: 5422–5428.
11. Youssef NH, Elshahed MS (2008) Diversity rankings among bacterial lineages in soil. The ISME Journal 3: 305–313.
12. Felske A, Wolterink A, Van Lis R, De Vos WM, Akkermans AD (2000) Response of a soil bacterial community to grassland succession as monitored by 16S rRNA levels of the predominant ribotypes. Applied and Environmental Microbiology 66: 3998–4003.
13. Hartmann M, Niklaus PA, Zimmermann S, Schmutz S, Kremer J, et al. (2014) Resistance and resilience of the forest soil microbiome to logging-associated compaction. The ISME Journal 8: 226–224.
14. Bachar A, Al-Ashhab A, Soares MI, Sklarz M, Angel R, et al. (2010) Soil microbial abundance and diversity along a low precipitation gradient. Microbial Ecology 60: 453–461.
15. Janssen PH (2006) Identifying the dominant soil bacterial taxa in libraries of 16S rRNA and 16S rRNA genes. Applied and Environmental Microbiology 72: 1719.
16. Priest FG (2009) Genus I. *Paenibacillus*. In: De Vos P, Garrity GM, Jones D, Krieg NR, Ludwig W, et al., editors. Bergey's Manual of Systematic Bacteriology The *Firmicutes*. 2nd ed. New York, USA: Springer. pp. 1422.
17. Rodríguez-Díaz M, Lebbe L, Rodelas B, Heyrman J, De Vos P, et al. (2005) *Paenibacillus wynnii* sp. nov., a novel species harbouring the *nifH* gene, isolated from Alexander Island, Antarctica. International Journal of Systematic and Evolutionary Microbiology 55: 2093–2099.
18. Kishore KH, Begum Z, Pathan AAK, Shivaji S (2010) *Paenibacillus glacialis* sp. nov., isolated from the Kafni glacier of the Himalayas, India. International Journal of Systematic and Evolutionary Microbiology 60: 1909–1913.
19. Nelson DM, Glawe AJ, Labeda DP, Cann IKO, Mackie RI (2009) *Paenibacillus tundrae* sp. nov. and *Paenibacillus xylanexedens* sp. nov., psychrotolerant, xylan-degrading bacteria from Alaskan tundra. International Journal of Systematic and Evolutionary Microbiology 59: 1708–1714.
20. Kim B-C, Lee K, Kim M, Kim E-M, Rhee M-S, et al. (2009) *Paenibacillus pinihumi* sp. nov., a cellulolytic bacterium isolated from the rhizosphere of *Pinus densiflora*. The Journal of Microbiology 47: 530–535.
21. Elo S, Suominen I, Kämpfer P, Juhanoja J, Salkinoja-Salonen M, et al. (2001) *Paenibacillus borealis* sp. nov., a nitrogen-fixing species isolated from spruce forest humus in Finland. International Journal of Systematic and Evolutionary Microbiology 51: 535–545.
22. Chan Q, Cornman RS, Birol I, Liao N, Chan S, et al. (2011) Updated genome assembly and annotation of *Paenibacillus larvae*, the agent of American foulbrood disease of honey bees. BMC Genomics 12: 450.
23. Qin X, Evans JD, Aronstein K, Murray KD, Weinstock GM (2006) Genome sequences of the honey bee pathogens *Paenibacillus larvae* and *Ascosphaera apis*. Insect Molecular Biology 15: 715–718.
24. Sirota-Madi A, Olender T, Helman Y, Ingham C, Brainis I, et al. (2010) Genome sequence of the pattern forming *Paenibacillus vortex* bacterium reveals potential for thriving in complex environments. BMC Genomics 11: 710.
25. Markowitz VM, Chen I-MA, Palaniappan K, Chu K, Szeto E, et al. (2012) IMG: the integrated microbial genomes database and comparative analysis system. Nucleic Acids Research 40: D115–D122.
26. Dsouza M, Taylor MW, Ryan J, MacKenzie A, Lagutin K, et al. (2014) *Paenibacillus darwinianus* sp. nov., isolated from gamma-irradiated Antarctic soil. International Journal of Systematic and Evolutionary Microbiology 64: 1406–1411.
27. Adlam LS, Balks MR, Seybold CA, Campbell DI (2010) Temporal and spatial variation in active layer depth in the McMurdo Sound Region, Antarctica. Antarctic Science 22: 45–52.
28. Niederberger TD, McDonald IR, Hacker AL, Soo RM, Barrett JE, et al. (2008) Microbial community composition in soils of Northern Victoria Land, Antarctica. Environmental Microbiology 10: 1713–1724.
29. Saul DJ, Aislabie JM, Brown CE, Harris L, Foght JM (2005) Hydrocarbon contamination changes the bacterial diversity of soil from around Scott Base, Antarctica. FEMS Microbiology Ecology 53: 141–155.
30. Aislabie J, Lau A, Dsouza M, Shepherd C, Rhodes P, et al. (2013) Bacterial composition of soils of the Lake Wellman area, Darwin Mountains, Antarctica. Extremophiles 17: 775–786.
31. Van Horn DJ, Van Horn ML, Barrett JE, Gooseff MN, Altrichter AE, et al. (2013) Factors controlling soil microbial biomass and bacterial diversity and community composition in a cold desert ecosystem: Role of geographic scale. PLoS ONE 8: e66103.
32. Tiao G, Lee CK, McDonald IR, Cowan DA, Cary SC (2012) Rapid microbial response to the presence of an ancient relic in the Antarctic Dry Valleys. Nature Communications 3: 660.
33. Lee CK, Barbier BA, Bottos EM, McDonald IR, Cary SC (2012) The inter-valley soil comparative survey: the ecology of Dry Valley edaphic microbial communities. The ISME Journal 6: 1046–1057.
34. Pointing SB, Chan Y, Lacap DC, Lau MCY, Jurgens JA, et al. (2009) Highly specialized microbial diversity in hyper-arid polar desert. Proceedings of the National Academy of Sciences USA 106: 19964–19969.
35. Aislabie JM, Jordan S, Barker GM (2008) Relation between soil classification and bacterial diversity in soils of the Ross Sea region, Antarctica. Geoderma 144: 9–20.
36. Aislabie J, Jordan S, Ayton J, Klassen JL, Barker GM, et al. (2009) Bacterial diversity associated with ornithogenic soil of the Ross Sea region, Antarctica. Canadian Journal of Microbiology 55: 21–36.
37. Bottos E, Woo A, Zawar-Reza P, Pointing S, Cary S (2014) Airborne bacterial populations above desert soils of the McMurdo Dry Valleys, Antarctica. Microbial Ecology 67: 120–128.
38. Montes MJ, Mercadé E, Bozal N, Guinea J (2004) *Paenibacillus antarcticus* sp. nov., a novel psychrotolerant organism from the Antarctic environment. International Journal of Systematic and Evolutionary Microbiology 54: 1521–1526.
39. Logan NA, De Clerck E, Lebbe L, Verhelst A, Goris J, et al. (2004) *Paenibacillus cineris* sp. nov. and *Paenibacillus cookii* sp. nov., from Antarctic volcanic soils and a gelatin-processing plant. International Journal of Systematic and Evolutionary Microbiology 54: 1071–1076.
40 Marshall BJ, Ohye DF (1966) *Bacillus macquariensis* n.sp., a psychrotrophic bacterium from Sub-Antarctic soil. Journal of General Microbiology 44: 41–46.
41 Xie C-H, Yokota A (2007) *Paenibacillus terrigena* sp. nov., isolated from soil. International Journal of Systematic and Evolutionary Microbiology 57: 70–72.
42 Lee F-L, Kuo H-P, Tai C-J, Yokota A, Lo C-C (2007) *Paenibacillus taiwanensis* sp. nov., isolated from soil in Taiwan. International Journal of Systematic and Evolutionary Microbiology 57: 1351–1354.
43. Vesth T, Lagesen K, Acar Ö, Ussery D (2013) CMG-Biotools, a free workbench for basic comparative microbial genomics. PLoS ONE 8: e60120.
44. Hyatt D, Chen G-L, LoCascio PF, Land ML, Larimer FW, et al. (2010) Prodigal: prokaryotic gene recognition and translation initiation site identification. BMC bioinformatics 11: 119.
45. Altschul SF, Madden TL, Schäffer AA, Zhang J, Zhang Z, et al. (1997) Gapped BLAST and PSI-BLAST: a new generation of protein database search programs. Nucleic Acids Research 25: 3389–3402.
46. Allen MA, Lauro FM, Williams TJ, Burg D, Siddiqui KS, et al. (2009) The genome sequence of the psychrophilic archaeon, *Methanococcoides burtonii*: The role of genome evolution in cold adaptation. The ISME Journal 3: 1012–1035.
47. Rabus R, Ruepp A, Frickey T, Rattei T, Fartmann B, et al. (2004) The genome of *Desulfotalea psychrophila*, a sulfate-reducing bacterium from permanently cold Arctic sediments. Environmental Microbiology 6: 887–902.
48. Aguilar PS, Hernandez-Arriaga AM, Cybulski LE, Erazo AC, de Mendoza D (2001) Molecular basis of thermosensing: a two-component signal transduction thermometer in *Bacillus subtilis*. The EMBO Journal 20: 1681–1691.
49. Rosario MML, Ordal GW (1996) CheC and CheD interact to regulate methylation of *Bacillus subtilis* methyl-accepting chemotaxis proteins. Molecular Microbiology 21: 511–518.
50. Sun G, Birkey SM, Hulett FM (1996) Three two-component signal-transduction systems interact for Pho regulation in *Bacillus subtilis*. Molecular Microbiology 19: 941–948.
51. Nakano MM, Zuber P, Glaser P, Danchin A, Hulett FM (1996) Two-component regulatory proteins ResD-ResE are required for transcriptional activation of *fnr* upon oxygen limitation in *Bacillus subtilis*. Journal of Bacteriology 178: 3796–3802.
52. Jers C, Kobir A, Søndergaard EO, Jensen PR, Mijakovic I (2011) *Bacillus subtilis* Two-component system sensory kinase DegS is regulated by serine phosphorylation in its input domain. PLoS ONE 6: e14653.
53. Blackman SA, Smith TJ, Foster SJ (1998) The role of autolysins during vegetative growth of *Bacillus subtilis* 168. Microbiology 144: 73–82.
54. Hesami S, Metcalf DS, Lumsden JS, MacInnes JI (2011) Identification of cold-temperature-regulated genes in *Flavobacterium psychrophilum*. Applied and Environmental Microbiology 77: 1593–1600.
55. Hopkins DW, Sparrow AD, Gregorich EG, Elberling B, Novis P, et al. (2009) Isotopic evidence for the provenance and turnover of organic carbon by soil microorganisms in the Antarctic dry valleys. Environmental Microbiology 11: 597–608.
56. Xu G, Gao Y, Lin Q, Li W, Chen L (2013) Characteristics of water-soluble inorganic and organic ions in aerosols over the Southern Ocean and coastal East Antarctica during austral summer. Journal of Geophysical Research: Atmospheres 118: 13303–13318.
57. Kawamura K, Semeré R, Imai Y, Fujii Y, Hayashi M (1996) Water soluble dicarboxylic acids and related compounds in Antarctic aerosols. Journal of Geophysical Research: Atmospheres 101: 18721–18728.
58. Teinilä K, Kerminen V-M, Hillamo R (2000) A study of size-segregated aerosol chemistry in the Antarctic atmosphere. Journal of Geophysical Research: Atmospheres 105: 3893–3904.

59. Graf S, Schmieden D, Tschauner K, Hunke S, Unden G (2014) The sensor kinase DctS forms a tripartite sensor unit with DctB and DctA for sensing C4-dicarboxylates in *B. subtilis*. Journal of Bacteriology 196: 1084–1093.
60. Mykytczuk NC, Foote SJ, Omelon CR, Southam G, Greer CW, et al. (2013) Bacterial growth at −15°C; molecular insights from the permafrost bacterium *Planococcus halocryophilus* Or1. The ISME journal 7: 1211–1226.
61. Riley M, Staley JT, Danchin A, Wang TZ, Brettin TS, et al. (2008) Genomics of an extreme psychrophile, *Psychromonas ingrahamii*. BMC Genomics 9: 210.
62. Wiegeshoff F, Beckering CL, Debarbouille M, Marahiel MA (2006) Sigma L is important for cold shock adaptation of *Bacillus subtilis*. Journal of Bacteriology 188: 3130–3133.
63. Helmann JD (2011) Regulation by alternative sigma factors. In: Storz G, Hengge R, editors. Bacterial stress responses. 2nd ed. Washington, DC: ASM Press. pp. 31–43.
64. Price CW, Fawcett P, Cérémonie H, Su N, Murphy CK, et al. (2001) Genome-wide analysis of the general stress response in *Bacillus subtilis*. Molecular Microbiology 41: 757–774.
65. Messner KR, Imlay JA (1999) The Identification of primary sites of superoxide and hydrogen peroxide formation in the aerobic respiratory chain and sulfite reductase complex of *Escherichia coli*. Journal of Biological Chemistry 274: 10119–10128.
66. Chattopadhyay M (2002) Low temperature and oxidative stress. Current Science 83: 109–109.
67. Methé BA, Nelson KE, Deming JW, Momen B, Melamud E, et al. (2005) The psychrophilic lifestyle as revealed by the genome sequence of *Colwellia psychrerythraea* 34H through genomic and proteomic analyses. Proceedings of the National Academy of Sciences USA 102: 10913–10918.
68. Nakano S, Küster-Schöck E, Grossman AD, Zuber P (2003) Spx-dependent global transcriptional control is induced by thiol-specific oxidative stress in *Bacillus subtilis*. Proceedings of the National Academy of Sciences USA 100: 13603–13608.
69. Helmann JD, Wu MFW, Gaballa A, Kobel PA, Morshedi MM, et al. (2003) The global transcriptional response of *Bacillus subtilis* to peroxide stress is coordinated by three transcription factors. Journal of Bacteriology 185: 243–253.
70. Bockheim J (1997) Properties and classification of cold desert soils from Antarctica. Soil Science Society of America Journal 61: 224–231.
71. Kempf B, Bremer E (1995) OpuA, an osmotically regulated binding protein-dependent transport system for the osmoprotectant glycine getaine in *Bacillus subtilis*. Journal of Biological Chemistry 270: 16701–16713.
72. Wood JM (2011) Osmotic stress. In: Storz G, Hengge R, editors. Bacterial stress responses. 2nd ed. Washington DC: ASM Press. pp. 133–156.
73. Médigue C, Krin E, Pascal G, Barbe V, Bernsel A, et al. (2005) Coping with cold: The genome of the versatile marine Antarctica bacterium *Pseudoalteromonas haloplanktis* TAC125. Genome Research 15: 1325–1335.
74. Hoffmann T, Boiangiu C, Moses S, Bremer E (2008) Responses of *Bacillus subtilis* to hypotonic challenges: physiological contributions of mechanosensitive channels to cellular survival. Applied and Environmental Microbiology 74: 2454–2460.
75. Campbell I, Claridge G (1987) Antarctica: soils, weathering processes and environment. The Netherlands: Elsevier Science. 368 p.
76. McKay CP, Nienow JA, Meyer MA, Friedmann EI (1993) Continuous nanoclimate data (1985–1988) from the Ross desert (McMurdo Dry Valleys) cryptoendolithic microbial ecosystem. Antarctic Meteorology and Climatology: Studies Based on Automatic Weather Stations. Washington, DC: AGU. pp. 201–207.
77. Phadtare S, Alsina J, Inouye M (1999) Cold-shock response and cold-shock proteins. Current Opinion in Microbiology 2: 175–180.
78. Graumann PL, Marahiel MA (1999) Cold shock response in *Bacillus subtilis*. Journal of Molecular Microbiology and Biotechnology 1: 203–209.
79. Ayala-del-Río HL, Chain PS, Grzymski JJ, Ponder MA, Ivanova N, et al. (2010) The genome sequence of *Psychrobacter arcticus* 273-4, a psychroactive Siberian permafrost bacterium, reveals mechanisms for adaptation to low-temperature growth. Applied and Environmental Microbiology 76: 2304–2312.
80. Gao H, Yang ZK, Wu L, Thompson DK, Zhou J (2006) Global transcriptome analysis of the cold shock response of *Shewanella oneidensis* MR-1 and mutational analysis of its classical cold shock proteins. Journal of Bacteriology 188: 4560–4569.
81. Kaan T, Homuth G, Mäder U, Bandow J, Schweder T (2002) Genome-wide transcriptional profiling of the *Bacillus subtilis* cold-shock response. Microbiology 148: 3441–3455.
82. Aldridge P, Hughes KT (2002) Regulation of flagellar assembly. Current Opinion in Microbiology 5: 160–165.
83. Guttenplan SB, Shaw S, Kearns DB (2013) The cell biology of peritrichous flagella in Bacillus subtilis. Molecular Microbiology 87: 211–229.
84. Szurmant H, Ordal GW (2004) Diversity in chemotaxis mechanisms among the bacteria and archaea. Microbiology and Molecular Biology Reviews 68: 301–319.
85. Ranea JAG, Grant A, Thornton JM, Orengo CA (2005) Microeconomic principles explain an optimal genome size in bacteria. Trends in Genetics 21: 21–25.
86. van Nimwegen E (2003) Scaling laws in the functional content of genomes. Trends in Genetics 19: 479–484.
87. Ranea JAG, Buchan DWA, Thornton JM, Orengo CA (2004) Evolution of protein superfamilies and bacterial genome size. Journal of Molecular Biology 336: 871–887.
88. Klappenbach JA, Dunbar JM, Schmidt TM (2000) rRNA operon copy number reflects ecological strategies of bacteria. Applied and Environmental Microbiology 66: 1328–1333.
89. Barrett JE, Virginia RA, Parsons AN, Wall DH (2005) Potential soil organic matter turnover in Taylor Valley, Antarctica. Arctic, Antarctic, and Alpine Research 37: 108–117.
90. Aislabie J, Bockheim J, Mcleod M, Hunter D, Stevenson B, et al. (2012) Microbial biomass and community structure changes along a soil development chronosequence near Lake Wellman, southern Victoria Land. Antarctic Science 24: 154–164.
91. Ayton J, Aislabie J, Barker GM, Saul D, Turner S (2010) Crenarchaeota affiliated with group 1.1b are prevalent in coastal mineral soils of the Ross Sea region of Antarctica. Environmental Microbiology 12: 689–703.
92. Hirsch P, Gallikowski CA, Siebert J, Peissl K, Kroppenstedt R, et al. (2004) *Deinococcus frigens* sp. nov., *Deinococcus saxicola* sp. nov., and *Deinococcus marmoris* sp. nov., low temperature and draught-tolerating, UV-resistant bacteria from continental Antarctica. Systematic and Applied Microbiology 27: 636–645.
93. Wilson K (1987) Preparation of genomic DNA from bacteria. Current Protocols in Molecular Biology: 2.4. 1–2.4. 5.
94. Pearson WR, Wood T, Zhang Z, Miller W (1997) Comparison of DNA sequences with protein sequences. Genomics 46: 24–36.
95. Zerbino DR, Birney E (2008) Velvet: algorithms for de novo short read assembly using de Bruijn graphs. Genome Research 18: 821–829.
96. Swain MT, Tsai IJ, Assefa SA, Newbold C, Berriman M, et al. (2012) A post-assembly genome-improvement toolkit (PAGIT) to obtain annotated genomes from contigs. Nature Protocols 7: 1260–1284.
97. Boetzer M, Henkel CV, Jansen HJ, Butler D, Pirovano W (2011) Scaffolding pre-assembled contigs using SSPACE. Bioinformatics 27: 578–579.
98. Angiuoli SV, Gussman A, Klimke W, Cochrane G, Field D, et al. (2008) Toward an online repository of Standard Operating Procedures (SOPs) for (meta) genomic annotation. OMICS A Journal of Integrative Biology 12: 137–141.
99. Albertsen M, Hugenholtz P, Skarshewski A, Nielsen KL, Tyson GW, et al. (2013) Genome sequences of rare, uncultured bacteria obtained by differential coverage binning of multiple metagenomes. Nature Biotechnology 31: 533–538.
100. Pagani I, Liolios K, Jansson J, Chen I-MA, Smirnova T, et al. (2012) The Genomes OnLine Database (GOLD) v.4: status of genomic and metagenomic projects and their associated metadata. Nucleic Acids Research 40: D571–D579.

9

Human Valacyclovir Hydrolase/Biphenyl Hydrolase-Like Protein Is a Highly Efficient Homocysteine Thiolactonase

Judit Marsillach[1,2☯], Stephanie M. Suzuki[1☯], Rebecca J. Richter[1☯], Matthew G. McDonald[3], Peter M. Rademacher[3¤a], Michael J. MacCoss[2], Edward J. Hsieh[2¤b], Allan E. Rettie[3], Clement E. Furlong[1,2*]

1 Department of Medicine (Division of Medical Genetics), University of Washington, Seattle, Washington, United States, **2** Department of Genome Sciences, University of Washington, Seattle, Washington, United States, **3** Department of Medicinal Chemistry, University of Washington, Seattle, Washington, United States

Abstract

Homocysteinylation of lysine residues by homocysteine thiolactone (HCTL), a reactive homocysteine metabolite, results in protein aggregation and malfunction, and is a well-known risk factor for cardiovascular, autoimmune and neurological diseases. Human plasma paraoxonase-1 (PON1) and bleomycin hydrolase (Blmh) have been reported as the physiological HCTL detoxifying enzymes. However, the catalytic efficiency of HCTL hydrolysis by Blmh is low and not saturated at 20 mM HCTL. The catalytic efficiency of PON1 for HCTL hydrolysis is 100-fold lower than that of Blmh. A homocysteine thiolactonase (HCTLase) was purified from human liver and identified by mass spectrometry (MS) as the previously described human biphenyl hydrolase-like protein (BPHL). To further characterize this newly described HCTLase activity, BPHL was expressed in *Escherichia coli* and purified. The sequence of the recombinant BPHL (rBPHL) and hydrolytic products of the substrates HCTL and valacyclovir were verified by MS. We found that the catalytic efficiency (k_{cat}/K_m) of rBPHL for HCTL hydrolysis was 7.7 × 10^4 $M^{-1}s^{-1}$, orders of magnitude higher than that of PON1 or Blmh, indicating a more significant physiological role for BPHL in detoxifying HCTL.

Editor: Fernando Rodrigues-Lima, University Paris Diderot-Paris 7, France

Funding: This work was supported by National Institutes of Health (http://www.nih.gov/) to CEF (P42 ES04696 and R01 ES09883), MJM (P41 GM103533), and AER (P01 GM32165); Drug Metabolism, Transport and Pharmacogenetics Research Fund within the School of Pharmacy at the University of Washington (http://sop.washington.edu/) to AER; and National Institute of Environmental Health Sciences (http://www.niehs.nih.gov/) to SMS (training grant T32 ES007032). The funders had no role in study design, data collection and analysis, decision to publish, or preparation of the manuscript.

Competing Interests: The authors have declared that no competing interests exist.

* Email: clem@uw.edu

☯ These authors contributed equally to this work.

¤a Current address: Global Blood Therapeutics, South San Francisco, California, United States of America
¤b Current address: Merck Research Laboratories, Palo Alto, California, United States of America

Introduction

Elevated homocysteine (Hcy) levels have been associated with cardiovascular disease [1,2], diabetic retinopathy [3], birth defects, osteoporosis, renal failure [4] and Alzheimer' disease (AD) [5,6]. Severe hyperhomocysteinemia is rare, but mild hyperhomocysteinemia is common in the general population. Hcy, a homologue of the amino acid cysteine, is generated from methionine following demethylation and can be converted back to methionine or to cysteine. In humans, Hcy serves as a precursor for maintaining the levels of methionine and cysteine. If methionine is needed, Hcy is methylated through two enzymes from the methionine and folate cycles with the aid of the co-factor vitamin B12. On the other hand, if there is an excess of methionine, Hcy is converted to cysteine via transsulfuration, which requires vitamin B6. Although Hcy can be synthesized in all human organs, most of it is detoxified in the liver and kidneys, organs that possess all the enzymes involved in the methionine metabolism [7]. Hcy is also converted to homocysteine thiolactone (HCTL) through error-editing reactions catalyzed by methionyl tRNA-synthetase. Homocysteine thiolactonase (HCTLase) activities hydrolyze HCTL to Hcy. Both Hcy and HCTL have been linked to many disease processes, however, HCTL appears to be the more toxic [4,8–10] and has been shown to be neurotoxic in mice [9,10].

HCTL is a reactive metabolite that modifies proteins by forming an isopeptide bond with the ε-amino groups of lysine residues [11]. Homocysteinylated proteins have the potential to misfold and aggregate which can activate the adaptive immune system and increase clot resistance to lysis, promoting atherosclerotic lesion formation. Homocysteinylated proteins may also lose biological activity [4,11]. Therefore, *in vivo* homocysteinylation of proteins has been suggested as the mechanism by which elevated Hcy levels contribute to cardiovascular, neurological and autoimmune diseases [11]. HCTL adducts have been described on several plasma proteins, such as albumin, hemoglobin γ-globulin, high-density lipoproteins (HDLs), low-density lipoproteins (LDLs) and fibrinogen [12,13]. Deficits in the enzymes metabolizing Hcy and HCTL can result in increased physiological concentrations of these metabolites and associated detrimental effects. Methylenetetrahydrofolate reductase (MTHFR), methionine synthase and cystathionine beta-synthase (CBS) are responsible for recycling Hcy back to amino acids. MTHFR is required to form 5-methyl tetrahydrofolate, a metabolite necessary for the conversion of Hcy to methionine. CBS is required to convert Hcy to cysteine. Deficiencies in any of these enzymes can result in increased levels of Hcy and HCTL (reviewed in [1,14]).

Human plasma paraoxonase-1 (PON1), a polymorphic high-density lipoprotein (HDL)-associated enzyme, synthesized primarily in liver and secreted to associate with HDL, hydrolyzes a number of lactones, including HCTL [15–17]. We have shown previously [18,19] that the physiological significance of PON1 in detoxification of organophosphorus (OP) insecticides is highly dependent on the catalytic efficiency of hydrolysis of a given OP substrate. For example, while the catalytic efficiency of $PON1_{R192}$ is ≈9-times greater than that of $PON1_{Q192}$ for hydrolysis of paraoxon, it provides little if any protection against paraoxon exposure whereas PON1 protects against exposures to chlorpyrifos oxon and diazoxon [18]. The reported low substrate affinity and very low specific activity for hydrolysis of HCTL by PON1 [15,16] have raised doubts about the physiological relevance of this activity [15,20]. However, studies with $PON1^{-/-}$ mice provided some evidence that HCTL as opposed to Hcy is neurotoxic *in vivo* and that PON1 could protect against the neurotoxicity of HCTL [9]. For HCTLase measurement, two substrates are commonly used, HCTL and γ-thiobutyrolactone (GTBL), the latter with the rationale that HCTL is unstable and/or because of albumin interference with the Ellman assay [21]. Despite the very low catalytic efficiency of PON1 for hydrolyzing HCTL [8,15,20], PON1 has been referred to as plasma HCTLase [22].

Another HCTLase has been purified from human placenta and identified as bleomycin hydrolase (Blmh) [8]. This enzyme, first studied for its ability to hydrolyze the anti-cancer drug bleomycin, has been detected in testis, skeletal muscle, pancreas, spleen, and at lower levels in thymus, prostate, ovary, small intestine, heart, brain, placenta and lung with very low levels of expression observed in liver, kidney, colon and peripheral blood leukocytes [23]. The catalytic efficiency of human Blmh for hydrolysis of HCTL was reported to be ≈100-fold greater than that of PON1, even though it was not saturated at 20 mM HCTL [8]. HCTL administered to yeast or mice was shown to be toxic, with yeast Blmh (referred to as BLH1) preventing HCTL toxicity in yeast [8] and mouse Blmh protecting mice from HCTL toxicity [10]. Yeast *BLH1* (also known as *LAP3*) mutants and $Blmh^{-/-}$ mice exhibited increased sensitivity to HCTL and $Blmh^{-/-}$ mice demonstrated deficits in spatial memory [24] as well as increased susceptibility to HCTL toxicity [10]. Blmh has also been associated with AD, based in part on its ability to process β-amyloid peptides [24,25]. HCTLase activity attributed to Blmh differed significantly between brains of AD patients and controls [5].

Biphenyl hydrolase-like protein (BPHL), also called valacyclovir (VC) hydrolase or valacyclovirase, is a 32,543 Da protein highly expressed in human liver and kidney and at lower levels in the intestine, heart and skeletal muscle [26,27]. BPHL hydrolyzes and activates the antiviral prodrug esters VC and valganciclovir, which are used in the management of herpes simplex, herpes zoster (shingles) and herpes B [28]. First cloned from breast carcinoma cells, BPHL, a member of the alpha/beta hydrolase fold family, is a serine hydrolase distantly related to other members of the serine hydrolase family [26,27]. BPHL has very low activity against standard esterase substrates such as *p*-nitrophenyl acetate/butyrate but higher activity for hydrolysis of a number of amino ester prodrugs [29]. Kim and coworkers have expressed and purified recombinant BPHL (rBPHL) from *Escherichia coli* (*E.coli*) [28]. Lai et al. reported a crystal structure of BPHL, which revealed a serine hydrolase consensus sequence of GSXSG and a catalytic triad at S122-D227-H255 [30]. A physiological function of BPHL has not yet been described [30,31].

In this study, we report the purification of a human hepatic enzyme that demonstrates high HCTLase activity and was identified as BPHL by mass spectrometry (MS). When expressed in *E. coli*, the purified rBPHL exhibited high HCTLase activity. Both native and recombinant purified BPHLs have much higher catalytic efficiencies for HCTL hydrolysis than described for purified plasma PON1 or Blmh, making BPHL a much more likely candidate for protecting against HCTL toxicity under physiological conditions.

Materials and Methods

Enzyme assays

HCTLase activity was measured using a modification of the Ellman cholinesterase method [32]. A fresh solution of 5,5'-dithiobis-(2-nitrobenzoic acid) (DTNB) (10.3 mM) was prepared in 100 mM $NaPO_4$, pH 7.0, and was added at a final concentration of 0.32 mM to a solution of 3 mM L-HCTL, in Dulbecco's PBS, with the volume dependent on the number of assays to be carried out. The substrate/DTNB solution (200 μL) was added immediately to enzyme samples (5–20 μL) in 96-well microplates. The assay was monitored at 405 nm for 4 min at 25°C for column fractions or at 37°C for kinetic analyses using a SPECTRAmax PLUS Microplate Spectrophotometer (Molecular Devices, Sunnyvale, CA). Inhibitors were added to the assays at levels indicated in the figures. GTBLase activity was measured by preparing a 28 mM solution of GTBL in Dulbecco's PBS. Just prior to use, a fresh solution of DTNB (10.3 mM) was prepared and added as above. The substrate and DTNB solution (200 μL) were added immediately to enzyme samples (5–20 μL) in 96-well microplates. The assay was monitored at 405 nm for 4 min at 37°C. Activities are expressed in Units/mL or Units/mg protein, based on the molar extinction coefficient of 14.15 $mmol/L^{-1}cm^{-1}$ for the hydrolysis of DTNB to 2-nitro-5-thiobenzoic acid (NTB) [33]. Only the initial linear rates of hydrolysis were used for calculations. Clopidogrel thiolactone, prasugrel thiolactone and the exo-thiol metabolites were a generous gift of Eli Lilly. Unless otherwise noted, all other reagents were obtained from Sigma-Aldrich (St. Louis, MO).

Albumin and DTNB

To examine the basis of reported albumin interference of the Ellman assay [21], we monitored the rate of appearance of NTB in the absence of added substrate. Both purified and recombinant serum albumin exhibited a rate of appearance of NTB, characteristic of the enzymatic hydrolysis of thiolactones and thio-esters (Figure 1). Thus, it is important to be cognizant of this reaction when measuring rates of formation of free sulfhydryl groups in samples containing albumin or other proteins with free sulfhydryl groups. The assay of albumin with DTNB alone has apparent activity similar to that observed with PON1 in the presence of HCTL, so controls must be performed when using DTNB and serum or plasma samples to correct for the "albumin interference". A concentration of 10 mM EDTA did not affect the rates of the DTNB/serum assay.

Purification of human liver HCTLase

Liver tissue was obtained from the Human Liver Bank maintained within the School of Pharmacy at the University of Washington [34]. De-identified samples were obtained from deceased, anonymous individuals with approval of the Institutional Review Board (IRB) at the University of Washington. A 33 g

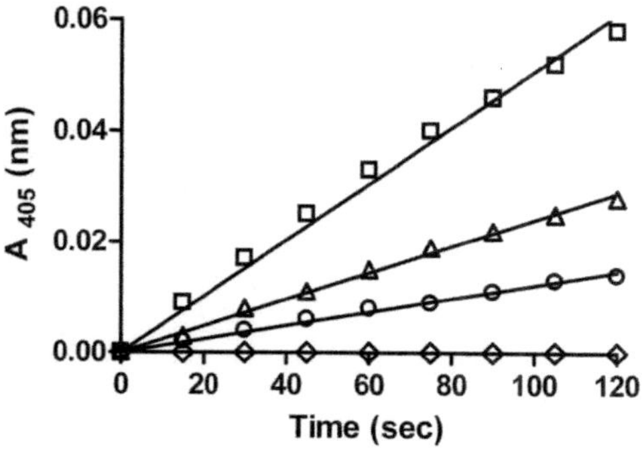

Figure 1. Reaction of human serum albumin (HSA) with Ellman's reagent (DTNB). Recombinant HSA (500 µg from *Pichia pastoris*) □_□; human plasma (10 µL) Δ_Δ; antibody purified HSA (200 µg) ○_○; no protein control ◇_◇.

human liver sample (pooled from five individual donors) was homogenized in 76 mL of 20 mM Tris-HCl (pH 8.0) buffer on ice using a Polytron tissue homogenizer. The sample was centrifuged at 17,000 × g in a Sorvall RC5B centrifuge for 40 min at 4°C. The supernatant was centrifuged at 424,000 × g in a tabletop ultracentrifuge (TL-100, Beckman Coulter, Brea, CA) for 30 minutes at 4°C, using a fixed-angle rotor (TLA 100.3, Beckman Coulter).

The soluble supernatant fraction (58 mL) was loaded onto a 25 mL DEAE (diethylaminoethyl) Sepharose Fast Flow (GE Healthcare, Piscataway, NJ) column (2.8 × 5.5 cm) equilibrated with 20 mM Tris-HCl, pH 8.0. The HCTLase activity did not bind to the column. The flow-through fractions with HCTLase activity were adjusted to pH 7.0 with 0.1N HCl, then loaded onto a 20 mL column of CM (carboxymethyl) Sepharose Fast Flow (1.8 × 8 cm) (GE Healthcare) that had been equilibrated with 20 mM Tris-HCl, pH 7.0 (CM buffer). The column was washed with 6 resin bed volumes (rbv's) of CM buffer, then eluted with 20 rbv's of a 0–600 mM NaCl linear gradient in CM buffer. Fractions with HCTLase activity were pooled and loaded onto a 2 mL ceramic hydroxyapatite (HA) column (1 × 4 cm, HA type II 40 µm, BioRad, Hercules, CA) that was equilibrated with 5 mM KPO_4, pH 6.9. The HA column was washed with equilibration buffer until the A_{280} reached baseline level, then eluted with a 25 rbv gradient from 5–200 mM KPO_4, pH 6.9. HCTLase-containing fractions were pooled and concentrated to 0.5 mL with a 10,000 molecular weight cut off (MWCO) Amicon Ultra centrifugal filter (Millipore, Billerica, MA) at 1,500 × g, loaded onto a 25 mL Superdex 200 (GE Healthcare) (1.2 × 32 cm) column equilibrated with 20 mM Tris-HCl (pH 7.0), 100 mM NaCl (equilibration buffer) and eluted with the same buffer. HCTLase-containing fractions were pooled, diluted 1:2 with 20 mM Tris-HCl, pH 7.0 and loaded onto a 0.5 mL CM Sepharose column (0.5 × 3 cm) equilibrated with CM buffer. The column was eluted with a gradient from 0–600 mM NaCl in 20 rbv's of CM buffer. The fractions containing HCTLase activity were pooled and concentrated to 0.2 mL on a 10,000 MWCO Amicon Ultra concentrator at 1,500 × g, then loaded onto a 5 mL Superdex 200 column (0.8 × 22 cm) in 20 mM Tris-HCl, pH 8.0. Fractions containing HCTLase activity were pooled and stored at 4°C. Protein concentrations were measured with the BCA Protein Assay (Thermo Fisher Scientific Inc., Waltham, MA).

In-gel activity and Coomassie Blue staining

Human liver purified BPHL and rBPHL (described below) were in-gel activity stained using an adapted protocol from Korenovsky et al. [35]. Both enzymes were first buffer exchanged into deionized water using Zeba spin desalting columns (Pierce Chemical, Rockford, IL) and then surface loaded on a 4.5 × 8 cm native IsoGel agarose isoelectric focusing (IEF) gel with a pH range of 6–10.5 (Lonza, Walkersville, MA). The anode solution used was 10 mM acetic acid (Thermo Fisher Scientific), and Cathode Fluid 10 was used as the cathode solution (SERVA Electrophoresis GmbH, Heidelberg, Germany). The IEF gel was run at 4°C with the voltage increased up to 60 V/cm over a period of 50 min (250 V maximum). Lysed human red blood cells provided a visible hemoglobin marker to ensure that the focusing of the proteins was completed. Following electrophoresis, half of the gel was used for HCTLase activity stain and the other half for Coomassie Blue stain. For the activity stain, the gel was washed with deionized water for 5 min, then equilibrated with reaction buffer [0.1 M sodium acetate, 20 mM glycine (Sigma-Aldrich), 4 mM cupric sulfate, 24% (w/v) sodium sulfate and 3 mM magnesium chloride (remaining reagents were from J.T. Baker Chemicals-Avantor Performance Materials, Center Valley, PA)] without substrate for 2 min. Then, the gel was incubated with 3.8 mM HCTL (Sigma-Aldrich) in reaction buffer with shaking for 1 h at room temperature. Following incubation with substrate, the gel was washed in deionized water for 5 min followed by a 3 M ammonium sulfate (Sigma-Aldrich) wash for 5 min. The gel was developed with 0.16% w/v dithio-oxamide in deionized water (Fluka, Sigma-Aldrich) until appearance of brown activity bands of copper- thio-oxamide was observed For Coomassie Blue staining, the gel was fixed for 15 min in a solution of 36% (v/v) methanol, 6% (w/v) trichloroacetic acid and 3.6% (w/v) sulfosalicylic acid (all from Sigma-Aldrich). After rinsing with deionized water, the gel was stained with Coomassie Blue stain solution [0.1%(w/v) Coomassie brilliant blue (Amresco, Solon, OH), 25% (v/v) ethanol and 9% (v/v) acetic acid (both from Thermo Fisher Scientific)] overnight at room temperature and de-stained in 25% (v/v) ethanol with 9% (v/v) acetic acid for 3 h.

Inhibition assays

Human liver homogenates were incubated with specific inhibitors for BPHL, Blmh or PON1. Briefly, 1 g of frozen human liver was thawed and homogenized on ice using a Tissue-Tearor hand-held homogenizer (Cole-Parmer, Vernon Hills, IL) in 1.5 mL of 20 mM Tris pH 8.0 supplemented with 1 mM $CaCl_2$ (to preserve PON1 activity). Liver homogenates were centrifuged as described above. The obtained supernatants were incubated for 10 min with BPHL inhibitors valacyclovir (20 mM) or L-proline benzyl ester (10 mM); Blmh inhibitors E-64 (50 µM) or iodoacetamide (2 mM); and PON1 inhibitors EDTA (10 mM) or 2-hydroxyquinoline (200 µM). Then, HCTL hydrolysis was monitored as described above but using 10 mM HCTL as substrate.

Cloning of human liver BPHL

A clone of human liver BPHL (BC106901) was obtained from Source Bioscience (Nottingham, UK). The 770 bp BPHL gene was amplified with primers containing NdeI restriction sites [(forward) CAG CAG CCA TGG GCA TGC CCA GGA ATC TGC TT] and NcoI [CAG CAG CAT ATG TCA TTG TAG GAA GTC TTC TGC (reverse)]. The resulting PCR product was cloned using the Invitrogen/Life Sciences TOPO TA Cloning Dual Promoter Kit, following the manufacturer's protocol. A pET15 expression vector (EMD Millipore, Billerica, MA) and the

TOPO-BPHL clone were digested with NdeI and NcoI, then run on a 1.2% double tier FlashGel DNA Cassette (Lonza). The digested pET15 and the BPHL DNA fragment were eluted according to the manufacturer's protocol. Following DNA elution, they were ligated with T4 Ligase (NEB, Ipswich, MA), then transformed into XL10-Gold competent subcloning cells (Agilent, Santa Clara, CA). The pET15-BPHL insert was sequenced to verify the fidelity of the sub-cloning protocol. The sequence-verified construct was transformed into Rosetta-Gami 2(DE3) expression cells (EMD Millipore).

Expression of rBPHL

Two 25°C-overnight cultures (600 mL and 800 mL) of Rosetta-Gami 2(DE3) cells transformed with the pET15-BPHL were inoculated respectively into two bubble jars containing 6 L and 8 L of LB media at 25°C with carbenicillin at 50 µg/mL. Cells were induced at an A_{600} of 0.6 with 1 mM IPTG and grown for an additional 12 h. Cells were harvested by centrifugation at 4,800 × g for 20 min at 4°C. Aliquots (0.1 g) of induced and uninduced cells were solubilized with BPER II Protein Extraction Reagent (Pierce Chemical) for analysis of HCTLase activity. The remaining induced cells (19 g) were frozen at -20°C.

Purification of rBPHL

Frozen, induced *E. coli* cells (9.6 g) were thawed on ice, then lysed and extracted with 2 volumes (w/v) of B-PER II containing 20 Units/mL of Benzonase (EMD Millipore) and Protease Inhibitor Cocktail Set III, EDTA-Free (EMD Millipore) diluted 1:200. The lysed, resuspended cells were gently mixed and incubated at room temperature for 30 min. Insoluble particulate was removed by centrifugation at 12,000 × g for 20 min at 4°C. The soluble supernatant fraction was loaded onto a 10 mL DEAE Sepharose Fast Flow column (1.5 × 6 cm) equilibrated with 20 mM Tris-HCl (pH 8.0) buffer. The HCTLase activity did not bind to the DEAE resin. The pool of the flow-through fractions containing HCTLase activity was adjusted to pH 6.9 with 0.1 N HCl and loaded onto a 10 mL ceramic HA column (1.2 × 8 cm) then washed with 5 mM KPO_4 (pH 6.9) buffer, and eluted with a 20 rbv 5–225 mM KPO_4 gradient. The active fractions were pooled and concentrated to 1.0 mL with a 10,000 MWCO Amicon Ultra centrifugal filter at 1,500 × g, then loaded onto a 75 mL Superdex 200 (1.8 × 45 cm) column equilibrated with 20 mM Tris-HCl (pH 8.0), 100 mM NaCl and eluted with the same buffer. Fractions containing HCTLase activity were pooled for gel analysis.

Mass spectrometric analysis of the purified proteins

The purified liver BPHL and rBPHL were analyzed by high-resolution MS. Five µg of the purified hepatic protein were loaded to a denaturing NuPAGE 4–12% Bis-Tris gel (Life Technologies, Grand Island, NY). After staining the gel with Imperial Protein Stain Solution (Thermo Fisher Scientific), three bands were excised for protein sequencing by MS. The excised bands were reduced in-gel with 10 mM dithiothreitol (DTT) and alkylated with 55 mM iodoacetamide (IAA). The carbamidomethylation step was followed by in-gel digestion with 2 µg of porcine trypsin (Promega, Madison, WI), with shaking at 37°C overnight. The generated tryptic peptides were extracted with 5% formic acid (v/v): acetonitrile (1:2), evaporated to near dryness (CentriVap vacuum concentrator, Labconco SpeedVac, Kansas City, MO) and reconstituted in 20 µL of 2% acetonitrile, 0.1% formic acid.

Five µg of the purified rBPHL were digested in solution. The surfactant RapiGest SF (Waters Corporation, Milford, MA) was added to a final concentration of 0.1% and the sample was reduced with 5 mM DTT and alkylated with 15 mM IAA. Then, the samples were digested with 1 µg of porcine trypsin, shaking at 37°C for 2 h. After the digestion, hydrochloric acid was added to a final concentration of 100 mM to hydrolyze the surfactant. The white precipitate formed was separated by centrifugation.

Both extracted peptides and peptides obtained by in-solution digestion were separated by nanoflow chromatography in a Waters nanoACQUITY UltraPerformance liquid chromatography (UPLC) system (Waters Corporation). A 20 cm, 75 µm fused I.D. silica capillary column was used, with a 5 µm I.D. tip pulled using a P-2000 CO_2 laser puller (Sutter Instrument Company, Novato, CA). The column was packed in-house to the stated length with Jupiter 4 µm Proteo 90Å C12 reversed-phase resin (Phenomenex, Torrance, CA). The peptides were eluted using a 50 min gradient of 2–32% acetonitrile gradient in 0.1% formic acid, at 250 nL/min, and analyzed by an LTQ-Orbitrap (Thermo Fisher Scientific). The SEQUEST algorithm [36] was used for database searches of the tandem MS output, with a 57.021464 Da static carbamidomethyl modification added to cysteine residues.

Mass spectrometric analysis of the products of HCTL hydrolysis

HCTLase (1 or 10 ng; for VC or HCTL, respectively) was pre-incubated in PBS buffer at 37°C and shaken at 70 rpm in a water bath for 2 min prior to addition of substrate (VC or HCTL at 100 µM final concentrations) in a 200 µL reaction volume. Reactions were quenched at time points of t = 0 and t = 5 min with the addition of 20 µL of a 15% aqueous $ZnSO_4$ solution on ice, then centrifuged to remove protein and buffer salts. Supernatants were analyzed by LC-MS/MS.

Reaction products were analyzed with a Micromass Quattro Micro API, Tandem Quadrupole Mass Spectrometer (Waters Corporation) equipped with a Shimadzu HPLC system, consisting of two LC-10AD pumps, an SCL-10Avp controller and an SIL-10ADvp autosampler (Shimadzu Scientific Instruments, Inc., Columbia, MD). The MS was run in positive electrospray ionization mode at a source temperature of 120°C and a desolvation temperature of 330°C. The cone voltage was set at 20 volts and the collision energy was set to 20 eV. The following mass transitions were monitored in separate ion channels: *m/z* 117 > 89 (for HCTL) and *m/z* 136 > 89 (for Hcy) for the thiolactonase assay, and *m/z* 326 > 152 (for VC) and *m/z* 226 > 152 (for acyclovir) for the VC assay. Hcy and HCTL were separated on a Zorbax CN 4.6 × 250mm HPLC column (Dupont Instruments, Cincinnati, OH) using an isocratic mix of 10% methanol and 90% aqueous (0.05% formic acid) with a flow rate of 0.7 mL/min. Acyclovir and VC were separated using the same column, solvents and flow rate, but with an isocratic mix of 20% methanol and 80% aqueous (0.05% formic acid).

Results

Purification of human liver HCTLase

Purification steps for human liver HCTLase are shown in Figure 2 A–E (DEAE column trace not shown) and summarized in Table 1. Human liver HCTLase was purified by >100-fold with 5 chromatographic column steps. Three major bands were observed following the final Superdex 200 column. The three bands (Figure 3A, lane 8) were excised for analysis and digested in-gel with trypsin. The extracted peptides were analyzed by liquid chromatography coupled with tandem MS (LC-MS/MS). The peptide mass spectra generated from each band were searched against a human International Protein Index (IPI) peptide database (3/25/09). The higher molecular weight band

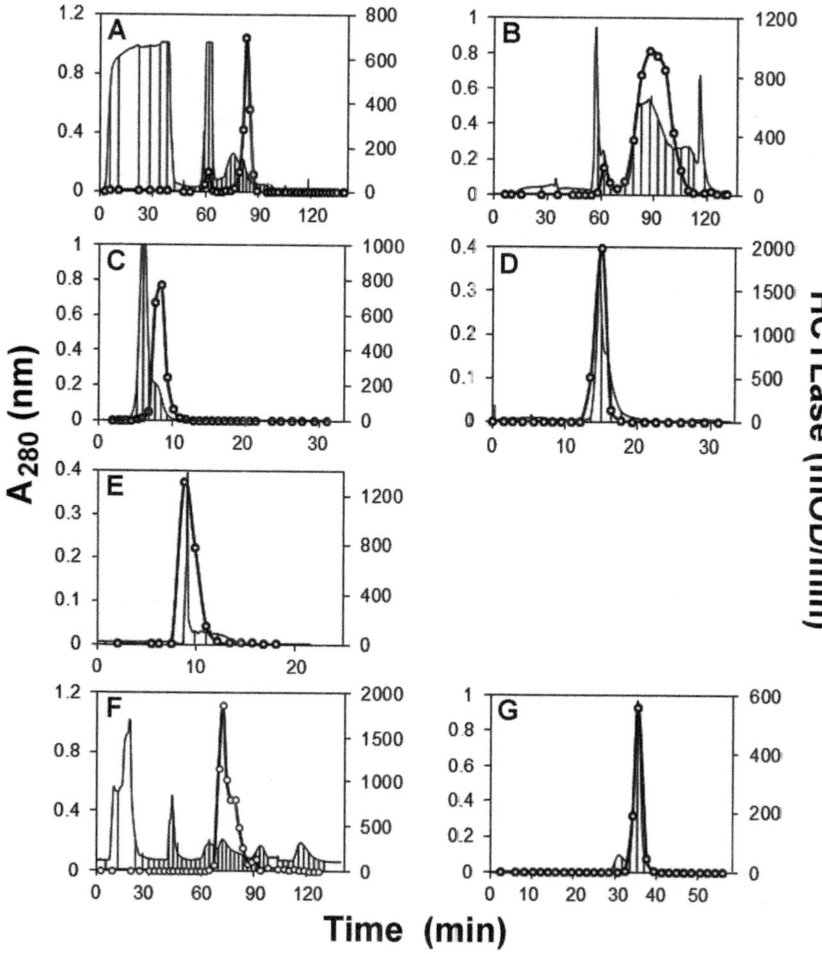

Figure 2. Protein absorbance (A_{280}) and HCTLase activity of column fractions from the chromatographic purification of human liver HCTLase (A–E) and recombinant HCTLase (F, G). A_{280} nm, solid lines; HCTL activity, solid lines with open circles. (A) Carboxymethyl (CM) column 1, (B) ceramic hydroxyapatite (HA) column, (C) Superdex 200 column 1, (D) CM column 2, (E) Superdex 200 column 2; (F) HA column, (G) Superdex 200 column. Ticks represent fraction changes.

(≈44 kDa) was identified as acetyl-Coenzyme A acyltransferase 1, with a sequence coverage of 55.2%. The best match for the band at ≈34 kDa was hydroxyacyl-Coenzyme A dehydrogenase, with a 56.4% sequence coverage. The 30 kDa molecular weight protein was identified as BPHL, with sequence coverage of 56.6%. BPHL has a basic isoelectric point of 9.4 [37], consistent with the lack of binding to the anion exchange resin (DEAE Sepharose) and its binding to the cation exchange resin (CM Sepharose). Human liver HCTLase was ≈30 kDa based on elution from the calibrated gel filtration column (Superdex 200). In addition, the activity stain and Coomassie Blue IEF gels confirmed that only the band corresponding to BPHL showed HCTLase activity (Figure 4).

Table 1. Purification of human liver HCTLase.

Purification Step	Volume, mL	Total Protein, mg	Total HCTLase activity, Units*	Specific activity, Units/mg	Fold Purification	Yield, %
Liver extract	58.0	435.0	118.3	0.27	–	100
DEAE	58.0	336.4	93.1	0.28	1.0	77.7
CM I	32.0	13.4	73.1	5.50	20.0	62.0
HA	8.0	8.3	48.5	5.80	21.4	41.0
Superdex I	3.5	1.5	41.7	27.90	102.7	34.7
CM II	2.5	0.8	27.6	34.10	125.5	23.4
Superdex II	2.2	0.7	23.6	34.60	127.0	20.0

DEAE indicates diethylaminoethyl; CM, carboxymethyl; and HA, hydroxyapatite.
* Units of HCTLase activity are μmol NTB produced per minute measured at $\approx K_m$ concentration of substrate with the partially purified liver BPHL.

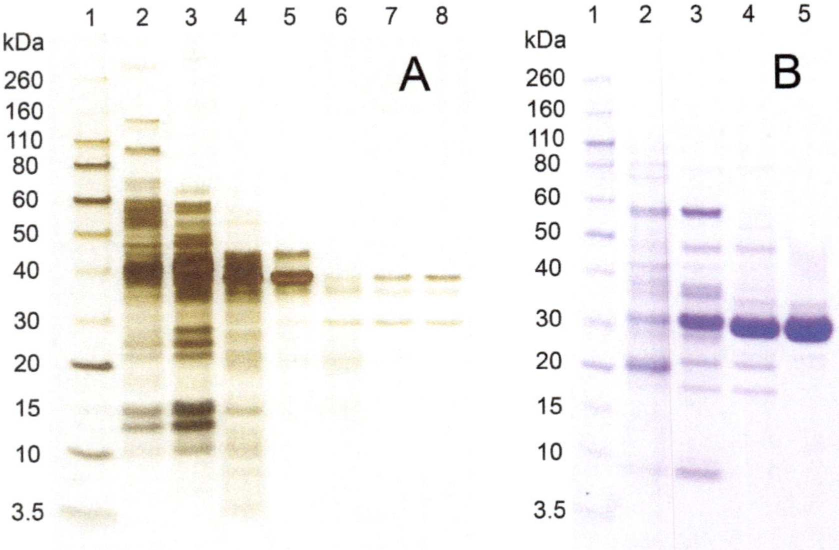

Figure 3. SDS-PAGE analysis of pooled column fractions from (A) human liver and (B) recombinant HCTLase purification. (A) Lane 1, molecular weight markers (kDa); lane 2, liver extract; lane 3, diethylaminoethyl (DEAE) pool; lane 4, CM-1 pool; lane 5, ceramic hydroxyapatite (HA) pool; lane 6, Superdex 200 (#1) pool; lane 7, CM-2 pool; lane 8, Superdex 200 (#2) pool. Gel was stained with silver stain kit (Pierce Chemical). (B) Lane 1, molecular weight markers (kDa); lane 2, *E. coli* extract; lane 3, DEAE pool; lane 4, HA pool; lane 5, Superdex 200 pool. Gel was stained with Imperial Protein Stain (Pierce Chemical).

Characterization of purified human liver BPHL

Human liver BPHL had a specific activity of 34.6 µmol/min/mg when assayed at 3 mM substrate, a sub-saturating concentration (Table 1). The substrate-dependence of HCTL hydrolysis by the partially purified human liver BPHL revealed a K_m value for HCTL hydrolysis of 3.92 mM (Figure 5A). A more detailed characterization of substrate hydrolysis was carried out with the highly purified recombinant BPHL (rBPHL) as described below.

Inhibition of human liver HCTLase activity

To determine whether liver BPHL is more important in detoxifying HCTL than Blmh or PON1, we incubated human liver homogenates with specific inhibitors for BPHL, Blmh and PON1 and assayed the residual HCTLase activity. The results obtained are summarized in Table 2. The BPHL specific inhibitors valacyclovir and L-proline benzyl ester decreased human liver HCTLase activity by 74% and 88%, respectively. In contrast, Blmh specific inhibitor E-64, a cysteine protease inhibitor, and the two PON1 specific inhibitors EDTA and 2-hydroxyquinoline did not decrease human liver HCTLase activity. Interestingly, iodoacetamide, a reported Blmh inhibitor [8], decreased human liver HCTL hydrolysis by ~50%. We confirmed that iodoacetamide is a BPHL inhibitor using rBPHL. Following a 10 min incubation of 2 mM iodoacetamide with rBPHL, the HCTLase activity was reduced by 50% (data not shown).

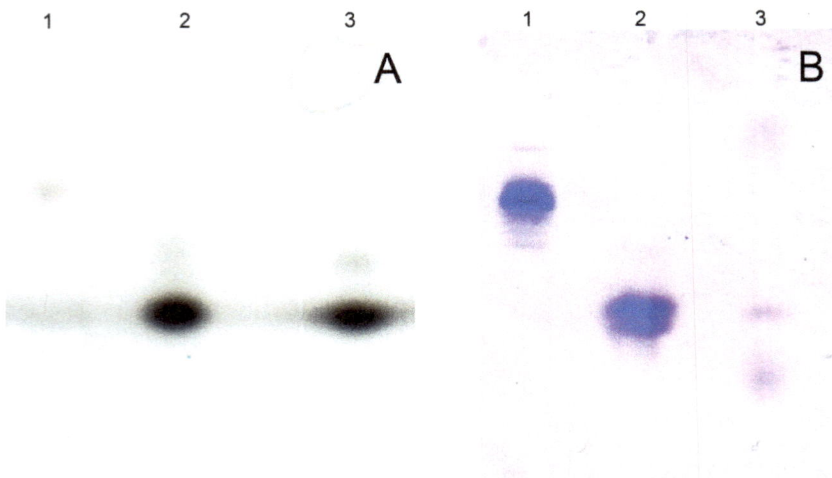

Figure 4. IEF activity and Coomassie stain of liver BPHL and rBPHL. (A) HCTLase activity stain of rBHPL (lane 2) and human liver BPHL (lane 3). Hemoglobin shows a faint band as well (lane 1). (B) Coomassie Blue stain of hemoglobin, rBPHL and human liver BPHL (lanes 1–3, respectively).

Purification of rBPHL

Following identification of BPHL as the most likely candidate responsible for the HCTLase activity, BPHL from a cDNA clone was expressed in *E. coli* and purified as described in the methods section (Figure 2 F-G, DEAE column trace not shown). Highly purified rBPHL, greater than 90% pure as assessed by gel electrophoresis, was obtained after three column chromatographic steps (Table 3). An SDS-PAGE analysis of the purification is shown in Figure 3B. Purified rBPHL had the same molecular weight as human BPHL. To confirm the identity of the purified recombinant protein, it was digested with trypsin and the peptides analyzed by LC-MS/MS. The obtained spectra were searched against the peptide database described above. Human BPHL was the most abundant protein, with a 51.9% sequence coverage.

Characterization of rBPHL activity

Purified rBPHL has high HCTLase activity, with a specific activity of 230 µmol/min/mg when assayed at 3 mM substrate (sub-saturating concentration) (Table 3). The K_m value for HCTL is 3.18 mM, with a V_{max} of 454 µmol/min/mg (Figure 5B), a turnover number of 246.24 s^{-1} (k_{cat}) and a catalytic efficiency of 7.7×10^4 M^{-1} s^{-1} (k_{cat}/K_m). The pH optimum of rBPHL was pH 8 (Figure 5C). The ability of rBPHL to hydrolyze HCTL and VC was confirmed by MS (Figure 6 A–D). The BPHL hydrolytic products of HCTL and VC showed the expected reaction products Hcy and acyclovir (Figure 6).

It was surprising to find that GTBL, an alternate substrate used interchangeably for the HCTLase activity of PON1 [21], was not a substrate for BPHL (Table 4). In addition, EDTA, which irreversibly inhibits human PON1, had no effect on BPHL activity at a concentration of 10 mM. Other BPHL substrates, VC and L-proline benzyl ester, the latter reported to be the best substrate for human valacyclovirase [30], inhibited the HCTLase activity of BPHL (Figure 7). Since rBPHL has thiolactonase activity, other PON1 lactone substrates were tested as substrates for rBPHL. Neither clopidogrel or prasugrel thiolactones nor the PON1 lactone substrates dihydrocoumarin and 2-coumaranone were substrates for rBPHL. The substrate specificity for BPHL is shown in Table 4.

Discussion

The link between homocysteinemia and atherosclerosis was established by McCully in 1969 [38]. A number of earlier related studies on hyperhomocysteinemia and hyperhomocysteineuria are cited in the McCully 1969 paper [38] and discussed together with more recent research in excellent reviews [14,39]. Several biochemical mechanisms appear to contribute to disease risk associated with hyperhomocysteinemia. As pointed out in the recent review by Joseph et al. [14], the relationship of hyperhomocysteinemia and cardiovascular disease is very complex and that the aim of their review was to "*...allow researchers to deconstruct this complex field into separate areas that, when*

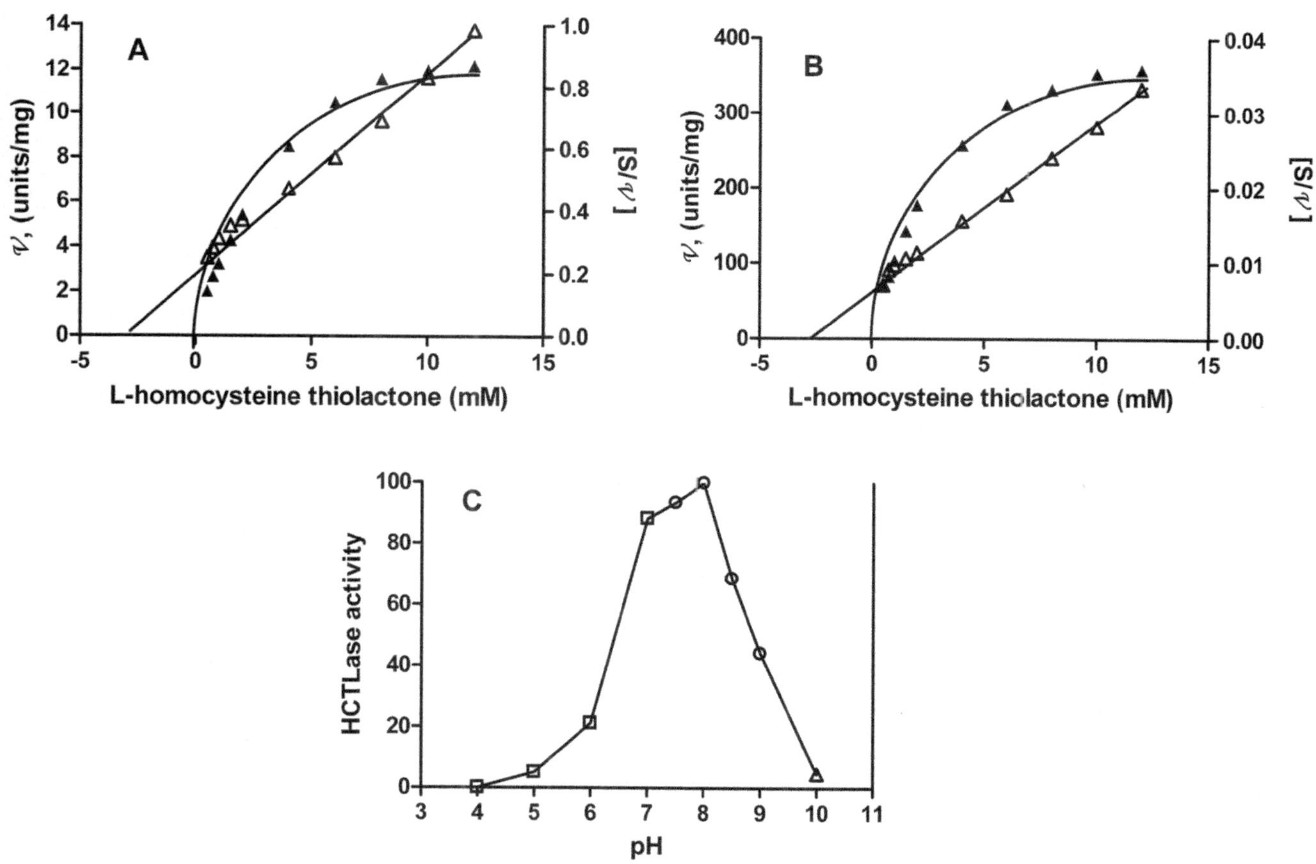

Figure 5. Substrate dependence and pH optimum of HCTL hydrolysis by BPHL. Substrate dependence of HCTL hydrolysis by (A) purified human liver HCTLase (BPHL) and (B) rBPHL. Closed symbols, velocity (v) vs substrate concentration [S]; open symbols, [S]/v vs [S]. K_m for HCTL hydrolysis by purified liver BPHL = 3.92 mM and for rBPHL K_m = 3.18 mM. (C) HCTLase activity of rBPHL (% maximal activity) as a function of pH. Open squares, □_□ 50 mM citrate/Na_2HPO_4; open circles ○_○, 50 mM Tris-HCl; open triangle, △_△ 50 mM Na_2PO_4.

Table 2. Inhibition of human liver HCTL activity.

Inhibitor name	% HCTL activity*
No inhibitor	100
BPHL inhibitors	
Valacyclovir (20 mM)	25.9
L-Proline benzyl ester (10 mM)	12.0
Blmh inhibitors	
E-64 (50 μM)	97.5
Iodoacetamide (2 mM)	57.9 †
PON1 inhibitors	
EDTA (10 mM)	141.5
2-hydroxyquinoline (200 μM)	107.0

* HCTLase activity was measured at 10 mM concentration of substrate.
† Iodoacetamide inhibited rBPHL to the same extent.

addressed adequately, may lead to findings that elucidate the overall link between hyperhomocysteinemia and cardiovascular disease and allow the design of appropriate clinical trials..." This report describes the characterization of the highly efficient HCTLase activity of BPHL, a protein previously known only for its role in drug metabolism [28,29] and provides a more complete understanding of the detoxication pathways for HCTL.

One risk associated with elevated levels of plasma Hcy is related to the formation of disulfide bonds with free cysteine residues on proteins. For example, disulfide bond derivatization of annexin A2 complex by Hcy prevents tissue plasminogen activator binding to A2, resulting in decreased fibrinolytic activity and inhibition of angiogenesis [40]. A second mechanism of protein modification involves the covalent addition of Hcy to the ε-amino groups of lysine residues by HCTL, which can cause protein misfolding, aggregation, loss of function and generation of autoimmune responses [4,11]. A number of experiments have examined the question of whether Hcy or HCTL is the more toxic species. Studies with yeast strains with mutations in yeast Blmh demonstrated increased sensitivity to HCTL [8] and studies with *Blmh*$^{-/-}$ mice also demonstrated increased sensitivity to HCTL [10]. Thus, detoxication of HCTL is an important component in understanding the relationship of hyperhomocysteinemia to vascular and other diseases.

The cardioprotective HCTLase activity had for a number of years been attributed to PON1 [16], then later to Blmh [8]. However, the data reported here strongly suggest that the previously unrecognized HCTLase activity of BPHL is likely to be physiologically more relevant than either PON1 or Blmh for hydrolyzing HCTL and reducing the risk of vascular and other HCTL-associated diseases.

In this regard, it is informative to compare the total HCTL detoxication capacity of plasma PON1 (K_m, 23 mM) with that of liver HCTLase (K_m, 3.18 mM). Since many of the HCTLase activities reported for plasma PON1 were carried out with labeled 1 mM HCTL, it makes sense to compare activities at this concentration of substrate. Rates for lower HCTL concentrations are easily extrapolated since 1 mM is below the K_m value for both proteins. The following extrapolations are from data reported by Jakubowski [16] and from this study. Assuming an average individual has ≈3 L of plasma with ≈0.0057 μmol HCTL hydrolyzed/min/mL plasma (at 1 mM HCTL) [16] the HCTL detoxication capacity for plasma PON1 of an individual would be ≈17.1 μmol/min/individual. Based on the data from Table 1, 33 g of human liver yielded 118.3 Units of HCTLase (measured at 3 mM HCTL; Units = μmol hydrolyzed/min), which would equate to ≈1.195 U/g liver tissue at 1 mM HCTL. Based on an average weight of a human liver of 1,414 g, [41] the total HCTLase Units in a liver would be ≈1,690 μmol/min/liver or

Table 3. Purification of recombinant human BPHL (HCTLase).

Purification Step	Volume, mL	Total Protein, mg	Total HCTLase activity, Units*	Specific activity, Units/mg	Fold Purification	Yield, %
Bacterial extract	18.0	410.0	3802	10.4	-	100
DEAE	50.0	64.5	1284	19.9	1.9	34
HA	30.0	10.1	911	90.1	8.7	24
Superdex	6.5	3.0	857	230.4	22.1	23

DEAE indicates diethylaminoethyl; and HA, hydroxyapatite.
* Units of HCTLase activity are μmol NTB produced per minute measured at ≈K_m concentration of substrate.

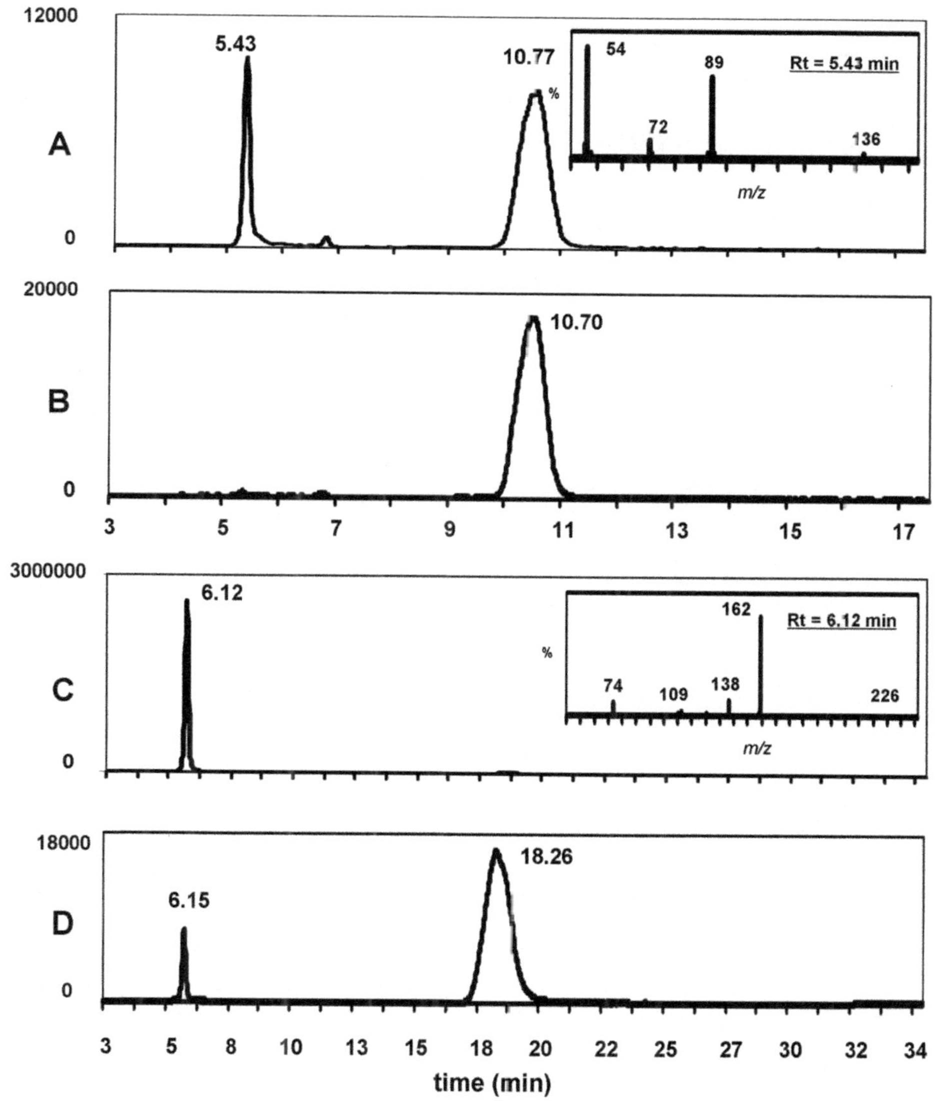

Figure 6. LC-MS chromatograms showing rBPHL-mediated HCTL cleavage at (A) t = 5 min and (B) t = 0 of reaction time, and valacyclovir ester cleavage at (C) t = 5 min and (D) t = 0. Chromatograms A and B represent the sum of two MRM channels monitoring the following transitions: *m/z* 117 > 89 (for HCTL, Rt = 10.7 min) and *m/z* 136 > 89 (for Hcy, Rt = 5.4 min). The inset shows the daughter ion spectrum (for *m/z* 136) of the product peak at Rt = 5.4 min, and is consistent with the spectrum obtained from commercial Hcy. Chromatograms C and D also represent the sum of two MRM channels monitoring the following transitions: *m/z* 326 > 152 (for VC, Rt = 18.3 min) and *m/z* 226 > 152 (for acyclovir, Rt = 6.1 min). The inset shows the daughter ion spectrum (for *m/z* 226) of the product peak at Rt = 6.1 min, and is consistent with the spectrum obtained from commercial acyclovir. The small product peak at t = 0 in chromatogram D reflects the extremely rapid metabolism of VC by rBPHL, which generated a detectable amount of acyclovir in even the very short period of time (~10 sec) between initiation and quenching of the enzymatic reaction.

98.8-times the detoxication capacity of plasma PON1. This approximation does not take into account the level of liver PON1 (which in mice represents a small percentage of total PON1) or BPHL levels in kidney and other tissues [26].

It is also useful to compare the catalytic efficiencies of HCTL hydrolysis by Blmh (1×10^3 M^{-1} s^{-1}) [8] and PON1 with that of BPHL (7.7×10^4 M^{-1} s^{-1}) (this study). Zimny and coworkers reported that the catalytic efficiency of HCTL hydrolysis by Blmh was ≈100-fold higher than that of PON1 [8] indicating that the catalytic efficiency of HCTL hydrolysis by BPHL is almost four orders of magnitude higher than that of PON1 and nearly two orders of magnitude higher than that of Blmh.

Earlier Northern blot expression analyses had shown high expression levels of BPHL in liver and kidney with lower levels in intestine, heart and skeletal muscle [26]. The primary band observed from liver tissue was 1.8 kb with both the 1.8 kb band and a 2.4 kb band visible from kidney tissue. Interestingly, a different expression pattern was observed for Blmh, where low to moderate levels of expression were observed in spleen, prostate, ovary, small intestine, heart, brain, placenta and lung. Elevated levels of expression of Blmh mRNA were observed in testis, skeletal muscle and pancreas with very low expression levels observed in liver, kidney, colon and peripheral blood leukocytes [23]. Thus, tissue expression patterns of BPHL and Blmh are somewhat complementary. In addition, in this study we have shown that the majority of liver HCTLase activity is attributable to BPHL (Table 2).

Table 4. BPHL substrates.

Compound	Specific Activity, Units/mg
L-homocysteine thiolactone	454.0 * †
γ−thiobutyrolactone	<0.001 †
Dihydrocoumarin	<0.001 †
2-coumaranone	<0.001 †
Phenyl acetate	<0.001 †
DL-N-acetyl homocysteine thiolactone	<0.001 †
Clopidogrel thiolactone	<0.02 †
Prasugrel thiolactone	<0.02 †
Valacylovir	103.7 ‡
L-Phe benzyl ester	358.3 ‡
L-Proline benzyl ester	5148.4 ‡
3-Amino-3-phenylpropionic acid ethyl ester	0.6 ‡
L-Ala benzyl ester	1367.7 ‡
β-Ala benzyl ester	1.8 ‡
L-Gly benzyl ester	849.4 ‡
L-Leu benzyl ester	635.0 ‡
L-Val benzyl ester	156.3 ‡
L-Met methyl ester	156.2 ‡
L-Phe benzyl ester	358.3 ‡
L-Tyr benzyl ester	41.4 ‡
L-Trp methyl ester	35.3 ‡
L-Ser methyl ester	484.3 ‡
L-Phe ethyl ester	75.3 ‡

* V_{max} from Figure 4B.
† from this study.
‡ from [27].

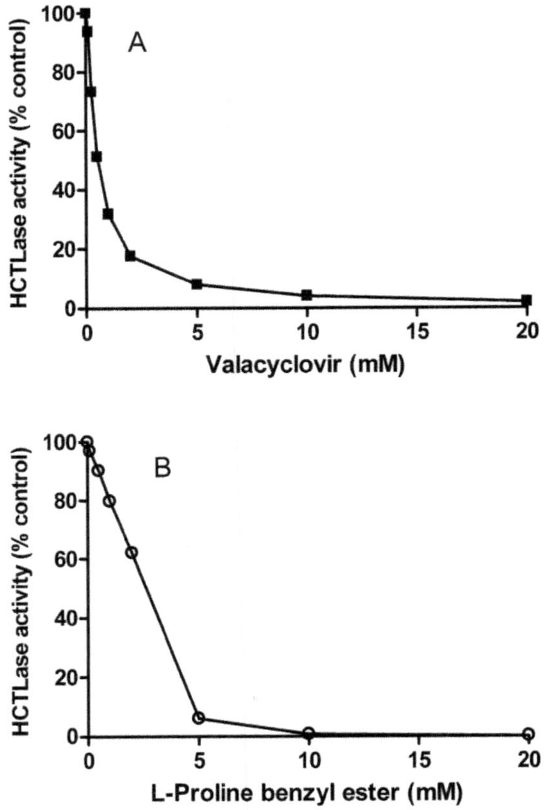

Figure 7. Concentration-dependent inhibition of rBPHL HCTLase by (A) VC and (B) L-proline benzyl ester.

This study began with the aim of identifying the thiolactonase activity responsible for the second metabolic step in the bioactivation of the platelet inhibitor clopidogrel to the active exo-thiol since it was clear that PON1 was not the physiologically relevant activity for this second step of bioactivation [42,43]. It has since been shown that both the first and second steps are carried out by cytochrome(s) P450 [43]. PON1 generates an endo-thiol from clopidogrel thiolactone that is not a pharmacologically active metabolite. In the search for a liver thiolactonase that would hydrolyze the intermediate thiolactone metabolite of clopidogrel, we chose HCTL as a convenient thiolactonase substrate. Following purification of the liver HCTLase activity, gel electrophoretic analysis revealed several protein bands of 30–40 kDa (Figure 3A), all too small to be either PON1 or Blmh. MS analysis identified BPHL as the most likely HCTLase candidate. Expression and purification of rBPHL from *E. coli* verified that BPHL did indeed have very high HCTLase activity. BPHL differs from PON1 and Blmh, previously identified HCTLases, in its substrate specificity and more significantly in its high catalytic efficiency. When human liver homogenates were incubated with specific inhibitors for BPHL, Blmh or PON1, HCTLase activity was significantly reduced following incubation with BPHL inhibitors. This finding demonstrates that BPHL is a more relevant liver HCTLase enzyme than the two previously reported Blmh and PON1. The rBPHL did not detectably hydrolyze the thiolactone metabolites of the drugs clopidogrel or prasugrel (another platelet inhibitor) (Table 4).

It is also interesting to compare the substrate specificities of the three identified HCTL hydrolases. Since BPHL metabolized the thiolactone HCTL, we tested it with several other PON1 substrates. The rBPHL failed to hydrolyze dihydrocoumarin, 2-coumaranone, phenyl acetate and DL-*N*-acetyl homocysteine thiolactone (Table 4). GTBL is often used as a surrogate for measuring HCTL hydrolysis; however, BPHL does not hydrolyze GTBL. Many investigators cite problems of albumin interference with the Ellman assay and degradation of HCTL at alkaline pH as a rationale for using GTBL, which has been used as a lactonase substrate for analysis of PON1 activity [3,5,21,22,44]. GTBL is used in a commercially-available assay expressly for measuring serum HCTLase, using Ellman's reagent. However, BPHL has no GTBLase activity. Blmh, an aminopeptidase, also catalyzes the hydrolysis of HCTL but not the hydrolysis of GTBL. In previous studies, typical esterase (*p*-nitrophenyl acetate, *p*-nitrophenyl butyrate) and peptidase substrates (amides such as Lys-*p*-nitroanilide) were tested as BPHL substrates [29]. BPHL had almost no activity for any of these substrates except for *p*-nitrophenyl butyrate, for which it had only low activity. The results from this study show BPHL to be an efficient HCTLase but not a general lactonase. Studies by Kim et al [28,29], provide additional information on the substrate specificity of BPHL. It is worth noting the structural similarities between HCTL and VC as both possess a free amino group on the carbon alpha to the carbonyl that is neutralized by the enzyme's active site residue D123. Lai et al. described BPHL as a specific alpha-amino acid ester hydrolase [30].

Prior to the studies reported here, a physiological activity of BPHL was unknown [30,31], but due to its localization to the liver and kidney using Northern blots, BPHL was hypothesized to be involved in detoxification and drug metabolism [26]. The discovery of BPHL's role in converting the prodrug VC to acyclovir identified one such pathway [28]. It is interesting that both Blmh [23] and BPHL [28] were first characterized as drug metabolizing enzymes. The role of PON1 in the metabolism of some drugs has also been characterized (reviewed in [45]). Since proteins did not evolve to metabolize drugs, the high HCTLase activity of BPHL reported here provides evidence that BPHL is most likely a primary detoxication pathway for HCTL and its high degree of sequence conservation across species would indicate that this HCTLase activity of BPHL is a universally shared function.

Acknowledgments

The authors thank Dr. Wan-Fen Li for helpful suggestions and editing of the manuscript, and Thomas Bukowski for his help with the native IEF activity stain gels.

Author Contributions

Conceived and designed the experiments: JM SMS RJR MJM AER CEF. Performed the experiments: JM SMS RJR MGM PMR EJH. Analyzed the data: JM SMS RJR MGM PMR EJH AER CEF. Wrote the paper: JM SMS RJR MJM AER CEF.

References

1. Lentz SR (2005) Mechanisms of homocysteine-induced atherothrombosis. J Thromb Haemost 3: 1646–1654.
2. Wald DS, Law M, Morris JK (2002) Homocysteine and cardiovascular disease: evidence on causality from a meta-analysis. BMJ 325: 1202.
3. Barathi S, Angayarkanni N, Pasupathi A, Natarajan SK, Pukraj R, et al. (2010) Homocysteinethiolactone and paraoxonase: novel markers of diabetic retinopathy. Diabetes Care 33: 2031–2037.
4. Yilmaz N (2012) Relationship between paraoxonase and homocysteine: crossroads of oxidative diseases. Arch Med Sci 8: 138–153.
5. Suszynska J, Tisonczyk J, Lee HG, Smith MA, Jakubowski H (2010) Reduced homocysteine-thiolactonase activity in Alzheimer's disease. J Alzheimers Dis 19: 1177–1183.
6. Seshadri S (2006) Elevated plasma homocysteine levels: Risk factor or risk marker for the development of dementia and Alzheimer's disease. Journal of Alzheimer's Disease 9: 393–398.
7. Finkelstein JD (1998) The metabolism of homocysteine: pathways and regulation. Eur J Pediatr 157 Suppl 2: S40–44.
8. Zimny J, Sikora M, Guranowski A, Jakubowski H (2006) Protective mechanisms against homocysteine toxicity: the role of bleomycin hydrolase. J Biol Chem 281: 22485–22492.
9. Borowczyk K, Shih DM, Jakubowski H (2012) Metabolism and neurotoxicity of homocysteine thiolactone in mice: evidence for a protective role of paraoxonase 1. J Alzheimers Dis 30: 225–231.
10. Borowczyk K, Tisonczyk J, Jakubowski H (2012) Metabolism and neurotoxicity of homocysteine thiolactone in mice: protective role of bleomycin hydrolase. Amino Acids 43: 1339–1348.
11. Jakubowski H (1999) Protein homocysteinylation: possible mechanism underlying pathological consequences of elevated homocysteine levels. FASEB J 13: 2277–2283.
12. Jakubowski H (2000) Homocysteine thiolactone: metabolic origin and protein homocysteinylation in humans. J Nutr 130: 377S–381S.
13. Jakubowski H (2002) The determination of homocysteine-thiolactone in biological samples. Anal Biochem 308: 112–119.
14. Joseph J, Handy DE, Loscalzo J (2009) Quo Vadis: Whither Homocysteine Research? Cardiovascular Toxicology 9: 53–63.
15. Billecke S, Draganov D, Counsell R, Stetson P, Watson C, et al. (2000) Human serum paraoxonase (PON1) isozymes Q and R hydrolyze lactones and cyclic carbonate esters. Drug Metab Dispos 28: 1335–1342.
16. Jakubowski H (2000) Calcium-dependent human serum homocysteine thiolactone hydrolase. A protective mechanism against protein N-homocysteinylation. J Biol Chem 275: 3957–3962.
17. Bayrak A, Bayrak T, Demirpence E, Kilinc K (2011) Differential hydrolysis of homocysteine thiolactone by purified human serum (192)Q and R-192 PON1 isoenzymes. Journal of Chromatography B-Analytical Technologies in the Biomedical and Life Sciences 879: 49–55.
18. Costa LG, Cole TB, Jarvik GP, Furlong CE (2003) Functional genomic of the paraoxonase (PON1) polymorphisms: effects on pesticide sensitivity, cardiovascular disease, and drug metabolism. Annu Rev Med 54: 371–392.
19. Li WF, Costa LG, Richter RJ, Hagen T, Shih DM, et al. (2000) Catalytic efficiency determines the in-vivo efficacy of PON1 for detoxifying organophosphorus compounds. Pharmacogenetics 10: 767–779.
20. La Du BN (2000) Historical Perspectives. In: Costa LG, Furlong, C.E., editor. Paraoxonase (PON1) in Health and Disease. Boston/Dordrecht/London: Kluwer Academic Publishers. pp.1–25.
21. Kosaka T, Yamaguchi M, Motomura T, Mizuno K (2005) Investigation of the relationship between atherosclerosis and paraoxonase or homocysteine thiolactonase activity in patients with type 2 diabetes mellitus using a commercially available assay. Clin Chim Acta 359: 156–162.
22. Jakubowski H (2010) The role of paraoxonase 1 in the detoxification of homocysteine thiolactone. Adv Exp Med Biol 660: 113–127.
23. Bromme D, Rossi AB, Smeekens SP, Anderson DC, Payan DG (1996) Human bleomycin hydrolase: molecular cloning, sequencing, functional expression, and enzymatic characterization. Biochemistry 35: 6706–6714.
24. Montoya SE, Aston CE, DeKosky ST, Kamboh MI, Lazo JS, et al. (1998) Bleomycin hydrolase is associated with risk of sporadic Alzheimer's disease. Nat Genet 18: 211–212.
25. Kajiya A, Kaji H, Isobe T, Takeda A (2006) Processing of amyloid beta-peptides by neutral cysteine protease bleomycin hydrolase. Protein Pept Lett 13: 119–123.
26 Puente XS, Lopez-Otin C (1995) Cloning and expression analysis of a novel human serine hydrolase with sequence similarity to prokaryotic enzymes involved in the degradation of aromatic compounds. J Biol Chem 270: 12926–12932.
27. Puente XS, Pendas AM, Lopez-otin C (1998) Structural characterization and chromosomal localization of the gene encoding human biphenyl hydrolase-related protein (BPHL). Genomics 51: 459–462.
28. Kim I, Chu XY, Kim S, Provoda CJ, Lee KD, et al. (2003) Identification of a human valacyclovirase: biphenyl hydrolase-like protein as valacyclovir hydrolase. J Biol Chem 278: 25348–25356.
29. Kim I, Song X, Vig BS, Mittal S, Shin HC, et al. (2004) A novel nucleoside prodrug-activating enzyme: substrate specificity of biphenyl hydrolase-like protein. Mol Pharm 1: 117–127.
30. Lai L, Xu Z, Zhou J, Lee KD, Amidon GL (2008) Molecular basis of prodrug activation by human valacyclovirase, an alpha-amino acid ester hydrolase. J Biol Chem 283: 9318–9327.
31. Long JZ, Cravatt BF (2011) The metabolic serine hydrolases and their functions in mammalian physiology and disease. Chemical Reviews 111: 6022–6063.
32. Ellman GL, Courtney KD, Andres V Jr, Feather-Stone RM (1961) A new and rapid colorimetric determination of acetylcholinesterase activity. Biochem Pharmacol 7: 88–95.
33. Eyer P, Worek F, Kiderlen D, Sinko G, Stuglin A, et al. (2003) Molar absorption coefficients for the reduced Ellman reagent: reassessment. Anal Biochem 312: 224–227.
34. Lin YS, Dowling AL, Quigley SD, Farin FM, Zhang J, et al. (2002) Co-regulation of CYP3A4 and CYP3A5 and contribution to hepatic and intestinal midazolam metabolism. Mol Pharmacol 62: 162–172.
35. Korenovsky A, Laev H, Mukherjee S, Mahadik SP (1990) Quantitative analyses of plasma cholinesterase isozymes in haloperidol-treated rats. Biol Psychiatry 27: 871–883.
36. Eng JK, McCormack AL, Yates JR (1994) An approach to correlate tandem mass spectral data of peptides with amino acid sequences in a protein database. J Am Soc Mass Spectrom 5: 976–989.
37. Burnette TC, Harrington JA, Reardon JE, Merrill BM, de Miranda P (1995) Purification and characterization of a rat liver enzyme that hydrolyzes valaciclovir, the L-valyl ester prodrug of acyclovir. J Biol Chem 270: 15827–15831.
38. McCully KS (1969) Vascular pathology of homocysteinemia: implications for the pathogenesis of arteriosclerosis. Am J Pathol 56: 111–128.
39. McCully KS (2005) Hyperhomocysteinemia and arteriosclerosis: historical perspectives. Clin Chem Lab Med 43: 980–986.
40. Jacovina AT, Deora AB, Ling Q, Broekman MJ, Almeida D, et al. (2009) Homocysteine inhibits neoangiogenesis in mice through blockade of annexin A2-dependent fibrinolysis. J Clin Invest 119: 3384–3394.
41. Molina DK, DiMaio VJ (2012) Normal organ weights in men: part II-the brain, lungs, liver, spleen, and kidneys. Am J Forensic Med Pathol 33: 368–372.
42. Camps J, Joven J, Mackness B, Mackness M, Tawfik D, et al. (2011) Paraoxonase-1 and clopidogrel efficacy. Nat Med 17: 1041–1042; author reply 1042–1044.
43. Dansette PM, Rosi J, Debernardi J, Bertho G, Mansuy D (2012) Metabolic activation of prasugrel: nature of the two competitive pathways resulting in the opening of its thiophene ring. Chem Res Toxicol 25: 1058–1065.
44. Koubaa N, Hammami S, Nakbi A, Ben Hamda K, Mahjoub S, et al. (2008) Relationship between thiolactonase activity and hyperhomocysteinemia according to MTHFR gene polymorphism in Tunisian Behcet's disease patients. Clin Chem Lab Med 46: 187–192.
45. Furlong CE (2008) Paraoxonases: An Historical Perspective. In: Paragh G, editor. The Paraoxonases: Their Role in Disease Development and Xenobiotic Metabolism. Dodrecht: Springer. pp. 3–31.

10

New Insights into Regulation of Proteome and Polysaccharide in Cell Wall of *Elsholtzia splendens* in Response to Copper Stress

Tingting Liu, Chaofeng Shen, Yi Wang, Canke Huang, Jiyan Shi*

Institute of Environmental Science and Technology, College of Environmental and Resource Sciences, Zhejiang University Hangzhou, P.R. China

Abstract

Background and Aims: Copper (Cu) is an essential micronutrient for plants. However, excess amounts of Cu are toxic and result in a wide range of harmful effects on the physiological and biochemical processes of plants. Cell wall has a crucial role in plant defense response to toxic metals. To date, the process of cell wall response to Cu and the detoxification mechanism have not been well documented at the proteomic level.

Methods: An recently developed 6-plex Tandem Mass Tag was used for relative and absolute quantitation methods to achieve a comprehensive understanding of Cu tolerance/detoxification molecular mechanisms in the cell wall. LC–MS/MS approach was performed to analyze the Cu-responsive cell wall proteins and polysaccharides.

Key Results: The majority of the 22 up-regulated proteins were involved in the antioxidant defense pathway, cell wall polysaccharide remodeling, and cell metabolism process. Changes in polysaccharide amount, composition, and distribution could offer more binding sites for Cu ions. The 33 down-regulated proteins were involved in the signal pathway, energy, and protein synthesis.

Conclusions: Based on the abundant changes in proteins and polysaccharides, and their putative functions, a possible protein interaction network can provide new insights into Cu stress response in root cell wall. Cu can facilitate further functional research on target proteins associated with metal response in the cell wall.

Editor: Raffaella Balestrini, Institute for Sustainable Plant Protection, C.N.R., Italy

Funding: The work was supported by the National Natural Science Foundation of China (11179025, 21177109) and Program for New Century Excellent Talents in University (NCET-11-0455). The funders had no role in study design, data collection and analysis, decision to publish, or preparation of the manuscript.

Competing Interests: The authors have declared that no competing interests exist.

* Email: shijiyan@zju.edu.cn

Introduction

Copper (Cu) is an essential micronutrient for plants because it is a structural and catalytic component of several proteins or enzymes involved in various physiological metabolisms [1]. However, excess Cu in plants is harmful; it induces phytotoxicity [2] and disturbs the physiological metabolisms, including carbohydrate metabolism, protein metabolism, mineral nutrition, cell respiration, and photosynthesis [3]. Given its redox properties, Cu^{2+} can catalyze the generation of reactive oxygen species (ROS) and Fenton reactions that damage cellular components or interfere with cellular transport processes [4]. Plants cannot avoid the stress conditions from Cu-contaminated areas. Thus, several resistance mechanisms for Cu tolerance have been developed [5], including exclusion, compartmentalization, chelation, and binding to organic ligands, which enable maintenance of Cu homeostasis for plant survival even in a highly contaminated environment [6].

Plant cell wall is the primary site for signal perception and defense response, which is significant in responding to environmental stresses [7]. The cell walls of plants that grow on heavy metal-contaminated areas provide physical barriers against toxic heavy metals and actively participate in plant defense response. The cell wall serves as a repository for high contents of metals; it senses stress signals and transmits them to the cell interior, thereby affecting the cell fate decision [8]. The main chemical components of a plant cell wall include polysaccharides and proteins. Polysaccharides represent up to 90% of the plant cell wall and constitute three different kinds of polymers, namely, cellulose, hemicelluloses, and pectins. Cell wall proteins (CWPs) of plants comprise less than 10% of the cell wall dry weight but have major roles in cell wall structure, metabolism, cell enlargement, signal transduction, defense responses, and many other physiological events [9].

Certain plants can accumulate high content of metal ions in their cell wall through various compounds. Cell wall polysaccharides possess crucial roles in heavy metal binding and accumulation. Binding of metal ions depends on the number of functional groups in the cell wall polysaccharides. Previous studies confirmed that the essential capacity of the cell wall for binding metal ions depends mainly on the abundance of polysaccharides in the carboxyl groups [10]. Current information about the regulation of cell wall polysaccharides under metal stress remains limited.

Abiotic stress induces expression of stress-responsive proteins, which lead to cellular adaptation in plant growth under stress conditions. Therefore, studies have been carried out to identify various CWPs involved in stress response, particularly those of regulatory or targeting function proteins [11–17]. Most reports have focused on leaf cell wall proteomes under abiotic stress, such as salt in tobacco [11]; wounding in *Medicago* [12]; dehydration in chickpea, rice, and maize [9,14,15]; zinc toxicity in *Hordeum* [18]; nickel toxicity in barley [19]; manganese toxicity in cowpea [20,21]; boron deficiency in *Lupinus albus* [22]; and flooding in soybean [23]. Few studies have focused on the root cell wall proteomes under abiotic stress [24–26]. All of these studies broadened our understanding of the complicated regulation of apoplast proteins. An in-depth proteome study of the cell wall can provide extensive information on the intrinsic mechanism of stress response attributed to a possible relationship between protein abundance and plant stress tolerance [27].

Elsholtzia splendens is a plant that can survive in a highly Cu-polluted area. This species can accumulate more than 1000 mg/kg of Cu in the body [28], and the major proportion (70%) of Cu ions is stored in the root cell wall [29]. Nevertheless, little is known about the defense and detoxification mechanisms in the root cell wall of *E. splendens*. A recently developed 6-plex Tandem Mass Tag (TMT) was used for relative and absolute quantitation methods to gain a comprehensive understanding of Cu tolerance/detoxification molecular mechanisms. LC–MS/MS approach was also conducted to analyze Cu-responsive CWPs and polysaccharides in this study. This proteomic study of the apoplast is the first in response to Cu. The results will significantly expand knowledge on the apoplast complexicity in metal-tolerant plants and develop a more comprehensive understanding of the involved cellular reactions in cell defense mechanisms.

Results

Isolation and purification of CWPs

Root CWPs from *E. splendens* were isolated with sucrose gradients and washed according to the method of Feiz et al. [30]. The activity of G'DH, a typical marker enzyme used to detect contamination of cytoplasmic proteins, was measured to validate the purification method [31]. The G6PDH activity in $CaCl_2$-extracted CWPs, NaCl-extracted CWPs, and cytosolic protein fractions were 8.36, 5.78, and 107.55 mU/mg protein, respectively (Figure 1). The G6PDH activity in $CaCl_2$- or NaCl-extracted CWPs was 10 times lower than that in cytosolic proteins ($p<0.01$). These results indicate that the contamination of intracellular proteins was minimal.

The identified proteins were classified into different gene ontology (GO) terms according to their cellular compartment(s), function(s), and biological processes (Table S1). Some of these identified proteins showed a compartment label (apoplast, cell wall, extracellular region, and plant-type cell wall) that is coherent with secreted proteins. The same proteins can also be attributed to various compartments in the GO classification. Three different computer programs (TargetP, Predotar, and WoLF PSORT) were used to predict the protein subcellular localization (Table S1). A protein was secreted when at least two out of the three algorithms predicted extracellular localization. Based on the results, 15 proteins were predicted to be secreted proteins in the fractions (Table S1).

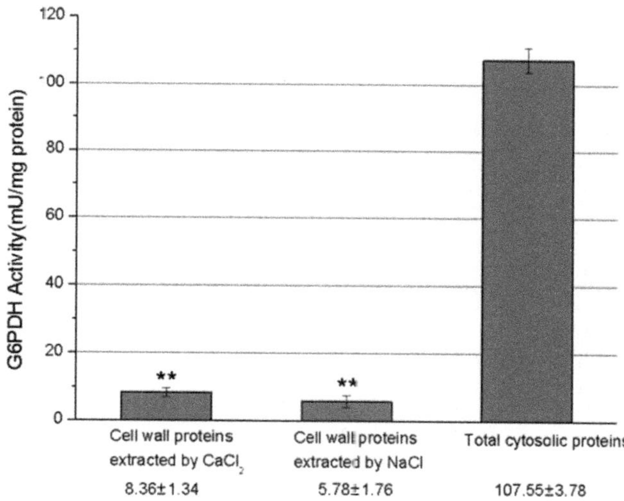

Figure 1. Evaluation of purity by the G6PDH activity in the cell wall proteins isolated from the roots of *Elsholtzia splendens.* The activity of G6PDH was assayed in $CaCl_2$-extracted cell wall proteins, NaCl-extracted cell wall proteins and total soluble proteins. One unit of G6PDH activity is defined as 1 µmol of NADPH turnover per min/mg protein. Results are presented as mean ±SE of G6PDH activity from three biological replicates. The asterisks indicate significant differences in the G6PDH activity of $CaCl_2$-extracted cell wall proteins, NaCl-extracted cell wall proteins compared with that of total soluble cytosolic proteins (**$p<0.01$).

Cu-induced changes in root CWP contents

Plants were grown on a medium supplemented with different $CuSO_4$ concentrations to determine changes in CWP abundance under Cu stress (Figure 2A). Cu toxicity led to a strong increase in CWP contents at low Cu concentrations. The CWP contents increased 1.39-fold at 25 µM Cu and 1.96-fold at 50 µM Cu. The highest content of the CWPs was observed at 50 µM Cu concentrations. The plant root began to decompose at 100 µM Cu (Figure 2C). The increase of Cu contents in the cell wall could proportionally increase with the Cu supply in the medium (slope = 0.3, $R^2 = 0.98$) at Cu concentrations below 75 µM. The Cu content in the root cell wall continued to increase significantly (slope = 0.45, $R^2 = 0.99$) above 75 µM Cu. The highest value of Cu content in the root cell wall reached 1.41 mg/g with the root cell wall dry weight at 100 µM Cu. These findings indicate an increase in Cu binding to cell walls with Cu supply in the medium (Figure 2A). Polypeptide separation by SDS-PAGE (Figure 2B) revealed about 30 polypeptide bands. Band intensities increased with the increase in Cu concentration below 50 µM (Figure S1). The staining intensity was the lowest among all the cell wall polypeptides at 100 µM Cu concentration.

Cu-induced remodeling in root cell wall polysaccharide

Evidence generally suggests that cell wall polysaccharides are crucial sites for Cu retention in plants. Therefore, changes in the contents of root cell wall polysaccharide of *E. splendens* were measured. Treatments with 25, 50, 75, and 100 µM Cu significantly increased the expression of CWPs and polysaccharides compared with controls (Figure 3). Cu supply enhanced the total sugar contents in various cell wall fractions. Sugar contents were higher in hemicellulose than in pectin and cellulose. Sugar contents in pectin increased 1.22-fold with 25 µM Cu. The major components of cell wall fractions are galacturonic acids. The content of galacturonic acids was considerably higher than that in

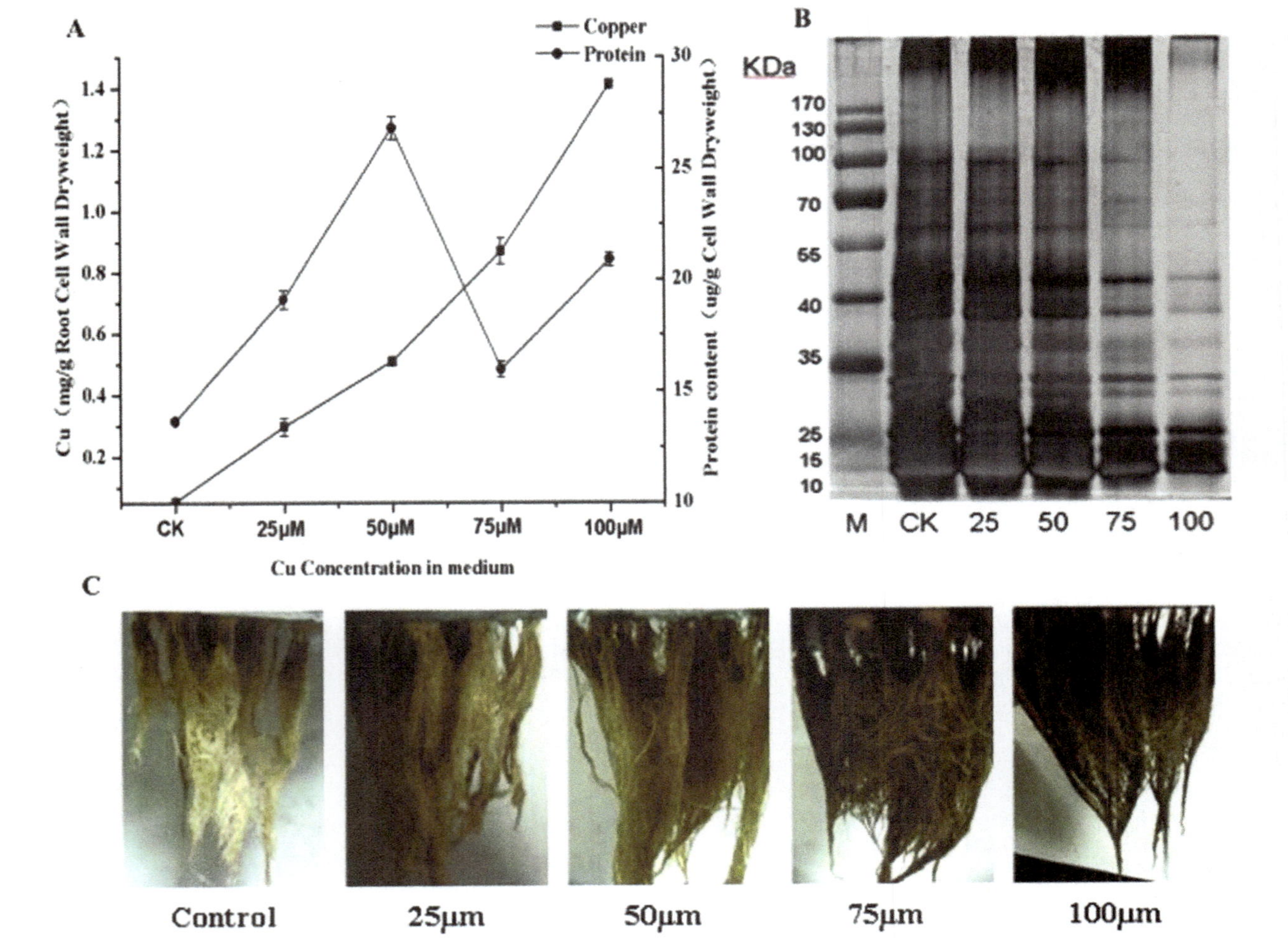

Figure 2. Effect of different copper concentration on the cell wall protein and copper ions in *Elsholtzia splendens*'s root cell. A: The content of copper and protein every root cell wall dry weight. B: SDS-PAGE of root cell wall protein under different copper stress. C: Root growth patterns of control and copper-stressed plants during different copper concentration. Data presented are mean ±SE (n = 20),*Significant mean differences from control at $p = 0.05$ in multiple comparison by LSD test.

the control during Cu treatment, particularly at higher Cu supply. The galacturonic acid values rapidly increased in pectin (1.28-fold), hemicellulose (1.22-fold), and cellulose (1.79-fold) at 100 μM Cu. A conserved disaccharide residue known as 2-keto-3-deoxyoctonic acid (KDO) is attached to C-3 of the backbone of rhamnogalacturonan II in pectin. The KDO concentration in pectin increased during Cu exposure. Thus, the highest concentration of KDO (1.49-fold) was found in the 100 μM Cu-treated group.

The results indicated that excess Cu affected the intracellular distribution of polysaccharides in the root cell wall. Given this reason, we focused on the relationship between wall polysaccharide composition and CWPs with increasing Cu intensity. The dendrogram proved the connection between the contents of polysaccharides and proteins as matched with different Cu concentrations (Figure 4). This finding suggests that Cu induced changes in the contents of root cell wall polysaccharides and proteins, which altered the intracellular distribution patterns in assigning effects on cellular activities.

Cu-responsive CWP identification using TMT LC-MS/MS

This study aimed to gain a better understanding of the metabolic processes and molecular mechanisms involved in Cu-responsive CWPs. A TMT-based shotgun quantitation approach was used to obtain an overall view of the proteome changes associated with Cu stress in the root cell wall of *E. splendens*. Unique proteins were successfully identified based on homology searching with a common protein confidence cutoff of 95% and relative quantitative information in control and Cu-treated samples, as indicated by the peak areas of the different TMT tags (Table 1, Table S2). A threshold of 1.5-fold change and a $p < 0.05$ were set as stringent criteria for significant differences between control and treatment. Fifty five proteins were classified as differentially expressed upon Cu stress using the criteria; 22 and 33 of these proteins showed increasing trends and decreasing abundance, respectively (Table 1, Table 2).

Functional classification of identified CWPs

Database search using Uni-Prot/NCBI accession numbers was conducted to obtain detailed information of differentially expressed proteins. These proteins were categorized into biological process and molecular function using an in-house Perl script according to the extracted GO terms from InterPro (www.ebi.ac.uk/interpro/) or Pfam (pfam.sanger.ac.uk/) (Figure 5). Functional classifications of the 55 differentially expressed proteins (Table 1, Table 2) were classified based on the GO terms that correspond to

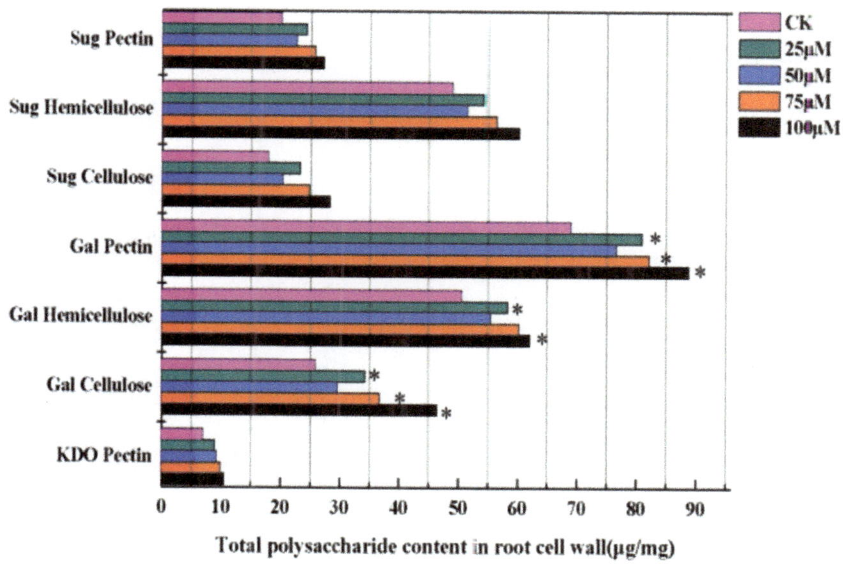

Figure 3. Effect of different copper concentrations on the contents of root cell wall polysacchride. Sug: sugar, Gal: galacturonic acids, KDO: 2-Keto-3-deoxyoctonic acid. Data presented are mean ±SE (n = 3). *Significant mean difference from control at $p = 0.05$ in multiple comparison by LSD test.

the biological process. The top three categories were those involved in metabolic processes (23.24%), cellular processes (16.20%), and response to stimuli (14.79%) (Figure 5A). These three categories were classified according to the GO molecular annotation into binding (41.38%), catalytic activity (31.03%), and transport activity (11.49%) (Figure 5B). The observed diversity in biological function and processes of the differentially expressed CWPs demonstrated that the response of the root cell wall of *E. splendens* to Cu stress was a complex process. Many physiological and biochemical changes were altered to counteract the adverse conditions.

Specific enzymes constituted by some small differentially expressed CWPs participated in various metabolic processes to counteract Cu stress. Classification pathways according to the

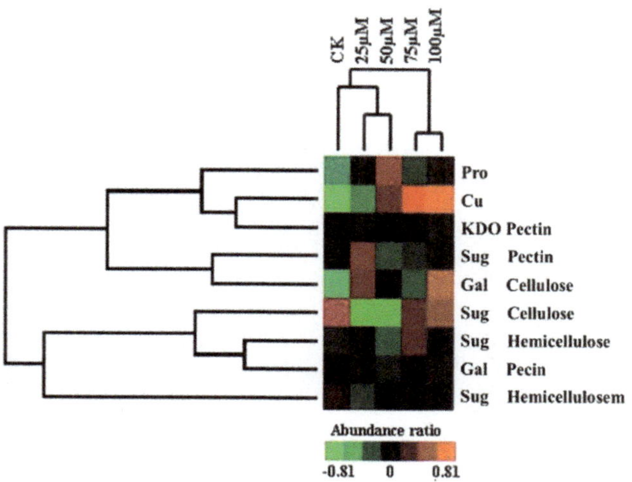

Figure 4. Hierarchical cluster result of cell wall polysaccharide abundance ratio using the average linkage distance between clusters is shown. The color weighting represents normalized levels of each variable from the high (red) to the low (green).

KEGG database searches categorized the cell wall enzymes into 12 classes (Table 3, Table S2). The more important category pathways were involved in oxidative phosphorylation, phenylalanine biosynthesis and metabolism, ascorbate and aldarate metabolism, and citrate cycle. All these pathways suggested the relatively altered abundance of the enzymes in various metabolic pathways that probably re-optimized the metabolic processes to combat against Cu stress.

Discussion

Cell wall polysaccharide remodeling in root under Cu stress

Cell walls provide a physical barrier to plants grown in Cu-contaminated soil, and cell wall polysaccharides have a crucial role in metal binding and accumulation [7]. The binding ability depends on the number of functional groups in polysaccharides, particularly for abundant carboxyl groups [7]. In the present study, the contents of the various cell wall polysaccharides significantly increased with the increase in Cu concentration. The expression of cell wall polysaccharides, which contain more carboxyl groups, provided more binding sites for Cu ions. Excess Cu also affected the intracellular distribution of polysaccharide in the root cell walls (Figure 3). Similar situations were also described in rice roots under Cd stress [45] and *Arabidopsis* roots under Al stress [46]. The starch and sucrose metabolism-related proteins (Table 3) – Chloroplast NAD-MDH (No. 1, Table 1) – were also markedly up-regulated in response to Cu stress based on our observations. These findings suggest that Cu or Cu-induced oxidative stress may activate some crucial enzymes in the glyconeogenesis pathway, which can alter the contents of cell wall polysaccharides.

Increasing physiological, biochemical, and molecular evidence also showed that the remodeling of the binding properties of root apoplasts was attributed to Cu resistance. Enhancement at the level of low-methylesterified pectins, where the polysaccharides bound more trivalent metal ions, is one of the most remarkable alterations. Pectin methylesterases (PMEs), known as pectinesterases, catalyzed

Table 1. Root cell wall proteins with significant increased in expression level under 50 μM Cu treatment identified by LC-ESI-MS/MS-based proteomics using SIEVE ($p<0.05$ and fold chang >1.5).

Serial No.	Protein name	gi number[a]	Percent coverage [b] (%)	Fold Change[c]	*p* value[d]	Species identified[e]	MW(kDa)	p*I*
1	Chloroplast NAD-MDH	3256066	8.68	1.50	0.00000	Arabidopsis thaliana	49.9	8.2
2	D-glyceraldehyde-3-phosphate Dehydrogenase(GAPDH)	51849658	17.26	1.52	0.00000	Periploca sepium	25.3	6.2
3	40S ribosomal protein S19	74231038	9.32	1.52	0.00271	Solanum peruvianum	15.7	5.5
4	Phosphoglycerate kinase	21272	5.08	1.53	0.00963	Spinacia oleracea	55.1	6.1
5	Putative malate dehydrogenase	50080249	15.29	1.55	0.00346	Oryza sativa Japonica Group	41.6	5.0
6	Cyc07	1856971	6.12	1.56	0.02389	Catharanthus roseus	35.6	6.0
7	2,3-bisphosphoglycerate-independent phosphoglycerate mutase	168588	1.61	1.57	0.00000	Zea mays	68.8	5.2
8	Non-symbiotic hemoglobin	11095158	13.75	1.58	0.01599	Medicago sativa	23.0	9.0
9	P0 ribosomal protein	1143507	6.83	1.59	0.00000	Lupinus luteus	41.0	9.8
10	Putative 40S ribosomal protein s12	643074	7.75	1.60	0.03853	Fragaria x ananassa	21.2	4.7
11	Unknown	255648032	11.18	1.62	0.02313	Glycine max	21.3	6.4
12	Mitochondrial chaperonin	415733	7.16	1.68	0.01830	Brassica napus	74.5	5.4
13	Maturase K	33332509	4.13	1.69	0.01270	Pitcairnia xanthocalyx	48.8	6.9
14	Pectin methylesterase	1222552	2.23	1.70	0.00633	Solanum lycopersicum	75.5	9.5
15	Porin-like protein	47847590	3.61	1.71	0.01348	Oryza sativa Japonica Group	36.1	6.4
16	Pollen allergen Que a 1 isoform	167472851	7.50	1.72	0.00062	Quercus alba	21.3	8.8
17	Putative 1,4-benzoquinone reductase	53749369	5.37	1. 73	0.00000	Oryza sativa Japonica Group	49.9	8.2
18	Thioredoxin h	13624884	9.32	1.89	0.02335	Pisum sativum	25.3	6.2
19	Predicted protein	162680354	5.16	1.91	0.00523	Physcomitrella patens subsp	24.8	8.9
20	Hypothetical protein Osl_31140	125563499	2.51	1.99	0.00035	Oryza sativa Indica Group	16.1	4.7
21	Hypothetical protein Osl_07053	218190658	1.47	2.70	0.00000	Oryza sativ a Indica Group	37.3	5.8
22	60s acidic ribosomal protein P2	551267	7.89	3.31	0.00000	Parthenium argentatum	56.7	5.4

[a] gi no., gene identification number as in GenBankTM.
[b] Coverage., sequence coverage.
[c] Fold change.
[d] p value, indicates the significance of up- or down- regulation of spots according to the F-test through analysis of variance ($p<0.05$).
[e] Species identified by Mascot search using NCBI.

Table 2. Root cell wall proteins with significant decreased in expression level under 50 μM Cu treatment identified by LC-ESI-MS/MS-based proteomics using SIEVE ($p<0.05$ and fold chang >1.5).

Serial No.	Protein name	gi number[a]	Percent coverage [b] (%)	Fold Change[c]	*p* value[d]	Species identified[e]	MW(kDa)	p*I*
23	Predicted protein	162671613	1.87	−2.68	0.00000	Physcomitrella patens subsp	71.6	4.9
24	Peroxidase	17066703	2.71	−2.59	0.00000	Capsicum annuum	39.5	6.9
25	Os07g0223100	113610742	18.75	−2.28	0.03704	Oryza sativa Japonica Group	6.1	6.0
26	Mitochondrial phosphate translocator	1842188	2.47	−2.25	0.00466	Betula pendula	45.3	9.4
27	Predicted protein	549049082	3.53	−2.23	0.01662	Populus trichocarpa	30.5	8.9
28	Actin	288300146	16.80	−2.17	0.00166	Cleistogenes songorica	46.2	5.5
29	Ubiquitin fusion protein	149391425	36.04	−2.08	0.02095	Oryza sativa Indica Group	16.6	10.0
30	Heat shock protein Hsp70	124360342	2.59	−1.98	0.02106	Medicago truncatula	80.9	6.4
31	Putative nascent polypeptide associated complex alpha chain	20160782	6.93	−1.97	0.00728	Oryza sativa Japonica Group	26.9	4.4
32	Hypothetical protein SORBIDRAFT_09g020360	241945088	8.50	−1.88	0.00003	Sorghum bicolor	49.1	5.2
33	Plasma membrane-type calcium ATPase	3335060	1.08	−1.86	0.03138	Arabidopsis thaliana	125.3	8.9
34	Cytochome b5	2695711	11.94	−1.83	0.01678	Olea europaea	18.3	5.5
35	Unknown	118482646	3.48	−1.82	0.00000	Populus trichocarpa	41.8	4.9
36	ATP synthase beta subunit	8452631	4.35	−1.80	0.04389	Cichorium intybus	18.1	6.2
37	Peptidyl-prolyl cis-trans isomerase	118104	7.56	−1.76	0.00000	Zea mays L	21.8	9.0
38	Ferredoxin–NADP+ reductase-like protein	7267299	8.61	−1.73	0.00081	Arabidopsis thaliana	48.3	8.9
39	Putative protein	5262759	0.82	−1.71	0.04231	Arabidopsis thaliana	124.3	8.5
40	Conserved hypothetical protein	223517971	3.65	−1.71	0.00000	Ricinus communis	24.2	4.2
41	Unnamed protein product	219742878	7.50	−1.70	0.01691	Glycine max	38.2	5.6
42	Unknown	255634488	7.50	−1.69	0.01879	Glycine max	38.0	5.4
43	Ferredoxin-NADP Reductase	6634773	7.59	−1.69	0.00157	Oryza sativa	50.9	9.2
44	Predicted protein	222844395	1.20	−1.68	0.00000	Populus trichocarpa	72.6	9.2
45	60S ribosomal protein L25	310935	14.94	−1.66	0.00002	Nicotiana tabacum	24.8	8.8
46	Vacuolar H(+)-ATPase subunit-like protein	7801655	2.27	−1.64	0.00000	Arabidopsis thaliana	58.0	5.9
47	Small G protein	974780	4.06	−1.61	0.01530	Beta vulgaris subsp. vulgaris	26.3	10.1
48	Elongation factor 1 subunit alpha	162036292	8.22	−1.58	0.02150	Cenchrus americanus	30.2	6.6
49	2-dehydro-3-deoxy phosphor heptonate aldolase	2546988	5.61	−1.52	0.01476	Morinda citrifolia	65.5	9.3
50	RAS-related GTP-binding protein	20756	5.31	−1.52	0.00000	Pisum sativum	26.7	8.3
51	Aspartic proteinase nepenthesin-1 precursor, putative	223535787	3.05	−1.52	0.00000	Ricinus communis	46.6	6.5
52	Hypothetical protein	147774267	0.73	−1.52	0.00000	Vitis vinifera	24.1	4.9
53	Hypothetical protein VITISV_025412	147822728	8.42	−1.52	0.00000	Vitis vinifera	21.9	7.3
54	Predicted protein	222855760	6.33	−1.51	0.00185	Populus trichocarpa	41.8	5.0
55	Hypothetical protein VITISV_023718	147801802	1.87	−1.51	0.00000	Vitis vinifera	71.1	7.8

[a] gi no., gene identification number as in GenBankTM.
[b] Coverage., sequence coverage.
[c] Fold change.
[d] p value, indicates the significance of up- or down- regulation of spots according to the F-test through analysis of variance ($p<0.05$).
[e] Species identified by Mascot search using NCBI.

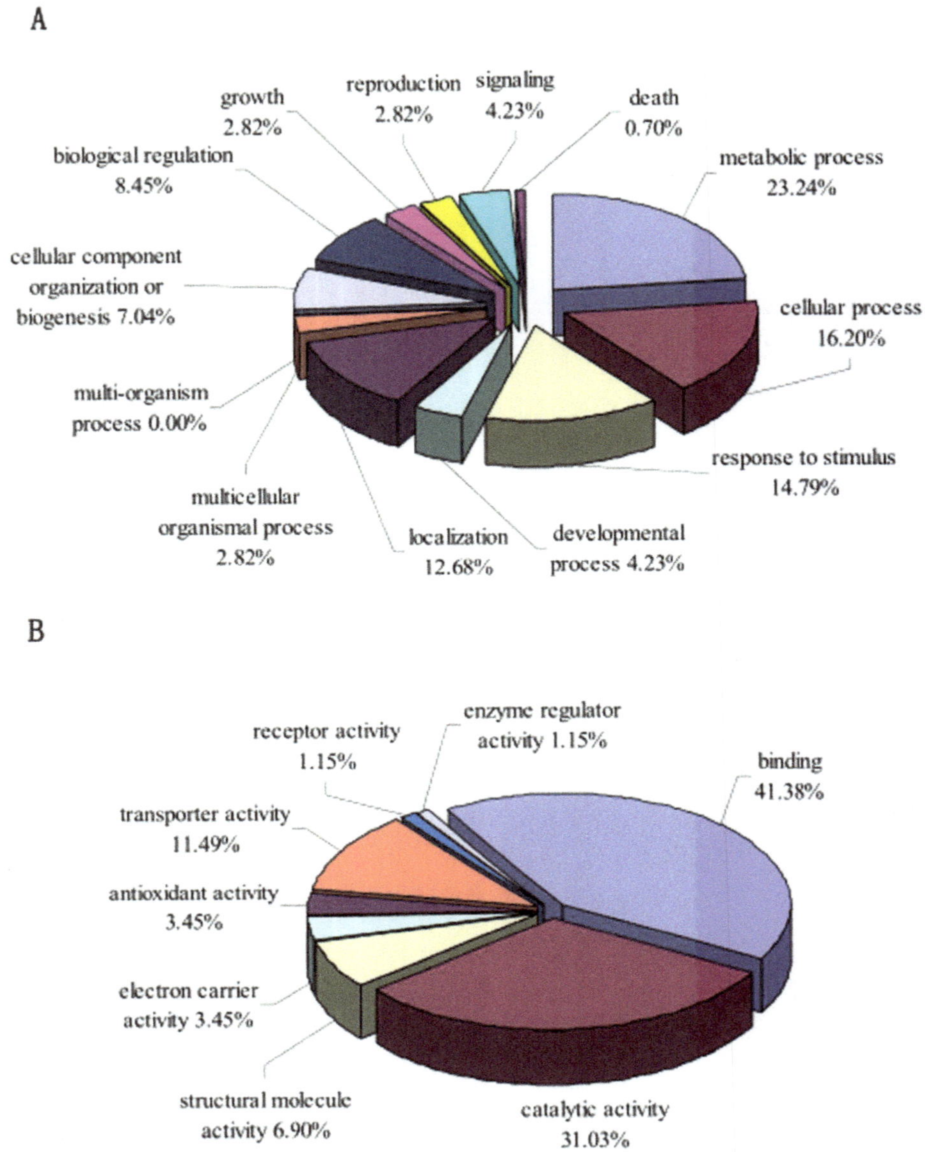

Figure 5. Functional cataloguing of 55 differentially expressed cell wall proteins in *Elsholtzia splendens*'s root based on GO annotation. The pie charts show the distribution of 55 differentially expressed cell wall proteins on of the Cu -responsive proteins into their functional classes in percentage. A: Biological Process Ontology, B: Molecular Function Ontology.

the demethylesterification of the homogalacturonan pectin domain in the cell wall. Demethylesterification of the pectin increases the abundance of free carboxylic acid groups on the galacturonic acid residue [47]. A flax PME has been recently implicated in wall remodeling following Cd treatment [48]. Interestingly, this study identified PMEs (No. 14, Table 1) as Cu-induced differentially expressed proteins; the protein had changes in abundance of 1.7-fold ($p \leq 0.5$). Catalyzed ROS by excess Cu^{2+} can also damage or cause degradation of essential complex molecules in cell wall polysaccharide. Generating hydroxyl radicals from H_2O_2 possesses a direct role in cell wall loosening through polysaccharide cleavage [49,50]. Hydroxyl radicals (OH) may cause non-enzymic scission of polysaccharides in vivo (e.g., in plant cell walls) [50]. Basing on these findings, we concluded that the chemical composition and distribution of plant cell wall polysaccharides are a factor in the outcome of plant–metal interaction, which elucidated a possible role in a novel Cu-resistance mechanism.

Cu-induced oxidative stress and antioxidant defense in cell wall

Cu is a redox active metal that catalyzes ROS production, such as superoxide ($O_2^{\cdot}$), hydrogen peroxide (H_2O_2), and hydroxyl radicals ($OH\cdot$) through Habere Weiss and Fenton reactions [51]. ROS can act as signaling molecules for stress response. However, ROS can cause damage to many cellular components above a certain threshold. Most Cu-tolerance mechanisms are primarily involved in protecting the cellular structure. An important method is the control of the level of ROS or the limitation of damage caused by ROS. In the current study, significant changes in the abundance of some antioxidant and defense-related proteins suggested that ROS can be involved in a Cu-induced oxidative

Table 3. Category of differentially expressed cell wall proteins refers to the entry on the classification of pathways from KEGG database.

No.	Pathway	Enzyme	Enzyme ID	Change Folds	Protein Name
1	Phenylpropanoid biosynthesis??? phenylalanine metabolism	Lactoperoxidase	ec:1.11.1.7	−2.59	Peroxidase
				−1.69	Ferredoxin-NADP Reductase
				−1.66	Predicted protein
2	Phenylalanine, tyrosine and tryptophan biosynthesis	Synthase	ec:2.5.1.54	−1.52	Hypothetical protein VITISV_025412
				2.70	Hypothetical protein OsI_07053
3	Nitrogen metabolism	(6S)-tetrahydrofolate aminomethyltransferase dehydratas	ec:2.1.2.10	1.72	Pollen allergen Que a 1 isoform
			ec:4.2.1.1	−1.83	Cytochome b5
4	One carbon pool by folate	(6S)-tetrahydrofolate aminomethyltransferase	ec:2.1.2 10	1.72	Pollen allergen Que a 1 isoform
5	Starch and sucrose metabolism	Synthase	ec:2.4.1.34	1.50	Chloroplast NAD-MDH
6	Oxidative phosphorylation	ATPase	ec:3.6.3.6	−1.76	Peptidyl-prolyl cis-trans isomerase
		Dehydrogenase	ec:1.3.5.1	1.53	Phosphoglycerate kinase
7	Ascorbate and aldarate metabolism	Oxidase	ec:1.1.3.8	−1.64	Vacuolar H(+)-ATPase subunit-like protein
8	Glyoxylate and dicarboxylate metabolism	(Si)-synthase	ec:2.3.3.1	−1.86	Plasma membrane-type calcium ATPase
9	Sphingolipid metabolism	Phosphodiesterase	ec:3.1.4.12	−1.52	2-dehydro-3-deoxyphosphoheptonate aldolase
10	Citrate cycle (TCA cycle)	(Si)-synthase	ec:2.3.3.1	−1.86	Plasma membrane-type calcium ATPase
		dehydratas	ec:1.3.5.1	1.53	Phosphoglycerate kinase
11	Glycine, serine and threonine metabolism	(6S)-tetrahydrofolate aminomethyltransferase	ec:2.1.2.10	1.72	Pollen allergen Que a 1 isoform
12	Methane metbolisma	Lactoperoxidase	ec:1.11.1.7	−2.59	Peroxidase
				−1.69	Ferredoxin-NADP Reductase
				−1.66	Predicted protein

stress response. Analysis of *E. splendens* root CWPs confirmed the associated gene expression for enzymes involved in ROS scavenging and oxidative phosphorylation (Table 3), such as peroxidase (No. 24, Table 2), peptidyl-prolyl cis-trans isomerase (No. 37, Table 2), and phosphoglycerate kinase (No. 4, Table 1). Among these proteins, peptidyl-prolyl cis-trans isomerase, as a protein chaperone, possess complementary and sometimes overlapping roles in protecting the proteins [15].

Excess Cu generates oxidative stress, thereby hindering some important metabolic process, such as up-regulation of antioxidant and stress-related regulatory proteins, that help maintain cellular homeostasis [52]. Maturase K (No. 13, Table 1) changed in abundance of 1.69-fold. Maturase K has an important role in plant growth, cell division, and expansion as well as in protecting metabolic processes against H_2O_2 and other toxic derivatives of oxygen. Hydroquinone formation was accompanied by the oxidation of two moles of NADPH and the presence of an inducible 1,4-benzoquinone reductase. Thus, putative 1,4-benzoquinone reductase (No. 17, Table 1) plays an essential role in antioxidant defense pathway [53]. Studies also reported about wall-bound malate dehydrogenase (No. 5, Table 1) [54,55], which can regenerate NAD(P)H that is needed by cell wall peroxidases for free radical generation associated with lignin polymerization. Thioredoxin (No. 18, Table 1) is also involved in redox regulation by reducing disulfides on the target protein for detoxifying lipid hydroperoxides or repairing oxidized proteins and relaying the signal to mitogen-activated protein kinase pathway of stress signaling [56]. Phosphoglycerate kinase (No. 4, Table 1) can also interact with cytosolic catalase and has a role in relieving oxidative stress. Interestingly, down-regulation of key metabolic enzymes revealed that oxidant protection conferred by these proteins was also regulated during Cu treatment. Ferredoxin-NADP reductase (No. 43, Table 2) and ferredoxin-NADP reductase-like protein (No. 38, Table 2), known to sequester highly reactive Fe^{3+} and prevent formation of toxic $\cdot$OH species, were also identified [57]. Vacuolar H(+)-ATPase subunit-like protein (No. 46, Table 2) was involved in ascorbate and aldarate metabolism. L-ascorbic acid is characterized by plant tissues, and ascorbate is one of the most important antioxidant molecules [58]. Furthermore, 2-dehydro-3-deoxy-phosphorheptonate aldolase synthase (No. 49, Table 2) may be involved in aromatic pathways in the secondary metabolites, which are known to act as defense responses to abiotic stress [59].

Cu stress-activated signaling pathways

The communication between the cytoskeleton and the CWPs is one of the most characterized features of cellular mechanisms that enable cells to respond effectively to various extracellular signals. Several candidate components involved in signal transduction were identified in this study, including Hsp70, small G protein, and cytochrome. The biosynthesis and accumulation of HSP proteins (No. 30, Table 2) can generally contribute to the protection and repair of cells under stress [60]. These proteins may be involved in cell wall biogenesis [61]. Small G proteins (No. 47, Table 2) transduced signals from receptors to control a wide range of cellular functions, particularly for regulating Ca^{2+} channel expression at the cell surface [62]. These proteins are clustered into distinct families but all act as molecular switches, which are active in their GTP-bound form but inactive when GDP-bound [63]. RAS-related GTP-binding protein (No. 50, Table 2) also has a primarily role in modulating cellular functions that involve actin cytoskeleton (No. 28, Table 2), such as establishing cell polarity and morphology [64]. Cytochrome has been identified in the stem cell wall of *Medicago sativa* [65] and germinating embryos of *Oryza sativa* [66]. This first proteomic study showed that cytochromes (No. 6, Table 1; No. 34, Table 2) were Cu-responsive proteins in the root cell wall of *E. splendens*. More preliminary evidence suggested that porins can form aqueous transmembrane channels for transporting solutes and macromolecules across the extracellular surface [67] [68]. These pores are regulated by ATP and GTP with a gating mechanism that modulates the pore size and ion selectivity [54]. In this case, Cu resistance mechanism may be induced by active apoplastic permeability because of up-regulated porin-like proteins (No. 15, Table 1).

Cu stress-activated energy pathways and protein synthesis

ATP is an essential metabolite in cell walls associated with energy conversion [69]. The plasma membrane calcium ATPase (PMCA) (No. 33, Table 2) or Ca^{2+} pump transports Ca^{2+} ions out of the cells using the stored energy in ATP. Control of Ca^{2+} concentration is significant in the cytosol [70]. PMCA down-regulation at the protein levels was observed during the experiment, which may represent an adaptive mechanism to facilitate removal of Ca^{2+} in the maintenance of calcium homeostasis in abiotic stress. The vacuolar H(+)-ATPase (No. 46, Table 2) functions as a primary proton pump that generates electrochemical gradients of protons across the transmembrane region [71], which provides the primary driving force for transporting numerous ions and metabolites against their electrochemical gradients [72]. Glyceraldehyde-3-phosphate dehydrogenase (GAPDH) (No. 2, Table 1), as a classical cellular enzyme involved in glycolysis, is also differentially expressed in response to Cu exposure, where GAPDH is specifically targeted at the cell wall [73]. Mitochondrial phosphate translocator (No. 26, Table 2), involved in ATP/ADP transportation, is also identified in the cell wall. Amino acids and amino acid-derived molecules can be chelated with metal ions by high affinity ligands [74]. Pollen allergen Que a 1 isoform (No. 16, Table 1), which is involved in glycine, serine, and threonine metabolism (Table 3), was up-regulated in the root cell wall of *E. splendens*, whereas aspartic proteinase nepenthesin-1 precursor (No. 51, Table 2) was down-regulated by Cu stress. Different expression levels of amino acid-related proteins in the cell wall elucidated their possible role in Cu tolerance.

Other Cu stress-responsive proteins and their potential functions

The 2,3-bisphosphoglycerate independent phosphoglycerate mutase (No. 7, Table 1) was up-regulated in the study, which was also differentially expressed in rice anthers under cold stress [75]. Previous evidence indicated that the expression of non-symbiotic hemoglobins (nsHbs) from cotton is induced with Verticillium wilt fungus [76]. The proteins also had a role in the defense responses against pathogen invasions in *Arabidopsis* [76]. Our studies showed that the nsHbs (No. 8, Table 1) were up-regulated in the cell wall of *E. splendens* root under Cu stress, although their physiological function is yet to be determined. A significant amount of the unidentified proteins were classified as hypothetical or predictable. These proteins were annotated in databases as unknown, hypothetical, or putative proteins because of the theoretical translation of open reading frame sequences [77]. The roles of these proteins remain to be investigated.

Overview of regulated CWP functions

This study provides insights into the functional role of the cell wall of *E. splendens* under Cu stress, where several CWPs were identified with possible roles in Cu tolerance/detoxification. Cu stress conditions can alter the composition (polysaccharide and protein) of the cell wall both qualitatively and quantitatively. Approximately 40% of the differentially expressed CWPs showed higher abundance in response to Cu stress involved in antioxidant defense, cell wall polysaccharide remodeling, and metabolism process. Up to 60% of the CWPs were in low abundance in response to Cu stress that is involved in signal, energy, and protein synthesis. Polysaccharide analysis confirmed the cell wall remodeling under Cu stress. The amount, composition, and distribution of the cell wall polysaccharides are consequential for plant adaptation to enhance Cu ion levels. Proteome analysis of cell wall confirmed that most proteins were associated with an antioxidant defense response. Hsp 70, small G protein, and RAS-related GTP-binding protein also have essential roles in signal transduction across the cell wall and through the cytoskeleton. Literature provides fundamental information about the role of polysaccharide composition of plant cell wall in metal tolerance and complementary evidence on continuous crosstalk between CWPs and the cytoskeleton. Therefore, knowledge has been expanded on plant stress-related signaling pathways in the cell wall. Cu regulation of these proteins may also not solely respond on abundance changes. Post-translational modification and dynamics are interesting subjects for future investigation. The present study is the first cell wall proteome and polysaccharide investigation of plants in response to Cu and important for the understanding of the plant cell wall response to environmental heavy metal stresses. We propose a possible protein interaction network (Figure 6) that provides new insights into Cu stress response in the root cell wall and facilitates further functional research of target proteins that are associated with Cu response based on the abundant changes in these proteins, as well as polysaccharides and their putative functions.

Materials and Methods

Ethics statement

No specific permissions were required for collecting *E. splendens* seeds from deposited Cu-mining soil in Zhuji County, Zhejiang Province, China. *E. splendens* is neither endangered nor protected. Authors maintained the population at sustainable levels. All plant work was conducted according to relevant national and international guidelines.

Plant material and Cu treatment

E. splendens seeds were collected from plants that grew on deposited Cu-mining soil in Zhuji County, Zhejiang Province, China. The seeds were washed with deionized water and soaked in distilled water to germinate in a controlled dark condition (25°C)

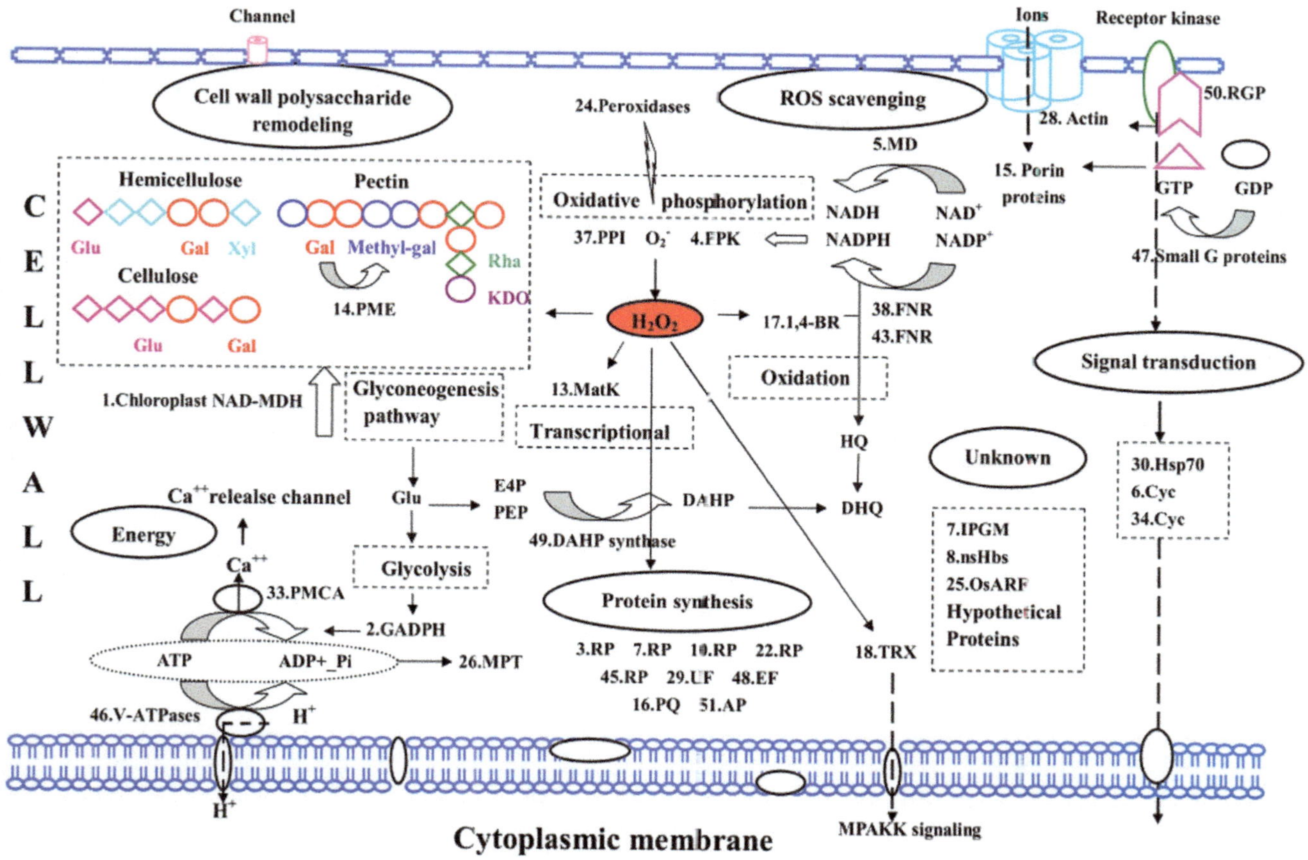

Figure 6. Pathways involved in cell defense, signaling, and cell wall remodeling under copper stress in the root cell wall of *Elsholtzia splendens*. Proteins identified in this study are displayed on the corresponding metabolic pathways and the number indicates the protein identification number. Gal, galacturonic acids; Methyl-gal, methylated galacturonic acids; KDO, 2-keto-3-deoxyoctonic acid; Rha, rhamnose; PME, Pectin methylesterase; Xyl, xylanase; Glu, glucose; MD, malate dehydrogenase; Trx, thioredoxin; MAPKK, mitogen-activated protein kinase kinase; PPI, peptidyl-prolyl cis-trans isomerase; PPK, phosphoglycerate kinase; MatK, maturase K; 1,4-BR, putative 1,4-benzoquinone reductase; oxidation; FNR, ferredoxin-NADP reductase; HQ, hydroquinone; DHQ, Durohydroquinone; DAHP, 3-deoxy-D-arabino-heptulosonic acid-7-phosphate; E4P, perythrose-4-phosphate; PEP, phosphoenolpyruvate; RGP, RAS-related GTP-binding protein; PLP, porin-like proteins; Cyc, cytochrome; PMCA, plasma membrane calcium ATPase; V-ATPases, vacuolar H(+)-ATPases; MPT, mitochondrial phosphate translocator; GADPH glyceraldehyde-3-phosphate dehydrogenase; RP, ribosomal protein; UF, ubiquitin fusion protein; EF, elongation factor 1 subunit alpha; PQ, Pollen allergen Que a 1 isoform; AP, aspartic proteinase nepenthesin-1 precursor; nsHbs. non-symbiotic hemoglobin; OsARF, Os07022З100; iPGM, 2,3-bisphosphoglycerate-independent phosphoglycerate mutase.

for 48 h. The solution was changed to one-quarter-strength complete nutrient solution after germination. At the fourth leaf stage, uniform seedlings were transferred into vessels filled with full-strength nutrient solution that contained macronutrients (in mM): 1.0 $Ca(NO_3)_2$, 0.5 $MgSO_4$, and 0.5 K_2HPO_4; and micronutrients (in μmol/L): 27.0 Fe(III)–EDTA, 23.0 H_3BO_3, 0.8 $CuSO_4$, 0.5 Na_2MoO_4, 0.5 $ZnSO_4$, and 4.5 $MnSO_4$. The pH was adjusted to 5.8 with 0.1 M HCl and NaOH. Nutrient solutions were renewed every 3 d and aerated continuously through 0.2 μm filters. Plants were grown in a hydroponic nutrient solution chamber with a 16 h, 25°C day and an 8 h, 20°C night regimen at 60% to 70% relative humidity. Light conditions in the growth chamber were fixed at 5 μmol to 10 μmol photons m^{-2} s^{-1}. Plants were exposed to various concentrations of Cu (25, 50, 75, and 100 μM) after 28 d of growth, which was added as sulfate for 48 h. Each treatment (15 plants) was conducted in triplicate, and the control plants (CK) were grown in pure nutrient solution for comparison. Plant roots were rinsed with distilled water at harvest and then immersed in 5 mmol/L of $Ca(NO_3)_2$ for 20 min to remove the putative adsorbed Cu^{2+} [32]. Roots were separated, pooled, and rinsed with deionized water; plant parts were washed, immediately frozen in liquid nitrogen, and stored at −80°C for analysis [33].

Cell wall preparation

Cell wall purification and CWP extraction were prepared as described by Feiz et al. [34] with slight modifications. All procedures were conducted at 4°C unless mentioned otherwise. Fifteen independent root preparations were pooled to yield one biological replicate. Root tissues (4 g) were homogenized in pre-cold extraction buffer [5 mM acetate, 0.4 M sucrose, 1 mM phenylmethylsulfonyl fluoride (PMSF), pH 4.6, 125 mL] with a chilled mortar and pestle. After adding 0.4 g of polyvinylpolypyrrolidone (PVPP), the mixture was incubated at 4°C for 30 min with stirring. Cell walls were separated from soluble cytoplasmic fluid by centrifuging the homogenate at 1,000×*g* for 15 min. The pellet was washed by suspension in 125 mL of 5 mM acetate buffer at pH 4.6 with 0.6 M sucrose. The mixture was centrifuged at 1,000×*g* for 15 min. The pellet was further purified by resuspension in 125 mL of 5 mM acetate buffer at pH 4.6 with

1 M sucrose and centrifuged at 1,000×*g* for 15 min. The residue was washed thoroughly with 750 mL of 5 mM acetate buffer at pH 4.6 while filtered on a layer of Miracloth (Merck, Darmstadt, Germany). The supernatant was discarded, and the final pellet was freeze-dried overnight. The freeze-dried cell wall materials were then stored at −80°C for further use. Most intracellular proteins were removed from the cell walls given the advantage of sucrose gradients and extensive washing with low ionic strength acidic buffer. PVPP was treated with acid to increase polymerization and to remove metal ions and contaminants.

Cell wall composition extraction and analysis

Isolation of cell wall material. Cell wall materials were extracted according to Zhong and Lauchli (1993) [35]. The pectin fraction was extracted twice with 0.5% ammonium oxalate buffer that contained 0.1% $NaBH_4$ (pH 4) in boiling water bath for 1 h each and pooled the supernatants. Pellets were subsequently subjected to triple extractions with 4% KOH that contained 0.1% $NaBH_4$ at room temperature for 24 h, followed by similar extraction with 24% KOH that contained 0.1% $NaBH_4$. The supernatants from the 4% and 24% KOH extractions were collected and thus yielded the hemicellulose fractions. The remaining pellet from the 24% KOH extraction was then lyophilized, weighed, and considered to be the cellulose fraction [36].

Protein extraction and digestion. A portion (2 g) of the roots was homogenized with 1 mL of PBS (pH 7.6) that contained 65 mM K_2HPO_4, 2.6 mM KH_2PO_4, 400 mM NaCl, and 3 mM NaN_3 in a mortar and pestle. The homogenate was centrifuged twice at 15,000×*g* for 10 min, and the supernatant was collected as the total soluble protein. CWPs were extracted from the root cell wall fraction in two successive steps. The first step used $CaCl_2$ solution (5 mM sodium acetate buffer, pH 4.6, 0.2 M $CaCl_2$, and 1 mM PMSF; $CaCl_2$ can efficiently extract CWPs that exert weak electrostatic interactions with other cell wall components [37]). The second step involved two extractions with NaCl solution (5 mM sodium acetate buffer, pH 4.6, 1 M NaCl, and 1 mM PMSF; NaCl was also used to extract the strong ionically bound proteins) [34]. For each extraction, the sample was incubated with vortexing at 4°C, and the supernatant was collected after centrifugation at 4,000×*g* for 15 min. The supernatants from all the extraction steps were pooled and concentrated to 4 mL by centrifugation at 1,500×*g* at 10°C using a 3 kDa molecular weight cut-off Amicon Spin Tube (Millipore, MA). The protein sample was buffer-exchanged with ultrapure water.

CWP extracts were dissolved in 100 μL of lysis buffer (7 M urea, 2 M thiourea, 5 mM EDTA, 10 mM DTT, and 1 mM PMSF). The supernatant was transferred to a new tube, reduced with 10 mM DTT for 1 h at 56°C, and alkylated with 55 mM iodoacetamide for 45 min at room temperature in darkness. The protein was precipitated with four volumes of pre-chilled acetone for 30 min at −20°C. The pellet was dissolved in 0.5 M TEAB after centrifugation and sonicated for 5 min. The centrifugation step was repeated, and the supernatant was collected. Approximately 100 μg of proteins from each sample was digested with trypsin (Promega) overnight at 37°C in a 1:20 trypsin-to-protein mass ratio.

Analysis of Cu, protein, and cell wall polysaccharide content. Cell wall fraction was dried at 70°C to a constant weight and then digested at 145°C for 24 h in an acid mixture of HNO_3:$HClO_4$ (3:1, v:v). Cu concentrations were measured by Elmer flame atomic absorption spectrometry (AAS-3600).

Protein contents were determined by the Bradford method [38] with bovine serum albumin as the standard. CWPs were dissolved with 400 μL of SDS sample lysis buffer by boiling for 5 min and then loaded onto a 12% acrylamide mini-gel (5 cm to 8 cm) for 1D SDS-PAGE.

The contents of total sugars, galacturonic acids, and KDO were determined colorimetrically by sulfate-phenol [39], hydroxydiphenyl [40], and thiobarbituric acid [41], respectively.

Labeling TMT reagents

Peptide was desalted by Strata XC18 SPE column (Phenomenex) and vacuum-dried after trypsin digestion. Peptide was reconstituted in 0.5 M TEAB and processed according to the protocol for 6-plex TMT reagent kits. In brief, two units of TMT reagent (defined as the amount of reagent required to label 100 μg of protein) was thawed and reconstituted in 41 μL of acetonitrile. Peptide samples from 50 μM Cu were treated for 48 h, and CK samples were labeled with TMT tags 126, 128, and 130 as well as TMT tags 127, 129, and 131. Both samples were incubated at room temperature for 2 h. The peptide mixtures were then pooled, desalted, and dried by vacuum centrifugation.

LC–ESI-MS/MS analysis by Q Exactive

The labeled peptide was resuspended in buffer A (2% ACN, 0.1% FA) and centrifuged at 20,000×*g* for 2 min. The supernatant was transferred into a sample tube and loaded onto an Acclaim PepMap 100 C18 trap column (75 μm×2 cm; Dionex) by EASY nLC1000 nano UPLC (Thermo). The peptide was then eluted onto an Acclaim PepMap RSLC C18 analytical column (50 μm×15 cm; Dionex). An 85 min gradient program was run at 300 nL/min, which started from 3% to 35% B (80% ACN, 0.1% FA), followed by 5 min linear gradient to 90% B, and maintained at 90% B for 5 min.

The peptides were subjected to NSI source, followed by tandem mass spectrometry (MS/MS) in Q Exactive (Thermo) coupled online to the UPLC. Intact peptides were detected in the Orbitrap at a resolution of 70,000. Peptides were selected for MS/MS using 27% NCE with 12% stepped NCE; ion fragments were detected in the Orbitrap at a resolution of 17,500. A data-dependent procedure that alternated between one MS scan followed by 20 MS/MS scans was applied to the top 20 precursor ions above a threshold ion count of 3E4 in the MS survey scan with 5.0 s dynamic exclusion. The applied electrospray voltage was 1.8 kV. Automatic gain control was used to prevent overfilling of the ion trap; 1E5 ions were accumulated for generation of MS/MS spectra. The m/z scan range was 350 Da to 2000 Da for MS scans.

Proteomic database search

The instrument data file (.raw) was merged and transformed to an.mgf file by Proteome Discoverer (ver. 1.3.0.339; Thermo). Peptide and protein identifications were performed using the Mascot search engine (ver. 2.3.02; Matrix Science). Derived protein sequences from plants in the NCBI were collected, and a database containing 1,596,443 sequences was created. In the Mascot search engine version 2.3.02 software, all parameters were set as follows. Database searching was restricted to tryptic peptides. Carbamidomethyl (C), TMT 6-plex (N-term), and TMT 6-plex (K) were selected as fixed, and deamidated (NQ), Gln→pyro-Glu (N-term Q), and oxidation (M) were selected as variable modifications, where two missed cleavages were allowed with precursor error tolerance at 10 ppm and fragment deviation at 0.02 Da. The Mascot search results were quantified using Mascot 2.3.02 with the following criteria: protein ratio type = median, minimum unique peptides = 1, peptide threshold type = at least homolog. Peptides were not quantified if peptide score was too low or the deviation was too large. The final ratios of protein were then normalized by taking the median of all the proteins quantified. The complete list of identified peptides was encoded in an Excel (Microsoft) database for grouping of results into proteins and calculation of ratios and coefficient of variation.

To obtain the differentially expressed proteins, we selected the proteins according to the following principles. (a) The expression change of protein folds is more than 1.5. (b) The protein quantitative T-test p value is less than 0.05. A total of 55 differentially expressed proteins were obtained in our bioinformatics analysis data.

Bioinformatics studies and statistical analysis

Data were the average of at least three independent experimental replicates. One-way ANOVA (Duncan's test) and LSD test with $p<0.05$ as the significance level were performed to analyze the differences between the control and treatments. All data were presented as the mean value ± standard error and were analyzed using SPSS statistical software package (version 16.0).

The clustering of content abundance profiles was performed using Cluster 3.0. Euclidean metric was used for computing the distance between points and cluster centroids, which is a typical choice for K-means clustering analysis. The output was visualized using TreeView after hierarchical clustering.

GO information was used to categorize the biological processes of identified proteins from TMT data sets. GO annotations were extracted from the Uni-Prot database and were matched with corresponding gene locus identifiers embedded in the NCBI RefSeq database. Proteins were then classified based on their biological process using the Web Gene Ontology Annotation Plot [42] (http://wego.genomics.org.cn/cgi-bin/wego/index.pl). Functionally analyzed Cu stress-induced CWPs in biosynthesis pathways were identified through the Kyoto Encyclopedia of Genes and Genomes (KEGG) [43]. These pathways were classified into hierarchical categories according to the KEGG website (http://www.genome.jp/kegg/). Subcellular locations were predicted using TargetP (http://www.cbs.dtu.dk/services/TargetP/), Predotar (http://urgi.versailles.inra.fr/predotar/predotar.html), and WoLF PSORT (http://wolfpsort.org/) (http://www.genscript.com/psort/wolf_psort.html)software applications as previously described [44].

Supporting Information

Figure S1 SDS–PAGE of root cell wall proteins under different copper stress.
(DOC)

Table S1 Protein subcellular localization and go annotation of 55 significantly differentially expressed proteins.
(XLS)

Table S2 Details of 55 significantly differentially expressed proteins identified by TMT using ANOVA including their *p* –value.
(XLS)

Author Contributions

Conceived and designed the experiments: TTL JYS. Performed the experiments: TTL CFS. Analyzed the data: TTL CKH JYS. Contributed reagents/materials/analysis tools: YW. Wrote the paper: TTL.

References

1. Maksymiec W (1998) Effect of copper on cellular processes in higher plants. Photosynthetica 34: 321-342.
2. Teisseire H, Guy V (2000) Copper-induced changes in antioxidant enzymes activities in fronds of duckweed (*Lemna minor*)Plant science 153: 65-72.
3. Vinit-Dunand F, Epron D, Alaoui-Sossé B, Badot PM (2002) Effects of copper on growth and on photosynthesis of mature and expanding leaves in cucumber plants. Plant science 163: 53-58.
4. Bona E, Marsano F, Cavaletto M, Berta G (2007) Proteomic characterization of copper stress response in *Cannabis sativa* roots. Proteomics 7: 1121-1130.
5. Hall J (2002) Cellular mechanisms for heavy metal detoxification and tolerance. Journal of Experimental Botany 53: 1-11.
6. Cobbett C, Goldsbrough P (2002) Phytochelatins and metallothioneins: roles in heavy metal detoxification and homeostasis. Annual review of plant biology 53: 159-182.
7. Krzesłowska M (2011) The cell wall in plant cell response to trace metals: polysaccharide remodeling and its role in defense strategy. Acta Physiologiae Plantarum 33: 35-51.
8. Jarvis MC (2009) Plant cell walls: supramolecular assembly, signalling and stress. Structural Chemistry 20: 245-253.
9. Zhu J, Chen S, Alvarez S, Asirvatham VS, Schachtman DP, et al. (2006) Cell wall proteome in the maize primary root elongation zone. I. Extraction and identification of water-soluble and lightly ionically bound proteins. Plant physiology 140: 311-325.
10. Vorwerk S, Somerville S, Somerville C (2004) The role of plant cell wall polysaccharide composition in disease resistance. Trends in plant science 9: 203-209.
11. Dani V, Simon WJ, Duranti M, Croy RR (2005) Changes in the tobacco leaf apoplast proteome in response to salt stress. Proteomics 5: 737-745.
12. Soares NC, Francisco R, Vielba JM, Ricardo CP, Jackson PA (2009) Associating wound-related changes in the apoplast proteome of Medicago with early steps in the ROS signal-transduction pathway. Journal of Proteome Research 8: 2298-2309.
13. Jamet E, Canut H, Boudart G, Pont-Lezica RF (2006) Cell wall proteins: a new insight through proteomics. Trends in plant science 11: 33-39.
14. Bhushan D, Pandey A, Choudhary MK, Datta A, Chakraborty S, et al. (2007) Comparative proteomics analysis of differentially expressed proteins in chickpea extracellular matrix during dehydration stress. Molecular & Cellular Proteomics 6: 1868-1884.
15. Pandey A, Rajamani U, Verma J, Subba P, Chakraborty N, et al. (2010) Identification of extracellular matrix proteins of rice (*Oryza sativa* L.) involved in dehydration-responsive network: a proteomic approach. Journal of Proteome Research 9: 3443-3464.
16. Floerl S, Druebert C, Majcherczyk A, Karlovsky P, Kües U, et al. (2008) Defence reactions in the apoplastic proteome of oilseed rape (*Brassica napus var. napus*) attenuate Verticillium longisporum growth but not disease symptoms. Bmc Plant Biology 8: 129.
17. Goulet C, Goulet C, Goulet MC, Michaud D (2010) 2-DE proteome maps for the leaf apoplast of Nicotiana benthamiana. Proteomics 10: 2536-2544.
18. Brune A, Urbach W, Dietz K-J (1994) Zinc stress induces changes in apoplasmic protein content and polypeptide composition of barley primary leaves. Journal of Experimental Botany 45: 1189-1196.
19. Blinda A, Koch B, Ramanjulu S, Dietz K (1997) De novo synthesis and accumulation of apoplastic proteins in leaves of heavy metal-exposed barley seedlings. Plant, Cell & Environment 20: 969-981.
20. Fecht-Christoffers MM, Braun H-P, Lemaitre-Guillier C, VanDorsselaer A, Horst WJ (2003) Effect of manganese toxicity on the proteome of the leaf apoplast in cowpea. Plant physiology 133: 1935-1946.
21. Fecht-Christoffers MM, Führs H, Braun H-P, Horst WJ (2006) The role of hydrogen peroxide-producing and hydrogen peroxide-consuming peroxidases in the leaf apoplast of cowpea in manganese tolerance. Plant physiology 140: 1451-1463.
22. Alves M, Francisco R, Martins I, Ricardo C (2006) Analysis of *Lupinus albus* leaf apoplastic proteins in response to boron deficiency. Plant and Soil 279: 1-11.
23. Komatsu S, Kobayashi Y, Nishizawa K, Nanjo Y, Furukawa K (2010) Comparative proteomics analysis of differentially expressed proteins in soybean cell wall during flooding stress. Amino Acids 39: 1435-1449.
24. Kataoka T, Furukawa J, Nakanishi T (2003) The decrease of extracted apoplast protein in soybean root tip by aluminium treatment. Biologia plantarum 46: 445-449.
25. Kong F-J, Oyanagi A, Komatsu S (2010) Cell wall proteome of wheat roots under flooding stress using gel-based and LC MS/MS-based proteomics approaches. Biochimica et Biophysica Acta (BBA)-Proteins and Proteomics 1804: 124-136.
26. Zhou L, Bokhari SA, Dong C-J, Liu J-Y (2011) Comparative proteomics analysis of the root apoplasts of rice seedlings in response to hydrogen peroxide. Plos One 6: e16723.
27. Hossain Z, Nouri M-Z, Komatsu S (2011) Plant cell organelle proteomics in response to abiotic stress. Journal of Proteome Research 11: 37-48.

28. Yang MJ, Yang XE, Römheld V (2002) Growth and nutrient composition of *Elsholtzia splendens Nakai* under copper toxicity. Journal of plant nutrition 25: 1359-1375.
29. Lou L, Shen Z, Li X (2004) The copper tolerance mechanisms of *Elsholtzia haichowensis*, a plant from copper-enriched soils. Environmental and Experimental Botany 51: 111-120.
30. Feiz L, Irshad M, Pont-Lezica RF, Canut H, Jamet E (2006) Evaluation of cell wall preparations for proteomics: a new procedure for purifying cell walls from Arabidopsis hypocotyls. Plant Methods 2: 10.
31. Kwon H-K, Yokoyama R, Nishitani K (2005) A proteomic approach to apoplastic proteins involved in cell wall regeneration in protoplasts of Arabidopsis suspension-cultured cells. Plant and Cell Physiology 46: 843-857.
32. Harrison S, Lepp N, Phipps D (1979) Uptake of Copper by excised roots II. Copper desorption from the free space. Zeitschrift für Pflanzenphysiologie 94: 27-34.
33. Li F, Shi J, Shen C, Chen G, Hu S, et al. (2009) Proteomic characterization of copper stress response in *Elsholtzia splendens* roots and leaves. Plant Molecular Biology 71: 251-263.
34. Feiz L, Irshad M, Pont-Lezica RF, Canut H, Jamet E (2006) Evaluation of cell wall preparations for proteomics: a new procedure for purifying cell walls from Arabidopsis hypocotyls. Plant Methods 2.
35. Zhong H, Lauchli A (1993) Changes of cell wall composition and polymer size in primary roots of cotton seedlings under high salinity. Journal of Experimental Botany 44: 773-778.
36. Yu M, Shen R, Liu J, Chen R, Xu M, et al. (2009) The role of root border cells in aluminum resistance of pea (*Pisum sativum*) grown in mist culture. Journal of Plant Nutrition and Soil Science 172: 528-534.
37. Jamet E, Boudart G, Borderies G, Charmont S, Lafitte C, et al. (2008) Isolation of plant cell wall proteins. 2D PAGE: Sample Preparation and Fractionation: Springer. pp. 187-201.
38. Bradford MM (1976) A rapid and sensitive method for the quantitation of microgram quantities of protein utilizing the principle of protein-dye binding. Analytical biochemistry 72: 248-254.
39. Dubois M, Gilles KA, Hamilton JK, Rebers P, Smith F (1956) Colorimetric method for determination of sugars and related substances. Analytical chemistry 28: 350-356.
40. Dronnet V, Renard C, Axelos M, Thibault J-F (1996) Heavy metals binding by pectins: selectivity, quantification and characterisation. Progress in Biotechnology 14: 535-540.
41. Mohnen D (2008) Pectin structure and biosynthesis. Current opinion in plant biology 11: 266-277.
42. Ye J, Fang L, Zheng H, Zhang Y, Chen J, et al. (2006) WEGO: a web tool for plotting GO annotations. Nucleic acids research 34: W293-W297.
43. Kanehisa M, Goto S (2000) KEGG: kyoto encyclopedia of genes and genomes. Nucleic acids research 28: 27-30.
44. Day A, Fénart S, Neutelings G, Hawkins S, Rolando C, et al. (2013) Identification of cell wall proteins in the flax (*Linum usitatissimum*) stem. Proteomics.
45. Xiong J, An L, Lu H, Zhu C (2009) Exogenous nitric oxide enhances cadmium tolerance of rice by increasing pectin and hemicellulose contents in root cell wall. Planta 230: 755-765.
46. Yang JL, Zhu XF, Peng YX, Zheng C, Li GX, et al. (2011) Cell wall hemicellulose contributes significantly to aluminum adsorption and root growth in Arabidopsis. Plant physiology 155: 1885-1892.
47. Pelloux J, Rusterucci C, Mellerowicz EJ (2007) New insights into pectin methylesterase structure and function. Trends in plant science 12: 267-277.
48. Paynel F, Schaumann A, Arkoun M, Douchiche O, Morvan C (2009) Temporal regulation of cell-wall pectin methylesterase and peroxidase isoforms in cadmium-treated flax hypocotyl. Annals of Botany 104: 1363-1372.
49. Liszkay A, Kenk B, Schopfer P (2003) Evidence for the involvement of cell wall peroxidase in the generation of hydroxyl radicals mediating extension growth. Planta 217: 658-667.
50. Fry S, Dumville J, Miller J (2001) Fingerprinting of polysaccharides attacked by hydroxyl radicals in vitro and in the cell walls of ripening pear fruit. Biochem J 357: 729-737.
51. Thounaojam TC, Panda P, Mazumdar P, Kumar D, Sharma G, et al. (2012) Excess copper induced oxidative stress and response of antioxidants in rice. Plant Physiology and Biochemistry 53: 33-39.
52. Ahsan N, Lee D-G, Lee S-H, Kang KY, Lee JJ, et al. (2007) Excess copper induced physiological and proteomic changes in germinating rice seeds. Chemosphere 67: 1182-1193.
53. Spain JC, Wyss O, Gibson DT (1979) Enzymatic oxidation of p-nitrophenol. Biochemical and Biophysical Research Communications 88: 634-641.
54. Gross GG (1977) Cell wall-bound malate dehydrogenase from horseradish. Phytochemistry 16: 319-321.
55. Li Z-C, McClure JW, Hagerman AE (1989) Soluble and bound apoplastic activity for peroxidase, β-D-glucosidase, malate dehydrogenase, and nonspecific arylesterase, in barley (*Hordeum vulgare* L.) and oat (*Avena sativa* L.) primary leaves. Plant physiology 90: 185-190.
56. Apel K, Hirt H (2004) Reactive oxygen species: metabolism, oxidative stress, and signal transduction. Annu Rev Plant Biol 55: 373-399.
57. Jensen KA, Houtman CJ, Ryan ZC, Hammel KE (2001) Pathways for extracellular Fenton chemistry in the brown rot basidiomycete Gloeophyllum trabeum. Applied and Environmental Microbiology 67: 2705-2711.
58. Howard L, Talcott S, Brenes C, Villalon B (2000) Changes in phytochemical and antioxidant activity of selected pepper cultivars (Capsicum species) as influenced by maturity. Journal of Agricultural and Food Chemistry 48: 1713-1720.
59. Keith B, Dong X, Ausubel FM, Fink GR (1991) Differential induction of 3-deoxy-D-arabino-heptulosonate 7-phosphate synthase genes in Arabidopsis thaliana by wounding and pathogenic attack. Proceedings of the National Academy of Sciences 88: 8821-8825.
60. Timperio AM, Egidi MG, Zolla L (2008) Proteomics applied on plant abiotic stresses: role of heat shock proteins (HSP). Journal of proteomics 71: 391-411.
61. Wang SB, Hu Q, Sommerfeld M, Chen F (2004) Cell wall proteomics of the green alga *Haematococcus pluvialis* (Chlorophyceae). Proteomics 4: 692-708.
62. Béguin P, Nagashima K, Gonoi T, Shibasaki T, Takahashi K, et al. (2001) Regulation of Ca2+ channel expression at the cell surface by the small G-protein kir/Gem. Nature 411: 701-706.
63. Rittinger K, Walker PA, Eccleston JF, Nurmahomed K, Owen D, et al. (1997) Crystal structure of a small G protein in complex with the GTPase-activating protein rhoGAP. Nature 388: 693-697.
64. Hall A (1994) Small GTP-binding proteins and the regulation of the actin cytoskeleton. Annual review of cell biology 10: 31-54.
65. Watson BS, Lei Z, Dixon RA, Sumner LW (2004) Proteomics of *Medicago sativa* cell walls. Phytochemistry 65: 1709-1720.
66. Zhang H, Lian C, Shen Z (2009) Proteomic identification of small, copper-responsive proteins in germinating embryos of *Oryza sativa*. Annals of Botany 103: 923-930.
67. Odou M, Singer E, Romond M, Dubreuil L (1998) Isolation and characterization of a porin-like protein of 45 kilodaltons from Bacteroides fragilis. Fems Microbiology Letters 166: 347-354.
68. Massari P, Ho Y, Wetzler LM (2000) Neisseria meningitidis porin PorB interacts with mitochondria and protects cells from apoptosis. Proceedings of the National Academy of Sciences 97: 9070-9075.
69. Chivasa S, Ndimba BK, Simon WJ, Lindsey K, Slabas AR (2005) Extracellular ATP functions as an endogenous external metabolite regulating plant cell viability. The Plant Cell Online 17: 3019-3034.
70. Carafoli E, Lim D (2009) Plasma Membrane Calcium ATPase. Handbook of Neurochemistry and Molecular Neurobiology: Springer. pp. 581-596.
71. Nelson H, Nelson N (1990) Disruption of genes encoding subunits of yeast vacuolar H (+)-ATPase causes conditional lethality. Proceedings of the National Academy of Sciences 87: 3503-3507.
72. Sze H, Ward JM, Lai S (1992) Vacuolar H+-translocating ATPases from plants: structure, function, and isoforms. Journal of bioenergetics and biomembranes 24: 371-381.
73. Petersen J, Brinkmann H, Cerff R (2003) Origin, evolution, and metabolic role of a novel glycolytic GAPDH enzyme recruited by land plant plastids. Journal of molecular evolution 57: 16-26.
74. Sharma SS, Dietz K-J (2006) The significance of amino acids and amino acid-derived molecules in plant responses and adaptation to heavy metal stress. Journal of Experimental Botany 57: 711-726.
75. Imin N, Kerim T, Rolfe BG, Weinman JJ (2004) Effect of early cold stress on the maturation of rice anthers. Proteomics 4: 1873-1882.
76. Qu Z-L, Zhong N-Q, Wang H-Y, Chen A-P, Jian G-L, et al. (2006) Ectopic expression of the cotton non-symbiotic hemoglobin gene GhHbd1 triggers defense responses and increases disease tolerance in Arabidopsis. Plant and Cell Physiology 47: 1058-1068.
77. Zhou X, Li J, Cheng W, Liu H, Li M, et al. (2010) Gene structure analysis of rice ADP-ribosylation factors (OsARFs) and their mRNA expression in developing rice plants. Plant Molecular Biology Reporter 28: 692-703.

11

Gene Expression and Pathway Analysis of Effects of the CMAH Deactivation on Mouse Lung, Kidney and Heart

Deug-Nam Kwon[1], Byung-Soo Chang[2], Jin-Hoi Kim[1]*

1 Department of Animal Biotechnology, Konkuk University, Seoul, South Korea, **2** Department of Cosmetology, Hanseo University, Seosan, Chungnam, South Korea

Abstract

Background: N-glycolylneuraminic acid (Neu5Gc) is generated by hydroxylation of CMP-Neu5Ac to CMP-Neu5Gc, catalyzed by CMP-Neu5Ac hydroxylase (CMAH). However, humans lack this common mammalian cell surface molecule, Neu5Gc, due to inactivation of the CMAH gene during evolution. CMAH is one of several human-specific genes whose function has been lost by disruption or deletion of the coding frame. It has been suggested that CMAH inactivation has resulted in biochemical or physiological characteristics that have resulted in human-specific diseases.

Methodology/Principal Findings: To identify differential gene expression profiles associated with the loss of Neu5Gc expression, we performed microarray analysis using Illumina MouseRef-8 v2 Expression BeadChip using the main tissues (lung, kidney, and heart) from control mice and CMP-Neu5Ac hydroxylase (Cmah) gene knock-out mice, respectively. Out of a total of 25,697 genes, 204, 162, and 147 genes were found to be significantly modulated in the lung, kidney, and heart tissues of the Cmah null mouse, respectively. In this study, we examined the gene expression profiles, using three commercial pathway analysis software packages: Ingenuity Pathways Analysis, Kyoto Encyclopedia of Genes and Genomes analysis, and Pathway Studio. The gene ontology analysis revealed that the top 6 biological processes of these genes included protein metabolism and modification, signal transduction, lipid, fatty acid, and steroid metabolism, nucleoside, nucleotide and nucleic acid metabolism, immunity and defense, and carbohydrate metabolism. Gene interaction network analysis showed a common network that was common to the different tissues of the Cmah null mouse. However, the expression of most sialytransferase mRNAs of Hanganutziu-Deicher antigen, sialy-Tn antigen, Forssman antigen, and Tn antigen was significantly down-regulated in the liver tissue of Cmah null mice.

Conclusions/Significance: Mice bearing a human-like deletion of the Cmah gene serve as an important model for the study of abnormal pathogenesis and/or metabolism caused by the evolutionary loss of Neu5Gc synthesis in humans.

Editor: Qing-Yuan Sun, Institute of Zoology, Chinese Academy of Sciences, China

Funding: This work was supported by the next generation of Biogreen 21 (PJ009625) from the Rural Development Administration (RDA), Republic of Korea. The funders had no role in study design, data collection and analysis, decision to publish, or preparation of the manuscript.

Competing Interests: The authors have declared that no competing interests exist.

* Email: jhkim541@konkuk.ac.kr

Introduction

Xenotransplantation using pig organs has the potential to solve the increasing shortage of donor organs available for allotransplantation [1]. Over the last two decades, there have been considerable advances in our understanding of the immunological and physiological hurdles to xenotransplantation. Many of the obstacles to xenotransplantion result from mismatches in receptor-ligand or enzyme-substrate interactions between the pig tissue and the recipient's blood and immune system [2]. The initial cause of failure for pig cardiac and renal xenografts is thought to be antibody-mediated injury to the endothelium, leading to the development of microvascular thrombosis. Factors contributing to the development of thrombotic microangiopathy include anti-non-Gal antibodies, natural killer cells or macrophage activity, and inherent coagulation dysregulation between pigs and primates [3]. To address these problems, several researchers have produced gene knockouts or transgenic animals to correct these mismatches. However, the combination of these modifications into an "idealized" transgenic animal has yet to be reported.

Neu5Gc is produced from Neu5Ac through enzymatic hydroxylation of the N-acetyl residue of free Neu5Ac, CMP-Neu5Ac, or glycoconjugate-linked Neu5Ac [4,5]. Neu5Gc, also called Hanganutziu-Deicher (H-D) antigen, is expressed on the endothelial cell surface of all mammals, with the exception of humans, and is a target for non-Galα 1,3 Gal antibodies [6]. As consequence of inactivation of the CMP-Neu5Ac hydroxylase (Cmah) gene during evolution, humans have lost the ubiquitous mammalian cell surface molecule Neu5Gc [7,8,9]. Conversely, Neu5Gc produced by CMAH activity is one of the non-Gal xenoantigens of secondary importance to α-1, 3-galactosyltransferase (GGTA1), for pig-to-human xenotransplantation [10]. Similar to the galactose α1,3 galactose (α-Gal), Neu5Gc is immunogenic in humans as it is responsible for the expression of Neu5Gc, a key non-Gal antigen [11]. Very recently, we have developed a biallelic CMAH knock-out in pigs [12]. However, these data raise the possibility of an alternate pathway in metabolism and the immune system, which might contribute to acute immune rejection of xenografts.

The silencing of the CMAH gene expression resulted in a number of genetic and biochemical changes to the biosynthesis of sialic acids, which may have contributed to several unique aspects of human biology in health and disease [13,14]. Cmah null mice show Nue5Ac accumulation, a characteristic present in humans. The Cmah null mice also exhibit many of the problems common in humans, including diminished acoustic sensitivity and startle response threshold, which resulted in hearing loss and delays to skin healing [15,16]. Recently, microarray analysis has been shown to be a powerful tool for the analysis of gene expression and is particularly suited for the identification of transcription factor target genes. In addition, network-assisted analysis of DNA chip data is an emerging area in which network-related approaches are developed and utilized in order to study human diseases or traits. To identify the transcriptional alterations caused by the human-specific loss of Neu5Gc, we examined expression profiling of the main organs, which included the lung, kidney, and heart harvested from wild type (WT) and Cmah null mice. Here, we report the use of network-assisted transcript profiling to various diseases and discuss the options relating to practical applications.

Materials and Methods

Animal ethics

All animal experiments were approved and performed under the guidelines of the Konkuk University Animal Care and Experimentation Community [IACUC approval number: KU12045]. Cmah <tm1Ykoz> knockout mice were kindly provided by RIKEN (Japan). All lines were maintained on a congenic C57Bl/6J background. The mice were allowed to eat and drink ad libitum and were fed with standard mouse chow (Cargill Agri Purina, Inc., Seongnam-Si, Korea). Twelve weeks old wild type and Cmah null male mouse in this study were used.

Immunohistochemistry (IHC)

For IHC, the tissues were fixed in neutral buffer with 10% formalin and then embedded on slides. Endogenous peroxidase activity was blocked using 3% hydrogen peroxide. The samples were then pretreated with Borg Decloaker, and blocked in background Sniper solution. After washing, the samples were incubated with specific primary antibodies for CMAH (Santa Cruz; Texas, USA, 1:100) and Neu5Gc (Sialix; San Diego, CA, USA; 1:200) at 4°C overnight. After the incubation, the samples were washed and incubated with horseradish peroxidase-conjugated secondary antibody. Samples were then stained with ImmPACT^{TM} DAB peroxidase substrate (Vector Laboratories; CA, USA) to visualize the signal. Samples were also stained with Hematoxylin QS to provide background information for reference. The samples were mounted using VECTORSHIELD HardSet mounting medium (Vector Laboratories; CA, USA) and observed using fluorescence microscopy (Olympus; Japan).

Microarray analysis

WT and Cmah null mice with same age (12 weeks) and genetic background (C57BL/6J) were used (n = 3 per each group) for microarray analysis. Total RNA was extracted and purified from the lung, kidney, and heart of WT and Cmah null mice using RNeasy columns (Qiagen; Valencia, CA, USA) according to the manufacturer's protocol. The RNA quality was verified using an Agilent Bioanalyzer 2100 (Agilent Technologies; Palo Alto, CA, USA) using the RNA 6000 Pico Assay. Generation of double-stranded cDNA, preparation and labeling of cRNA, hybridization to Mouse Ref-8 v2.0 Expression BeadChip (Illumina, Inc.; San Diego, CA, USA), washing, and scanning were all performed according to the standard Illumina protocol. Arrays were scanned using the Illumina Bead Array Reader Confocal Scanner. Microarray data have been deposited in NCBI's Gene Expression Omnibus and are accessible through GEO Series accession number GSE59964 (Gene expression profile of the Cmah gene depletion on mouse liver, lung, kidney, and heart).

Raw data preparation and statistical analysis

The quality of hybridization and overall chip performance were monitored by visual inspection of both internal quality control checks and the raw scanned data. The raw data were extracted using the software provided by the manufacturer (Illumina GenomeStudio v2009.2 [Gene Expression Module v1.5.4]). Array data were filtered by a detection p-value < 0.05 in at least 50% of the samples. Selected gene signal values were logarithm-transformed and normalized using the quantile method [17]. Comparative analysis between the wild-type group and Cmah null mice group was carried out based on fold-change in expression levels.

Gene ontology (GO) analysis

GO analysis of the significant probe list was performed using PANTHER (http://www.pantherdb.org/), using text files containing the Gene ID list and accession numbers of the Illumina probe ID. All data analysis and visualization of differentially expressed genes were conducted using R 2.4.1 (www.r-project.org). In addition, the DAVID Functional Annotation Bioinformatics Microarray Analysis tools (http://david.abcc.ncifcrf.gov/) were used to study the biological function of the regulated genes [18].

Kyoto Encyclopedia of Genes and Genomes (KEGG) pathway analysis

KEGG is a collection of online databases dealing with genomes, enzymatic pathways, and biological chemicals [19]. The PATHWAY database records networks of molecular interactions in the cell that includes organism-specific network maps (http://www.genome.jp/kegg/). A total of 9 pathways, involving 49 genes, were collected from KEGG.

Interactome/network analysis of differentially expressed genes

Significantly affected or differentially expressed genes were subjected to a comprehensive search to identify their biological functions. Gene interaction networks, bio functions, and pathway analysis were generated by Ingenuity Pathway Analysis (IPA) (Ingenuity Systems; Mountain View, CA, USA), which assists with microarray data interpretation via grouping of differentially expressed genes into known functions, pathways, and networks primarily based on human and rodent studies. The identified genes were mapped to genetic networks available from the Ingenuity database and were then ranked by score. The significance was set at a p-value of 0.05.

Pathway Studio analysis

To identify molecular pathways, we arranged the data by relation type, using Pathway Studio 9.0 software (Ariadne Genomics; Rockville, MD, USA). This program integrates relevant information among the imported genes, consequently allowing identification of biological pathways, gene regulation networks, and protein interaction maps.

Quantitative real-time polymerase chain reaction (RT-qPCR)

The total RNAs obtained from each tissues (lung, kidney, heart, and liver) used for microarray analysis was reverse-transcribed with the QuantiTect Reverse Transcription Kit (Qiagen; Valencia, CA, USA), according to the manufacturer's recommendations. To assess gene expression, quantitative real-time polymerase chain reaction (RT-qPCR) was performed on an ABI ViiA^{TM} 7 system (Applied Biosystems; Foster City, CA, USA) with SYBR Green as the fluorescence detection method (Bio-Rad; Hercules, CA, USA). Specific primer sets for RT-qPCR were listed in Table S1 and S2. The mouse glyceraldehyde-3-phosphate dehydrogenase (Gadph) gene was used as an internal control to normalize the RT-qPCR efficiency and for quantification of gene expression. After normalization with Gadph expression, we compared the relative expression of each mRNA sample in the Cmah null mice with those of the controls. The RT-qPCR was performed in triplicate for each sample.

Results

Disruption of the Cmah gene completely abolishes Neu5Gc production

The production of Cmah null mice has been described previously [15,16]. The Cmah null mouse was confirmed by PCR and the lack of CMAH protein expression was confirmed by western blotting analysis (data not shown). When Cmah null mice from Cmah$^{+/-}$ × Cmah$^{+/-}$ crosses were examined and weighed, homozygous Cmah null mice were indistinguishable from their wild type or heterozygous littermates in all respects.

IHC was used to determine expression patterns for Cmah and Neu5Gc in the lung, kidney, and heart tissues of control and Cmah null mice at 12 weeks of age, using specific antibodies for CMAH and Neu5Gc. As shown in Figure 1, both the CMAH protein and Neu5Gc epitope were highly expressed in the control mice tissues, whereas the Cmah null mice were completely deficient for both Cmah and Neu5Gc in the lung, kidney, and heart. These observations suggested that the lung, kidney and heart tissues of the Cmah null mouse may contain possible molecular targets associated with xenoantigens relevant to animal-to-human xenotransplantation or human species-specific metabolic syndrome.

Cmah null mice show reduced sialyltransferase-related gene expression

Neu5Gc, also called the H-D antigen, is produced by the CMAH enzyme and is one of the non-Gal xenoantigens for pig-to-human xenotransplantation [10]. Therefore, we examined the sialyltransferase (ST3Gal1–4 and ST6Gal1) expression pattern in the livers of control and Cmah null mice. Gadph gene expression was simultaneously assayed as an endogenous invariant control for data normalization. Compared with the H-D antigen gene expression in the control mouse (1±0.079; 1±0.024; 1±0.056; 1±0.061; 1±0.061; 1±0.005), RT-qPCR analysis showed that Cmah null mice had low ST6Gal1 (0.53±0.012), ST3Gal1 (0.20±0.015), ST3Gal2 (0.39±0.004), ST3Gal3 (0.21±0.012), ST3Gal4 (0.70±0.0016), and ST3Gal6 (0.54±0.079) expression levels, whereas the ST3Gal5 expression level was slightly elevated (1.13±0.079) (Figure 2). Interestingly, the H-D antigen mRNA expression in Cmah null mice was not completely lost, suggesting that the H-D antigen mRNA expression may be compensated when one gene does not function for some reason. Since ST3GAL1, 3, and 4 catalyze the addition of Neu5Ac to the nonreducing terminal of the galactose (Gal) residue of glycans, our observations indicate that down-regulation of ST3GAL1, 3, and 4 and ST6GAL1 expression in Cmah null mice may reduce cytotoxic T cell numbers and therefore attenuate resistance to pathogens and tumorigenic lesions.

Expression levels of the Sialy-Tn antigen-related genes ST6GalNac2 (0.018±0.003), ST6GalNac3 (0.018±0.021), ST6GalNac4 (0.017±0.012), and ST6GalNac6 (0.28±0.023), were significantly lower in the Cmah null mice than in the control mice (1±0.024, 1±0.030, 1±0.052, and 1±0.179, respectively; Figure. 2A). Expression levels of GBGT1 (0.56±0.015) for the Forssman antigen, and of GalNT2 (0.83±0.063), GalNT4 (0.68±0.065), and GalNT6 (0.67±0.031) for the Tn antigen in Cmah null mice were also significantly lower in comparison to those of the control mice (1±0.058, 1±0.066, and 1±0.202, respectively). However, GalNT7 expression for the Tn antigen, unlike other family members, was significantly increased. Considering that the GalNT7 enzyme shows exclusive specificity for partially GalNAc-glycosylated acceptor substrates and shows no activity with non-glycosylated peptides, increased GalNT7 expression in the Cmah null mouse may promote the initiation of the O-glycosylation step.

We next analyzed the binding of natural xenoreactive antibodies in human serum using control (WT), Cmah heterozygote (mKO), and Cmah homozygote (Cmah dKO or Cmah null)) mice. As shown in Figure S1, most of the serum samples from healthy volunteers with different blood groups (A, B, O, and AB) showed the presence of naturally occurring IgM and IgG antibodies that were bound to thymocytes from WT, Cmah-mKO, Cmah-dKO mice. Of note, IgG binding to Cmah dKO mouse-derived thymocytes was significantly reduced in all tested blood groups compared to WT- or Cmah mKO-derived thymocytes. Whereas, there was no significant IgM antibody in the A, O, and AB blood types.

Microarray analysis of gene expression profiles in the lung, kidney, and heart of Cmah null mice

Following microarray analysis using the Illumina MouseRef-8 v2 Expression BeadChip (composed of 25,697 filtered probe sets) using the main tissues (lung, kidney, and heart) from the control and Cmah null mouse groups (n = 3, respectively), all hybridization spots on the image were quantified. The fluorescence intensity data were converted into $\log_{10}$ values. Genes with significantly different expression levels ($P<0.05$) in the lung, kidney, and heart from the Cmah null mouse, relative to the control mouse, were extracted for further analysis. The results showed that 204, 162, and 147 genes were differentially expressed, with a more than 1.5-fold change in the lung, kidney, and heart of Cmah null mice when compared with age- and sex-matched control mice, respectively (Figure 3A). In the lung, 101 genes were up-regulated and 103 genes were down-regulated (File S1). In the kidney, 108 genes were up-regulated and 54 genes were down-regulated (File S2). In the heart, 78 genes were up-regulated and 69 genes were down-regulated (File S3). Even though 2 up-regulated genes (Entpd4 and Dex3y) and 4 down-regulated genes (Nr1d1, Sybl1, LOC10047427, and Cfd) overlapped in all tissues (Table S3), overall, the overlapping between 2 different organs (lung vs. heart, kidney vs. heart, and lung vs. kidney) and 3 different organs (lung vs. kidney vs. heart) was minimal.

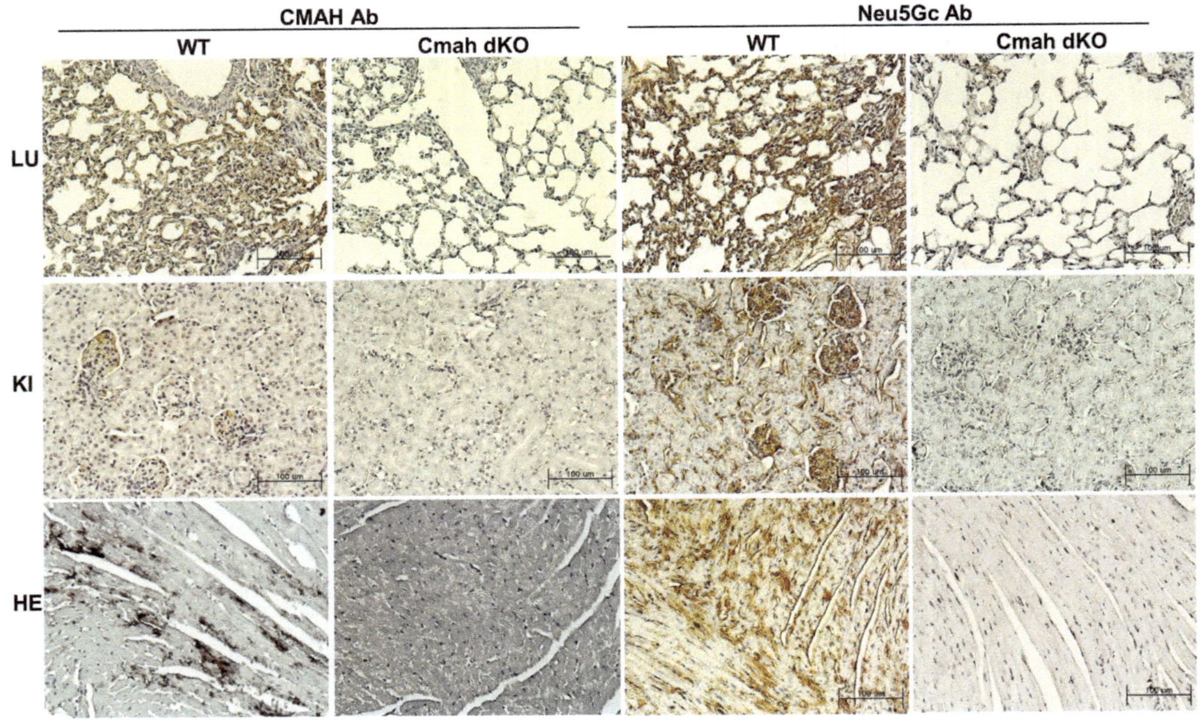

Figure 1. Immunohistochemistry in Cmah-dKO mouse-derived tissues (lung, kidney, and heart) for detection of CMAH and Neu5Gc. WT mice expressed both Cmah and Neu5Gc epitopes. Cmah-dKO mice were completely deficient for both CMAH and Neu5Gc epitopes in the lung, kidney, and heart tissues. Bar: 100 μm. LU; lung, KI; kidney, HE; heart.

GO enrichment, KEGG, and IPA pathway analysis of differentially expressed genes

To understand their biological roles, the genes with significant changes in expression detected in Cmah null mice by microarray analysis were assigned to establish GO classification categories using the PANTHER and DAVID tools. There are 3 ontologies in GO: cellular component, molecular function, and biological process [20]. In GO classification, the top 6 enriched biological processes from the GO analysis were categorized according to their functional role, as shown in Table 1 and Files S1, S2, and S3. These biological groups covered 5 subcategories (protein biosynthesis, protein complex assembly, protein folding, protein modification, and proteolysis) for protein metabolism and modification, 3 subcategories (cell communication, cell surface receptor-mediated signal transduction, and intracellular signaling cascade) for signal transduction, 4 subcategories (fatty acid metabolism, lipid and fatty acid transport, lipid metabolism, and steroid metabolism) for lipid, fatty acid, and steroid metabolism, 10 subcategories (antioxidation and free radical removal, B-cell- and antibody-mediated immunity, blood clotting, complement-mediated immunity, cytokine/chemokine-mediated immunity, detoxification, interferon-mediated immunity, macrophage-mediated immunity, stress response, and T-cell-mediated immunity) for immunity and defense, 6 subcategories (carbohydrate transport, gluconeogenesis, glycogen metabolism, glycolysis, pentose-phosphate shunt, and tricarboxylic acid pathway) for carbohydrate metabolism, and 3 subcategories (mRNA transcription, purine metabolism, pyrimidine metabolism) for nucleoside, nucleotide, and nucleic acid metabolism. Although expression profiles were completely different for each tissue of the Cmah null mice, some genes were shared between tissues in a biological group (Fkbp5, Cfd, Dbp, and Entpd4).

The results generated using the PANTHER classification system is shown in Figure 3B and Table 1. Each gene was assigned to one or more biological groups, according to the function of its proteins. The classification was based on 31 biological groups; the 6 groups that contained the majority of the differentially expressed genes were protein metabolism and modification (33, 24, and 16 genes in the lung, kidney, and heart, respectively), signal transduction (26, 13, and 14 genes in the lung, kidney, and heart), lipid, fatty acid, and steroid metabolism (16, 24, and 8 genes in the lung, kidney, and heart), nucleoside, nucleotide, and nucleic acid metabolism (16, 15, and 13 genes in the lung, kidney, and heart), immunity and defense (15, 14, and 9 genes in the lung, kidney, and heart), and carbohydrate metabolism (12, 11, and 8 genes in the lung, kidney, and heart) (Figure 3B). In addition, GO classification by molecular function showed that differentially expressed genes were assigned 29 different functions (viral protein, cell junction protein, isomerase, chaperone, lyase, ion channel, synthase/synthetase, cell adhesion molecule, extracellular matrix, select calcium binding protein, defense/immunity protein, phosphatase, transfer/carrier protein, membrane traffic protein, protease, ligase, cytoskeletal protein, signaling molecule, transporter, hydrolase, oxidoreductase, miscellaneous function, kinase, transferase, receptor, select regulatory molecule, transcription factor, nucleic acid binding, and molecular function unclassified), while 42 (24 down-regulated genes and 18 up-regulated genes), 23 (5 down-regulated genes and 18 up-regulated genes), and 28 (15 down-regulated genes and 13 up-regulated genes) genes had multiple functions in the lung, kidney, and heart, respectively (Files S1, S2, and S3). The GO classification by cellular component using the DAVID tool showed that 34 differentially expressed genes were mapped to cell fraction, membrane fraction, plasma membrane, etc.; 32 genes mapped to

A

Antigen	Gene	WT	dKO
H-D antigen	ST6Gal1	1±0.079	0.53±0.012**
	ST3Gal1	1±0.024	0.20±0.015***
	ST3Gal2	1±0.056	0.39±0.004**
	ST3Gal3	1±0.061	0.21±0.012**
	ST3Gal4	1±0.061	0.70±0.016*
	ST3Gal5	1±0.195	1.13±0.079
	ST3Gal6	1±0.005	0.54±0.079**
Sialy-Tn antigen	ST6GalNac2	1±0.024	0.18±0.003***
	ST6GalNac3	1±0.030	0.18±0.021***
	ST6GalNac4	1±0.052	0.17±0.012**
	ST6GalNac6	1±0.179	0.28±0.023*
Forssman antigen	GBGT1	1±0.015	0.56±0.015***
Tn antigen	GalNT2	1±0.058	0.83±0.063**
	GalNT3	1±0.233	1.10±0.078
	GalNT4	1±0.066	0.68±0.065**
	GalNT6	1±0.202	0.67±0.031**
	GalNT7	1±0.063	1.34±0.032***

*P<0.05, **P<0.01, ***P<0.001

B

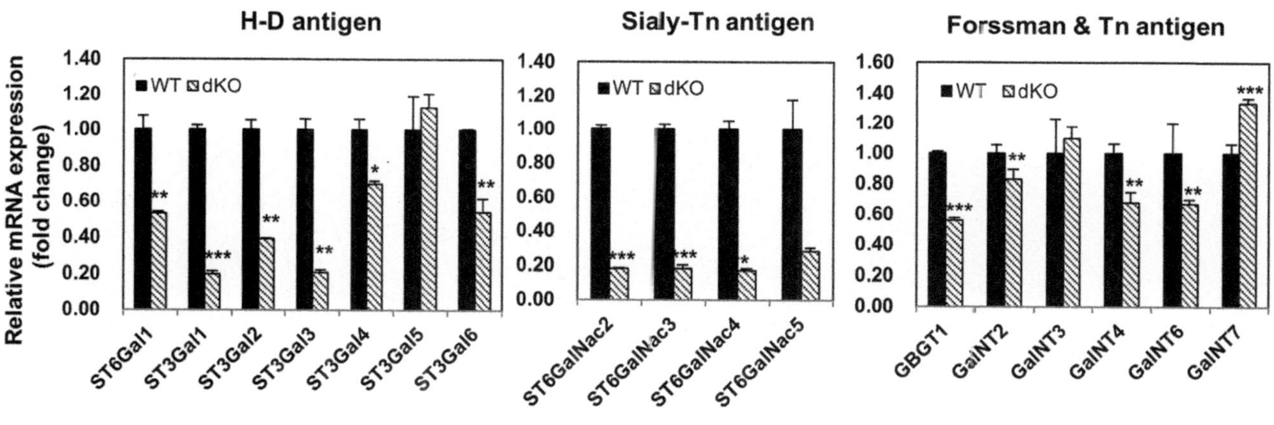

Figure 2. Sialyltransferase gene expression levels in WT and Cmah-dKO mice. A. Comparison of sialyltransferase gene expression in the liver of WT and Cmah-dKO mice by RT-qPCR. B. RT-qPCR analysis in liver of WT and Cmah-dKO mice. All RT-qPCR were conducted in triplicate and normalized with mouse *Gadph* gene expression. The data were presented as mean ± SD. *P<0.05, **P<0.01, ***P<0.001.

extracellular region, insoluble fraction, organelle membrane, etc.; and 36 genes mapped to extracellular matrix, cytoplasmic vesicle, organelle lumen, etc. in the lung, kidney, and heart, respectively (data not shown).

The molecular pathways associated with the differentially expressed genes from the lung, kidney, and heart of Cmah null mice were identified using KEGG pathway analysis. KEGG pathways are manually drawn maps representing well-known molecular interaction and reaction networks. Differentially expressed genes from the lung, kidney, and heart of Cmah null mice were associated with the following pathways (P<0.05): for the lung, retinol metabolism and glycerolipid metabolism; for the kidney, drug metabolism, arachidonic acid metabolism, complement and coagulation cascades, retinol metabolism, fatty acid metabolism, and circadian rhythm; for the heart, focal adhesion (Table 2). These data indicate that deletion of the Cmah gene leads to both up- and down-regulation of gene expression in the lung, kidney, and heart tissues, which may regulate the metabolism and signaling pathway within these tissues.

We then performed IPA, which identified 3 gene interaction networks identified in the Cmah null tissues from the uploaded gene lists, based on the literature contained in the IPA knowledge base (P<0.05) (Table S4). The analysis revealed one network associated with lipid metabolism and small molecular biochemistry that was common to all tissue types (Figure 4). The network was merged and clustered around several central genes, including the up- and down-regulated genes identified from the microarray data (40.63% of the total nodes in the lung, 51.43% of the total nodes in the kidney, and 41.18% of the total nodes in the heart). This network contained 5 up-regulated genes (Cap1, Gpr182, Myl2, Myl3, and Prg2) and 8 down-regulated genes (Cel, Cela1, Clps, Cpa1, Ctrl, Plip, Pnliprp1, and Zg16) in the lung, whereas all the

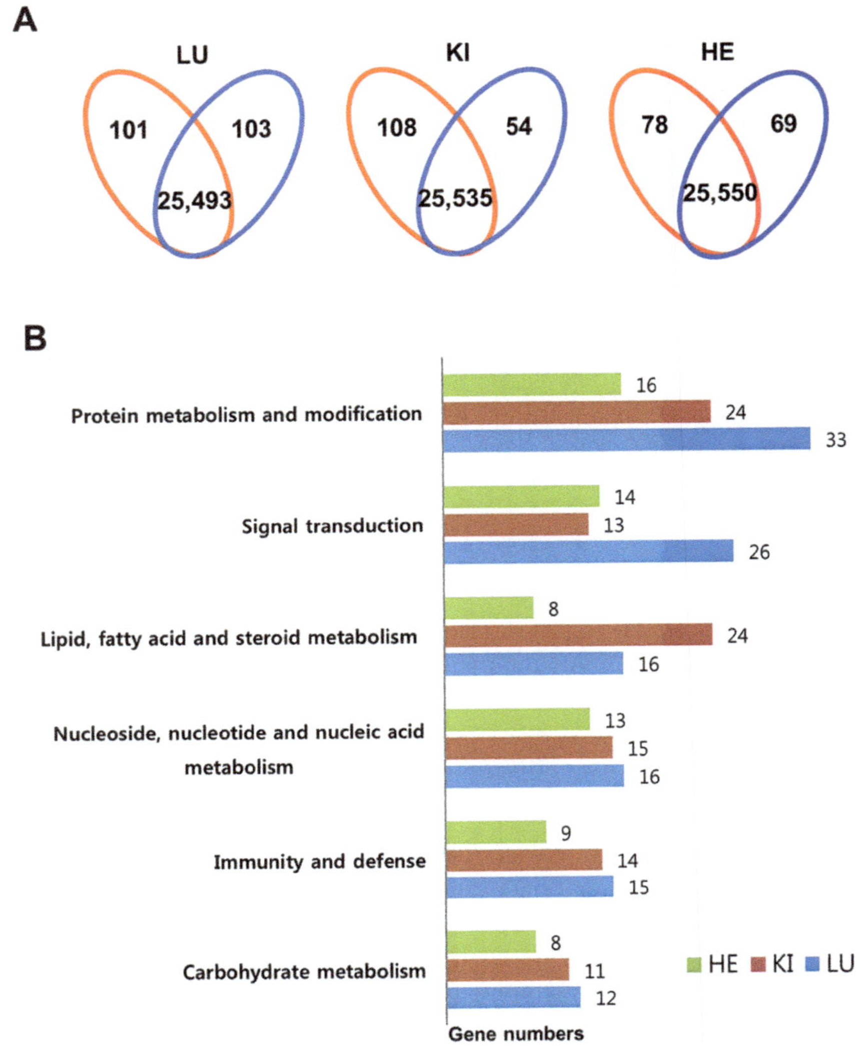

Figure 3. Gene expression profile in the lung, kidney, and heart of Cmah null mice. A. Venn diagram showing differential expression of genes in the lung, kidney, and heart of Cmah null mice. Numbers in red and blue Venn diagram present up- and down- regulated genes, respectively. LU, lung; KI, kidney; HE, heart. B. The differentially up- or down-regulated genes were classified according to biological process in the lung, kidney, and heart of Cmah-dKO mice. LU, lung; KI, kidney; HE, heart. X axis of bar graph indicates the gene numbers.

genes (Abcc3, Ahsg, Akr1c3, Apoa1, Apoc1, Azgp1, Cfd, Cyp27b1, Cyp4a14, Cyp4a22, Hdc, Hp, Hpx, Gtp, Ly6a, Mat1a, Prlr, and Serpina1) were up-regulated in the kidney. Similarly, the network in the heart was composed of 9 up-regulated genes (Adh1c, Dbp, Fkbp5, Gck, Herpud1, Per2, Ptgds, Scgb1a1, and Tef) and 5 down-regulated genes (Alas2, Chad, Egr1, Scd, and Snca).

IPA pathway analysis identified putative disease and disorders in lung, kidney, and heart of Cmah null mouse, respectively. These are as follows: inflammatory response, cardiovascular disease, developmental disorder, and skeletal and muscular diseases for lung, gastrointestinal disease, nutrition disease, and cancer for kidney, neurological disease, cancer, cardiovascular disease, and skeletal and muscular diseases for heart (Table 3). Taken together, the predicted interacting molecular networks in the Cmah null tissues from the IPA suggest that loss of Neu5Gc production has affected the signaling pathways that involved in human disease and disorders. Therefore, this result provides further information for more detailed evaluation of the potential effects of the loss of function of CMAH.

RT-qPCR validation

The microarray expression data were validated by RT-qPCR analysis using genes selected for analysis of KEGG pathway in the lung, kidney, and heart of Cmah null mice. As shown in Figure 5A, the expression levels of Pnlip (pancreatic lipase), Cel (carboxyl ester lipase), Pnliprp1 (pancreatic lipase related protein 1), and Dgat2 (diacylglycerol O-acyltransferase 2) for glycerolipid metabolism in the lung, Cyp4a12b (cytochrome P450, family 4, subfamily a, polypeptide 12B), Adh1 (alcohol dehydrogenase 1), and Cyp4a14 (cytochrome P450, family 4, subfamily a, polypeptide 14) for fatty acid metabolism in the kidney, Actb (actin, beta), Myl7 (myosin, light chain 7), Itgb6 (integrin beta 6), and Pik3r1 (phosphatidylinositol 3-kinase, regulatory subunit, polypeptide 1) for regulation of actin cytoskeleton in the heart corresponded to the microarray results. This shows that the data sets obtained from the microarray analysis accurately reflect the differential gene

Table 1. Top 6 biological processes overrepresented in the lung, kidney, and heart of Cmah null mice.

Biological group	Functional role	Lung	Kidney	Heart	Biological group	Functional role	Lung	Kidney	Heart
Protein metabolism and modification (P = 1.26E-07 in LU, P = 0.019 in KI, P = 0.042 in HE)	Protein biosynthesis	Rps3(−1.58)	Rpl23(−1.62), Eif4ebp2(−1.56)	Rpl23(−1.71)	Immunity and defense (P = 0.022 in LU, P = 6.44E-06 in KI, P = 0.045 in HE)	Cytokine/chemokine mediated immunity	Ccl11(−1.52), Gab3(1.59)		Tnfaip2(−1.97), Ccl9(1.67)
	Protein complex assembly	Hspa1a(1.65)		Hspa8(1.55)		Detoxification		Abcc3(−3.41), Gsta2(−1.76)	Gsta3(1.89), Gpx3(−1.60)
	Protein folding	Fkbp5(−1.94), Dnajb6(1.52), Hspb1(1.70), etc.	Fkbp5(1.53)	Hspa8(1.55), Eif4ebp2 (−1.56), Fkbp5(2.05)		Interferon-mediated immunity	Gbp6(1.54), Gbp10(1.56), Gbp3(1.75), etc.	Oasl2(1.57), Ifit3(1.65), Irf7(1.92)	Gbp2(1.56)
	Protein modification	A4galt(−1.66), Mapk9(1.63), Bmx(1.86), etc.	Wnk1(−1.66), Pdk4(−1.66), Cml2(1.57), Cml3(1.61)	Wnk1(−1.64), Map3k6(1.53), Man2a1(1.69), etc.		Macrophage-mediated immunity	Sdc4(−1.55), Gbp6(1.54), Gab3(1.59), etc.	Asgr1(1.56), Fcgr4(1.80)	Scara3(−2.01), Tnfaip2(−1.97), Gbp2(1.56)
	Proteolysis	Ela2a(−12.87), Try4(−3.12), Cfd(−2.67), Ctrl(−2.61), Cpb1(−2.31), etc.	Cfd(−3.01), Itih1(1.71), Serpina1e(4.72), etc.	Cfd(−4.45), Mmp14(−1.85), Usp2(1.69), Ela2a(4.32), etc.		Stress response	U4606(−4.81), Plunc(−2.61), Hspb1(1.70), Bmx(1.86), etc.	Hp(2.19)	Gpx3(−1.60), Ier3(1.51), Lyz2(1.67), etc.
Signal transduction (P = 0.023 in LU, P = 0.049 in KI, P = 0.043 in HE)	Cell communication	Adam23(−1.76), Per1(−1.52), Cap1(2.77), etc.	Foxq1(−2.17), Per1(−1.58), Foxa3(1.53), Cap1(2.31), etc.	Chad(−2.11), Itgb6(−1.84), Igf2(−1.77), Per2(3.47), etc.		T-cell mediated immunity	Fkbp5(−1.94), Gab3(1.59), Bmx(1.86), etc.	Fkbp5(1.53), Azgp1(2.09)	Sqstm1(1.51), Ab1(1.53), Fkbp5(2.05), etc.
	Cell surface receptor mediated signal transduction	Tff2(−3.65), Cd8b1(−1.65), Edn1(1.64), Gpr182(2.07), etc.	Prlr(−2.41), Tob2(−1.53), Foxa3(1.53), Lrg1(1.66), etc.	Rrad(−1.69), P2ry1(1.66), Ccl9(1.67), etc.	Carbohydrate metabolism (P = 0.0003 in LU, P = 0.007 in KI, P = 0.0003 in HE)	Carbohydrate transport	Slc35b1(1.59)	Slc2a2(−1.52), Pck1(1.55)	Adipoq(−1.64)
	Intracellular signaling cascade	Fkbp5(−1.94), Mapk9(1.63), Bmx(1.86), etc.	Prlr(−2.41), Wnk1(−1.66), Fkbp5(1.53), Cap1(2.31), etc.	Wnk1(−1.64), Pik3r1(1.87), Fkbp5(2.05), etc.		Gluconeogenesis			Adipoq(−1.64)
Lipid, fatty acid and steroid metabolism (P = 1.61E-05 in LU, P = 2.70E-06 in KI, P = 4.74E-05 in HE)	Fatty acid metabolism	Cyp2a5(−2.58), Dgat2(−1.68), Acoxl(−1.53), etc.	Cyp4a14(−2.33), Cyp2a5(1.66), Cyp4a12a(3.28), etc.	Scd1(−2.41), Scd4(−1.97), Adipoq(−1.64), Ptgds(2.01)		Glycogen metabolism	Amy2(−2.63)		

Table 1. Cont.

Biological group	Functional role	Lung	Kidney	Heart	Biological group	Functional role	Lung	Kidney	Heart
	Lipid and fatty acid transport		Apoc3(−1.74), Apoa2(1.95), Apoc1(2.09)			Glycolysis	6430537H07Rik(−1.52)		Pfkm(1.68), Gck(2.68)
	Lipid metabolism	Pnlip(−5.43), Clps(−4.93), Pnliprp1(−3.07), Cel(−2.36), etc.	Lip1(−1.67)	Pnlip(1.74), Clps(1.76), Ptgds(2.01)		Pentose-phosphate shunt	Tkt(1.81)		
	Steroid metabolism		Ugt1a2(−4.03), Cyp27b1(−2.12), Rdh7(1.58), etc.	Gpx3(−1.60), Ela1(1.63)		Tricarboxylic acid pathway	Mdh1b(−1.51)		Idh1(1.59)
Immunity and defense (P = 0.022 in LU, P = 6.44E-06 in KI, P = 0.045 in HE)	Antioxidation and free radical removal		Sepp1(1.87)	Gpx3(−1.60)	Nucleoside, nucleotide and nucleic acid metabolism (P = 0.0002 in LU, P = 4.43E-05 in KI, P = 2.15E-07 in HE)	mRNA transcription	Dbp(−4.22), Nr1d1(−2.26), Fos(−2.03), Per1(−1.52), etc.	Foxq1(−2.17), Dbp(−1.66), Per2(1.51), Foxa3(1.53), etc.	Egr1(−2.75), Fos(−1.56), Dbp(2.74), Per2(3.47), etc.
	B-cell- and antibody-mediated immunity	Vpreb3(−1.70), Cd69(−1.51), Gab3(1.59), etc.	Fcgr4(1.80)			Purine metabolism	Entpd4(1.80)	Entpd4(2.12)	Entpd4(1.65)
	Blood clotting	Anxa8(−1.55)	F2(1.56), Asgr1 (1.56), Fga(1.87)	Angptl4(−1.94), Itgb6(−1.84), Vwf(−1.54)		Pyrimidine metabolism	Nme5(−1.78)	Upp2(1.59)	
	Complement-mediated immunity	Cfd(−2.67)	Cfd(−3.01), C4b(1.89), C3(1.79), etc.	Cfd(−4.45)					

Table 2. KEGG pathways significantly associated with genes altered in the lung, kidney, and heart of Cmah null mice.

Tissues	Term	Count	P value	Genes
Lung	Retinol metabolism	6	0.0006147	Alah1a1, Dgat2, Cyp26b1, Cyp2a5, Adh7, Aldh1a7
	Glycerolipid metabolism	4	0.0121528	Pnlip, Cel, Pnliprp1, Dgat2
Kidney	Drug metabolism	9	0.0000009	Cyp2d9, Gsta2, Fmo5, Ugt2b36, Adh1, Ugt1a2, Cyp2a5, Cyp2d26, Cyp2e1
	Arachidonic acid metabolism	5	0.0107399	Cyp4a12b, Cyp2j13, Cyp4a12a, Cyp2e1, Cyp4a14
	Complement and coagulation cascades	10	0.0000001	Mbl1, Fga, Serpina1b, C3, Serpina1a, C4b, Serpina1d, F2, Serpina1e, Cfd
	Retinol metabolism	7	0.0000671	Cyp4a12b, Ugt2b36, Cyp4a12a, Adh1, Ugt1a2, Cyp2a5, Cyp4a14
	Fatty acid metabolism	4	0.0112964	Cyp4a12b, Cyp4a12a, Adh1, Cyp4a14
	Circadian rhythm	3	0.0077886	Nr1d1, Per2, Per1
Heart	Focal adhesion	8	0.0010197	Actb, Vwf, Myl7, Col4a1, Itgb6, Col1a1, Pik3r1, Chad

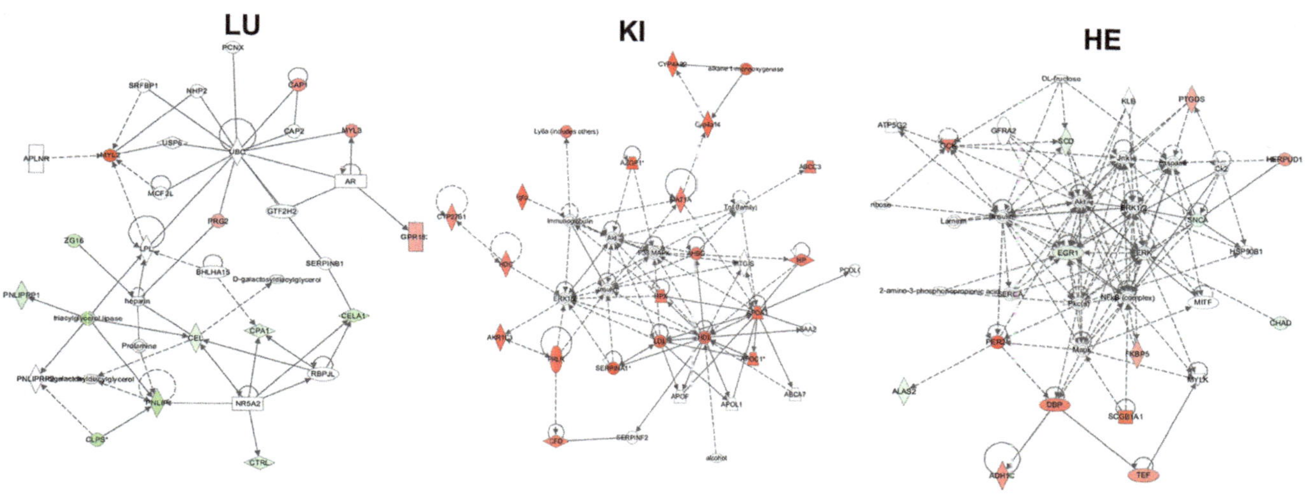

Tissues	Associated Network Functions	Molecules	Score	Focus molecules
Lung	Lipid Metabolism, Small Molecule Biochemistry, Vitamin and Mineral Metabolism	APLNR, AR, BHLHA15, CAP1, CAP2, CEL, CELA1, CLPS, CPA1, CTRL, GPR182, GTF2H2, heparin, LPL, MCF2L, MYL2, MYL3, NHP2, NR5A2, PCNX, PNLIP, PNLIPRP1, PNLIPRP2, PRG2, Protamine RBPJL, SERPINB1, SRFBP1, triacylglycerol lipase, UBC, USP6, ZG16	30	13
Kidney	Lipid Metabolism, Small Molecule Biochemistry, Vitamin and Mineral Metabolism	ABCA7, ABCC3, AHSG, AKR1C3, Akt, alcohol, alkane 1-monooxygenase, APOA1, APOC1, APOF, APOL1, AZGP1, CFD CYP27B1, Cyp4a14, CYP4A22, ERK1/2, HDC, HDL, HP, HPX, gip, Immunoglobulin, Insulin, LDH, Ly6a, MAT1A, P38 MAPK, PCOLCE2, PRLR, PTGIS, SAA2, SERPINA1, SERPINF2, Tnf	48	18
Heart	Lipid Metabolism, Small Molecule Biochemistry, Moleculare Transport	2-amino-3-phosphonopropionic acid, ADH1C, Akt, ALAS2, ATP5G2, caspase, CHAD, Ck2, DBP, DL-fructose, EGR1, ERK, ERK1/2, FKBP5, GCK, GFRA2, HERPUD1, HSP90B1, Insulin, Jnk, KLB, Laminin, Mapk, MYLK, NFkB, PER2, Pkc, PTGDS, ribose, SCD, SCGB1A1, SERCA, SNCA, TEF	34	14

Figure 4. Networks predicted by Ingenuity Pathway Analysis in the Cmah-dKO mice. Upper: The common network identified was lipid metabolism and small-molecule biochemistry in the lung (LU), kidney (KI), and heart (HE), respectively. The network is displayed graphically as nodes (genes). The node color intensity indicates the expression of genes, with red representing up-regulation and green representing down-regulation. Solid lines and dotted lines indicate direct relationship and indirect relationships, respectively. Bottom: The table includes the molecules, score, and focus molecules in identified common networks (lipid metabolism and small molecule biochemistry) from each tissue by IPA (score 30, 13 focus molecules in LU; score 48, 18 focus molecules in KI; score 34, 14 focus molecules in HE). Red and blue denote up-regulated and down-regulated genes among the differentially expressed genes.

Table 3. Disease and disorders predicted by Ingenuity Pathway Analysis in lung, kidney, and heart of Cmah null mice.

Tissues	Lung (molecules)	Kidney (molecules)	Heart (molecules)
Disease and disorders	Inflammatory response (10)	Gastrointestinal disease (13)	Neurological disease (11)
	Cardiovascular disease (3)	Nutrition disease (7)	Cancer (9)
	Developmental disorder (5)	Cancer (21)	Cardiovascular disease (3)
	Skeletal and muscular diseases (6)		Skeletal and muscular diseases (12)

expression in the lung, kidney, and heart between the control and Cmah null mice (Figure 5A).

Putative Cmah interaction network analysis

Heterogeneous high-throughput biological data has become readily available for a number of diseases. However, the amount of data points generated by such experiments does not facilitate manual integration of the information, which is required to design the most optimal therapy for a disease. In this study, we examined a novel computational workflow for designing a therapy strategy using the Pathway Studio Software. Using subnetwork enrichment analysis (Figure 5B), we identified eight CMAH downstream-localized proteins (ERBB2, CMAS, CYB5R3, CTNNB1, IL4, IL17A, IL6, and IFNG), six small molecules (Fe^{2+}, NSC2921, sialate, CMP-N-glycolylneuraminate, N-glycolylneuraminate, CMP-N-acetylneuraminate), three cell processes (immune reactivity, B-cell activation, cell growth), and two diseases (metabolic disease and Duchenne muscular dystrophy). In addition, we identified four CMAH upstream-localized diseases (neoplasm, malaria, Mdx dystrophy, and nematode infections), three small molecules (NaCl, NADH, and ferrosulfate), eight functional classes (serotonin receptor, ncRNA, DNA-directed RNA polymerase,

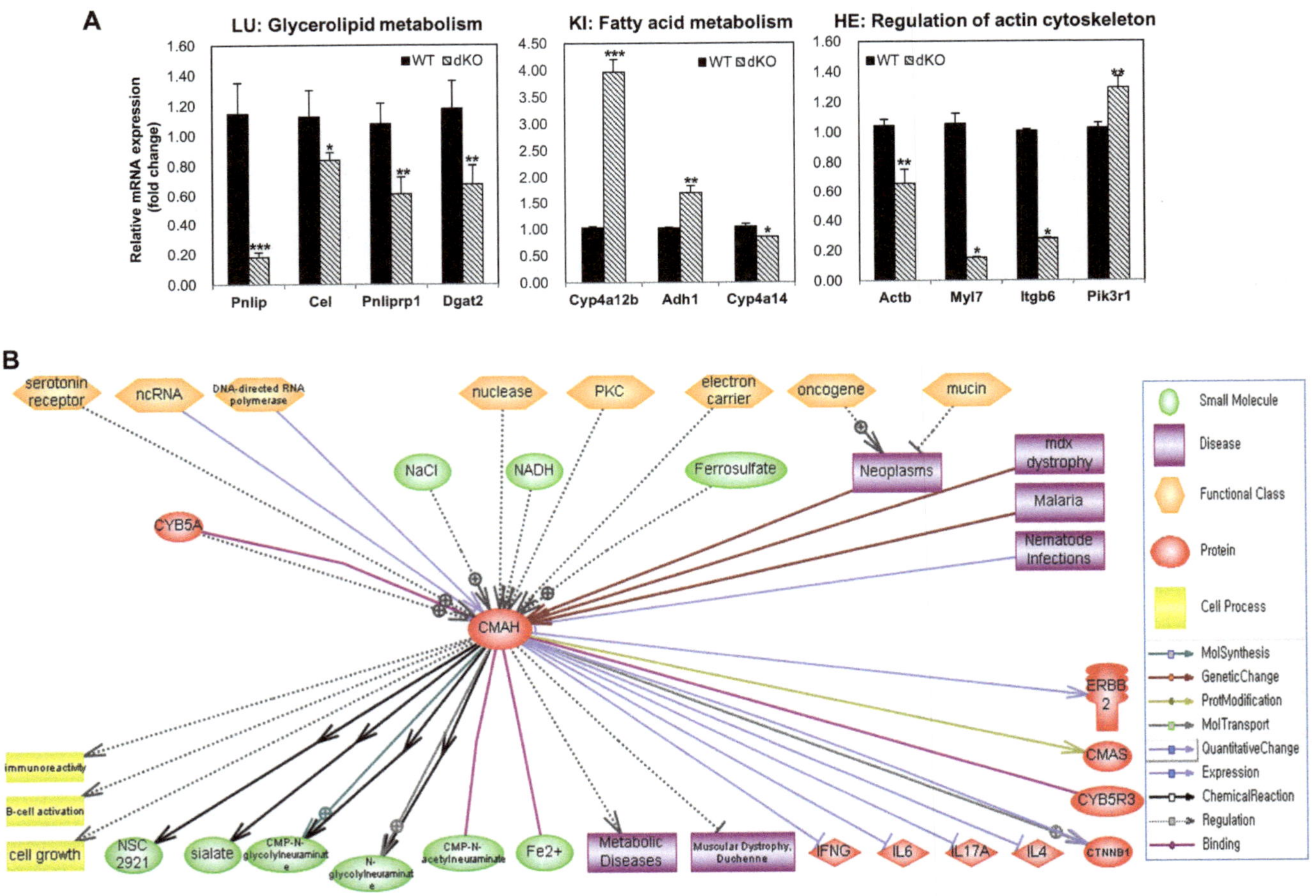

Figure 5. Schematic representation of biological networks was constructed by CMAH function. A. Validation of microarray gene expression by RT-qPCR. Gene expression levels of selected genes were determined by RT-qPCR using WT and Cmah-dKO derived lung, kidney, and heart tissues. Measurements were performed in triplicate, after which the calculated mean expression level was corrected for Gadph gene expression levels. Error bars indicate standard deviations. B. Functional correlation and interaction of both the signaling proteins and Cmah gene were reconstituted into a network model using the Pathway Studio software (Ariadne Genomics Inc.; Rockville, MD, USA). All genes are shown by their gene symbols. Direct and indirect regulation is indicated by colored lines and dotted gray lines, respectively.

nuclease, PKC, electron carrier, oncogene, and muscin), and one protein (CYB5A). As shown in Figure 5B, the data set obtained from a representative potential signaling pathway accurately predicted previously reported metabolic disease models, such as human type II diabetes [21] and human Duchenne muscular dystrophy syndrome [16]. Taken together, this analysis suggests that regulation analysis of both Cmah and other proteins provide a valuable tool to evaluate human-specific diseases, caused by loss of CMAH function during human evolution.

Discussion

Microarray analysis was selected to uncover the underlying molecular mechanisms necessary for animal-to-human xenotransplantation, as well as the metabolic conditions arising from the evolutionary loss of CMAH. The microarray expression data were validated by RT-qPCR analysis of selected genes from the lung, kidney, and heart of Cmah null mice. Overall, the decrease in expression of glycosyltransferases suggests a corresponding reduction in the cell surface sialylation and carbohydrate modification in the Cmah null mouse tissues. From the regulatory network identified from the Cmah null mice, we were able to show that many of the transcription factors and target genes associated with CMAH activity were linked in our network.

Recently, several studies have linked CMAH deactivation to hearing loss [15], skin healing delay [15], a human-like muscular dystrophy phenotype following combined mutation of the Dmd gene [16], abnormal B cell proliferation and antibody production, and type 2 diabetes-like syndromes [22]. To elucidate the role of CMAH in an animal model, a conventional knockout mouse has been constructed and made available to the scientific community [22]. We have also recently reported the first successful CMAH knockout in pigs [12]. First, we compared the expression profiles of Cmah$^{+/+}$ and Cmah $^{-/-}$ mice littermates and between the Cmah null mouse and pig models. In this study, GalNT7 expression for Tn antigen in Cmah null mouse, unlike other family, significantly increased whereas, the expression of this gene in CMAH null pigs was significantly decreased. Moreover, ST6GalNac2 expression for Sialyl-Tn antigen and GalNT2, GalNT3, and GalNT4 expression for Tn antigen in CMAH null pigs were not changed compared to the control [12], whereas the expression of these genes in Cmah null mice was significantly decreased (Figure 2). While considerable differences in sialyltransferase expression between the Cmah null mouse and pig models were observed, their sialyltransferase expression profiles were strikingly similar overall. With respect to expression of H-D antigens, ST6GalNac3, ST6GalNac4, and ST6GalNac6 expression for sialyl-Tn antigen and GalNT6 expression for Tn antigen were all significantly down-regulated in both the Cmah null mouse and CMAH-null pig, suggesting a common expression signature for both models. This observation also suggests that the potential for Cmah null mice to serve as model for human-specific disease is becoming a reality, whereas that they have limitations because of their differences in gene expression and physiology compared to humans.

GO is an international standardized classification system for determining gene function, which supplies a set of controlled vocabulary to comprehensively describe the property of genes and gene products [23]. There are 3 ontologies in GO: cellular component, molecular function, and biological process [20]. In this study, analysis of molecular function showed that 204, 162, and 147 genes were assigned to 29 different functions based on gene background, while 42, 23, and 28 genes were involved in multiple functions in the lung, kidney, and heart, respectively (Files S1, S2, and S3). GO enrichment for gene background, based on the cellular component, showed that 34, 32, and 36 altered genes were mapped in the lung, kidney, and heart to GO terms in the database. As for the KEGG pathway analysis, it is worth noting that the following pathways were significantly (P value< 0.0121528–0.0000001) enriched in the Cmah null mouse (Table 2): for the heart, focal adhesion; for the lung, retinol metabolism and glycerolipid metabolism; for the kidney, drug metabolism, arachidonic acid metabolism, complement and coagulation cascades, retinol metabolism, fatty acid metabolism, and circadian rhythm. The IPA further identified interacting modules involved in lipid metabolism and apoptosis signaling networks, which include fatty acids and their surrogate genes Pnlip (pancreatic lipase), Pnliprp1 (pancreatic lipase-related protein 1), and Pnliprp2 (pancreatic lipase-related protein 2). These significantly enriched pathways may imply that alterations in glycerolipid metabolism and fatty acid metabolism are involved in the pathogenesis of potentially fatal conditions, such as obesity, type II diabetes, and cancer. The results of the GO annotation and KEGG pathway analyses can therefore provide direction for future research. This study identified a common network in the three main organs, as the lung, kidney and heart, which are associated with lipid metabolism and small-molecule biochemistry.

The microarray analysis revealed distinct expression profiles for the lung, kidney, and heart from Cmah$^{+/+}$ and Cmah $^{-/-}$ littermates. The microarray data were analyzed using the Ingenuity System Database software that includes the Ingenuity Knowledge Base (IKB) and the Global Molecular Network (GMN). These databases integrate published findings on biologically meaningful genetic or molecular gene/gene product interactions and identify functionally related gene networks [24]. The differential gene expression values (Cmah$^{+/+}$ versus Cmah $^{-/-}$ litter mates) were entered into the IPA to determine the most highly regulated networks of gene interactions and to highlight the biological processes that are relevant to each of the treatments. Only networks with a score of 30 or higher were selected for further analysis. Networks describing the relationships between a subset of genes, their neighboring genes, and symbols representing the functional categories of the molecules are presented in Figure 4. Focus genes are denoted by red symbols for up-regulated genes and green symbols indicate down-regulated genes. Grey and open symbols are intermediate genes, placed in the network by the Ingenuity Software and shown in the literature to interact with genes in this dataset. Genes in gray have been shown to interact with the colored gene products that appear in this scheme.

The lung network (left of Figure 4) showed a score of 30 and contains 13 differentially regulated genes. Among them, 10 and 13 genes were involved lipid metabolism (P value: 3.21E-07–4.83E-02) and small-molecule biochemistry (3.21E-07–4.83E-02), respectively. The 5 genes (Cel, Lpl, Pnlip, Pnliprp1, and Pnliprp2) in this network have a central role in glycerolipid metabolism. These genes (Pnlip, Pnliprp1, and Pnliprp2) encode key lipolytic enzymes and are known to alter lipid metabolism. Pnliprp1, and Pnliprp2 encode two novel pancreatic lipase-related proteins, referred to as Pnlip-related proteins 1 and 2. Both these proteins have an amino acid sequence identity of 68% to Pnlip [25]. Pnliprp2 shows lipolytic activity that is marginally dependent on the presence of colipase, whereas the function of Pnliprp1 remains unclear [26]. Overall, these functionally related lipases play key roles in directing the regulation of fatty acid turnover and signaling. The most common symptoms associated with lipase deficiency are muscle spasms, acne, arthritis, gallbladder stress, formation of gallstones, bladder problems, and cystitis [27]. Therefore, our study has also provided the first evidence that a lower expression of

Pnlip in Cmah null mice may be closely associated with above mentioned disease symptoms. The kidney network (middle of Figure 4), with the highest score (48), includes 18 genes. Among them, 10 and 13 genes were involved lipid metabolism (P value: 3.94E-087–4.82E-02) and small-molecule biochemistry (3.94E-08–4.82E-02), respectively. The heart network (right of Figure 4) received a score of 34 and contains 14 differentially regulated genes. Among them, 9 and 10 genes were involved lipid metabolism (P value: 1.14E-05–4.56E-02) and small-molecule biochemistry (1.14E-05–4.99E-02), respectively. Taken together, these data suggested that the gene interacting network predicted in the Cmah null tissues could be critical in understanding the role of evolutionary loss of CMAH function in human metabolism disorders.

PathwayAssist is a software application developed for navigation and analysis of biological pathways, gene regulation networks, and protein interaction maps [28]. Using Pathway Studio, we identified a number of protein interaction pathways with important roles following the evolutionary loss of CMAH function (Figure 5B). Comparing previously reported diseases caused by loss of CMAH function, we were able to accurately predict metabolic diseases, including human type II diabetes and human Duchenne muscular dystrophy syndrome, caused by CMAH inactivation. Therefore, our observations suggested that accuracy and efficiency in the interpretation of CMAH function is improved by Pathway Studio analysis.

Herein we identified 8 disease and disorders identified in the Cmah null tissues, based on the literature contained in the IPA knowledge base ($P<0.05$) (Table 3). Of them, specific disorders in Cmah null mice were closely associated with cardiovascular disease (Ptgds, Scd, and Egrl)-, skeletal and muscular disease (Adh1c, Dbp, Dkk3, Egr1, Herpud1, Scd, Snca, Ptgds, Timp4, Myl4, Alas2, and Fkbp5)-, and cancer (Cfd, Egr1, Scd, Timp4, Per2, Scgb1a1, Spon2, Slc46a3, and Gck)-associated genes in heart (Table 3). In conclusion, these findings suggest that CMAH inactivation could affect a variety of signaling pathways accompanying orchestrated gene expression changes. Therefore, our results argue that evolutionary loss of Neu5Gc affect complex regulation of cellular signaling pathways that involved in human diseases.

Supporting Information

Figure S1 Binding of natural xenoreactive antibodies in human sera to thymocytes from WT, Cmah-mKO (+/−; hetero) and Cmah-dKO (−/−; homo) mice. Indirect immunofluorescence staining of thymocytes from WT, Cmah-mKO, and Cmah-dKO mice was used to detect xenoreactive antibody levels in healthy human serum samples (blood group: A, B, O, and AB). Detection of IgM or IgG binding was achieved by further incubating the cells with DyLight 649-conjugated Monkey anti-Human IgM or DyLight 488 Monkey anti-Human IgG Abs. Histogram profiles show differences in binding of natural antibodies present in human sera to the different thymocytes. Green lines indicate binding to thymocytes from WT; pink lines, binding to thymocytes from Cmah-mKO; sky blue lines, binding to thymocytes from Cmah-dKO, and gery lines indicate negative control WT thymocytes stained with secondary antibody alone. (TIF)

Table S1 Primer sets used for detection of sialyltransferases mRNA expression.
(DOCX)

Table S2 Primer sets for real-time RT-qPCR.
(DOCX)

Table S3 Common up- or down-regulated genes in lung, kidney, and heart of Cmah null mouse.
(DOCX)

Table S4 Networks predicted by Ingenuity Pathway Analysis in lung, kidney, and heart of Cmah-null mice.
(DOCX)

File S1 UP or down-regulated genes in lung of Cmah-null mice.
(XLSX)

File S2 UP or down-regulated genes in kidney of Cmah-null mice.
(XLSX)

File S3 UP or down-regulated genes in heart of Cmah-null mice.
(XLSX)

Author Contributions

Conceived and designed the experiments: DNK JHK. Performed the experiments: DNK BSC. Analyzed the data: DNK BSC. Contributed reagents/materials/analysis tools: DNK BSC. Contributed to the writing of the manuscript: DNK JHK.

References

1. Sandrin MS, McKenzie IF (1994) Gal alpha (1,3)Gal, the major xenoantigen(s) recognised in pigs by human natural antibodies. Immunol Rev 141: 169–190.
2. Griesemer A, Yamada K, Sykes M (2014) Xenotransplantation: immunological hurdles and progress toward tolerance. Immunol Rev 258: 241–258.
3. Ekser B, Rigotti P, Gridelli B, Cooper DK (2009) Xenotransplantation of solid organs in the pig-to-primate model. Transpl Immunol 21: 87–92.
4. Schauer R, Wember M (1971) Hydroxylation and O-acetylation of N-acetylneuraminic acid bound to glycoproteins of isolated subcellular membranes from porcine and bovine submaxillary glands. Hoppe Seylers Z Physiol Chem 352: 1282–1290.
5. Buscher HP, Casals-Stenzel J, Schauer R, Mestres-Ventura P (1977) Biosynthesis of N-glycolylneuraminic acid in porcine submandibular glands. Subcellular site of hydroxylation of N-acetylneuraminic acid in the course of glycoprotein biosynthesis. Eur J Biochem 77: 297–310.
6. Miwa Y, Kobayashi T, Nagasaka T, Liu D, Yu M, et al. (2004) Are N-glycolylneuraminic acid (Hanganutziu-Deicher) antigens important in pig-to-human xenotransplantation? Xenotransplantation 11: 247–253.
7. Irie A, Koyama S, Kozutsumi Y, Kawasaki T, Suzuki A (1998) The molecular basis for the absence of N-glycolylneuraminic acid in humans. J Biol Chem 273: 15866–15871.
8. Chou HH, Takematsu H, Diaz S, Iber J, Nickerson E, et al. (1998) A mutation in human CMP-sialic acid hydroxylase occurred after the Homo-Pan divergence. Proc Natl Acad Sci U S A 95: 11751–11756.
9. Hayakawa T, Satta Y, Gagneux P, Varki A, Takahata N (2001) Alu-mediated inactivation of the human CMP- N-acetylneuraminic acid hydroxylase gene. Proc Natl Acad Sci U S A 98: 11399–11404.
10. Chen G, Qian H, Starzl T, Sun H, Garcia B, et al. (2005) Acute rejection is associated with antibodies to non-Gal antigens in baboons using Gal-knockout pig kidneys. Nat Med 11: 1295–1298.
11. Ezzelarab M, Ayares D, Cooper DK (2005) Carbohydrates in xenotransplantation. Immunol Cell Biol 83: 396–404.
12. Kwon DN, Lee K, Kang MJ, Choi YJ, Park C, et al. (2014) Production of biallelic CMP-Neu5Ac hydroxylase knock-out pigs. Sci Rep 3: 1981.
13. Varki A (2010) Colloquium paper: uniquely human evolution of sialic acid genetics and biology. Proc Natl Acad Sci U S A 107 Suppl 2: 8939–8946.
14. Varki NM, Strobert E, Dick EJ Jr, Benirschke K, Varki A (2011) Biomedical differences between human and nonhuman hominids: potential roles for uniquely human aspects of sialic acid biology. Annu Rev Pathol 6: 365–393.
15. Hedlund M, Tangvoranuntakul P, Takematsu H, Long JM, Housley GD, et al. (2007) N-glycolylneuraminic acid deficiency in mice: implications for human biology and evolution. Mol Cell Biol 27: 4340–4346.

16. Chandrasekharan K, Yoon JH, Xu Y, deVries S, Camboni M, et al. (2010) A human-specific deletion in mouse Cmah increases disease severity in the mdx model of Duchenne muscular dystrophy. Sci Transl Med 2: 42ra54.
17. Bolstad BM, Irizarry RA, Astrand M, Speed TP (2003) A comparison of normalization methods for high density oligonucleotide array data based on variance and bias. Bioinformatics 19: 185–193.
18. Huang da W, Sherman BT, Lempicki RA (2009) Systematic and integrative analysis of large gene lists using DAVID bioinformatics resources. Nat Protoc 4: 44–57.
19. Kanehisa M, Goto S (2000) KEGG: kyoto encyclopedia of genes and genomes. Nucleic Acids Res 28: 27–30.
20. Ashburner M, Ball CA, Blake JA, Botstein D, Butler H, et al. (2000) Gene ontology: tool for the unification of biology. The Gene Ontology Consortium. Nat Genet 25: 25–29.
21. Kavaler S, Morinaga H, Jih A, Fan W, Hedlund M, et al. (2011) Pancreatic beta-cell failure in obese mice with human-like CMP-Neu5Ac hydroxylase deficiency. FASEB J 25: 1887–1893.
22. Naito Y, Takematsu H, Koyama S, Miyake S, Yamamoto H, et al. (2007) Germinal center marker GL7 probes activation-dependent repression of N-glycolylneuraminic acid, a sialic acid species involved in the negative modulation of B-cell activation. Mol Cell Biol 27: 3008–3022.
23. Harris MA, Clark J, Ireland A, Lomax J, Ashburner M, et al. (2004) The Gene Ontology (GO) database and informatics resource. Nucleic Acids Res 32: D258–261.
24. Jimenez-Marin A, Collado-Romero M, Ramirez-Boo M, Arce C, Garrido JJ (2009) Biological pathway analysis by ArrayUnlock and Ingenuity Pathway Analysis. BMC Proc 3 Suppl 4: S6.
25. Cao H, Hegele RA (2003) DNA polymorphisms of lipase related genes. J Hum Genet 48: 443–446.
26. Reboul E, Berton A, Moussa M, Kreuzer C, Crenon I, et al. (2006) Pancreatic lipase and pancreatic lipase-related protein 2, but not pancreatic lipase-related protein 1, hydrolyze retinyl palmitate in physiological conditions. Biochim Biophys Acta 1761: 4–10.
27. Zhang G, He P, Tan H, Budhu A, Gaedcke J, et al. (2013) Integration of metabolomics and transcriptomics revealed a fatty acid network exerting growth inhibitory effects in human pancreatic cancer. Clin Cancer Res 19: 4983–4993.
28. Nikitin A, Egorov S, Daraselia N, Mazo I (2003) Pathway studio–the analysis and navigation of molecular networks. Bioinformatics 19: 2155–2157.

Heart Mitochondrial Proteome Study Elucidates Changes in Cardiac Energy Metabolism and Antioxidant PRDX3 in Human Dilated Cardiomyopathy

Esther Roselló-Lletí[1], Estefanía Tarazón[1], María G. Barderas[2], Ana Ortega[1], Manuel Otero[3], Maria Micaela Molina-Navarro[1], Francisca Lago[3], Jose Ramón González-Juanatey[3], Antonio Salvador[4], Manuel Portolés[5], Miguel Rivera[1]*

1 Cardiocirculatory Unit, Health Research Institute Hospital La Fe, Valencia, Spain, **2** Department of Vascular Physiopathology, Hospital Nacional de Parapléjicos, SESCAM, Toledo, Spain, **3** Cellular and Molecular Cardiology Research Unit, Department of Cardiology and Institute of Biomedical Research, University Clinical Hospital, Santiago de Compostela, Spain, **4** Cardiology Service, Hospital La Fe, Valencia, Spain, **5** Cell Biology and Pathology Unit, Health Research Institute Hospital La Fe, Valencia, Spain

Abstract

Background: Dilated cardiomyopathy (DCM) is a public health problem with no available curative treatment, and mitochondrial dysfunction plays a critical role in its development. The present study is the first to analyze the mitochondrial proteome in cardiac tissue of patients with DCM to identify potential molecular targets for its therapeutic intervention.

Methods and Results: 16 left ventricular (LV) samples obtained from explanted human hearts with DCM (n = 8) and control donors (n = 8) were extracted to perform a proteomic approach to investigate the variations in mitochondrial protein expression. The proteome of the samples was analyzed by quantitative differential electrophoresis and Mass Spectrometry. These changes were validated by classical techniques and by novel and precise selected reaction monitoring analysis and RNA sequencing approach increasing the total heart samples up to 25. We found significant alterations in energy metabolism, especially in molecules involved in substrate utilization (ODPA, ETFD, DLDH), energy production (ATPA), other metabolic pathways (AL4A1) and protein synthesis (EFTU), obtaining considerable and specific relationships between the alterations detected in these processes. Importantly, we observed that the antioxidant PRDX3 overexpression is associated with impaired ventricular function. PRDX3 is significantly related to LV end systolic and diastolic diameter ($r = 0.73$, p value$<$ 0.01; $r = 0.71$, p value<0.01), fractional shortening, and ejection fraction ($r = -0.61$, p value<0.05; and $r = -0.62$, p value$<$ 0.05, respectively).

Conclusion: This work could be a pivotal study to gain more knowledge on the cellular mechanisms related to the pathophysiology of this disease and may lead to the development of etiology-specific heart failure therapies. We suggest new molecular targets for therapeutic interventions, something that up to now has been lacking.

Editor: Vincenzo Lionetti, Scuola Superiore Sant'Anna, Italy

Funding: This work was supported by grants from the NIH "Fondo de Investigaciones Sanitarias del Instituto Carlos III," (RD12/0042/0003; FIS Project PI10/00275; PI13/00100) and Health Research Institute Hospital La Fe (Project 07/2013). The funders had no role in study design, data collection and analysis, decision to publish, or preparation of the manuscript.

Competing Interests: The authors have declared that no competing interests exist.

* Email: miguelrivera492@gmail.com

Introduction

Heart failure (HF), a major and growing public health problem, is a current worldwide pandemic with an unacceptable high level of morbidity and mortality in industrialized countries and with no curative treatment currently available. Dilated cardiomyopathy (DCM), one of the most frequent causes of HF, is a severe pathology of unknown etiology characterized by impaired systolic function with increased ventricular mass, volume, and wall thickness [1,2]. The mechanisms underlying the development of this cardiomyopathy are multiple, complex, and not well understood.

Mitochondria are the major energy production sites within cells [3]. Cardiac energy deficits have been reported in the failing heart, with convincing evidence of the important effect of mitochondrial dysfunction in the development and progression of HF in human and animal models resulting from its central role in energy production, metabolism, calcium homeostasis, oxidative stress, and cell death [4–8]. Some studies identify mitochondria as both the target and origin of major pathogenic pathways that cause myocardial dysfunction [9]. Nevertheless, the mitochondria-specific role and the proteins contributing to HF are unclear. In earlier studies, this organelle has been studied using experimental models and classic biochemical methods [10–12]. These studies usually focused on only one particular protein rather than the whole cardiac mitochondrial proteome, despite the fact that methods designed to enrich and purify the mitochondria represent one of the most long-standing examples of proteome subfractionation [13–15]. Thus, characterization of the mitochondrial proteome could provide new insight into cardiac dysfunction and suggest new molecular targets for the therapeutic intervention of DCM. However, the mitochondrial proteome has not been analyzed in pathological human hearts.

Here, we isolate mitochondria from left ventricular (LV) samples of explanted human hearts with DCM and use a proteomic approach to investigate the variations in mitochondrial protein expression. Our results identify the overexpression of several proteins involved mainly in energy metabolism but also in stress response and protein synthesis in dilated human hearts. We focus on seven representative mitochondrial proteins with different expressions in control (CNT) and diseased hearts validated by different classical techniques as well as novel and precise selected reaction monitoring (SRM) analysis and RNA sequencing (RNAseq) approach. We find that some proteins involved in the different components of cardiac energy metabolism and protein biosynthesis could have an important role in this cardiomyopathy. LV dysfunction is directly related with the antioxidant PRDX3 expression in DCM.

Materials and Methods

Ethics statement

The project was approved by the Ethics Committee of Hospital La Fe, Valencia, and all participants gave their written, informed consent. The study was conducted in accordance with the guidelines of the Declaration of Helsinki [16].

Tissue sources

The experiments were performed using LV samples from explanted human hearts from Caucasian patients with DCM undergoing cardiac transplantation. Clinical history, hemodynamic study, electrocardiography, and Doppler echocardiography data were available from all of these patients. Non-ischemic DCM was diagnosed when patients had LV systolic dysfunction (ejection fraction, <40%) with a dilated non-hypertrophic LV (LV diastolic diameter, >55 mm) on echocardiography. Moreover, none of the patients had existing primary valvular disease or a familial history of DCM. All of the patients were functionally classified according to the New York Heart Association (NYHA) criteria and were receiving medical treatment following the guidelines of the European Society of Cardiology [17].

Non-diseased donor hearts were used as CNT samples. The hearts were initially considered for transplantation but were subsequently deemed unsuitable either because of blood type or size incompatibility. The cause of death was cerebrovascular or motor vehicle accident. All donors had normal LV function and no history of myocardial disease or active infection at the time of the transplantation.

Transmural samples were taken from the region around the apex of the left ventricle and stored at 4°C for a maximum of 6 h from the time of coronary circulation loss. The samples were stored at −80°C until the mitochondrial isolation was performed. Of the 25 heart samples, 16 were used in the proteomic analysis (DCM, n = 8; CNT, n = 8). The 25 heart samples were used in the validation to improve the numerical base with a higher number of patients (DCM, n = 17; CNT, n = 8).

Mitochondrial isolation and proteomic analysis

Mitochondrial isolation was performed using standard homogenization, protease digestion, and differential centrifugation methods as previously described by Imahashi *et al.* [15]. The isolation was made from 50 mg of left ventricular tissue, obtaining a final protein concentration of 10 μg/μl.

2D-DIGE

Protein samples were precipitated using a 2-D Clean-Up Kit (GE Healthcare) as per the manufacturer's protocol. Samples were labeled using CyDye DIGE Fluor Minimal Dyes (GE Healthcare) according to the manufacturer's recommendations. The samples were resuspended in 20 μL of labeling buffer containing 7 M urea, 2 M thiourea, 4% CHAPS, and 30 mM Tris. The pH was then checked in every sample to ensure an optimal labeling reaction (pH 8.0–9.0). Aliquots (50 μg protein in 10 μL) were separated into individual tubes and a pooled internal standard was generated by mixing of equal amounts of all samples included in the experiment. A total of 400 pmol of the appropriate CyDye was added to each sample according to the experimental protocol. This experimental design included 2 experimental groups and 8 samples per group, and thus 16 samples were labeled with either Cy3 or Cy5. The labeling reactions were developed for 30 min on ice in the dark and then quenched by incubation with 1 μL of 10 mM lysine for 10 min in the dark. The labeled samples were combined according to the experimental design and loaded onto the same immobilized pH-gradient (IPG) strip. We randomized the CyDye assignments and sample combinations in each gel, which also included one aliquot of the pooled internal standard labeled with Cy2.

Two-dimensional electrophoresis

The combined samples were diluted with a rehydration solution containing 8 M urea, 4% CHAPS, 20 mM DTT, and 1% IPG buffer (pH 3–11 NL; GE Healthcare) to a final volume of 450 μL. The IPG strips (24 cm; pH 3–11 NL) were allowed to rehydrate overnight in a re-swelling tray (GE Healthcare). After the strips were passively rehydrated, they were placed in an Ettan IPGphor Manifold ceramic tray (GE Healthcare) and isoelectric focusing (IEF) was performed in an Ettan IPGphor 3 unit (GE Healthcare) at 20°C according to the following program: 1-h gradient to 1000 V, 1.5-h gradient to 1500 V, 2-h gradient to 3500 V, 4-h constant at 3500 V, 2-h gradient to 6000 V, and, finally, constant at 6000 V to constitute a total of 60 kV/h. After IEF, the strips were equilibrated for 20 min in 1.5 M Tris (pH 8.8) buffer containing 6 M urea, 30% glycerol, 2% sodium dodecyl sulfate (SDS), and bromophenol blue to which 1% DTT had been added, followed by equilibration for 20 min more in the same buffer to which 2.5% iodoacetamide had been added. SDS–polyacrylamide gel electrophoresis (SDS-PAGE) was performed overnight as described by Laemmli [18] by using the Ettan DALT*six* Electrophoresis System (GE Healthcare) at 1 W/gel.

Image acquisition and analysis

After SDS-PAGE, the gels were scanned with a Typhoon 9400 fluorescence gel scanner (GE Healthcare, Piscataway, NJ) using appropriate individual excitation and emission wavelengths, filters and photomultiplier (PTM) values that are sensitive for each of the Cy3, Cy5 and Cy2 dyes (PTM values: 480 nm, 490 nm, 500 nm, respectively). Relative protein quantification was performed on AS and healthy valves with DeCyder software v6.5 (GE Healthcare) and the multivariate statistical module EDA (Extended data analysis). The Differential in-gel analysis (DIA) module co-detected the 3 images of a gel (the internal standard and the two samples), measured the spot abundance in each image, and expressed these values as Cy3/Cy2 and Cy5/Cy2 ratios. These DIA datasets were then analysed using the Biological Variation Analysis module (BVA), which enabled the spot maps to be matched and the Cy3/Cy2 and Cy5/Cy2 ratios to be compared. Only protein spots with >1.5-fold differences in abundance were considered for the analysis. A statistical analysis was then carried out to determine the changes in protein species, with P-values below 0.05 accepted as significant when the Students t-test was applied. The gels were then re-stained with a silver staining kit (GE-Healthcare).

In-gel protein digestion

Protein spots were excised manually and then automatically digested using the Ettan Digester (GE Healthcare). We used the digestion protocol previously described by Schevchenko *et al.* [19] with minor variations: the gel plugs were reduced using 10 mM DTT (Sigma-Aldrich; St. Louis, MO, USA) in 50 mM ammonium bicarbonate (99% purity; Scharlau) and alkylated using 55 mM iodoacetamide (Sigma-Aldrich) in 50 mM ammonium bicarbonate. The gel pieces were then rinsed with 50 mM ammonium bicarbonate in 50% methanol (gradient, high-performance liquid chromatography [HPLC] grade; Scharlau) and acetonitrile (gradient, HPLC grade; Scharlau) and dried in a Speedvac. Modified porcine trypsin (sequencing grade; Promega, Madison, WI, USA) at a final concentration of 20 ng/µL in 20 mM ammonium bicarbonate was added to the dry gel pieces, and the digestion was allowed to proceed at 37°C overnight. Finally, 60% aqueous acetonitrile and 0.5% trifluoroacetic acid (99.5% purity; Sigma-Aldrich) were added to extract peptides.

MALDI-MS(/MS) and database searching

A total of 0.5 µL of each digestion solution was deposited using the thin-layer method onto a 384 Opti-TOF 123×81 mm MALDI plate (Applied Biosystems) and allowed to dry at RT. The same volume of matrix (3 mg/mL α-cyano-4-hydroxycinnamic acid (Sigma-Aldrich) in 60% acetonitrile/0.5% trifluoroacetic acid) was applied on every sample in the MALDI plate. MALDI-MS(/MS) data were obtained in an automated analysis loop by using a 4800 Plus MALDI TOF/TOF Analyzer (Applied Biosystems). Spectra were acquired in the reflector positive-ion mode by using an Nd:YAG 355-nm wavelength laser at 200 Hz laser frequency, and 1,000–2,000 individual spectra were averaged. The spectra were acquired uniformly by using fixed laser intensity. For the MS/MS 1-kV analysis mode, precursors were accelerated to 8 kV in Source 1 and selected using a relative resolution of 200 (FWHM) and metastable suppression. Fragment ions generated by collision with air in a CID chamber were further accelerated by 15 kV in Source 2. The mass data were automatically analyzed using the 4000 Series Explorer Software, Version 3.5.3 (Applied Biosystems). The MALDI-TOF mass spectra were internally calibrated using 2 trypsin autolysis ions with m/z = 842.510 and 2211.105. In the case of MALDI-MS/MS, calibrations were performed using the fragment-ion spectra obtained for Glub-fibrinopeptide (4700 Cal Mix; Applied Biosystems). MALDI-MS and MS/MS data were combined in the GPS Explorer Software Version 3.6 to search a non-redundant protein database (Swiss-Prot 2012_08) by using the Mascot software, version 2.2 (Matrix Science) [20], featuring 50 ppm precursor tolerance, 0.6-Da MS/MS fragment tolerance, carbamidomethyl cysteine as the fixed modification, and oxidized methionine as the variable modification, and allowing for 1 missed cleavage. The MALDI-MS (/MS) spectra and database search results were manually inspected in detail using the aforementioned software. In the case of the combined MS and MS/MS data, identifications were accepted when the confidence interval (CI%) calculated using the GPS software was ≥95%. Because Protein Scores and Ion Scores obtained from distinct searches cannot be directly compared, the GPS software calculates the CI% to combine the results of MS and MS/MS database searches. This coefficient value refers to a <5% probability of the observed match being a random event. In the case of the PMF spectra, identifications were accepted when the CI% was ≥99%.

Gel electrophoresis and western blot analysis

Protein samples for the detection of pyruvate dehydrogenase E1 component subunit α, somatic form (ODPA), electron transfer flavoprotein-ubiquinone oxidoreductase (ETFD), dihydrolipoyl dehydrogenase (DLDH), delta-1-pyrroline-5-carboxylate dehydrogenase (AL4A1), ATP synthase subunit α (ATPA), elongation factor Tu (EFTU) and thioredoxin-dependent peroxide reductase (PRDX3) were separated using Bis-Tris electrophoresis on 4–12% polyacrylamide gels under reducing conditions. After the electrophoresis, the proteins were transferred from the gel to a polyvinylidene difluoride membrane using an iBlot Dry Blotting System (Invitrogen Ltd., UK) for western blot analyses. The primary detection antibodies used were anti-pyruvate dehydrogenase E1-alpha subunit mouse monoclonal antibody (1:400), anti-EFTDH mouse monoclonal antibody (1:400), anti-lipoamide dehydrogenase rabbit monoclonal antibody (1:1000), anti-ATP5A mouse monoclonal antibody (1:300), anti-TUFM rabbit polyclonal antibody (1:2000), and anti-peroxiredoxin 3 mouse monoclonal antibody (1:1000) (all obtained from Abcam, Cambridge, UK); and anti-ALDH4A1 mouse monoclonal antibody (1:500) from Sigma-Aldrich. Anti-COX IV rabbit polyclonal antibody (1:200) (Thermo Scientific, Rockford, IL, USA) was used as a loading control.

The bands were visualized using an acid phosphatase-conjugated secondary antibody and nitro blue tetrazolium/5-bromo-4-chloro-3-indolyl phosphate (NBT/BCIP, Sigma-Aldrich) substrate system. Finally, the bands were digitalized using an image analyzer (DNR Bio-Imagining Systems, Israel) and quantified using the GelQuant Pro (v12.2) program.

Fluorescence microscopy

Human myocardial LV samples were fixed in 4% formalin, embedded in paraffin, cut into 5-µm sections, and mounted on superfrost glass slides. Sections were kept at 60°C overnight, deparaffinized with xylol followed by washing in 100%, 96%, 80%, and 70% ethanol. The samples were then blocked with phosphate buffered saline (PBS) containing 1% bovine serum albumin (BSA) for 15 min at RT. After blocking, the sections were incubated for 120 min at RT with the primary antibodies (described in the Western Blot Analysis section above) in the same buffer solution, and then with Alexa-conjugated secondary antibody (Invitrogen, USA) for 60 min at RT [21]. Finally, the sections were rinsed in PBS, mounted in Vectashield-conjugated 4′,6-diamidino-2-phenylindole (DAPI) for identifying the nucleous (Vector Laboratories, Burlingame, CA, USA), and examined under an Olympus BX50 fluorescence microscope (Tokyo, Japan). The images were processed by ImageJ (v. 1.4.3.67) software.

Immunocytochemistry and electron microscopy

Myocardial samples (size 1 mm^3) from the LV were fixed in a solution of 1.5% glutaraldehyde and 1% formaldehyde in 0.05 M cacodylate buffer (pH 7.4) for 1 h at 4°C. The samples were then post-fixed in 1% OsO_4 for 1 h at 4°C, dehydrated in ethanol, and embedded in Epon 812. Ultra-thin sections measuring 80 nm were obtained and mounted on nickel grids and counter-stained with 2% uranyl acetate for 20 min and 2.7% lead citrate for 3 min [22,23].

For immunogold labeling, ultra-thin sections were floated for 30 min on 0.1% BSA-Tris buffer (20 mM Tris-HCl, 0.9% NaCl [pH 7.4] containing 0.1% BSA, type V) and 2 h in a moist chamber at RT on sodium metaperiodate [24]. After being rinsed with bi-distilled water, the sections were incubated for 5 min with 3% H_2O_2. The grids were rinsed again with bi-distilled water and incubated separately in a moist chamber overnight at RT with the primary antibodies (described in the Western Blot Analysis section above) in the 0.1% BSA-Tris buffer. After being rinsed with 0.1% BSA-Tris buffer, the sections were incubated in a moist chamber for 1 h at 37°C with 0.1% BSA-Tris buffer (containing 0.05% Tween-20) and a goat anti-rabbit IgG-gold antibody (10 nm, 1:10 dilution; Sigma) for DLDH and EFTU and a goat anti-mouse IgG-gold antibody (5 nm, 1:10 dilution; Sigma) for ODPA, EFTD, AL4A1, ATPA, and PRDX3.

After rinses with 0.1% BSA-Tris buffer and bi-distilled water, the sections were air dried and counterstained first with uranyl acetate for 30 min and then with lead citrate for 5 s. Finally, the grids were air dried completely. For the electron microscopy observation, a Philips CM-100 was used, with magnifications ranging X4500–15000.

Selected reaction monitoring (SRM)

Protein samples were reduced by incubating them with 100 mM DTT (Sigma Aldrich) in 50 mM ammonium bicarbonate (99% purity; Scharlau) for 30 min at 37°C. After reduction, alkylation with 55 mM iodoacetamide (Sigma Aldrich) in 50 mM ammonium bicarbonate was conducted for 20 min at RT. Next, we added 50 mM ammonium bicarbonate, 15% acetonitrile (LCMS grade, Scharlau), and, finally, sequencing-grade modified porcine trypsin (Promega) at a final ratio of 1 μg trypsin: 50 μg protein. After digestion at 37°C overnight, 2% formic acid (99.5% purity; Sigma Aldrich) was added and samples were cleaned using Pep-Clean spin columns (Pierce) according to the manufacturer's instructions. Tryptic digests were dried in a Speedvac and resuspended in 2% acetonitrile/2% formic acid prior to MS analysis.

The LC-MS/MS system consisted of a TEMPO nano LC system (Applied Biosystems) combined with a nano LC Autosampler and coupled to a modified triple quadrupole (Applied Biosystems 4000 QTRAP LC/MS/MS System). Three replicate injections (4 μL containing 8 μg of protein) were made for each sample (except 2 samples with only 1 injection per sample) by using mobile phase A (2% ACN/98% water, 0.1% FA) at a flow rate of 10 μL/min for 5 min. Peptides were loaded onto a μ-Precolumn Cartridge (Acclaim Pep Map 100 C18; 5 μm, 100Å; 300 μm i.d. ×5 mm, LC Packings) to preconcentrate and desalt samples. Reversed-phase LC was performed on a C18 column (Onyx Monolithic C18; 150×0.1 mm i.d., Phenomenex) in a gradient of phase A and phase B (98% ACN/2% water, 0.1% FA). Peptides were eluted at a flow rate of 900 nL/min by following these steps: 2–15% B for 2 min, 15–30% B for 18 min, 30–50% B for 5 min, 50–90% B for 2 min, and, finally, 90% B for 3 min. The column was then regenerated with 2% B for another 15 min. Both TEMPO nano LC and 4000 QTRAP system were controlled using the Analyst Software, v.1.4.5. Theoretical SRM transitions were designed using MRMpilot software v1.1 (ABSciex), with the following settings: Enzyme = trypsin, missed cleavages = 0; modifications in peptide ≤3; charge states = +1 from 300 to 600 Da, +2 from 500 to 2000 Da, +3 from 900 to 3000 Da, +4 from 1600 to 4000 Da, +5 from 2400 to 10,000 Da; studied modification = - none; fixed modifications = carboxyamidomethylation; variable modifications = none; min. number of amino acids ≥5; max. number of amino acids ≤30; ignore multiple modification sites; 3 transitions per peptide (Table_S2 in File S1). A pool containing a mixture of all the samples was digested as described previously and analyzed in the 4000QTrap using a MIDAS acquisition method that included the theoretical transitions. Transitions were selected when the three co-eluting peaks (corresponding to the three transitions of the same peptide) had a signal-to-noise ratio over 5 and the MS/MS data matched the theoretical spectrum for that peptide.

The mass spectrometer was set to operate in the positive-ion mode with an ion-spray voltage of 2800 V and a nanoflow interface heater temperature of 150°C. Source gas 1 and curtain gas were set to 20 and 20 psi, respectively, and nitrogen was applied as both curtain and collision gases. Collision energy was optimized to obtain maximal transmission efficiency and sensitivity for each SRM transition. A total of 42 MRM transitions (3 per peptide) were monitored during the analysis of each sample and were acquired at unit resolution in both Q1 and Q3, with dwell times of 20 and 50 ms that resulted in a cycle time of 1.2303 s. The IntelliQuan algorithm included in the Analyst 1.4.5 software was used to calculate abundances based on the peak areas after integration.

To carry out this analysis we pooled the samples (four samples per pool). Correlations between proteins were evaluated by calculating Pearson product–moment correlation coefficient of every transition from every peptide analyzed for each protein with respect to every transition of the peptides of the other proteins. If the results of 1 out of the 3 assayed transitions were divergent from the other two, this transition was not considered. While evaluating correlations, when 2 or more peptides were measured, the most significant peptide was selected, while the second one was used as a qualifier for the correlation, which was rejected if this second peptide had divergent results. Pearson's correlation coefficient (r) was calculated as a mean of all coefficients from all transitions from the most significant peptide-to-peptide correlation. When any of the transition-to-transition correlations was not significant, but close to signification, the greater transition-to-transition p value of the most significant peptide-to-peptide correlation was considered.

RNA extraction

Heart samples were homogenized in TRIzol reagent in a TissueLysser LT (Qiagen, UK). All RNA extractions were performed using a PureLink Kit according to the manufacturer's instructions (Ambion Life Technologies, CA, USA). RNA was quantified using a NanoDrop1000 spectrophotometer (Thermo Fisher Scientific, UK), and the purity and integrity of the RNA samples were measured using an Agilent 2100 Bioanalyzer with an RNA 6000 Nano LabChip kit (Agilent Technologies, Spain). All samples showed a 260/280 ratio of >2.0 and an RNA integrity number of ≥9.

RNAseq

The RNA samples were isolated using a MicroPoly(A) Purist Kit (Ambion, USA). The total polyA-RNA samples were used to generate whole transcriptome libraries that were sequenced on a SOLiD 5500XL platform as per the manufacturer's recommendations (Life Technologies, CA). The amplified cDNA quality was analyzed using the Bioanalyzer 2100 DNA 1000 kit (Agilent Technologies, Spain), and the cDNA was quantified using the Qubit 2.0 Fluorometer (Invitrogen, UK). Whole transcriptome libraries were used to generate SOLiD templated beads by following the SOLiD Templated Bead Preparation guide. Bead quality was estimated based on WFA (workflow analysis) parameters. The samples were sequenced using the 50625 paired-end protocol, which generated 75 nt+35 nt (Paired-End) +5 nt (Barcode) sequences. Quality data were measured using the SETS software parameters (SOLiD Experimental Tracking System).

Computational analysis of RNAseq data

The initial whole transcriptome paired-end reads obtained from the sequencing were mapped against the latest version of the human genome (Version GRchr37/hg19) by using the Life Technologies mapping algorithm (http://www.lifetechnologies.com/). The aligned records were reported in the BAM/SAM format [25]. Bad quality reads (Phred score <10) were eliminated using the Picard Tools software [26].

The isoform and gene predictions were subsequently estimated using the cufflinks method [27], and the expression levels were calculated using the HTSeq software [28]. The Edge method was applied to analyze the differential expression between conditions [29]. This method relies on a Poisson model to estimate the RNAseq data variance for differential expression. We selected genes and isoforms that were calculated to exhibit p value<0.05 and fold-change >1.5.

Statistics

Data are presented as mean ± standard deviation (SD). The Kolmogorov–Smirnov test was used to analyze the normal distribution of the variables. Comparisons between 2 groups were performed using Student's t-test, while Pearson's correlation coefficient was calculated to analyze the association between variables. Analyses were considered significant when p value< 0.05. All statistical analyses were performed using SPSS software v. 20 for Windows (IBM SPSS Inc., Chicago, IL, USA).

Results

Patients' clinical characteristics

LV tissue samples were obtained from 17 patients with DCM (82% men; mean age, 54±11 years; ejection fraction, <40%). These patients had a New York Heart Association (NYHA) functional classification of III–IV and were previously diagnosed with significant comorbidities including hypertension, and diabetes mellitus. The patients' clinical and echocardiographic characteristics are summarized in Table 1. Eight non-diseased donor hearts were used as CNT samples (63% men; mean age, 55±8 years; ejection fraction, >50%).

Differentially expressed mitochondrial proteins in patients with HF of dilated etiology

The protein expressions of purified heart mitochondria from 8 DCM patients and 8 CNT were compared using two-dimensional differential gel electrophoresis (2D-DIGE). Each gel contained the mitochondrial proteome of DCM patients, CNT samples, and an internal standard. Gel images were imported into DeCyder Differential Analysis Software, which detected 1,171–1,418 protein spots. Reproducibility was tested by comparing the variation within the different gels in the same group. The t-test statistical analysis did not show significant differences. We focused on the identification of up- and downregulation of spot intensities where the fold change was ≥1.5 (p value<0.05). Considering these criteria, statistical analysis of the data in the DeCyder software revealed changes in the abundance of 19 protein spots corresponding to 17 mitochondrial proteins revealed by mass spectrometry. We encountered 16 significantly upregulated and 3 downregulated spots in DCM hearts (Figure_S1 in File S1). The tandem mass spectrometry (MS/MS) identification details are summarized in Table_S1 in File S1.

Table 2 shows that most of the proteins altered in the mitochondrial proteome of DCM patients are involved in cardiac energy metabolism, some implicated in substrate utilization such as ETFD or DLDH, while others are implicated in energy production such as ATPA. The remaining identified mitochondrial proteins were basically structural or implicated in the protein synthesis and stress response, such as coiled-coil-helix-coiled-coil-helix domain-containing protein 3, EFTU, and PRDX3, respectively. The majority of the altered proteins belong to the mitochondrial matrix or the inner membrane.

Validation of protein differential abundance and mRNA levels

A selection of representative proteins of each mitochondrial function and/or localization and involved in heart damage was validated using different techniques that compared its levels among DCM patients (n = 17) and CNT (n = 8). Therefore, using western blot techniques, we determined the levels of some relevant proteins involved in metabolism, including ODPA, ETFD, DLDH, AL4A1, ATPA; protein synthesis EFTU; and stress response PRDX3. As shown in Figures 1a-g, levels of all of the analyzed molecules were significantly increased in the pathological samples (ODPA, 152±19 vs. 100±18 au, p value<0.0001; ETFD, 157±35 vs. 100±37 au, p value<0.01; DLDH, 132±42 vs. 100±25 au, p value<0.05; AL4A1, 193±75 vs. 100±43 au, p value<0.01; ATPA, 149±36 vs. 100±25 au, p value<0.01; EFTU, 200±51 vs. 100±53 au, p value<0.0001; and PRDX3, 148±57 vs. 100±23 au, p value<0.01). These results coincided with those of the proteomic analysis.

We determined whether there was a relationship between these protein levels and the clinical characteristics shown in Table 1. The LV function parameters were completely available in 15 of 17 samples from DCM patients. We found significant relationships between PRDX3 levels and LV function (Fig. 1h). The results obtained showed that PRDX3 is significantly correlated with LV end systolic diameter, LV end diastolic diameter, fractional shortening, and ejection fraction (r = 0.73, p value<0.01; r = 0.71, p value<0.01; r = −0.61, p value<0.05; and r = −0.62, p value<0.05, respectively).

The immunofluorescence study findings were consistent with the increased levels observed by western blotting and proteomic analysis showing that the intensity of all validated proteins was higher in the dilated hearts than in the CNT samples. These proteins had a diffuse cytoplasmic distribution with a significantly higher percentage of fluorescence in the dilated group (ODPA, 21%, p value<0.01 [Fig. 2a]; ETFD, 44%, p value<0.01 [Fig. 2b]; DLDH, 24%, p value<0.05 [Fig. 2c]; AL4A1, 32%, p value<0.01 [Fig. 2d]; ATPA, 18%, p value<0.01 [Fig. 2e]; EFTU, 42%, p value<0.05 [Fig. 2f]; PRDX3, 29%, p value< 0.05 [Fig. 2g]).

Immunocytochemistry studies confirmed the previous results and the localization or distribution of these mitochondrial proteins. Figure 3 shows an increase in immunogold labeling in ODPA, ETFD, DLDH AL4A1, ATPA, EFTU, and PRDX3 in DCM hearts compared to CNT hearts. We observed similar localization of each protein in the pathological and CNT samples.

To validate the previous analyses and to evaluate the possible relationship between the altered proteins involved in the different mitochondrial processes, these molecules were monitorized by SRM (Table_S2 and S3 in File S1). Differential expression was confirmed in all three transitions per peptide (p value<0.05). We

Table 1. Clinical and echocardiographic characteristics of heart failure patients.

	DCM (n = 17)
Age (years)	54±11
Gender male (%)	82
BMI (kg/m^2)	26±5
Prior hypertension (%)	29
Diabetes mellitus (%)	29
NYHA class	3.3±0.4
Hemoglobin (mg/dL)	13±2
Hematocrit (%)	40±7
Total cholesterol (mg/dL)	144±43
Duration of disease (months)	85±69
Echo-Doppler study	
Ejection fraction (%)	24±7
Fractional shortening (%)	13±4
Left ventricular end systolic diameter (mm)	66±13
Left ventricular end diastolic diameter (mm)	75±13
Left ventricle mass (g)	484±132
Left ventricle mass index (g/cm^2)	255±80

Duration of disease from diagnosis of heart failure until heart transplant. BMI, body mass index; DCM, dilated cardiomyopathy; NYHA, New York Heart Association.

obtained a good relationship between the abundance of altered proteins involved in substrate utilization (ODPA vs. DLDH, r = 0.58, p value<0.05; ODPA vs. ETFD, r = 0.65, p value<0.01; DLDH vs. ETFD, r = 0.59, p value<0.05) and with other altered molecules implicated in the metabolic process (ODPA vs. AL4A1, r = 0.73, p value<0.001; DLDH vs. AL4A1, r = 0.58, p value< 0.05; ETFD vs. AL4A1, r = 0.76, p value<0.01) and protein synthesis (ODPA vs. EFTU, r = 0.66, p value<0.01; DLDH vs. EFTU, r = 0.74, p value<0.001; ETFD vs. EFTU, r = 0.67, p value<0.01). In addition, these molecules showed a relevant correlation with ATPA (ODPA vs. ATPA, r = 0.77, p value< 0.001; DLDH vs. ATPA, r = 0.74, p value<0.001; ETFD vs. ATPA, r = 0.61, p value<0.01; AL4A1 vs. ATPA, r = 0.78, p value<0.001; EFTU vs. ATPA, r = 0.89, p value<0.0001).

The mRNA differences between DCM patients and CNT were determined by RNAseq. The mRNA levels of the ODPA (*PDHA1*), ETFD (*ETFDH*), AL4A1 (*ALDH4A1*), ATPA (*ATP5A1*), and EFTU (*TUFM*) genes were increased in the DCM group (18-fold, p value<0.05; 62-fold, p value<0.01; 88-fold, p value<0.01; 79-fold, p value<0.01; and 25-fold, p value< 0.05, respectively) (Fig. 4a). These results also coincided with those of proteomic analysis demonstrating the same tendency between gene expression and protein levels. However, the DLDH (*DLD*) and PRDX3 (*PRDX3*) gene expressions did not reach statistical significance. Finally, we obtained a good relationship between the mRNA expression of altered molecules involved in substrate utilization (*PDHA1* vs. *ETFDH*, r = 0.60, p value<0.01) and with other altered genes implicated in the metabolic process (*PDHA1* vs. *ALDH4A1*, r = 0.50, p value<0.05; *ETFDH* vs. *ALDH4A1*, r = 0.62, p value<0.01) and protein synthesis (*PDHA1* vs. *TUFM*, r = 0.54, p value<0.05; *ETFDH* vs. *TUFM*, r = 0.70, p value< 0.01). In addition, these molecules showed a relevant correlation with *ATP5A1* (*ETFDH* vs. *ATP5A1*, r = 0.75, p value<0.001; *ALDH4A1* vs. *ATP5A1*, r = 0.62, p value<0.001; *TUFM* vs. *ATP5A1*, r = 0.70, p value<0.001) (Fig. 4b).

Discussion

Mitochondrial proteome of DCM human hearts

In the present study, we carried out a 2D-DIGE analysis of isolated cardiac mitochondria from the LV tissue of DCM patients to investigate mitochondrial cardiac protein expression changes. The mechanisms responsible for mitochondrial dysfunction in human hearts are poorly understood and the animal models used to elucidate it may not reflect the true pathophysiology of DCM. There are some studies based on proteomic approach in failing hearts [30–33], however this is the first study to analyze the mitochondrial proteome in pathological human hearts. As described so far, isolating and analyzing single organelles is a right approach to uncover essential information regarding their regulation and interplay [13].

Because the heart consumes more ATP than any other organ, it is particularly rich in mitochondria. During HF development, both energy demands and metabolism change show a remarkable decrease in oxidative phosphorylation and a shift toward glucose over fatty acid utilization [5]. Mitochondria are a major source and target of free radicals, and the collapse of the mitochondrial transmembrane potential can initiate the signaling cascades involved in apoptosis [34]. They also play an important role in antioxidant regeneration and are responsible for the majority of ATP production [35]. A recent study by Ahuja *et al.* showed that DCM is clearly associated with enhanced mitochondrial biogenesis and mtDNA deletion, making evident the essential role of mitochondria in this form of HF [36]. Thus, analysis of the mitochondrial proteome could provide new insights into cardiac dysfunction in DCM patients. Here, we identified 19 protein spots corresponding to 17 mitochondrial proteins altered in failing hearts (14 increased, 3 decreased). These changes comprise many aspects of mitochondrial function, including metabolism, transport, respiratory chain, stress response, protein synthesis, and cell death.

Table 2. Mitochondrial proteins differentially regulated in dilated cardiomyopathy *vs.* controls.

Spot	Accesion code	Protein name	Fold-change	*p*-value	Main Localization	Function
324	ETFD_HUMAN	Electron transfer flavoprotein-ubiquinone oxidoreductase	+1.74	0.047	Inner membrane	Metabolism/Transport
335	CH60_HUMAN	60 kDa heat shock protein	+1.95	0.023	Matrix	Stress response
344	ETFD_HUMAN	Electron transfer flavoprotein-ubiquinone oxidoreductase	+1.77	0.001	Matrix/Inner membrane	Metabolism/Transport
348	DLDH_HUMAN	Dihydrolipoyl dehydrogenase	+2.25	0.008	Matrix	Metabolism
361	AL4A1_HUMAN	Delta-1-pyrroline-5-carboxylate dehydrogenase	+1.79	0.002	Matrix	Metabolism
407	DLDH_HUMAN	Dihydrolipoyl dehydrogenase	+1.73	0.019	Matrix	Metabolism
472	ODO2_HUMAN	Dihydrolipoyllysine-residue succinyltransferase component of 2-oxoglutarate dehydrogenase complex	+1.51	0.0001	Matrix	Metabolism
601	ODPA_HUMAN	Pyruvate dehydrogenase E1 component subunit alpha, somatic form	+2.21	0.001	Matrix	Metabolism
604	ACADM_HUMAN	Medium-chain specific acyl-CoA dehydrogenase	+1.79	0.027	Matrix	Metabolism
	ODPA_HUMAN	Pyruvate dehydrogenase E1 component subunit alpha, somatic form			Matrix	Metabolism
614	KCRS_HUMAN	Creatine kinase S-type	−2.01	0.001	Inner membrane	Metabolism
689	EFTU_HUMAN	Elongation factor Tu	+2.41	0.008	Mitochondrion	Protein biosynthesis
732	MDHM_HUMAN	Malate dehydrogenase	−1.75	0.008	Matrix	Metabolism
861	CY1_HUMAN	Cytochrome c1, heme protein	+1.71	0.007	Inner membrane	Metabolism/Respiratory chain
	ECH1_HUMAN	Delta(3,5)-Delta(2,4)-dienoyl-CoA isomerase			Matrix	Metabolism
882	ATPA_HUMAN	ATP synthase subunit alpha	+2.23	0.01	Inner membrane	Metabolism/Respiratory chain
904	ATPA_HUMAN	ATP synthase subunit alpha	+1.93	0.049	Inner membrane	Metabolism/Respiratory chain
960	CHCH3_HUMAN	Coiled-coil-helix-coiled-coil-helix domain-containing protein 3	+1.79	0.006	Inner membrane	Structural
1019	PRDX3_HUMAN	Thioredoxin-dependent peroxide reductase	+1.73	0.031	Mitochondrion	Stress response
	NDUV2_HUMAN	NADH dehydrogenase [ubiquinone] flavoprotein 2			Inner membrane	Metabolism/Respiratory chain
1039	PRDX3_HUMAN	Thioredoxin-dependent peroxide reductase	+1.59	0.046	Mitochondrion	Stress response
1140	PRDX5_HUMAN	Peroxiredoxin-5	−1.89	0.023	Mitochondrion	Stress response

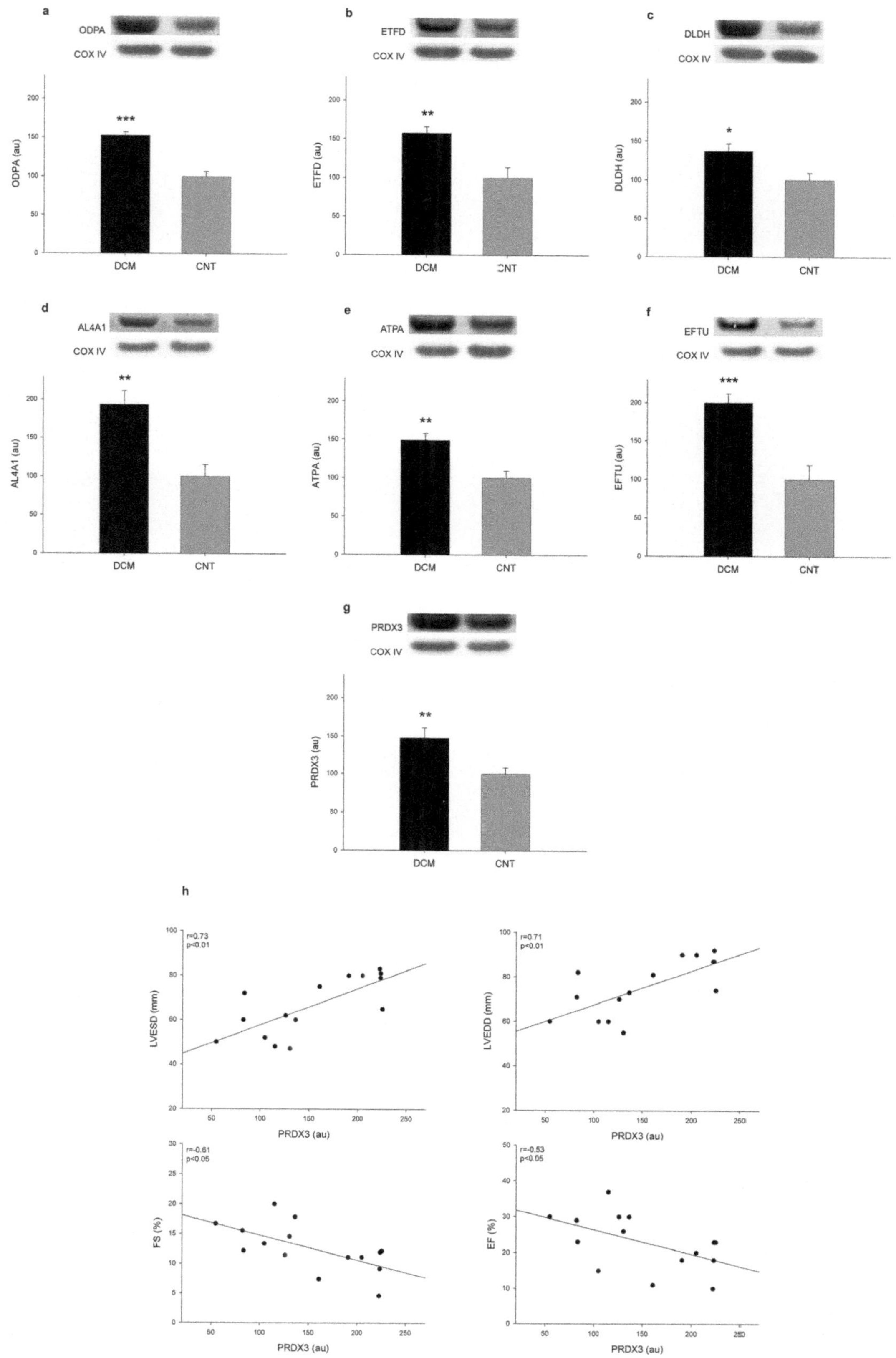

Figure 1. Mitochondrial protein overexpression in dilated human hearts and relationship between PRDX3 and left ventricular function. (a–g) The influence of dilated cardiomyopathy on the amount of each representative protein involved in cardiac energy metabolism (ODPA, ETFD, DLDH, AL4A1, and ATPA), protein biosynthesis (EFTU), and stress response (PRDX3) analyzed using western blotting techniques. As shown, all proteins were significantly increased in the DCM group (n = 17) compared with the CNT group (n = 8). The values from the controls were set to 100. Values were normalized to COX IV and finally to the CNT group. The data are expressed as mean+SEM in arbitrary units (optical density).

Images are representative of the results obtained for all of the patients with DCM and the CNT included in the study. (h) Scatter plots showing the relationship between PRDX3 protein levels and left ventricular function, specifically with fractional shortening, left ventricular end systolic diameter, and left ventricular end diastolic diameter. CNT, control; DCM, dilated cardiomyopathy; ODPA, pyruvate dehydrogenase E1 component subunit α, somatic form; ETFD, electron transfer flavoprotein-ubiquinone oxidoreductase; DLDH, dihydrolipoyl dehydrogenase; AL4A1, delta-1-pyrroline-5-carboxylate dehydrogenase; ATPA, ATP synthase subunit α; EFTU, elongation factor Tu; PRDX3, thioredoxin-dependent peroxide reductase; FA, fractional shortening; LVESD, left ventricular end systolic diameter; LVEDD, left ventricular end diastolic diameter. *p value<0.05, **p value<0.01, ***p value<0.0001.

Alterations in substrate utilization in DCM human hearts

Impaired cardiac energy metabolism is known to play a major role in HF [4,6]. Our results are in concordance with these studies, as we found that most of the proteins altered in the mitochondrial proteome of DCM patients are involved in cardiac energy metabolism, 12 of the 17 differentially regulated proteins identified (70%) are involved in this process. The metabolic machinery has three components: substrate utilization, energy production, and energy transfer and utilization. The main way that myocytes compensate for decreasing mitochondrial function is by increasing

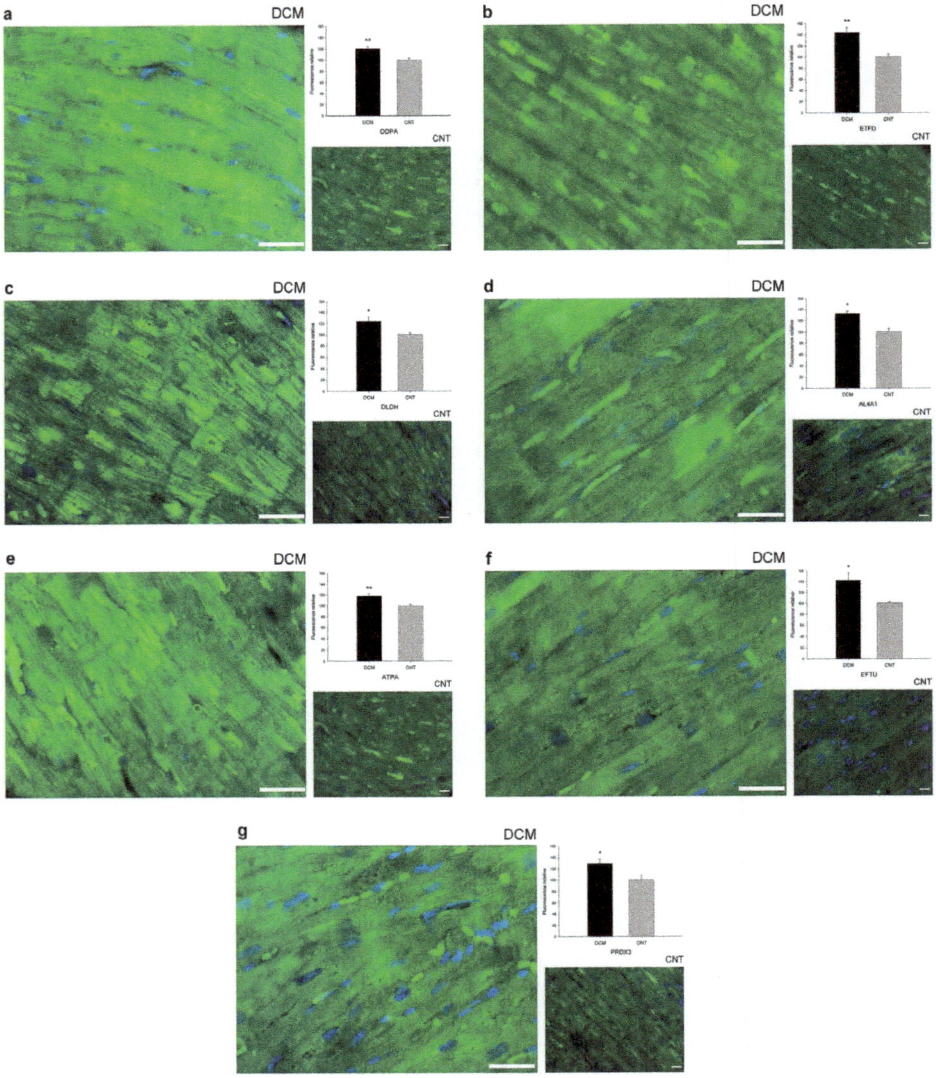

Figure 2. Mitochondrial protein overexpression in dilated human hearts according to immunofluorescence techniques. Influence of dilated cardiomyopathy on the amount of each representative protein involved in cardiac energy metabolism (ODPA, ETFD, DLDH, AL4A1, and ATPA), protein biosynthesis (EFTU), and the stress response (PRDX3). Immunofluorescence of (a) ODPA, (b) ETFD, (c) DLDH, (d) AL4A1, (e) ATPA, (f) EFTU, and (g) PRDX3 were significantly increased in patients with dilated cardiomyopathy compared with the control group. Here we show the nucleus co-stained with DAPI (blue). All of the micrographs are representative of the results obtained in four independent experiments for each group and protein studied, DCM (n = 4) and CNT (n = 4). The bar represents 100 μm. The bar graph shows the relative fluorescence intensity in dilated compared to control hearts. The data are expressed as mean ± SEM. CNT, control; DCM, dilated cardiomyopathy; ODPA, pyruvate dehydrogenase E1 component subunit α, somatic form; ETFD, electron transfer flavoprotein-ubiquinone oxidoreductase; DLDH, dihydrolipoyl dehydrogenase; AL4A1, delta-1-pyrroline-5-carboxylate dehydrogenase; ATPA, ATP synthase subunit α; EFTU, elongation factor Tu; PRDX3, thioredoxin-dependent peroxide reductase. *p value<0.05, **p value<0.01.

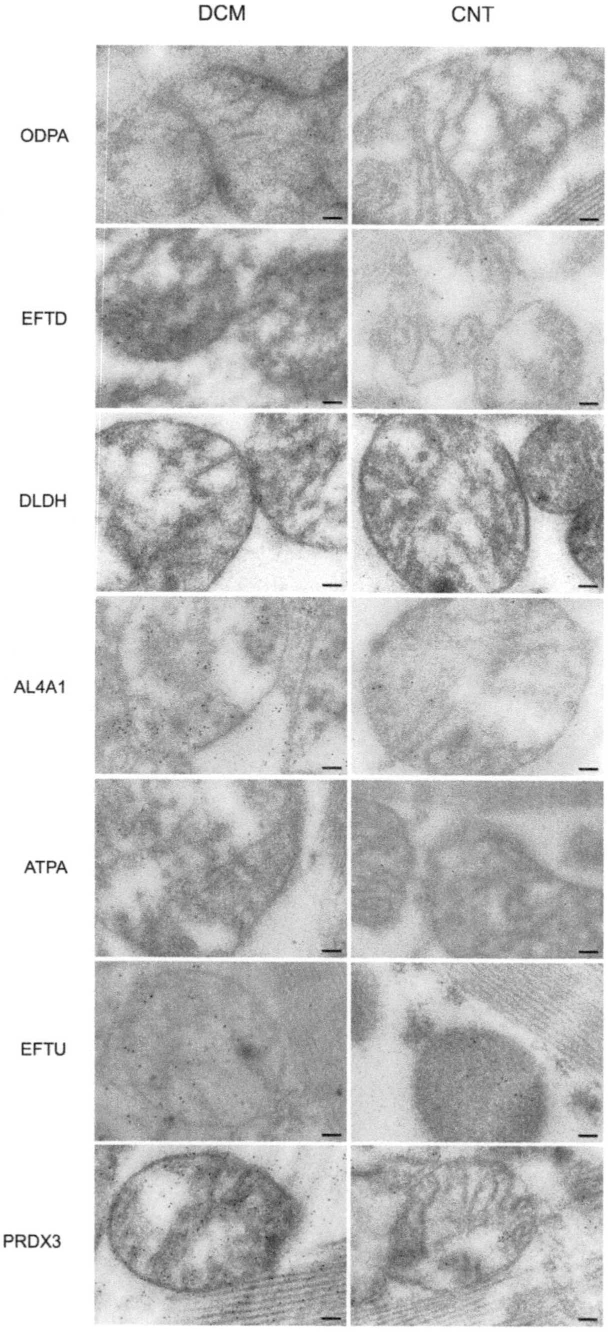

Figure 3. Mitochondrial protein localization and overexpression in dilated human hearts analyzed using transmission electron microscopy. Influence of dilated cardiomyopathy on the amount and localization of the representative proteins involved in cardiac energy metabolism (ODPA, ETFD, DLDH, AL4A1, and ATPA), protein biosynthesis (EFTU), and stress response (PRDX3). We observed an increase in immunogold labeling in all proteins studied in DCM hearts. We also confirmed the location of all proteins analyzed and observed a similar distribution of each protein upon comparing pathological with control samples. The bar represents 100 nm. CNT, control; DCM, dilated cardiomyopathy; ODPA, pyruvate dehydrogenase E1 component subunit α, somatic form; ETFD, electron transfer flavoprotein-ubiquinone oxidoreductase; DLDH, dihydrolipoyl dehydrogenase; AL4A1, delta-1-pyrroline-5-carboxylate dehydrogenase; ATPA, ATP synthase subunit α; EFTU, elongation factor Tu; PRDX3, thioredoxin-dependent peroxide reductase.

glycolysis-related protein expression [5]. Therefore, we focused on three overexpressed proteins involved in substrate utilization. The pyruvate dehydrogenase complex is a multi-enzyme system composed of multiple copies of 3 catalytic components, E_1 ($E_1\alpha$ and $E_1\beta$), E_2, and E_3. Specifically, ODPA catalyzes the overall conversion of pyruvate to acetyl-CoA and CO_2, and thereby links the glycolytic pathway to the tricarboxylic cycle [37,38]. We found that levels of this protein and its mRNA are significantly increased in DCM in concordance with the earlier study results of a canine HF model [39]. In addition, we found that ETFD was overexpressed. Electron transfer flavoproteins are heterodimeric proteins that transfer electrons between primary dehydrogenases and respiratory chains and link the oxidation of fatty acids and some amino acids to the mitochondrial respiratory system [40,41]. Finally, we also validated the overexpression of DLDH, a stable homodimer and essential component of the pyruvate dehydrogenase and glycine cleavage system as well as the α-ketoacid dehydrogenase complex [42]. This result is consistent with those published by our group and also by Li *et al.* in previous studies of total homogenate of LV tissue of DCM patients observing increased DLDH levels [43,44]. In addition, we found a good correlation between the protein levels and mRNA expression of these molecules and also with other altered proteins and its mRNA levels implicated in metabolic process and protein synthesis, specifically AL4A1 and EFTU, thereby interconnecting the alterations found in different processes.

Alterations in energy production in DCM human hearts

High myocardial energy production rates are required to maintain the constant demand of the working heart ATP and alterations in oxidative phosphorylation reduce cardiac function by providing an insufficient supply of ATP to cardiomyocytes [5]. Although the activity of electron transport chain complexes and ATP synthase (the complex responsible for ATP production) activity are known to be reduced in HF [45], reports on the individual levels of ATP synthase subunits in this syndrome are contradictory. While some authors observed that ATP synthase levels did not change [45] or diminish [46] in failing hearts, other studies revealed an increase in ATPA [47,48]. In the present work, we observed a significant overexpression of this protein. This lack of agreement between studies might be because of differences in the sample type or protocols and techniques used to detect these proteins. We also found a significant positive correlation of ATPA protein levels and mRNA expression with the overexpressed molecules involved in substrate utilization, highlighting the relationship between two principal components of the cardiac energy metabolism system. In other words, changes in some proteins involved in substrate utilization implicate modifications in specific components of oxidative phosphorylation. In addition, we found a good correlation between ATPA protein levels and mRNA expression with EFTU and *TUFM*, respectively. Thus, alterations in the energy production system are linked to higher activation of protein biosynthesis and oxidative damage in failing hearts since EFTU promotes the GTP-dependent binding of aminoacyl-tRNA to the A-site of ribosomes, and its downregulation increase reactive oxygen species [49].

Antioxidant PRDX3 overexpression is associated with impaired ventricular function

A large number of studies have reported that oxidative stress is important in the pathophysiology and development of HF via free radical production [50,51]. Reactive oxygen species play a key role in the onset and progression of coronary heart disease, tissue necrosis, and contractile dysfunction [52,53]. PRDX3 is a

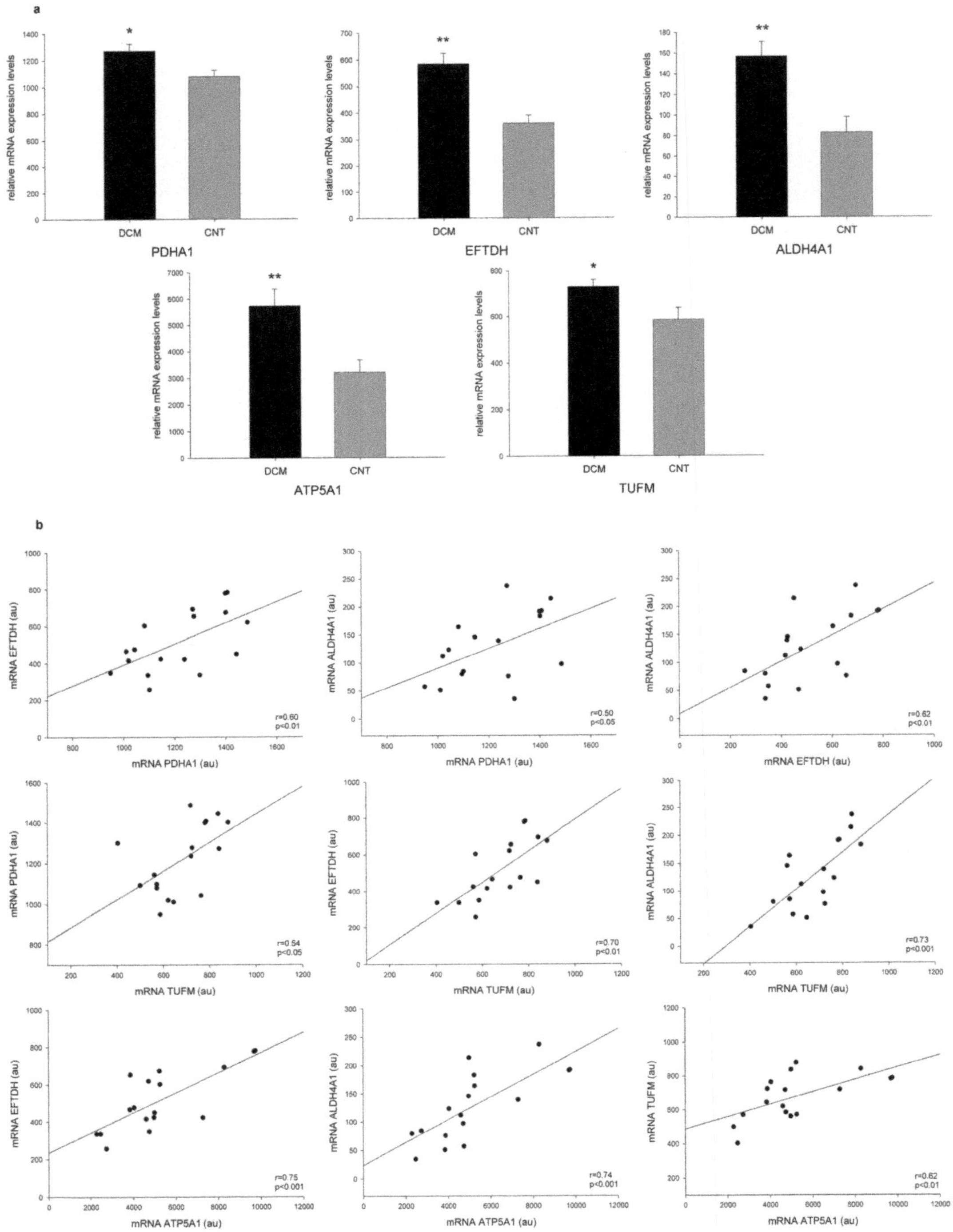

Figure 4. Levels of mRNA expression determined by RNAseq. (a) mRNA levels of the ODPA gene (*PDHA1*), ETFD gene (*ETFDH*), AL4A1 gene (*ALDH4A1*), ATPA gene (*ATP5A1*), and EFTU gene (*TUFM*) were increased in the dilated group compared to the control group. The data are expressed as mean ± SEM in mRNA relative expression. (b) The scatter plots show the relationship between the mRNA expression of molecules altered involved in substrate utilization (*PDHA1* and *ETFDH*) and also with other altered protein implicated in metabolic processes (*ALDH4A1*) and protein synthesis (*TUFM*). These genes showed a relevant correlation with *ATP5A1*, which is involved in energy production. CNT, control; DCM, dilated cardiomyopathy; ODPA, pyruvate dehydrogenase E1 component subunit α, somatic form; ETFD, electron transfer flavoprotein-ubiquinone oxidoreductase; AL4A1, delta-1-pyrroline-5-carboxylate dehydrogenase; ATPA, ATP synthase subunit α; EFTU, elongation factor Tu. *p value<0.05, **p value<0.01.

mitochondrial antioxidant protein that protects radical-sensitive enzymes against oxidative damage by a radical-generating system. Matsusshima *et al.* reported that PRDX3 overexpression protects the heart against post-myocardial infarct remodeling and failure in mice, reducing LV cavity dilation, dysfunction, fibrosis, and apoptosis [54]. These results are consistent with our findings, since we found a significant increase in the protein levels of PRDX3 in the cardiac tissue of DCM patients probably due to its specific role in the attenuation mechanisms of these failing hearts. Although, we did not observe the same tendency between protein levels and gene expression, likely because of a different removal mechanism of this protein in failing hearts; however, the most remarkable finding was the good correlation between this protein level and LV function, indicating that an increased PRDX3 level is associated with impaired ventricular function. Thus, our findings demonstrated once again that mitochondrial oxidative stress is key player in the pathogenesis of cardiac failure and showed for the first time a direct relationship between the level of this antioxidant and LV dysfunction in human cardiac tissue, suggesting that it could be a primary line of defense against this disease process. With the objective to evaluate the causality of this significant relationship, further studies need to be done.

It is noteworthy that our results are consistent and have been validated by different established techniques and by novel and precise SRM analysis and RNAseq approach. Despite these significant data, more work is needed to fully understand the causal role and which of the alterations observed are adaptive or maladaptive. A promising way to obtain new therapeutic options is to explore strategies based on gene therapy to clarify the mechanism leading to energetic derangement and, moreover, to restore ventricular function in DCM patients. For example, silencing or inducing overexpression of *PRDX3* through gene therapy with the creation of a murine DCM model would allow us to investigate whether LV is aggravated or improved and to develop etiology-specific therapies in HF. In this way, we suggest new molecular targets and all these experiments could be providing a new therapeutic approach.

Study limitations

A common limitation of studies that use cardiac tissues from end-stage failing human hearts is the fact that there is high variability in disease etiology and treatment. To make our study population etiologically homogeneous, we chose DCM patients who did not report any family history of the disease.

Moreover, our tissue samples were taken from the transmural left ventricle apex, so our findings could not be generalized to all layers and regions of the left ventricle. However, we want to emphasize the importance of having carried out this study in a significant number of samples from explanted human hearts from DCM patients undergoing cardiac transplantation. This has allowed us to extract the region of tissue that we wanted to analyze, which could not have been possible if the study had been made using biopsied tissues.

In summary, the present study is the first to analyze the mitochondrial proteome in cardiac tissue of DCM patients. We found significant and reproducible alterations in cardiac energy metabolism, especially in molecules involved in substrate utilization and energy production, obtaining a considerable relationship between the alterations detected in both processes. LV dysfunction is directly related with the antioxidant PRDX3 expression. This work provides new insight into the cellular mechanisms associated with the pathophysiology of DCM and could serve as a pivotal study to develop etiology-specific therapies in HF.

Supporting Information

File S1 Supporting figure and tables. Figure_S1: Representative two-dimensional DIGE gel. Table_S1: Additional data on MS protein identification of DCM spots with differential expression by MALDI-MS. Table_S2: Additional data on selected reaction monitoring (SRM) analysis. Table_S3: Additional data on selected reaction monitoring (SRM) analysis. Analyte peak area (counts) for each peptide in all samples.
(DOCX)

Acknowledgments

The authors thank the Transplant Coordination Unit (Hospital Universitario La Fe, Valencia, Spain) for their help in obtaining the samples. Furthermore, we are grateful to Lorena Gómez and Guillermo Esteban for the assistance in microscopy procedures.

Author Contributions

Conceived and designed the experiments: ERL MR AS. Performed the experiments: ERL ET AO MO. Analyzed the data: ERL MGB MR MP JRGJ FL MO. Contributed reagents/materials/analysis tools: ERL MGB MMMN. Contributed to the writing of the manuscript: ERL MR.

References

1. Jefferies JL, Towbin JA (2010) Dilated cardiomyopathy. Lancet 375: 752–762.
2. Olson EN (2004) A decade of discoveries in cardiac biology. Nat Med 10: 467–474.
3. Bindoff L (2003) Mitochondria and the heart. Eur Heart J 24: 221–224.
4. Maloyan A, Sanbe A, Osinska H, Westfall M, Robinson D, et al. (2005) Mitochondrial dysfunction and apoptosis underlie the pathogenic process in alpha-B-crystallin desmin-related cardiomyopathy. Circulation 112: 3451–3461.
5. Neubauer S (2007) The failing heart–an engine out of fuel. N Engl J Med 356: 1140–1151.
6. Neubauer S, Horn M, Cramer M, Harre K, Newell JB, et al. (1997) Myocardial phosphocreatine-to-ATP ratio is a predictor of mortality in patients with dilated cardiomyopathy. Circulation 96: 2190–2196.
7. Rosca MG, Hoppel CL (2010) Mitochondria in heart failure. Cardiovasc Res 88: 40–50.
8. Sanbe A, Tanonaka K, Kobayasi R, Takeo S (1995) Effects of long-term therapy with ACE inhibitors, captopril, enalapril and trandolapril, on myocardial energy metabolism in rats with heart failure following myocardial infarction. J Mol Cell Cardiol 27: 2209–2222.
9. Bayeva M, Gheorghiade M, Ardehali H (2013) Mitochondria as a therapeutic target in heart failure. J Am Coll Cardiol 61: 599–610.
10. Horstkotte J, Perisic T, Schneider M, Lange P, Schroeder M, et al. (2011) Mitochondrial thioredoxin reductase is essential for early postischemic myocardial protection. Circulation 124: 2892–2902.
11. Lal H, Zhou J, Ahmad F, Zaka R, Vagnozzi RJ, et al. (2012) Glycogen synthase kinase-3alpha limits ischemic injury, cardiac rupture, post-myocardial infarction remodeling and death. Circulation 125: 65–75.
12. Liesa M, Luptak I, Qin F, Hyde BB, Sahin E, et al. (2011) Mitochondrial transporter ATP binding cassette mitochondrial erythroid is a novel gene required for cardiac recovery after ischemia/reperfusion. Circulation 124: 806–813.
13. Agnetti G, Husberg C, Van Eyk JE (2011) Divide and conquer: the application of organelle proteomics to heart failure Circ Res 108: 512–526.
14. Deng WJ, Nie S, Dai J, Wu JR, Zeng R (2010) Proteome, phosphoproteome, and hydroxyproteome of liver mitochondria in diabetic rats at early pathogenic stages. Mol Cell Proteomics 9: 100–116.
15. Imahashi K, Schneider MD, Steenbergen C, Murphy E (2004) Transgenic expression of Bcl-2 modulates energy metabolism, prevents cytosolic acidification during ischemia, and reduces ischemia/reperfusion injury. Circ Res 95: 734–41.
16. Macrae DJ (2007) The Council for International Organizations and Medical Sciences (CIOMS) guidelines on ethics of clinical trials. Proc Am Thorac Soc 4: 176–178, discussion 178–179.
17. Swedberg K, Cleland J, Dargie H, Drexler H, Follath F, et al. (2005) Guidelines for the diagnosis and treatment of chronic heart failure: executive summary (update 2005): The Task Force for the Diagnosis and Treatment of Chronic Heart Failure of the European Society of Cardiology. Eur Heart J 26: 1115–1140.
18. Laemmli UK (1970) Cleavage of structural proteins during the assembly of the head of bacteriophage T4. Nature 227: 680–685.
19. Shevchenko A, Wilm M, Vorm O, Mann M (1996) Mass spectrometric sequencing of proteins silver-stained polyacrylamide gels. Anal Chem 68: 850–858.
20. Perkins DN, Pappin DJ, Creasy DM, Cottrell JS (1999) Probability-based protein identification by searching sequence databases using mass spectrometry data. Electrophoresis 20: 3551–3567.
21. Azorin I, Portoles M, Marin P, Lazaro-Dieguez F, Megias L, et al. (2004) Prenatal ethanol exposure alters the cytoskeleton and induces glycoprotein microheterogeneity in rat newborn hepatocytes. Alcohol Alcohol 39: 203–212.
22. Portoles M, Faura M, Renau-Piqueras J, Iborra FJ, Saez R, et al. (1994) Nuclear calmodulin/62 kDa calmodulin-binding protein complexes in interphasic and mitotic cells. J Cell Sci 107 (Pt 12): 3601–3614.

23. Reynolds WA (1963) The Effects of Thyroxine Upon the Initial Formation of the Lateral Motor Column and Differentiation of Motor Neurons in Rana Pipiens. J Exp Zool 153: 237–249.
24. Tomas M, Fornas E, Megias L, Duran JM, Portoles M, et al. (2002) Ethanol impairs monosaccharide uptake and glycosylation in cultured rat astrocytes. J Neurochem 83: 601–612.
25. Li H, Handsaker B, Wysoker A, Fennell T, Ruan J, et al. (2009) The Sequence Alignment/Map format and SAMtools. Bioinformatics 25: 2078–2079.
26. McKenna A, Hanna M, Banks E, Sivachenko A, Cibulskis K, et al. (2010) The Genome Analysis Toolkit: a MapReduce framework for analyzing next-generation DNA sequencing data. Genome Res 20: 1297–1303.
27. Trapnell C, Williams BA, Pertea G, Mortazavi A, Kwan G, et al. (2010) Transcript assembly and quantification by RNA-Seq reveals unannotated transcripts and isoform switching during cell differentiation. Nat Biotechnol 28: 511–515.
28. Anders S, Huber W (2010) Differential expression analysis for sequence count data. Genome Biol 11: R106.
29. Robinson MD, McCarthy DJ, Smyth GK (2010) edgeR: a Bioconductor package for differential expression analysis of digital gene expression data. Bioinformatics 26: 139–140.
30. de Weger RA, Schipper ME, Siera-de Koning E, van der Weide P, van Oosterhout MF, et al. (2011) Proteomic profiling of the human failing heart after left ventricular assist device support. J Heart Lung Transplant 30: 497–506.
31. Hammer E, Goritzka M, Ameling S, Darm K, Steil L, et al. (2011) Characterization of the human myocardial proteome in inflammatory dilated cardiomyopathy by label-free quantitative shotgun proteomics of heart biopsies. J Proteome Res 10: 2161–71.
32. Nishtala K, Phong TQ, Steil L, Sauter M, Salazar MG, et al. (2011) Virus-induced dilated cardiomyopathy is characterized by increased levels of fibrotic extracellular matrix proteins and reduced amounts of energy-producing enzymes. Proteomics 11: 4310–20.
33. Comunian C, Rusconi F, De Palma A, Brunetti P, Catalucci D, et al. (2011) A comparative MudPIT analysis identifies different expression profiles in heart compartments. Proteomics 11: 2320–8.
34. Green DR, Reed JC (1998) Mitochondria and apoptosis. Science 281: 1309–1312.
35. Kagan VE, Shvedova A, Serbinova E, Khan S, Swanson C, et al. (1992) Dihydrolipoic acid–a universal antioxidant both in the membrane and in the aqueous phase. Reduction of peroxyl, ascorbyl and chromanoxyl radicals. Biochem Pharmacol 44: 1637–1649.
36. Ahuja P, Wanagat J, Wang Z, Wang Y, Liem DA, et al. (2013) Divergent mitochondrial biogenesis responses in human cardiomyopathy. Circulation 127: 1957–1967.
37. Kato M, Wynn RM, Chuang JL, Tso SC, Machius M, et al. (2008) Structural basis for inactivation of the human pyruvate dehydrogenase complex by phosphorylation: role of disordered phosphorylation loops. Structure 16: 1849–1859.
38. Korotchkina LG, Patel MS (1995) Mutagenesis studies of the phosphorylation sites of recombinant human pyruvate dehydrogenase. Site-specific regulation. J Biol Chem 270: 14297–14304.
39. Heinke MY, Wheeler CH, Chang D, Einstein R, Drake-Holland A, et al. (1998) Protein changes observed in pacing-induced heart failure using two-dimensional electrophoresis. Electrophoresis 19: 2021–2030.
40. Gempel K, Topaloglu H, Talim B, Schneiderat P, Schoser BG, et al. (2007) The myopathic form of coenzyme Q10 deficiency is caused by mutations in the electron-transferring-flavoprotein dehydrogenase (ETFDH) gene. Brain 130: 2037–2044.
41. Goodman SI, Axtell KM, Bindoff LA, Beard SE, Gill RE, et al. (1994) Molecular cloning and expression of a cDNA encoding human electron transfer flavoprotein-ubiquinone oxidoreductase. Eur J Biochem 219: 277–286.
42. Brautigam CA, Chuang JL, Tomchick DR, Machius M, Chuang DT (2005) Crystal structure of human dihydrolipoamide dehydrogenase: NAD+/NADH binding and the structural basis of disease-causing mutations. J Mol Biol 350: 543–552.
43. Li W, Rong R, Zhao S, Zhu X, Zhang K, et al. (2012) Proteomic analysis of metabolic, cytoskeletal and stress response proteins in human heart failure. J Cell Mol Med 16: 59–71.
44. Rosello-Lleti E, Alonso J, Cortes R, Almenar L, Martinez-Dolz L, et al. (2012) Cardiac protein changes in ischaemic and dilated cardiomyopathy: a proteomic study of human left ventricular tissue. J Cell Mol Med 16: 2471–2486.
45. Rosca MG, Okere IA, Sharma N, Stanley WC, Recchia FA, et al. (2009) Altered expression of the adenine nucleotide translocase isoforms and decreased ATP synthase activity in skeletal muscle mitochondria in heart failure. J Mol Cell Cardiol 46: 927–935.
46. Agnetti G, Kaludercic N, Kane LA, Elliott ST, Guo Y, et al. (2010) Modulation of mitochondrial proteome and improved mitochondrial function by biventricular pacing of dyssynchronous failing hearts. Circ Cardiovasc Genet 3: 78–87.
47. Cieniewski-Bernard C, Mulder P, Henry JP, Drobecq H, Dubois E, et al. (2008) Proteomic analysis of left ventricular remodeling in an experimental model of heart failure. J Proteome Res 7: 5004–5016.
48. Yang J, Moravec CS, Sussman MA, DiPaola NR, Fu D, et al. (2000) Decreased SLIM1 expression and increased gelsolin expression in failing human hearts measured by high-density oligonucleotide arrays. Circulation 102: 3046–3052.
49. Zhang DX, Yan H, Hu JY, Zhang JP, Teng M, et al. (2012) Identification of mitochondria translation elongation factor Tu as a contributor to oxidative damage of postburn myocardium. J Proteomics 77: 469–479.
50. Keith M, Geranmayegan A, Sole MJ, Kurian R, Robinson A, et al. (1998) Increased oxidative stress in patients with congestive heart failure. J Am Coll Cardiol 31: 1352–1356.
51. Shiomi T, Tsutsui H, Matsusaka H, Murakami K, Hayashidani S, et al. (2004) Overexpression of glutathione peroxidase prevents left ventricular remodeling and failure after myocardial infarction in mice. Circulation 109: 544–549.
52. Brioschi M, Polvani G, Fratto P, Parolari A, Agostoni P, et al. (2012). Redox proteomics identification of oxidatively modified myocardial proteins in human heart failure: implications for protein function. PLoS One 7: e35841.
53. Ide T, Tsutsui H, Kinugawa S, Utsumi H, Kang D, et al. (1999) Mitochondrial electron transport complex I is a potential source of oxygen free radicals in the failing myocardium. Circ Res 85: 357–363.
54. Matsushima S, Ide T, Yamato M, Matsusaka H, Hattori F, et al. (2006) Overexpression of mitochondrial peroxiredoxin-3 prevents left ventricular remodeling and failure after myocardial infarction in mice. Circulation 113: 1779–1786.

13

Proteomics Indicators of the Rapidly Shifting Physiology from Whole Mountain Pine Beetle, *Dendroctonus ponderosae* (Coleoptera: Curculionidae), Adults during Early Host Colonization

Caitlin Pitt[1]*, Jeanne A. Robert[1], Tiffany R. Bonnett[1], Christopher I. Keeling[2], Jörg Bohlmann[2], Dezene P. W. Huber[1]*

1 Ecosystem Science and Management Program, University of Northern British Columbia, Prince George, BC, Canada, **2** Michael Smith Laboratories, University of British Columbia, Vancouver, BC, Canada

Abstract

We developed proteome profiles for host colonizing mountain pine beetle adults, *Dendroctonus ponderosae* Hopkins (Coleoptera: Curculionidae). Adult insects were fed in pairs on fresh host lodgepole pine, *Pinus contorta* Dougl. ex Loud, phloem tissue. The proteomes of fed individuals were monitored using iTRAQ and compared to those of starved beetles, revealing 757 and 739 expressed proteins in females and males, respectively, for which quantitative information was obtained. Overall functional category distributions were similar for males and females, with the majority of proteins falling under carbohydrate metabolism (glycolysis, gluconeogenesis, citric acid cycle), structure (cuticle, muscle, cytoskeleton), and protein and amino acid metabolism. Females had 23 proteins with levels that changed significantly with feeding ($p<0.05$, FDR<0.20), including chaperones and enzymes required for vitellogenesis. In males, levels of 29 proteins changed significantly with feeding ($p<0.05$, FDR<0.20), including chaperones as well as motor proteins. Only two proteins, both chaperones, exhibited a significant change in both females and males with feeding. Proteins with differential accumulation patterns in females exhibited higher fold changes with feeding than did those in males. This difference may be due to major and rapid physiological changes occurring in females upon finding a host tree during the physiological shift from dispersal to reproduction. The significant accumulation of chaperone proteins, a cytochrome P450, and a glutathione S-transferase, indicate secondary metabolite-induced stress physiology related to chemical detoxification during early host colonization. The females' activation of vitellogenin only after encountering a host indicates deliberate partitioning of resources and a balancing of the needs of dispersal and reproduction.

Editor: Daniel Doucet, Natural Resources Canada, Canada

Funding: This work was supported by the following: Natural Science and Engineering Research Council of Canada (NSERC) Postdoctoral Fellowship (http://www.nserc-crsng.gc.ca/index_eng.asp); Natural Science and Engineering Research Council of Canada (NSERC) Discovery Grant (http://www.nserc-crsng.gc.ca/index_eng.asp); Genome Canada (http://www.genomecanada.ca); Genome British Columbia (http://www.genomebc.ca); Genome Alberta (http://genomealberta.ca); Canada Research Chairs (CRC) Program (http://www.chairs-chaires.gc.ca); Canada Foundation for Innovation (CFI) (http://www.innovation.ca; and British Columbia Knowledge Development Fund (BCKDF) (http://www.aved.gov.bc.ca/researchandinnovation/Funding/BCKDF/). The funders had no role in the study design, data collection and analysis, decision to publish, or preparation of the manuscript. This work was supported in part by Genome Canada, Genome BC, and Genome Alberta as part of the Tria Project (www.thetriaproject.ca). Information for all proteins in the dataset is provided in the Supplemental Materials file.

Competing Interests: This work was supported in part by Genome Canada, Genome BC, and Genome Alberta as part of the Tria Project (www.thetriaproject.ca).

* Email: huber@unbc.ca (DH); pitt@unbc.ca (CP)

Introduction

The mountain pine beetle (MPB), *Dendroctonus ponderosae* Hopkins (Coleoptera: Curculionidae), is a member of the large subfamily Scolytinae (~6,000 species) contained within the extremely large (~60,000 species) and diverse weevil family [1–2]. Like all other weevils, MPB is an herbivore and it spends most of its life cycle closely associated with its host plant [3]. In the case of MPB, the host plant can be any of the pine species native to western North America [4] as well as novel pine hosts in an expanding range [5–6]. This intimate association with its host has resulted in a large number of traits that allow MPB to survive, and even to thrive, in the often-hostile environment of its host pine's tissues [7–8].

While MPB is an important ecological factor in western North American pine forests [9] – ranging from Mexico to central British Columbia, and from the west coast to the Black Hills of South Dakota – it has, in recent years, moved into an outbreak state of unprecedented dimensions. In British Columbia alone, this insect impacted up to 53% of merchantable pine by 2012 [10] causing tens of billions of dollars in economic damage. This outbreak is predicted to have long-term extreme effects on forest ecology and carbon cycling [11] and has devastated local silvicultural economies.

Recently, MPB has breached the historical barrier of the Canadian Rocky Mountains and has begun to spread in the lodgepole pine (*Pinus contorta* Douglas ex Louden) and hybrid lodgepole pine x jack pine (*Pinus banksiana* Lamb.) forests of northwestern Alberta. Since the MPB is known to be able to attack and kill almost every pine species [4] – and even occasionally spruces [12] – in its native range, it is not surprising that it has recently been found to be spreading and successfully reproducing in wild stands of pure jack pines [5]. The MPB is now considered an invasive pest that threatens boreal jack pine forests, and other pine species as well [6], ranging from central Alberta through the Great Lakes region and to the eastern seaboard of Canada and the United States of America.

We are interested in the critical host colonization phase of the beetle's life cycle when it must cooperatively invade and kill its host tree so that it can later successfully reproduce [3–4]. While most of the insect's life cycle is cryptic – spent under the bark of its brood tree – the adults can spend several days in a foraging dispersal flight [13] in the summer during which they find and colonize new host pines for reproduction. During the dispersal phase the insects – especially the females, which are the pioneering sex – search for new susceptible host trees in a mixed landscape that includes a variety of nonhosts [14–15] and non-susceptible individuals of the host species. Dispersing beetles do not feed, and so this phase is fraught with hazards including reduction in energy available for reproductive provisioning [16], encounters with predators or other antagonists, and potentially inclement weather. When insects find a suitable host, colonization is coordinated by aggregation and antiaggregation pheromones [17] that allow the insects and their symbiotic fungi [18–19] to overcome the complex and copious host defenses [20–22] and complete their reproductive cycle.

Host colonization begins immediately following the MPB's dispersal flight. We postulate that insect physiology must rapidly shift from supporting flight and host search towards survival when exposed to host defenses and commencing with reproductive activities. The host-colonizing beetles and their associated fungi must tolerate, detoxify or in other ways overcome host defenses and eventually kill the host prior to reproducing. Using expressed sequence tag (EST) data for the MPB [23], combined with a proteomics approach, we have observed a number of proteins that indicate a shift towards reproduction and may be involved in the stress physiology of insect survival during the initial period of host colonization. This investigation of the total and differentially expressed proteomes of adult male and female MPB during the host colonization phase of their life cycle will provide candidates for further study to better understand, and to perhaps better manage, this ecologically and economically important forest insect pest [5–6].

Materials and Methods

Beetles

Lodgepole pine bolts infested with MPB were field collected from an outbreak area near Penticton, British Columbia, in May 2010 (49°25′60″N; 119°27′30″W). Doug Bateman Logging collected the bolts from a company harvest; no further permission was required and no endangered or protected species were involved. Immediately after collection the ends of the bolts were sealed with molten paraffin wax to prevent drying during experiments. Bolts were placed in vented plastic containers and stored at ambient outdoor temperatures. In early July, young MPB adults began to emerge and were collected daily. Sex was determined by examining the 7th abdominal tergite [24]. Emerged adults were stored in Petri dishes with a lightly moistened Kimwipe at 4°C. Beetles were stored for a maximum of ten days prior to use.

Feeding treatments

On the 26th and 27th of July 2010, control beetles were placed individually in 1.5 mL tubes with perforated lids and kept in the dark for 24 hours at room temperature prior to flash-freezing in liquid nitrogen and storage at −80°C. Simultaneously, randomly paired female and male adults were placed into holes drilled into the phloem of freshly collected and waxed bolts of a healthy lodgepole pine tree collected near Prince George, BC (53°58′12″N; 122°52′18″W). Holes were spaced roughly 10 cm apart. The female was inserted first, followed by a male; a metal screen was stapled over the hole to prevent escape. Logs were set upright and beetles were left to feed at room temperature for 24 hours. Logs were then dissected to remove live individuals that had produced visible frass (boring dust and feces, indicating active phloem tissue colonization), and the sex of each recovered beetle was confirmed. Recovered beetles were immediately flash-frozen in liquid nitrogen and stored at −80°C.

Protein extraction

Eight frozen adult MPB were used per extraction and treatment (control females, fed females, control males, fed males) and each extraction and treatment was replicated four times with different beetles. Samples were thawed on ice for 5 minutes then homogenized with 500 µl TCA buffer (trichloroacetic acid, Sigma) containing 15% TCA and 1% DTT (dithiothreitol, Fisher Scientific) by weight [25]. The samples were homogenized six times for 1 min each time at 1,500 cycles per min (2000 Geno/Grinder, Spex SamplePrep) interspersed with 3 min incubation on ice. The samples were transferred to new tubes and incubated on ice for 30 min. Samples were centrifuged at 18,000xg for 10 min at 4°C and the supernatant was removed. The pellets were resuspended in 1 mL ice-cold acetone and incubated on ice for 5 min. The acetone wash was repeated four times. The washed pellets were air dried for 10 min to allow the acetone to evaporate completely, then suspended in 1 mL urea:thiourea buffer containing 6 M Urea (Fisher Scientific) and 1 M thiourea (Fisher Scientific) in 100 mM Tris-Cl pH 8.0 (Ultrapure, Invitrogen). A final centrifugation at 18,000xg for 10 min at room temperature removed any remaining insoluble debris. The supernatant containing total insect protein was removed and stored at −80°C prior to subsequent iTRAQ proteome analysis. Protein concentration and quality was confirmed by a Coomassie (Bradford) Protein Assay Kit (Fisher Scientific) and Experion Pro260 Chip analysis (Biorad).

iTRAQ analysis

Samples were processed at the University of Victoria Genome British Columbia Proteomics Centre (Victoria, British Columbia, Canada) for proteome quantification and analysis. Analysis used eight-plex isobaric tags for relative and absolute quantification (iTRAQ) [26–27]. Samples were analyzed as sets of two eight-plexes. The first multiplex included four biological replicates of control female-derived proteins and four biological replicates of fed female-derived proteins. The eight-plex of male protein samples followed in an identical manner. Experimental conditions were reported as follows: Protein concentrations were determined using a Bradford protein assay (Sigma). Samples (100 µg of each) were precipitated overnight in acetone at 4°C followed by resolubilzation in 0.5 M TEAB, 0.2% SDS. Proteins were reduced

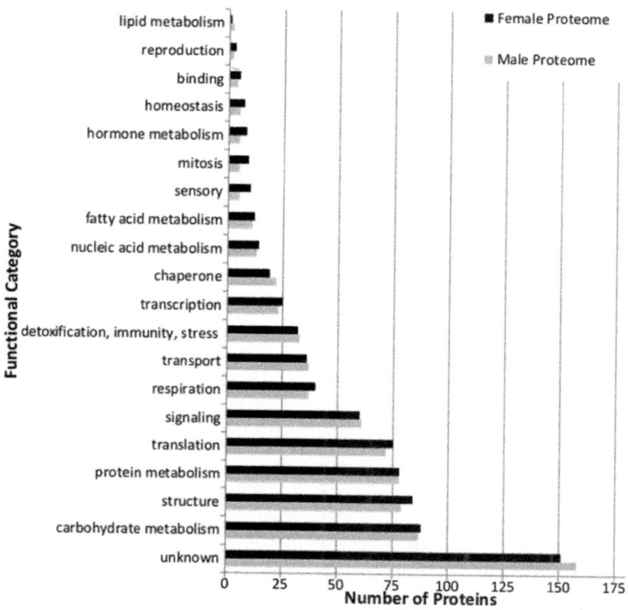

Figure 1. General functional proportions of all proteins. Detected proteins were divided into generalized functional groups based on gene ontology information from UNIprot.

with TCEP and alkylated with MMTS. Proteins were then in solution digested with trypsin (Promega) and labeled with the appropriate iTRAQ label. Labeled peptides were then combined and separated by strong cation exchange HPLC. SCX fractions containing peptides were then reduced in volume by speed-vac and analyzed by LC-MS/MS. The length of the reverse gradient used was 2 hours per HPLC strong cation exchange fraction. Samples were analyzed by reversed phase nanoflow (300 nL/min) HPLC with nano-electrospray ionization using a quadrupole time-of-flight mass spectrometer (QStar pulsar i, Applied Biosystems) operated in positive ion mode.

Data analysis

Data were analyzed using ProteinPilot Software v3.0 (Applied Biosystems) with the Paragon algorithm [28]. Analytical parameters were as follows: Cys alkylation: MMTS; digestion: trypsin; instrument: QSTAR ESI; and an Unused ProtScore threshold of >1.3 (95% protein confidence). Proteins were identified against the translated sequences from a MPB EST-derived transcriptome database [23].

Data were presented as ratios, normalized to one of the four biological control replicates for each sex, which was subsequently assigned a quantity of 1. Protein abundances were represented by average fold-change for each protein relative to control samples from untransformed iTRAQ ratios. Ratios were $\log_{10}$ transformed to obtain a normal distribution and a two-tailed independent samples t-test was conducted on the differences between starved control and fed treatment groups for each identified protein within each eight-plex. A Benjamini and Hochberg Correction (BH) was applied to the p-values to estimate the false discovery rate (FDR) [29–30]. Proteins were considered differentially expressed at $p < 0.05$ with an FDR<0.20 [31].

Proteins were manually annotated by their best BLASTx match to other insect proteins using their representative assembled EST sequences. Each quantitatively detected protein was investigated for its general functional role and subcellular location using UniProtKB [32] and the Panther Database [33]. The information was used to assign each protein to a general functional category (Tables S1 and S2 in File S1).

Results

Overall male and female proteome composition

Female and male samples yielded 757 and 739 proteins, respectively, for which quantitative information was obtained (Tables S1 and S2 in File S1). Proteins detected from females and males were designated by their EST library matches. 463 expressed proteins were shared between the sexes, leaving 294 and 276 proteins unique to the female and male proteomes, respectively.

All proteins were assigned to general functional categories using annotations provided by NCBI BLASTx and UNIprotKB (Figure 1). Besides those with unknown designations, the functional categories containing the most proteins were carbohydrate metabolism (11.6% in females, 11.8% in males; including enzymes involved in glycolysis, gluconeogenesis, and the citric acid cycle), structure (11.1% in females, 10.7% in males; proteins related to the cuticle, muscle, and cytoskeleton), protein and amino acid metabolism (10.3% in females, 9.7% in males), translation (9.9% in females and 10.6% in males; including ribosomal components and elongation factors), and signaling (7.9% in females, 8.3% in males; including enzymes affiliated with mediating cellular differentiation and development). In spite of the fact that only 61–63% of detected proteins were shared between the sexes, general proportions of proteins present in the functional groups differed little between females and males, perhaps suggesting common sources of precursors are being used for differing purposes.

The general functional distributions of the detected proteins unique to each sex were similarly compared (Figure 2). Females had more proteins involved in protein metabolism, respiration, and mitosis than did males. Males expressed isoprenoid biosynthesis-related enzymes not found in females, namely HMG-CoA synthase (HMG-S, h_cluster_01168-1+2) and a protein similar to geranylgeranyl pyrophosphate synthase (GGPPS, h_cluster_11996-1)(Table S2 in File S1). The most obvious difference between males and females in this experimental context was in carbohydrate metabolism, where females expressed more than double the complement of different enzymes involved in this process than did males.

Proteomic shifts following feeding

The proportion of proteins that showed differential shifts in accumulation between treatments and controls, and the significance levels of those changes are depicted in volcano plots (Figure 3). Most detected proteins fell below the level of significance ($p<0.05$) when comparing treatments for either males or females against the male or female controls. The fold-changes of proteins showing significant differences in fed males versus control males are of lower magnitude (with one exception, all are under +/−2-fold) than those of female treatments (Tables 1 and 2).

A Benjamini and Hochberg correction was applied to the p-values to control for false discovery at the 0.20 level [31]. In females, 23 proteins showed expression differences in accumulation between starved and fed treatments (Table 1). Of these, 18 increased with feeding (from 1.2-fold to a maximum of 8.9-fold change) and five decreased (from -1.4-fold to -2.5 -fold change). Proteins changing in quantity with treatment included the following categories: carbohydrate metabolism, chaperone, detoxification/stress response, hormone metabolism, nucleic acid

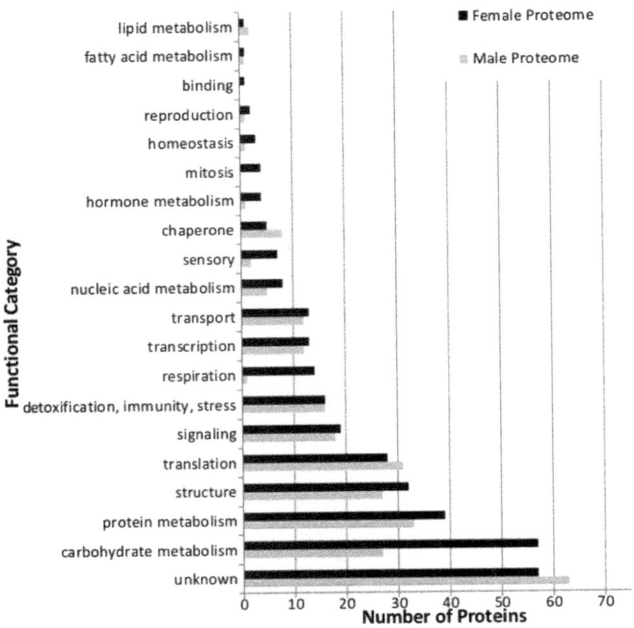

Figure 2. Functional distribution of detected proteins that were unique to each sex.

metabolism, protein metabolism, reproduction, respiration, sensory, signaling, structure, transcription, translation, and transport. Notably, vitellogenin proteins – glycolipoproteins that usually function as yolk precursors during oogenesis – exhibited some of the highest fold-change increases (7.8 and 8.9). Also increasing were two heat shock proteins and a cytochrome P450, identified as CYP6DE1. Those that showed significant decreases with feeding in females included a possible juvenile hormone binding protein, and proteins involved in proline biosynthesis and possibly odorant binding.

Of the proteins detected in males, 29 exhibited significant differences between fed and control treatments (Table 2). Twelve increased (ranging from 1.2-fold to 1.6-fold change) and 17 decreased (-1.1 to -2.2-fold change) with feeding. Proteins that changed with feeding treatment were classified in the following categories: carbohydrate metabolism, chaperone, detoxification/stress, fatty acid metabolism, homeostasis, protein metabolism, respiration, signaling, structure, translation, and transport. The heat shock proteins increasing in females also did so in males, as well as a glutathione S-transferase (GST, DpGSTs4). The fatty acid metabolism category had no representatives that showed differential changes in females related to the feeding treatment, and males lacked differentially expressed proteins in the transcription category.

Discussion

Overall male and female proteomes

Male and female protein functional category distributions were similar (Figure 1), despite only sharing ~60% identity of specific proteins between them. The overall similar pattern of functional categories present in males and females may reflect the similar physiological challenges in both sexes during host colonization. An EST study of *Ips pini* midguts showed similar category distributions, with carbohydrate metabolism being highly represented, more so than in insects that do not feed on phloem [34–35]. Microarray data also found carbohydrate metabolism to be well represented in midguts [35].

Digestion of plant cell walls is of obvious importance to phytophagous animals. Plant cell wall degrading enzymes (PCWDEs) have been identified in the transcriptome and genome of MPB and other beetles [23] [36] [37]. Several PCWDEs were present in the MPB proteomes, but quantities did not change with host colonization. Dispersing beetles may maintain constitutive expression of PCWDEs. Both sexes maintained proteins with annotation matches to endopolygalacturonase, secreted cellulase, cellulose 1-4-β-cellobiosidase, and amylase A. Females had three pectinesterase representatives. The presence of PCWDEs affirms their importance during host colonization and indicates that colonizing beetles are already primed for wood digestion upon arriving at a suitable host tree.

Distribution of 'unique' proteins

We identified differences between the functional category distributions of proteins uniquely presented in females or males. Because female MPB are the pioneering sex, an increase in carbohydrate metabolism and cell respiration may be necessary for producing the energy required to establish a maternal gallery, produce aggregation pheromone components, and increase provisioning to eggs while simultaneously overcoming host defences.

Males expressed proteins directly related to host colonization. One sub-category of constitutively expressed proteins unique to males were those involved in isoprenoid pheromone biosynthesis, including HMG-S. HMG-S is part of the mevalonate pathway, which synthesizes juvenile hormone (JH), a molecule that mediates numerous developmental changes related to differentiation and reproduction [38]. In *Ips pini*, HMG-S has been found to be inducible by JH, and the mevalonate pathway responds to feeding on host phloem [39–40]. Pheromone production is not exclusively dependent on phloem-derived precursors in some species; exposure to JH elicits pheromone production in *I. paraconfsus* and *I. pini* [41–43]. Several studies have demonstrated that JH induces the mevalonate pathway for synthesis of isoprenoid bark beetle pheromones such as frontalin, ipsenol and ipsdienol [44–46]. Male MPB produce the multifunctional isoprenoid-derived pheromone, frontalin, to signal that a host tree has succumbed to mass attack [47]. Frontalin is synthesized via the mevalonate pathway [44] [48–49], explaining the exclusive detection of HMG-S in host-colonizing males. Another isoprenoid pathway protein unique to the male proteome was similar to GGPPS, which functions in MPB frontalin synthesis [50]. Matching to the same full-length cDNA clone (DPO061_E16) as the protein described here, DponGGPPS transcripts were expressed primarily in the anterior midguts of feeding males with mates, with accumulation peaking after 24 hours. RNA interference of DponGGPPS reduced frontalin content to 9% of that found in controls [50].

Proteomic shifts following feeding

In total, 23 and 29 proteins of the identified proteins significantly changed in quantity following feeding in females and males, respectively (FDR<0.2). The lowest FDR obtained was 0.06. A less stringent FDR cutoff was used in this experiment (as in reference [31]) compared to that of the recently published MPB larval proteome [25]. We were trying to capture a very small window –the expression changes occurring during the shift from dispersal to host colonization (represented by 24 hours), whereas the study of MPB larval proteomes had sample collection times separated by months and by different larval instars. When

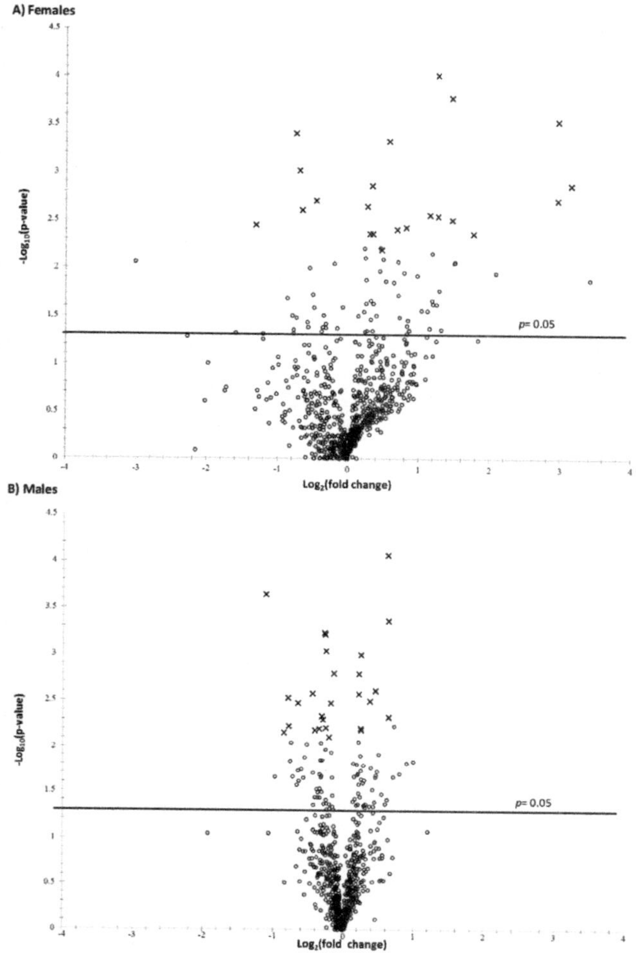

Figure 3. Volcano plots of iTRAQ protein quantification ratios in female (a) and male (b) iTRAQ runs. Points above the horizontal line represent proteins with significant ($p<0.05$) differences between starved and fed treatments, with points with FDR<0.2 marked as "x".

compared to the larvae, 54% and 60% of the female proteome was also present in the spring and fall libraries, respectively. Males had 54% and 58% similarity to spring and fall larvae. All categories were represented similarly in all four comparisons, with the highest percentages being for the unknown and carbohydrate metabolism proteins (15% and 14%, respectively), followed by those assigned to structure and translation (~11%).

Generally, females had proteomic shifts of greater magnitude than the males. Seven proteins related to translation showed increases in females following feeding (fed males had one increase and one decrease), which could suggest a higher investment in translational machinery and explain the greater fold change values.

A transcriptome analysis of this sample set has also been conducted [51], and all but 3 proteins listed in Tables 1 and 2 matched transcripts detected in that study (identified by the Genome Accession numbers to which they were mapped). Those that were also found to significantly change with treatment in the transcriptome study have a fold change value marked with an asterisk. Gene expression analysis revealed many more significantly changing genes, which could be indicative of differences in isolation efficiency. For example, proteins range greatly in size and solubility and occupy different subcellular compartments. Both studies used the same sample pool and thus the same temporal 'snapshot'; comparisons should bear in mind the time between transcription and translation and the influence of posttranscriptional and posttranslational modifications on the final protein product. That said, comparing these results where possible provides a more complete picture of what processes might be occurring.

Chaperones and host colonization

Two proteins with BLAST matches to insect 70 kDa heat shock proteins (HSPs) increased significantly with host colonization in both females (3.4- and 2.8-fold change) and males (1.6- and 1.4-fold change). Both HSPs were also found to increase on the transcriptome level (YQE07729, [51]). Many stressors illicit a HSP response, including heat, desiccation, radiation, and chemical compounds [52]. HSPs are chaperones that facilitate the repair of misfolded or damaged enzymes. HSPs also assist in the proper folding of nascent proteins and are thought to control regulators of cellular signals for homeostasis, proliferation and apoptosis [53]. The significant increase in these proteins in both sexes highlights the role of stress physiology during host colonization. In fact, both proteins also exhibited a significant increase in MPB larvae during springtime [25], suggesting that these HSPs are very important to feeding MPB.

Further indication of stress physiology in host-attacking males is the presence of calreticulin – a chaperone – that exhibited one of the higher significant increases of proteins detected in the fed male proteome (1.5-fold change). Calreticulin increased by the same magnitude in MPB larvae as they developed in spring [25]. The transcript of this protein also increased in host-exposed males (YQE10054, [51]) and microarray analysis of *Ips pini* midgut tissue also indicated an induction with feeding [35]. Insect calreticulin has also been shown to be part of the response to microorganisms [54]. In some insects calreticulin may mediate phagocytosis of foreign bodies via an association with hemocytes as a part of the innate immune system. Exposure to the host, which will also be more susceptible to other pathogens as the beetles overwhelm its defensive systems, may require an enhanced immune response against an increasingly microorganism-laden environment. While calreticulin is present, it is unclear why females would not also increase production of this protein, but they do arrive at the host tree first. Males could be arriving at a time where secondary infection by microorganisms is advancing, making an enhanced immune system important for their initial contact with the host environment.

Host colonization increases the abundance of detoxification enzymes

Several enzymes with potential functions in detoxification were differentially expressed in the female proteome. One upregulated protein was a cytochrome P450, a member of a ubiquitous group of enzymes known for their roles in detoxification, pheromone biosynthesis, and hormone metabolism [55]. First identified in the MPB EST library (AFI45031, [23]), CYP6DE1 increased by 1.8-fold with feeding. The corresponding transcript also showed an increase in the MPB transcriptome, although without statistical significance (YQE02814, [51]). CYP6DE1 was found to be decreasing in MPB larvae during the spring months; it was suggested that this was in response to host defenses waning as the tree succumbed to the attack [25].

The proteins with a relatively higher fold-change in fed males included one matching to the sigma GST (1.3-fold change), annotated DpGSTs4 [36]. GSTs are generally recognized as detoxification enzymes which conjugate toxins with glutathione,

Table 1. Proteins that showed significant changes in accumulation levels in females following feeding ($p<0.05$, FDR<0.2).

Translated EST contig ID[a]	EST/Genome Accession No.	BLASTx annotation	Subject	E-Value	Functional category	Fold change[b]	p-value
s_cluster_03564+2	GT386951/YQE05171	TSA: Dendroctonus ponderosae Contig451, mRNA sequence	n/a	n/a	?	7.8	0.0003
h_cluster_16459-2+2	GT400641/YQE10688	TSA: Dendroctonus ponderosae Contig584, mRNA sequence	n/a	n/a	?	2.4*	0.0001
h_cluster_01569+1	GT469028/YQE04735	No Hits	n/a	n/a	?	−1.4	0.0020
h_cluster_12831+2	GT371138/YQE11261	similar to CG15105-PA, isoform A	tca:656122	3E-83	?	−1.6	0.0025
h_cluster_00498-1+3	GT323752/YQE11922	similar to CG8036-PB, isoform B; K00615 transketolase	tca:660692	0	carbohydrate metabolism	2.4	0.0028
h_cluster_02588-1+3	GT346060/YQE03281	G6pd; glucose-6-phosphate dehydrogenase; K00036 glucose-6-phosphate 1-dehydrogenase	tca:662934	0	carbohydrate metabolism	1.3	0.0043
h_cluster_00930-1+1	GT352910/YQE07729	similar to heat shock 70 kD protein cognate; K09490 heat shock 70 kDa protein 5	tca:659147	4E-141	chaperone	3.4*	0.0043
h_cluster_00722-1+1	GT484212/YQE07729	similar to ENSANGP00000012893; K09490 heat shock 70 kDa protein 5	nvi:100118379	0	chaperone	2.8*	0.0002
h_cluster_00993-1+3	GT356404/YQE02814	similar to Probable cytochrome P450 6a14 (CYPVIA14), MPB CYP6DE1	tca:664473	9E-116	detoxification, immunity, stress response	1.8	0.0037
h_cluster_00871-1+1	GT375000/YQE10157	similar to CG14661-PA	tca:663762	1E-33	hormone metabolism	−2.5*	0.0035
h_cluster_03047+1	GT472768/YQE7451	similar to CG5840-PA, isoform A; K00286 pyrroline-5-carboxylate reductase	tca:664343	8E-99	protein metabolism	−1.6	0.0010
h_cluster_19511-3	GO487182/YQE09294	hypothetical protein (vitellogenin 6 precursor)	aag: AaeL_AAEL006138	2E-15	reproduction	8.9	0.0014
h_cluster_01826-1+1	GT391859/YQE09293	similar to vitellogenin	tca:660042	2E-99	reproduction	7.8*	0.0019
h_cluster_00532-1+3	GT449473/YQE03475	similar to AGAP008724-PA; K02265 cytochrome c oxidase subunit Vb	tca:656612	5E-50	respiration	−1.7	0.0004
h_cluster_00098-1+1	GT337131/YQE05782	similar to 40S ribosomal protein SA (p40) (34/67 kDa laminin receptor); K02998 small subunit ribosomal protein SAe	tca:658600	1E-99	translation	2.8	0.0031
h_cluster_00022+2	GT333266/YQE11710	similar to S3Ae ribosomal protein; K02984 small subunit ribosomal protein S3Ae	tca:656438	9E-121	translation	1.8	0.0062
h_cluster_07603+3	GT338829/YQE07229	similar to ribosomal protein S8; K02995 small subunit ribosomal protein S8e	tca:655685	2E-99	translation	1.6	0.0039
h_cluster_14581+1	GT489617/YQE02752	similar to ribosomal protein S6; K02991 small subunit ribosomal protein S6e	tca:656796	2E-101	translation	1.5	0.0005

Table 1. Cont.

Translated EST contig ID[a]	EST/Genome Accession No.	BLASTx annotation	Subject	E-Value	Functional category	Fold change[b]	p-value
h_cluster_09822-1	**GT339811/n/a**	**similar to Ribosomal protein L18A CG6510-PA; K02882 large subunit ribosomal protein L18Ae**	**ame:409832**	**2E-73**	**translation**	**1.4**	**0.0064**
h_cluster_00223+3	GT402293/YQE04450	similar to CG9282-PA; K02896 large subunit ribosomal protein L24e	tca:654912	5E-36	translation	1.3	0.0014
h_cluster_18699+2	GT490858/YQE07111	similar to ribosomal protein L31; K02910 large subunit ribosomal protein L31e	tca:663175	1E-50	translation	1.2	0.0043
h_cluster_02284-1+3	**GT403942/YQE01291**	**similar to Vacuolar ATP synthase catalytic subunit A, osteoclast isoform (V-ATPase A subunit 2) (Vacuolar proton pump alpha subunit 2) (V-ATPase 69 kDa subunit 2) (Isoform HO68); K02145 V-type H+-transporting ATPase subunit A**	**tca:663322**	**0**	**transport**	**2.2**	**0.0028**
h_cluster_17723-2-2	**GO487586/YQE00120**	**similar to Vacuolar ATP synthase catalytic subunit A, osteoclast isoform (V-ATPase A subunit 2) (Vacuolar proton pump alpha subunit 2) (V-ATPase 69 kDa subunit 2) (Isoform HO68); K02145 V-type H+-transporting ATPase subunit A**	**tca:663322**	**7E-105**	**transport**	**1.2**	**0.0022**

Plain text represents proteins whose levels changed following feeding that were unique to the female iTRAQ runs. Those in bold were also present in males. Bold, underlined contig Ids signify proteins whose levels also changed significantly in males. Fold change values marked with an asterisk also changed significantly in the RNA-seq transcriptome.

[a]Suffix (+1, +2, +3, −1, −2, −3) indicates translation frame. Contig sequences available in File S1.

[b]Fold change is relative to control (starved) females. Values marked with an asterisk also changed significantly in the transcriptome (51).

Table 2. Proteins that showed significant changes in accumulation levels in males following feeding ($p<0.05$, FDR<0.2).

Translated EST contig ID[a]	EST/Genome Accession No.	BLASTx annotation	Subject	E-Value	Functional category	Fold change[b]	p-value
h_cluster_00011-1+2	**GT413372/YQE02827**	**similar to CG6084-PA, isoform A**	**tca:658019**	**8E-144**	**?**	**1.6***	**0.0058**
f_cluster_c4596+1	n/a/YQE09455	No Hits	n/a	n/a	?	1.5	0.0004
h_cluster_06285-1+1	**GT360056/YQE02828**	**similar to CG6084-PA, isoform A**	**tca:657943**	**9E-124**	**?**	**1.2**	**0.0062**
h_cluster_00437-1+1	**GT424689/YQE03298**	**similar to cxpwmw03; K03522 electron transfer flavoprotein alpha subunit**	**tca:657777**	**1E-141**	**?**	**−1.1**	**0.0016**
h_cluster_08736-1+1	**GT327980/YQE04947**	**similar to alanine aminotransferase; K00814 alanine transaminase [EC:2.6.1.2]**	**api:100164899**	**4E-120**	**?**	**−1.2**	**0.0061**
h_cluster_09712-2	GT428841/YQE06399	No Hits	n/a	n/a	?	−1.2	0.0006
h_cluster_01584-2	GT321155/n/a	No Hits	n/a	n/a	?	−1.8	0.0069
s_cluster_03357+1	GT380648/YQE03583	No Hits	n/a	n/a	?	−2.2	0.0002
f_cluster_c11478+2	**n/a/YQE07184**	**similar to CG11661-PF, isoform F; K00164 2-oxoglutarate dehydrogenase E1 component [EC:1.2.4.2]**	**tca:662219**	**7E-163**	**carbohydrate metabolism**	**−1.1**	**0.0033**
h_cluster_10171-3-3	**GT377795/YQE07184**	**similar to CG4001-PB, isoform B; K00850 6-phosphofructokinase [EC:2.7.1.11]**	**tca:658327**	**2E-148**	**carbohydrate metabolism**	**−1.2**	**0.0046**
f_cluster_c577+3	**GT478810/YQE02618**	**similar to CG1516-PE, isoform E; K01959 pyruvate carboxylase subunit A [EC:6.4.1.1]; K01960 pyruvate carboxylase subunit B [EC:6.4.1.1]**	**tca:662701**	**0**	**carbohydrate metabolism**	**−1.7**	**0.0059**
h_cluster_00722-1+1	**GT484212/YQE07729**	**similar to ENSANGP00000012893; K09490 heat shock 70 kDa protein 5**	**nvi:100118379**	**0**	**chaperone**	**1.6***	**0.0047**
h_cluster_00159-1+2	**GT369839/YQE10054**	**similar to CG9429-PA: calreticulin**	**tca:660439**	**9E-163**	**chaperone**	**1.5***	**0.0001**
h_cluster_00930-1+1	**GT352910/YQE07729**	**similar to heat shock 70 kD protein cognate; K09490 heat shock 70 kDa protein 5**	**tca:659147**	**4E-141**	**chaperone**	**1.4***	**0.0025**
h_cluster_00442-1+3	**GT437135/YQE06755**	**similar to CG8938-PA, isoform A, DpGSTs4**	**tca:662754**	**6E-62**	**detoxification, immunity, stress response**	**1.3**	**0.0032**
h_cluster_01581-1+3	**GT330637/YQE07531**	**similar to AMP dependent coa ligase**	**tca:655052**	**0**	**fatty acid metabolism**	**−1.2***	**0.0009**
h_cluster_00955-1+1	**GT463172/YQE11663**	**similar to CG4703-PA; K00248 butyryl-CoA dehydrogenase [EC:1.3.99.2] (short-chain specific acyl-CoA dehydrogenase, mitochondrial)**	**tca:661684**	**0**	**fatty acid metabolism**	**−1.2**	**0.0050**
h_cluster_00668-1+1	**GT452594/YQE03432**	**similar to CG9360-PA (3-oxoacyl-[acyl-carrier-protein] reductase)**	**tca:658671**	**1E-63**	**fatty acid metabolism**	**−1.7**	**0.0030**
h_cluster_08348+3	GT323407/YQE08406	similar to Low molecular weight phosphotyrosine protein phosphatase 2 (Low molecular weight cytosolic acid phosphatase 2) (PTPase 2)	tca:657373	1E-63	protein metabolism	1.2	0.0027
h_cluster_02035+3	GT326460/YQE10953	similar to CG13340-PA (cytosol and leucine aminopeptidase)	tca:663523	0	protein metabolism	−1.3	0.0065

Table 2. Cont.

Translated EST contig ID[a]	EST/Genome Accession No.	BLASTx annotation	Subject	E-Value	Functional category	Fold change[b]	p-value
h_cluster_01760+1	**GT378898/YQE02609**	**similar to AGAP007123-PA (sarcosine dehydrogenase - mitochondrial)**	**tca:656849**	**0**	**protein metabolism**	**−1.6**	**0.0033**
h_cluster_01629-1+3	GT473612/YQE07881	similar to proteasome (prosome, macropain) subunit, alpha type 1; K02725 20S proteasome subunit alpha 6 [EC:3.4.25.1]	tca:655339	3E-121	protein metabolsim	1.2	0.0064
h_cluster_00049-2+1	GT317557/YQE01516	similar to ATP synthase delta chain, mitochondrial; K02134 F-type H+-transporting ATPase subunit delta [EC:3.6.3.14]	tca:656683	6E-51	respiration	−1.4	0.0026
h_cluster_02330-1+3	**GT376610/YQE08847**	**similar to CG2331-PA, isoform A**	**tca:661187**	**0**	**signaling**	**1.2**	**0.0010**
h_cluster_03460+2	GT328587/YQE07802	similar to CG14996-PB; calponin-like protein Chd64	tca:663982	3E-86	structure	1.4	0.0024
f_cluster_c3717+1	**GT360252/YQE04803**	**similar to CG4376-PB, isoform B; K05699 actinin alpha**	**tca:661042**	**0**	**structure**	**−1.2**	**0.0077**
f_cluster_c3347-3	n/a/n/a	similar to Kinesin light chain (KLC); K10407 kinesin light chain	tca:655195	2E-98	structure	−1.3	0.0062
h_cluster_13890-3	**GT322908/YQE05135**	**similar to ribosomal protein S24e; K02974 small subunit ribosomal protein S24e**	**tca:655928**	**2E-49**	**translation**	**1.2**	**0.0016**
h_cluster_06800+1	**GT334993/YQE03706**	**GL10774 gene product from transcript GL10774-RA**	**dpe: Dper_GL 10774**	**3E-117**	**translation**	**−1.2**	**0.0006**

Plain text represents proteins whose levels changed following feeding that were unique to the male iTRAQ runs. Those in bold were also present in females. Bold, underlined contig Ids signify proteins whose levels also changed significantly in females. Fold change values marked with an asterisk also changed significantly in the RNA-seq transcriptome.
[a]Suffix (+1, +2, +3, −1, −2, −3) indicates translation frame. Contig sequences available in File S1.
[b]Fold change is relative to control (starved) males. Values marked with an asterisk also changed significantly in the transcriptome (51).

thereby facilitating their excretion [56–57]. DpGSTs4's expression pattern is consistent with a possible role in MPB coping with its toxic host environment.

Host colonization and the initiation of reproduction

Proteins linked to reproductive physiology were shown to react quantitatively to feeding treatment in females. Two vitellogenin-matching proteins had significant and substantial fold change increases, specifically an 8.9-fold increase for vitellogenin-6 precursor and a 7.8-fold increase for vitellogenin. This pattern echoes the increases of vitellogenin transcripts in fed females (YQE09293, [51]). Vitellogenin, and its precursor, are phospholipoglycoprotein yolk precursors from the female's fat body that are taken up by maturing insect oocytes in a process controlled by JH [58]. Although it exceeded the 0.2 FDR cut-off, maternal exuperantia protein (Table S1 in File S1, h_cluster_01681-1+3) increased with feeding treatment (2.3-fold change, $p = 0.006$, FDR = 0.21). This protein – involved in promoting the transcription of reproductive genes – mediates the action of the bicoid gene product – a maternal effect gene vital for early larval development and the establishment of correct anatomical polarity [59]. A protein matching to hormone metabolism – a possible JH binding protein called soldier-specific protein 1 [60]– exhibited the largest fold-change decrease of the proteins detected in the female proteome (-2.5-fold change). The transcript for this protein was also found to decrease with feeding (YQE10157, [51]). JH is linked to pheromone production in some bark beetles and to reproduction and many other physiological processes in insects. That females wait until encountering the host tree and commencing with colonization activities to produce these reproductive proteins indicates a tradeoff between dispersal and reproduction.

This tradeoff and rapid upregulation of reproductive activity once dispersal is completed is also supported by indicators of females using host tissue as an energy source. Two proteins with BLASTx matches to components of a vacuolar H^+-transporting ATPase – a complex involved in a variety of cellular processes related to active transport systems – increased in females with feeding (2.2- and 1.2-fold change). Fed *Manduca sexta* larvae have active transporter systems in the midgut, and the supply of K^+ needed for digestion requires a gradient of H^+ facilitated by V-type H^+-transporting ATPases [61]. The increased production of these proteins in female MPB may similarly indicate the activation of gut tissue once feeding begins. Two carbohydrate metabolism proteins involved in glycolysis– transketolase and glucose-6-phosphate dehydrogenase – showed significant increases in females (2.4 and 1.3 fold-change) during early host colonization. This trend is corroborated, albeit not significantly, by the transcriptome data (YQE11922, YQE03281, [51]). Female MPB have been found to compensate for lower energetic stores after dispersal by obtaining energy at the breeding site (phloem) [16]. Here females are expressing proteins related to carbohydrate metabolism once they colonize a host, likely to process the phloem they consume during gallery excavation into energy. Combined with the production of reproductive proteins, this shift in energetics over just 24 hours demonstrates the speed and extent of the transition from flight dispersal to reproductive physiology in female MPB upon entering a host.

Male muscle tissue as an energy source

Male MPB exhibited changes in proteins related to muscle and cellular structure. Females have higher relative lipid stores from which to draw energy [62], which might explain why we did not observe such changes in females. A protein matching to calponin increased significantly in colonizing males (1.4-fold change). Proteins in this family regulate muscle contractions, inhibiting the movement of actin filaments over myosin in vertebrate smooth muscle [63]. Insects exclusively have striated muscle, but calponin has been speculated to be integral to visceral muscle function [64]. Actinin exhibited a decrease with feeding treatment (-1.2-fold). Actinin cross-links actin filaments [65–66]. A protein that matched the light chain of kinesin, which is involved in cytoskeletal structuring, also decreased with feeding treatment (-1.3-fold change). This change in the way contractions are being regulated and the reduction of both muscle and cellular structure proteins may be related to males no longer needing their flight muscles, and it is likely that they begin to access muscle tissue for energy during early host colonization. As males do not construct the parental gallery, they may not be benefiting from energy contained in host phloem as females do [16].

Conclusion

The proteomes of host colonizing males and females contained similar numbers of proteins, overall, some of which shifted significantly and rapidly following initial colonization of a pine host. In addition, the overall classification of proteins in both sexes resided in similar functional categories, but the proteins with quantities that significantly changed following feeding differed between the sexes.

Stress physiology played a role for both females and males during host colonization, as indicated by the significant accumulation of proteins such as various chaperones in both sexes as well as the CYP6DE1 in females and DpGSTs4 in males. Rapid accumulation of vitellogenin indicated that female reproductive physiology shifts dramatically during initial host colonization. Thus the mainly semelparous males and females – struggling in a highly variable and highly stressful environment during the initial hours of host colonization – rapidly shift towards and then prioritize reproductive physiology. MPB females conserve energy stores and then rapidly use those stores in their reproductive attempt, while males, conversely, consume much of their energy stores during the dispersal.

We detected indicators of different strategies between males and females during host colonization. Both sexes produced PCWDEs without feeding treatment, but had to encounter hosts before producing several chaperones and detoxification enzymes. However, males seemed to constitutively produce proteins related to pheromones, and hormones without host exposure. Females did not produce reproductive proteins until they colonized a host, at which point they also activated pathways relating to gut tissue transporters and carbohydrate metabolism. During colonization, males decreased proteins related to flight muscle, possibly to access energy. Females did not do this within the treatment time used here, suggesting that males and females employ different energetic strategies.

Not only were almost all of the differentially accumulating proteins discussed here detected in RNA-seq transcriptome analysis [51], but several transcripts with significant changes did so with the same patterns shown in our work. RNA-seq, microarray, and real-time PCR studies reveal the activity of genes through mRNA, but do not account for the effects of post-translational modification and regulation of the final protein products. This proteomics study provides another layer to the existing proteome, genome and transcriptome assets that have been developed for MPB [23] [25] [36] [51] [67] and provides new data and targets for ongoing exploration of the short, yet vital and perilous, host colonization phase of the MPB life cycle.

Supporting Information

File S1 Contains Tables S1, S2, and S3. Table S1 - Female iTRAQ Run Data, Table S2 - Male iTRAQ Run Data, Table S3-Untranslated Transcriptome Sequences.
(XLSX)

Acknowledgments

We thank collaborators within the Tria 1 and Tria 2 projects (http://www.thetriaproject.ca), a study of Mountain Pine Beetle System Genomics. We thank the staff of the UVic Genome BC Proteomics Centre for excellent technical support of the iTRAQ analysis. We also thank Doug Bateman for the MPB-infested logs, Sophia Unger for help in the lab, and members of the UNBC Forest Insect Research Group for their input.

Author Contributions

Conceived and designed the experiments: JAR DH. Performed the experiments: CP JAR TRB. Analyzed the data: CP TRB. Contributed reagents/materials/analysis tools: DPWH CIK JB. Wrote the paper: CP DPWH.

References

1. Jordal BH, Sequeira AS, Cognato AI (2011) The age and phylogeny of wood boring weevils and the origin of subsociality. Mol Phylogenet Evol 59: 708–724.
2. Oberprieler RG, Marvaldi AE, Anderson RS (2007) Weevils, weevils, weevils everywhere. Zootaxa 520: 491–520.
3. Rudinsky JA (1962) The ecology of the Scolytidae. Annu Rev Entomol 7: 327–348.
4. Wood SL, Bright DE (1992) Great Basin Naturalist Memoirs No. 13: A catalogue of Scolytidae and Platypodidae (Coleoptera), Part 2: Taxonomic Index, Volume A. Provo: Brigham Young University Press. pp. 159.
5. Cullingham CI, Cooke JEK, Dang S, Davis CS, Cooke BJ, et al. (2011) Mountain pine beetle host-range expansion threatens the boreal forest. Molec Ecol 20: 2157–2171.
6. Raffa KF, Powell EN, Townsend PA (2013) Temperature-driven range expansion of an irruptive insect heightened by weakly coevolved plant defenses. P Natl Acad Sci USA 110: 2193–2198.
7. Raffa KF, Berryman AA (1983) The role of host plant resistance in the colonization behavior and ecology of bark beetles (Coleoptera: Scolytidae). Ecol Monogr 53: 27–49.
8. Byers JA (1989) Chemical Ecology of Bark Beetles. Experientia 45: 271–283.
9. Amman G (1977) Role of the mountain pine beetle in lodgepole pine ecosystems: impact on succession. In: W.J . MattsonMattson, editor. The Role of Arthropods in Forest Ecosystems. New York: Springer-Verlag. pp. 3–18.
10. Walton A (12 April 2013) Provincial-level projection of the current mountain pine beetle outbreak: update of the infestation projection based on the Provincial Aerial Overview Surveys of Forest Health conducted from 1999 through 2012 and the BCMPB model (year 10). Victoria, BC: BC Ministry of Forests, Lands and Natural Resources Operations. Available: https://www.for.gov.bc.ca/ftp/hre/external/!publish/web/bcmpb/year10/BCMPB.v10.BeetleProjection.Update.pdf. Accessed 16 August 2013.
11. Kurz WA, Dymond CC, Stinson G, Rampley GJ, Neilson ET, et al. (2008) Mountain pine beetle and forest carbon: feedback to climate change. Nature 454: 987–990.
12. Huber DPW, Aukema BH, Hodgkinson RW, Lindgren BS (2009) Successful colonization, reproduction, and new generation emergence in live interior hybrid spruce, *Picea engelmannii x glauca*, by mountain pine beetle, *Dendroctonus ponderosae*. Agric For Entomol 11: 83–89.
13. Safranyik L, Linton DA, Silversides R, McMullen LH (1992) Dispersal of released mountain pine beetles under the canopy of a mature lodgepole pine stand. J Appl Entomol 113: 1–5.
14. Huber DPW, Borden JH (2003) Comparative Behavioural Responses of *Dryocoetes confuses* Swaine, *Dendroctonus rufipennis* (Kirby), and *Dendroctonus ponderosae* Hopkins (Coleoptera: Scolytidae) to Angiosperm Tree Bark Volatiles. Environ Entomol 32: 742–751.
15. Huber DPW, Gries R, Borden JH, Pierce HD Jr (2000) A survey of antennal responses by five species of coniferophagous bark beetles (Coleoptera: Scolytidae) to bark volatiles of six species of angiosperm trees. Chemoecology 10: 103–113.
16. Elkin CM, Reid ML (2005) Low energy reserves and resource allocation decisions affect reproduction by Mountain Pine Beetles, *Dendroctonus ponderosae*. Funct Ecol 19: 102–109.
17. Borden JH (1985) Aggregation pheromones. In: Kerkut GA, Gilbert, LI, editors. Comprehensive Insect Physiology, Biochemistry, and Pharmacology, Vol. 9. Oxford: Pergamon Press. pp. 257–285.
18. Raffa KF, Berryman AA (1982) Physiological differences between lodgepole pines resistant and susceptible to the mountain pine beetle and associated microorganisms. Environ Entomol 11: 486–492
19. DiGuistini S, Wang Y, Liao NY, Taylor G, Tanguay P, et al. (2011). Genome and transcriptome analyses of the mountain pine beetle-fungal symbiont *Grosmannia clavigera*, a lodgepole pine pathogen. P Natl Acad Sci USA 108: 2504–2509.
20. Seybold SJ, Huber DPW, Lee LC, Graves AD, Bohlmann J (2006). Pine monoterpenes and pine bark beetles: A marriage of convenience for defense and chemical communication. Phytochemistry Rev 5: 143–178.
21. Boone CK, Aukema BH, Bohlmann J, Carroll AL, Raffa KF (2011) Efficacy of tree defense physiology varies with bark beetle population density: a basis for positive feedback in eruptive species. Can J Forest Res 41: 1174–1188.
22. Hall DE, Yuen MMS, Jancsik S, Quesada AL, Dullat HK, et al. (2013) Transcriptome resources and functional characterization of monoterpene synthases for two host species of the mountain pine beetle, lodgepole pine (*Pinus contorta*) and jack pine (*Pinus banksiana*). BMC Plant Biol 13: doi: 10.1186/1471-2229-13-80.
23. Keeling CI, Henderson H, Li M, Yuen M, Clark EL, et al. (2012) Transcriptome and full-length cDNA resources for the mountain pine beetle, *Dendroctonus ponderosae* Hopkins, a major insect pest of pine forests. Insect Biochem Molec 42: 525–536. doi: 10.1016/j.ibmb.2012.03.010.
24. Lyon RL (1958) A Useful Secondary Sex Character in Dendroctonus Bark Beetles. Can Entomol 90: 582–584.
25. Bonnett TR, Robert JA, Pitt C, Fraser JD, Keeling CI, et al. (2012) Global and comparative proteomic profiling of overwintering and developing mountain pine beetle, *Dendroctonus ponderosae* (Coleoptera: Curculionidae), larvae. Insect Biochem Molec 42: 890–901.
26. Lippert D, Ralph SG, Phillips M, White R, Smith D, et al. (2009) Quantitative iTRAQ proteome and comparative transcriptome analysis of elicitor-induced Norway spruce (*Picea abies*) cells reveals elements of calcium signaling in the early conifer defense response. Proteomics 9: 350–367.
27. Ohlund LB, Hardie DB, Elliott MH, Camenzind AG, Smith DS, et al. (2011) Standard Operating Procedures and Protocols for the Preparation and Analysis of Plasma Samples Using the iTRAQ Methodology. In: Ivanov AR, Lazarev AV, editors. Sample Preparation in Biological Mass Spectrometry. New York: Springer Science. pp. 575–626.
28. Shilov IV, Seymour SL, Patel AA, Loboda A, Tang WH, et al. (2007) The Paragon Algorithm, a next generation search engine that uses sequence temperature values and feature probabilities to identify peptides from tandem mass spectra. Mol Cell Proteomics 6: 1638–1655.
29. Benjamini Y, Hochberg Y (1995) Controlling the false discovery rate: a practical and powerful approach to multiple testing. J Roy Stat Soc B Met 57: 289–300.
30. Hakimov HA, Walters S, Wright TC, Meidinger RG, Verschoor CP, et al. (2009) Application of iTRAQ to catalogue the skeletal muscle proteome in pigs and assessment of effects of gender and diet dephytinization. Proteomics 9: 4000–4016.
31. Kitteringham NR, Abdullah A, Walsh J, Randle L, Jenkins RE, et al. (2010) Proteomic analysis of Nrf2 deficient transgenic mice reveals cellular defence and lipid metabolism as primary Nrf2-dependent pathways in the liver. J Proteomics 73: 1612–1631.
32. The Uniprot Consortium (2011) Ongoing and future developments at the Universal Protein Resource. Nucl Acids Res 39: D214–D219.
33. Thomas PD, Kejariwal A, Campbell MJ, Mi H, Diemer K, et al. (2003) PANTHER: a browsable database of gene products organized by biological function, using curated protein family and subfamily classification. Nucl Acids Res 31: 334–341.
34. Eigenheer AL, Keeling CI, Young S, Tittiger C (2003) Comparison of gene representation in midguts from two phytophagous insects, *Bombyx mori* and *Ips pini*, using expressed sequence tags. Gene 316: 127–136.
35. Tittiger C, Keeling CI, Blomquist GJ (2005) Chapter Three Some insights into the remarkable metabolism of the bark beetle midgut. Recent Adv Phytochem 39: 57–78.
36. Keeling CI, Yuen MMS, Liao NY, Docking TR, Chan SK, et al. (2013) Draft genome of the mountain pine beetle, *Dendroctonus ponderosae* Hopkins, a major forest pest. Genome Biol 14: R27. doi: 10.1186/gb-2013-14-3-r27.
37. Pauchet Y, Wilkinson P, Chauhan R, ffrench-Constant RH (2010) Diversity of Beetle Genes Encoding Novel Plant Cell Wall Degrading Enzymes. PLOS ONE 5: e15635. doi:10.1371/journal.pone.0015635.
38. Bellés X, Martín D, Piulachs MD (2005) The Mevalonate Pathway and the Synthesis of Juvenile Hormone in Insects. Annu Rev Entomol 50: 181–199.
39. Bearfield JC, Keeling CI, Young S, Blomquist GJ, Tittiger C (2006) Isolation, endocrine regulation and mRNA distribution of the 3-hydroxy-3-methylglutaryl coenzyme A synthase (HMG-S) gene from the pine engraver, *Ips pini* (Coleoptera: Scolytidae). Insect Molec Biol 15: 187–195.
40. Keeling CI, Blomquist GJ, Tittiger C (2004) Coordinated gene expression for pheromone biosynthesis in the pine engraver beetle, *Ips pini* (Coleoptera: Scolytidae). Naturwissenschafen 91: 324–328.

41. Chen NM, Borden JH, Pierce HD Jr (1988) Effect of juvenile hormone analog, fenoxycarb, on pheromone production by *Ips paraconfusus* (Coleoptera: Scolytidae) J Chem Ecol 14: 1087–1098.
42. Byers JA, Birgersson G (1990) Pheromone Production in a Bark Beetle Independent of Myrcene Precursor in Host Pine Species. Naturwissenschaften 77: 385–387.
43. Seybold SJ, Quilici DR, Tillman JA, Vanderwel D, Wood DL, et al. (1995) De novo biosynthesis of the aggregation pheromone components ipsenol and ipsdienol by the pine bark beetles *Ips paraconfusus* Lanier and *Ips pini* (Say) (Coleoptera: Scolytidae). P Natl Acad Sci USA 92: 8393–8397.
44. Tittiger C, Blomquist GJ, Ivarsson P, Borgeson CE, Seybold SJ (1999) Juvenile hormone regulation of HMG-R gene expression in the bark beetle *Ips paraconfusus* (Coleoptera: Scolytidae): implications for male aggregation pheromone biosynthesis. Cell Mol Life Sci 55: 121–127.
45. Hall GM, Tittiger C, Blomquist GJ, Andrews GL, Mastick GS, et al. (2002) Male Jeffrey pine beetle, *Dendroctonus jeffreyi*, synthesizes the pheromone component frontalin in anterior midgut tissue. Insect Biochem Molec 32: 1525–1532.
46. Keeling CI, Bearfield JC, Young S, Blomquist J, Tittiger C (2006) Effects of juvenile hormone on gene expression in the pheromone-producing midgut of the pine engraver beetle, *Ips pini*. Insect Molec Biol 15: 207–216.
47. Lindgren BS, Borden JH (1989) Semiochemicals of the mountain pine beetle (*Dendroctonus ponderosae* Hopkins). In Proceedings: symposium on the management of lodgepole pine to minimize losses to the mountain pine beetle, July 12–14, 1988. G. D. Amman (compiler). Kalispell, Montana, USDA Forest Service, Intermountain Research Station, Ogden, Utah. pp. 83–88.
48. Barwaki LS, Franke W, Blomquist GJ, Seybold SJ (2003) Frontalin: De novo biosynthesis of an aggregation pheromone component by *Dendroctonus* spp. bark beetles (Coleoptera: Scolytidae). Insect Biochem Molec 33: 773–788.
49. Tittiger C, Barkawi LS, Bengoa CS, Blomquist GJ, Seybold SJ (2003) Structure and juvenile hormone-mediated regulation of the HMG-CoA reductase gene from the Jeffrey pine beetle, *Dendroctonus jeffreyi*. Mol Cell Endocinol 199: 11–21.
50. Keeling CI, Chiu CC, Aw T, Li M, Henderson H, et al. (2013) Frontalin pheromone biosynthesis in the mountain pine beetle, *Dendroctonus ponderosae*, and the role of isoprenyl diphosohate synthases. P Natl Acad Sci USA 110: 18838–18843.
51. Robert JA, Pitt C, Bonnett TR, Yuen MMS, Keeling CI, et al. (2013) Disentangling Detoxification: Gene Expression Analysis of Feeding Mountain Pine Beetle Illuminates Molecular-Level Host Chemical Defense Detoxification Mechanisms. PLOS ONE 8: e77777. doi:10.1371/journal.pone.0077777.
52. Feder ME, Hofmann GE (1999) Heat-shock proteins, molecular chaperones, and the stress response: evolutionary and ecological physiology. Annu Rev Physiol 61: 243–282.
53. Mayer MP, Bukau B (2005) Hsp70 chaperones: Cellular functions and molecular mechanism. Cell Mol Life Sci 62: 670–684.
54. Asgari S, Schmidt O (2003) Is cell surface calreticulin involved in phagocytosis by insect hemocytes? J Insect Physiol 49: 545–550.
55. Feyereisen R (1999) Insect P450 Enzymes. Annu Rev Entomol 44: 507–533.
56. Enayati AA, Ranson H, Hemingway J (2005) Insect glutathione transferases and insecticide resistance. Insect Molec Biol 14: 3–8.
57. Chahine S, O'Donnell MJ (2011) Interactions between detoxification mechanisms and excretion in Malpighian tubules of *Drosophila melanogaster*. J Exp Biol 214: 462–468.
58. Hartfelder K (2000) Insect juvenile hormone: from "status quo" to high society. Braz J Med Biol Res 33: 157–177.
59. Marcey D, Watkins WS, Hazelrigg T (1991) The temporal and spatial distribution pattern of maternal exuperantia protein: evidence for a role in establishment but not maintenance of bicoid mRNA localization. EMBO J 10: 4259–4266.
60. Hojo M, Morioka M, Matsumoto T, Miura T (2005) Identification of soldier caste-specific protein in the frontal gland of nasute termite *Natsutitermes takasagoensis* (Isoptera: Termitidae). Insect Biochem Molec 35: 347–354.
61. Beyenbach KW, Wiezorek H (2006) The V-type H+ ATPase: molecular structure and function, physiological roles and regulation. J Exp Biol 209: 577–589.
62. Graf M, Reid ML, Aukema BH, Lindgren BS (2012) Association of tree diameter with body size and lipid content of mountain pine beetles. Can Entomol 144: 467–477.
63. El-Mezgueldi M (1996) Calponin. Int J Biochem Cell B 28: 1185–1189.
64. Park K, Kwak IS (2008) Calponin Gene Expression in *Chironomus riparius* Exposed to Di(2-ethylhexyl) phthalate. Env Tox 24: 555–562.
65. Goll DE, Suzuki A, Temple J, Holmes GR (1972) Studies on purified α-actinin: I. Effect of temperature and tropomyosin on the α-actinin/F-actin interaction. J Molec Biol 67: 469–472.
66. Bullard B, Sainsbury GM (1977) The proteins in the Z line of insect flight muscle. Biochem J 161: 399–403.
67. Aw T, Schlauch K, Keeling CI, Young S, Bearfield JC, et al. (2010) Functional genomics of mountain pine beetle (*Dendroctonus ponderosae*) midguts and fat bodies. BMC Genomics 11: 215.

14

Gene Expression Differences among Three *Neurospora* Species Reveal Genes Required for Sexual Reproduction in *Neurospora crassa*

Nina A. Lehr[1], Zheng Wang[1,2], Ning Li[1], David A. Hewitt[3,4], Francesc López-Giráldez[1], Frances Trail[5,6], Jeffrey P. Townsend[1,2,7,8]*

1 Department of Ecology and Evolutionary Biology, Yale University, New Haven, Connecticut, United States of America, **2** Department of Biostatistics, Yale University, New Haven, Connecticut, United States of America, **3** Department of Botany, Academy of Natural Sciences, Philadelphia, Pennsylvania, United States of America, **4** Wagner Free Institute of Science, Philadelphia, Pennsylvania, United States of America, **5** Department of Plant Biology, Michigan State University, East Lansing, Michigan, United States of America, **6** Department of Plant, Soil and Microbial Sciences, Michigan State University, East Lansing, Michigan, United States of America, **7** Program in Computational Biology and Bioinformatics, Yale University, New Haven, Connecticut, United States of America, **8** Program in Microbiology, Yale University, New Haven, Connecticut, United States of America

Abstract

Many fungi form complex three-dimensional fruiting bodies, within which the meiotic machinery for sexual spore production has been considered to be largely conserved over evolutionary time. Indeed, much of what we know about meiosis in plant and animal taxa has been deeply informed by studies of meiosis in *Saccharomyces* and *Neurospora*. Nevertheless, the genetic basis of fruiting body development and its regulation in relation to meiosis in fungi is barely known, even within the best studied multicellular fungal model *Neurospora crassa*. We characterized morphological development and genome-wide transcriptomics in the closely related species *Neurospora crassa, Neurospora tetrasperma*, and *Neurospora discreta*, across eight stages of sexual development. Despite diverse life histories within the genus, all three species produce vase-shaped perithecia. Transcriptome sequencing provided gene expression levels of orthologous genes among all three species. Expression of key meiosis genes and sporulation genes corresponded to known phenotypic and developmental differences among these *Neurospora* species during sexual development. We assembled a list of genes putatively relevant to the recent evolution of fruiting body development by sorting genes whose relative expression across developmental stages increased more in *N. crassa* relative to the other species. Then, in *N. crassa*, we characterized the phenotypes of fruiting bodies arising from crosses of homozygous knockout strains of the top genes. Eight *N. crassa* genes were found to be critical for the successful formation of perithecia. The absence of these genes in these crosses resulted in either no perithecium formation or in arrested development at an early stage. Our results provide insight into the genetic basis of *Neurospora* sexual reproduction, which is also of great importance with regard to other multicellular ascomycetes, including perithecium-forming pathogens, such as *Claviceps purpurea*, *Ophiostoma ulmi*, and *Glomerella graminicola*.

Editor: Stefanie Pöggeler, Georg-August-University of Göttingen Institute of Microbiology & Genetics, Germany

Funding: This study was supported by a Gaylord Donnelley Environmental Fellowship of the Yale Institute of Biospheric studies to NL, and National Science Foundation (NSF) Grant MCB 0923797 to FT and JPT. The funders had no role in study design, data collection and analysis, decision to publish, or preparation of the manuscript.

Competing Interests: The authors have declared that no competing interests exist.

* Email: Jeffrey.Townsend@Yale.edu

Introduction

Many ascomycete fungi sexually reproduce by forming three-dimensional fruiting bodies that produce their sexual spores (ascospores) in sacs called asci. The genetics of fruit-body development in the filamentous fungi has been studied extensively in several fungi including *Aspergillus nidulans*, *Neurospora crassa*, *Sordaria macrospora*, *Fusarium graminearum*, and *Podospora anserina* [1–9]. Comparative genomics analyses suggest that meiosis machinery and ascospore production are generally conserved within ascomycetes [10–17]. Indeed, much of what we know about meiosis in plant and animal taxa has been deeply informed by studies of meiosis in *Saccharomyces* and, historically, *Neurospora* [18,19]. Although some genes involved in fungal fruiting body development have been identified by mutagenesis screens and characterized, the comparative genetic basis for the sexual cycle of these organisms in terms of interactions among gene networks has not been little explored [7,20]. Expression and comparative transcriptomics studies between *N. crassa* and *N. tetrasperma* have revealed key genes involved in asexual development and mating behaviors in these fungi [21,22]. However, inference and comparison of regulatory pathways for sexual development across these species has been difficult to achieve, partly due to the complex environmental stimuli relevant to sexual development and partly due to a lack of molecular detail regarding relevant gene interactions across sexual development. *Neurospora* species represent attractive models for elucidating the regulation of fruiting body development by transcriptional profiling and functional analysis due to their simple nutritional requirements, fast vegetative growth, and clearly recognizable stages during sexual development [23,24]. The most studied species is *N. crassa* [25], particularly well-known for its use in the original "one gene-one enzyme" experiments by Beadle and Tatum, for its eight-spored ordered asci which enabled centromere mapping of mutants, and for its experimentally tractable and intensively-studied genetic network for circadian rhythm [26,27].

The life histories of *Neurospora* spp. span the most common sexual strategies in the fungal kingdom, i.e. heterothallism (self-incompatibility with distinct mating types), pseudohomothallism (self-compatibility in which paired mating types coexist in one mycelium), and homothallism (self-compatibility regardless of mating type). Initiation of sexual reproduction is regulated by mating type genes and leads to cell fusion, nuclear pairing, nuclear fusion, meiosis, and the production of haploid ascospores. The determinant sequences for mating type, *mat A* and *mat a*, are at the same genetic locus in a given species but exhibit little to no similarity in nucleotide sequence; that is, they represent idiomorphs rather than alleles [28]. Heterothallic *Neurospora* species such as *N. crassa* and *N. discreta* [23] possess a bipolar mating system, and two strains with opposite mating types, *mat A* and *mat a*, must cross to initiate sexual development. Either mating type can produce female structures (protoperithecia with a trichogyne) as well as male reproductive structures (conidia) [29,30]. Pseudohomothallic species such as *N. tetrasperma* are generally self-compatible, even though they also require both mating types (*A* and *a*) to reproduce. In *N. tetrasperma*, both mating types are generally found within a single individual, although strains containing a single mating type exist in nature and can be isolated in lab. For *N. tetrasperma*, *mat A* and *mat a* nuclei associated in pairs in four heterokaryotic spores, packaged within an ascus for discharge [31]. Despite mating behavior differences that distinguish *N. tetrasperma* from *N. crassa* and *N. discreta*, phylogenetically *N. tetrasperma* and *N. crassa* are closely related and share the most recent common ancestor with *N. discreta* [32].

The genetic basis of sexual development has been the subject of many investigations, with unresolved controversy arising about the ancestral state of the life style of fungal sexual reproduction, heterothallism or homothallism, and its underlying genetic mechanism(s) [33–40]. For example, the heterothallic life style has been suggested to be ancestral within *Neurospora* [31], because the pseudohomothallic *N. tetrasperma* does not require a mating partner, but does require both *mat* idiomorphs to complete the sexual cycle. However, function of *mat* genes in *Neurospora* sexual development is not well understood, except their roles in heterothallic species such as *N. discreta* and *N. crassa* as regulators of pheromone expression to direct hyphal growth and fusion [4,41]. For heterothallic species as in *N. crassa*, the trichogynes, which originate from protoperithecia, change their direction of growth to approach conidia of the opposite mating type and fuse with them. Plasmogamy is followed by development of the perithecium, and later ascal development, karyogamy, and then formation of ascospores [42–44]. Two haploid nuclei of opposite mating type fuse in a young ascus resulting in a diploid zygote nucleus that immediately undergoes meioses, followed by a postmeiotic mitosis. The mature ascus delimits eight linearly arranged ascospores, each containing a single nucleus. After mitosis the ascospores become binucleate, gradually grow to their full size, and become pigmented.

Genomes of multiple species of *Neurospora* have been sequenced, and comparative genomic analyses have been focused on *N. crassa* and its closely related species [3,6,45]. Here we reveal candidate genes involved in fungal development and in the evolution of perithecia by comparisons of the gene expression levels within and across three *Neurospora* species. We assayed for large-scale differences in morphology and the transcriptomic landscape over the time course of sexual development, identifying putative genes involved in sexual development by comparative gene expression profiling. We also tested for knockout phenotypes of selected candidate genes by assaying knockout strains across sexual development for their ability to produce wild type perithecia. Our results provide insights into the links between gene expression and sexual development of *N. crassa* and related species, as well as contributing to our understanding of how fungi reproduce sexually.

Materials and Methods

2.1. Strains and culture conditions

Strains of complementary mating types *mat a* and *mat A* for *N. crassa* (FGSC4200, FGSC2489), *N. tetrasperma* (FGSC2509, FGSC2508) and *N. discreta* (FGSC8578, FGSC8579) were obtained from the Fungal Genetics Stock Center (FGSC) [46]. The strains were grown on Carrot Agar (CA), made as previously described [47]. The CA petri dish was covered with a cellophane membrane (Fisher Scientific Company) and plugs of agar with strains were deposited on the membrane and incubated at 26°C under constant artificial light from several Ecolux bulbs (F17T8.SP41-ECO, General Electric Company), which provided a net intensity of 14 μMol/m^2 S at the media surface. Conidia from the *mat a* strain on CA were collected and suspended in 2.5% Tween 60 (10^5–10^6 conidia/ml). Cultures of the *mat A* strain on CA were examined using a stereomicroscope for the formation of protoperithecia in 5–7 days, and areas with evenly distributed protoperithecia of a common size were delineated with a marker on the bottom of the plate to be harvested for stage-specific transcriptomics.

Crosses were performed by applying 2 ml of the suspension of *mat a* conidia in 2.5% Tween 60 (10^5–10^6 conidia/ml) to the surface of the *mat A* protoperithecia plates, at which point considerable disturbance to surface hyphae and other fungal tissues was unavoidable. Sexual development was monitored with a stereomicroscope until fully developed perithecia appeared [48]. Fungal material was harvested by scraping the surface with a razor blade in the areas, where protoperithecia or young perithecia similar in size were densely aggregated, right before the crossing and at 2, 24, 48 h after crossing. Sets of individual perithecia of similar morphological development were picked at 72, 96, and 144 h after crossing. For transcriptomic analysis, all tissues and perithecia were immediately and rapidly frozen in liquid nitrogen as they were sampled, then stored at −80°C.

2.2. Fixation and microscopy

Perithecium development was monitored for all three *Neurospora* species with a stereomicroscope over the time course of the sexual development. Cultures and crossings were performed as described in 2.1. Pieces of cellophane membrane (about 4 mm×2 mm) carrying 5–20 perithecia of similar size were cut from cultures and fixed in 1.5% formaldehyde and 0.025 M phosphate buffer for at least 48 h. The samples were embedded in resin and prepared for light microscopy as previously described [48]. Briefly, resin blocks were sectioned to a thickness of 1 to 2 µm using a glass knife and stained with 1% toluidine blue. A Leica DM LB microscope (Leica Microsystem Gmbh, Wetzlar, Germany) was used to capture images using a Zeiss AxioCam MRc color camera and AxioVision 4.8.2 (Göttingen, Germany). Image processing and annotation were performed using Adobe Photoshop CS3 (San Jose, CA).

To compare mature perithecia, scanning electron microscopy (SEM) was performed. Perithecia were collected, including the cellophane membrane they were growing on, and fixed immediately in 0.1 M sodium cacodylate buffer (pH 7.2) containing 2% glutaraldehyde over night. The samples were then washed with 0.1 M sodium cacodylate buffer, postfixed with 1% osmium tetroxide for 1 h, washed with distilled water, and dehydrated in an increasing series of ethanol. Then, all samples were critical point dried, mounted on an SEM stub, and sputter coated with gold. The samples were examined with a Scanning Electron Microscope (ISI SS-40).

2.3. RNA extraction, cDNA preparation, and transcriptomic sequencing

RNA was isolated from homogenized mycelia with TRI REAGENT (Invitrogen) and RNAeasy Kit (Qiagen) following protocol described by Clark et al. [49] and mRNA was purified using Dynabeads oligo(dT) magnetic separation (Invitrogen). The cDNA libraries for RNA sequencing were prepared according to the Illumina mRNA Sequencing Sample Preparation Guide. In brief, triplicates were prepared for each time point and pooled. Then, mRNA was purified from total RNA and 100 ng (9 µl) were fragmented with 10X fragmentation buffer (Ambion AM8740) and incubated at 70°C for 5 min prior to adding 1 ul stop buffer (Ambion). Fragmented mRNA was precipitated using 100% ethanol with glycogen (Ambion) at −80°C. Random hexamers (N_6, Invitrogen) were added to prime reverse transcription of the first strand cDNA separately for each sample, and to recover the second strand cDNA for all samples. After the ligation of standard adapters for Illumina sequencing, all samples were separated on a 2% low melting point agarose gel and processed cDNA fragments of lengths between 200 and 400 bp were selected by gel extraction and purified with Qiaquick gel extraction kit (Qiagen). The quantity of the samples was increased by a PCR using Pfx DNA polymerase (Invitrogen), and 15 cycles of PCR, each cycle comprising 98°C for 10 s, 65°C for 30 s, and 68°C for 30 s. The quantity and quality of the purified PCR products were checked at the Yale Center for Genome Analysis prior to sequencing. Single-end 35 bp reads of N_6-primed preparations were separately sequenced, each on eight lanes of an Illumina Genome Analyzer (Yale Center for Genome Analysis).

2.4. Data acquisition and analysis

The libraries were run on eight lanes of an Illumina Genome Analyzer, generating an average of 28 million single-end reads of 36 nucleotides each. Since transcriptomic tags also contain sequences that span exon junctions, the program Tophat v1.1.14 [50] was used to perform spliced alignments of the tags against the *N. crassa* OR74A genome (NC10; [51]), and those of *N. tetrasperma* FGSC 2508 (v2) and *N. discreta* FGSC 8579 (v1) obtained from JGI Genome Portal [52]. We scored results only for tags that mapped to a single unique location in the genome (–max-multihits option was set to 1) with less than three mismatches (–splice-mismatches option was set to 2). We used the default settings for all other Tophat options. We tallied tags aligning to exons of genes with the program HTSeq v0.4.5p6 (Unpublished; http://www-huber.embl.de/users/anders/HTSeq/doc/) and the gene structure annotation file for the reference genome. LOX v1.4 [53] was applied to the tallies for each sample for each gene to estimate gene expression levels and credible intervals across developmental stages. LOX provided relative gene expression levels standardized by the lowest sample, with credible intervals, for all three species (Table S1).

Instead of using simple sequence similarity for homolog identification, we applied a phylogenetic approach that is reliable in calling homologs among closely related genomes, with a cost of power to identify potential homologs for recently evolved gene families and genes experiencing multiple duplication events in their evolutionary history. A total of 2352 orthologous genes were selected by the BranchClust method [54] for *N. crassa*, *N. tetrasperma* and *N. discreta*. BranchClust uses the Reciprocal Best Blast hit method and phylogenetic trees to select putatively orthologous genes using a default threshold e-value of 10^{-4}; we only choose the complete families in those three species. Transcriptomic sequencing revealed expression levels for all orthologous gene triads across sexual development (Table S2). Data is also deposited as accession 239 at the Filamentous Fungal Gene Expression Database (FFGED; [55]) and as accession GSE41484 for *N. crassa* [56], GSE60256 for *N. tetrasperma*, and GSE60255 for *N. discreta* at the National Center for Biotechnology Information Gene Expression Omnibus (GEO; http://www.ncbi.nlm.nih.gov/geo). More orthologs can be called among these fungi using BLAST approaches, and we also identified more single copy ortholog for these species using a BLAST-based method. However, from our other evolutionary study targeting several gene families with these species, we found that phylogeny-based ortholog calls by BranchClust were more robust and reliable.

A comparative heat map was constructed based on the level of expression of each gene for all three species over the entire time course of sexual development. Gene expression levels for each ortholog were normalized row-by-row by subtraction of the mean and division by the standard deviation. To cluster observed gene expression hierarchically, rows only were clustered by iteratively agglomerating the most similar two gene expression profiles and by averaging the agglomeration for the next iteration (Unweighted Pair Group Method with Arithmetic mean, UPGMA) within which similarity was assessed by the Pearson correlation coefficient of expression between genes. The Functional Catalogue (FunCat: [57]; http://mips.helmholtz-muenchen.de/proj/funcatDB/) annotation scheme was used to group genes according to their functions. The statistically significant overrepresentation of gene groups in functional categories relative to the whole genome was determined by a hypergeometric distribution *P* value calculation, facilitated by the MIPS FunCat online web application.

2.5. Comparative gene expression analysis of *N. crassa*, *N. tetrasperma*, and *N. discreta*

The LOX estimates across developmental stages for each orthologous gene for each species were assembled to compare gene expression levels across development of *N. crassa*, *N. tetrasperma*, and *N. discreta*. For each developmental stage, two values were calculated using the upper bound of the 95% confidence interval (CI) or the lower bound of the 95% CI from LOX. If the expression of a gene in species 1 was higher than species 2, we calculated the difference between the lower bound of expression in species 1 and the upper bound of expression in species 2. Conversely, if the expression of a gene in species 2 was higher than species 1, we calculated the difference between the lower bound of species 2 and the upper bound of species 1. These differences constitute those of which we can be highly confident. *N. crassa* was compared with *N. tetrasperma*, *N. crassa* with *N. discreta*, and lastly *N. tetrasperma* with *N. discreta*. The calculation described above was performed for all time points (2, 24, 48, 72, 96 and 144 h) for each gene. Finally, for each comparison, all genes were prioritized according to magnitude of difference in a descending order and the first 130 genes with the biggest changes across all time points and common to all comparisons were subsequently selected for knockout and phenotyping. The data for the comparative gene expression analysis can be found in the Tables S3–S5.

2.6. Assessing phenotypes of knock out mutants

Knockout strains for the top 130 candidate genes were compared to wild-type (WT) strains and screened for defects in fruiting body formation. Knockouts of candidate genes were obtained for both mating types in *Neurospora crassa* from the Fungal Genetics Stock Center (FGSC). These knockouts had been preliminarily assayed for phenotypes by several high-throughput screens as part of a *Neurospora* knockout project [58,59], and in that project had not exhibited a mutant phenotype in asexual growth or development. We performed a detailed, controlled screen crossing the two mating types of each deletion strain, cultured on CA in triplicate. The mat A cultures on CA were fertilized with conidia from the mat a strain. Perithecium formation was monitored with stereo- and light microscopy (Nikon Diaphot 300) for the presence and for the orientation of a single beak, black coloration and the typical vase-shape. Mature perithecia were examined in squash mounts for the presence of asci with normal pores, ascospores with normal shapes and numbers, and normal paraphyses. Normal spore development and ascus firing were examined by checking the lid of the Petri dish for black ascospores.

2.7 Cosegregation experiments of knockout mutants showing a phenotype different from the wild type

Knockouts of candidate genes obtained from FGSC were produced using a high-throughput gene deletion strategy in *N. crassa* strains with deletion mutations of *mus-51* and/or *mus-52*, mutations required for nonhomologous end-joining DNA repair. [58,59]. A high rate of spontaneous mutations has been observed in in *Δmus-51* and/or *Δmus-52* strains, and cosegregation experiments were used to demonstrate that the intended knockout deletion was responsible for the mutant phenotype [60]. A hygromycin resistance cassette at the location of the deletion mutation provides a selectable marker. The KO strains showing phenotypes in sexual development were crossed with wild-type strains (FGSC2489 *mat A* or FGSC4200 *mat a*). Individual ascospore progeny were isolated for resistance to hygromycin. Their phenotypes were then examined on SCM medium. Cosegregation of hygromycin resistance and the observed phenotype is necessary evidence that the observed phenotype was result of the deletion of the specified gene.

Results

3.1 Perithecium development in *N. crassa*, *N. tetrasperma*, and *N. discreta* on CA follows a common time course

To monitor the sexual development of all three *Neurospora* species, we crossed wild-type strains of both mating partners of each *Neurospora* species. The observed perithecial development of *N. crassa, N. tetrasperma*, and *N. discreta* aligned with the common time course known for these species [56]. We observed that during perithecial development, grayish to yellowish gray protoperithecia darkened, then blackened once mature perithecia had formed, indicating the biosynthesis of melanin in the time course of perithecial development (Fig. 1). As expected, we also identified the fading of the orange pigmentation in the colony across the time course of sexual development, indicating a reduction in the asexual phase of the life cycle. Additionally, we found no obvious tissue differentiation between protoperithecia and perithecia within 24 h after crossing, except for a slight increase in size and a slight darkening in colour, indicating the biosynthesis of melanin as part of the successful fruiting body formation. Furthermore, the centrum parenchyma of thin-walled cells expanded with increasing perithecial size and differentiated filamentous structures and croziers formed after 48 h to 72 h. Asci containing developing ascospores were visible after 96 h, along with some narrow paraphyses. From 120 h to 144 h post crossing, a beak formed at the apex of the perithecium, and scanning electron microscopy of all three species revealed fully developed fruiting bodies with a beak on top of the perithecium (Fig. 2, panels A–C). Furthermore, squash mounts of perithecia 144 h after crossing revealed that the inside of the perithecium contained mature asci with ascospores (Fig. 2, panels D–F). A major phenotypic difference among species was the number of spores produced, which, as expected in *N. tetrasperma* was four, in contrast to eight in *N. crassa* and *N. discreta*. Additionally, *N. discreta* produced abundant conidia, which were visible associated with the perithecia after the sexual cycle has concluded (Fig. 2, panels C and F).

3.2. Transcriptional profiling of *N. crassa*, *N. tetrasperma*, and *N. discreta* reveals eight functional categories

To determine the gene expression of all three *Neurospora* species during the time course of sexual development, we performed Illumina next generation transcriptomic sequencing. We prepared samples by isolation of mRNA and performance of reverse transcription of genomic RNA with random hexamers (N_6), which were typically consistent in DNA concentration. Single-end 35-base reads of N_6-primed preparations were sequenced separately, each developmental stage of each species on one of eight lanes of an Illumina Genome Analyzer, at the Yale Center for Genomic Analysis. A total of 2352 single copy orthologous genes were selected by BranchClust [54] for *N. crassa*, *N. tetrasperma* and *N. discreta*. BranchClust uses the Reciprocal Best Blast hit method and phylogenetic trees to select putatively orthologous genes, with default threshold e-value of 10^{-4}. Our deep transcriptomic sequencing revealed expression in all three species across sexual development for all of these identified orthologous genes. To compare gene expression levels for these genes, we constructed a comparative heatmap with the Hierarchical Clustering Explorer Software [61] based on the level of expression of each gene for all three species over the entire time course of sexual development (Fig. 3) as described in section 2.4. Transcriptional profiling of *N. crassa*, *N. tetrasperma*, and *N. discreta* revealed eight clusters, which we then analyzed separately for functional category representation.

3.2.1 Among genes highly expressed in early stages of perithecium development, the ortholog set was enriched for genes involved in metabolic processes. Where previously identified, genes were assigned cellular or molecular functions by their Functional Catalogue (FunCat) [57] annotation. The statistical significance of overrepresentation of gene groups in functional categories relative to the whole genome was determined using the hypergeometric distribution, facilitated by the Munich Information Center for Protein Sequences (MIPS) FunCat online web application. Eight major clusters were identified based on expression patterns across sexual development for the identified

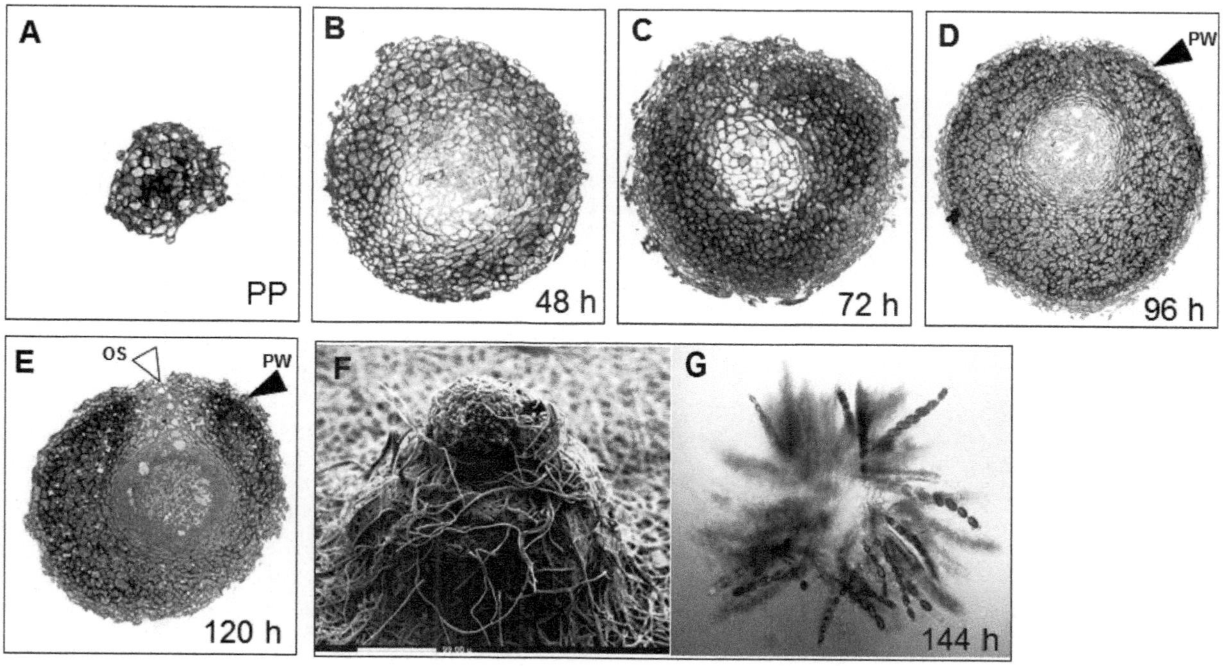

Figure 1. Sexual development of *N. crassa*. Cross sections of developing perithecia, from A) the protoperithecium (PP) through a time course of (B) 48 h, (C) 72 h, (D) 96 h, and (E) 120 h after fertilization. These images illustrate the development of several cell layers within the fruiting body such as the perithecium wall (PW, black arrowheads), composed of thick-walled cells, as well as the initial stages of formation of asci and ascospores and the ostiole (OS, white arrowheads) through which spores are released in a later stage. After (F) 144 h, the perithecium including its beak is fully developed, as shown by scanning electron microscopy (SEM). (G) A squash mount of a mature fruiting body, showing asci and ascospores.

orthologs. In cluster 1 (Fig. 3), 61 of the 671 proteins were unclassified. Of the classified proteins we identified, most were involved in metabolism, cellular transport, cell cycle and DNA processing, protein with binding function, cellular communication and transcription.

Genes in cluster 2 were mainly involved in metabolism, cellular transport, transport facilities and transport routes as well as

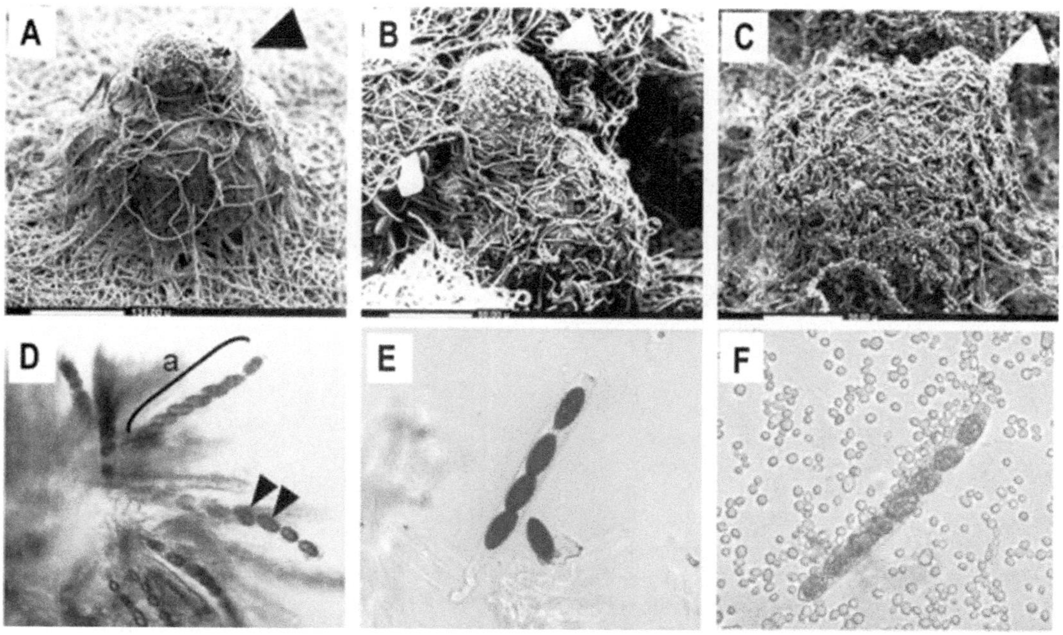

Figure 2. Key morphological characters of *N. crassa*, *N. tetrasperma* and *N. discreta* 144 h after crossing. Scanning electron micrographs, in which arrowheads indicate the perithecial beaks, of A) *N. crassa* (bar: 134 µm) B) *N. tetrasperma* (bar: 98 µm) and C) *N. discreta* (bar: 98 µm), and light micrographs of squash mounts of (D) *N. crassa* (a: ascus), (E) *N. tetrasperma*, and (F) *N. discreta* (with conidia). At 144 h, some spores are not fully mature.

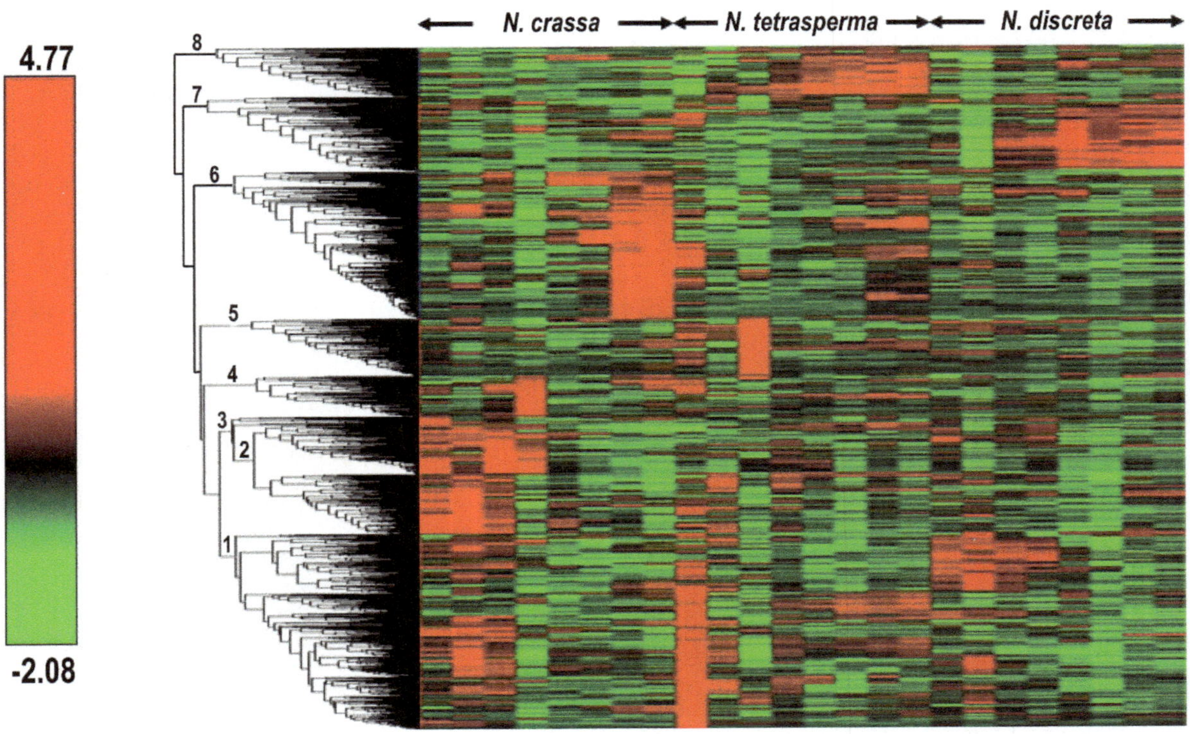

Figure 3. Comparative heat map of *N. crassa* (*N.c*), *N. tetrasperma* (*N.t*) and *N. discreta* (*N.d*) gene expression. Relative gene expression levels of the orthologous genes in all three species inferred by BranchClust, from Before Crossing (BC) to 144 h after crossing, were normalized row-by-row by subtraction of the mean and division by the standard deviation. Rows were hierarchically clustered by average linkage (UPGMA) applied to the Pearson correlation coefficient. Clusters 1–8 were analyzed for their function with FunCat. The scale of mean-centered relative gene expression ranges from −2.08–4.79, as indicated with the scale bar.

proteins with binding function, while only 16 of the proteins in cluster 2 were not classified. The metabolic genes were primarily involved in nucleotide/nucleoside/nucleobase metabolism, C-compound and carbohydrate metabolism, lipid, fatty acid and isoprenoid metabolism and metabolism of vitamins, cofactors, and prosthetic groups. The cellular transport-related genes were involved in transported compounds (substrates), protein transport, electron transport, vesicular transport such as the Golgi network and vacuolar/lysosomal transport. Genes encoding proteins with binding functions such as nucleic acid binding, metal binding and complex cofactor/cosubstrate binding. In cluster 3, only 19 out of 199 proteins were not classified, while the remaining proteins were mainly involved in metabolism, especially C-compound and carbohydrate metabolism. Genes in cluster 3 also tended to be involved in cellular transport, including cation and heavy metal ion transport as well as proteins with binding functions, e.g. nucleic acid binding, metal binding and RNA binding. Cluster 4 contained 139 proteins, which were involved mainly in metabolism, especially lipid, fatty acid and isoprenoid metabolism, cellular transport, and cell cycle including DNA restriction or modification. Overall, we observed an enrichment of genes involved in metabolic processes in early stages of perithecium development.

3.2.2 In later stages of the perithecium development, the ortholog set was enriched for genes involved in transcription, cell cycle, protein synthesis and cellular transport. Based on Functional Category (FunCat, [57]) annotations, cluster 5 comprises 198 proteins, of which the majority was involved in nucleotide/nucleoside/nucleobase metabolism, nucleotide/nucleoside/nucleobase metabolism and lipid, fatty acid, and isoprenoid metabolism, but also transcription such as RNA synthesis and RNA processing.

Interestingly, in clusters 6–8 there appears to be a species-specific expression in later stages of the development. Proteins of cluster 6 were mainly involved in transcriptional processes such as RNA synthesis, processing and modification, but also, cell cycle and DNA processing, as well as and metabolism. Cluster 7 contained 259 genes involved in transcription, protein binding, cell type differentiation and cellular transport. In cluster 8 only 13 genes were not identified. The cluster mainly comprised proteins involved in cell differentiation, cellular transport and transcription.

While in the early stages of the perithecium development, the fungi were observed to exhibit an enrichment of expressed genes involved in metabolic processes; in later stages of perithecium development, expressed genes from the ortholog set were enriched for genes involved in transcription, cell cycle, protein synthesis and cellular transport.

3.3 Comparative gene expression analyses revealed genes crucial for the successful development of fruiting bodies

Crossing of the *N. crassa* WT strains resulted in the successful formation of perithecia with perithecial beak (Fig. 4A). In order to identify genes that are required for the successful development of fruiting bodies, knockouts in *N. crassa* from the whole genome knockout project [58,59] of the 130 top candidate genes identified through pairwise comparison of orthologous gene expression in *N. crassa*, *N. tetrasperma*, and *N. discreta* were screened for mutant phenotypes in sexual development. Eight mutant phenotypes were observed that affected perithecium formation. Interestingly, development of NCU06874, encoding a HMG box (high mobility group)-containing protein was arrested after formation of protoperithecia on

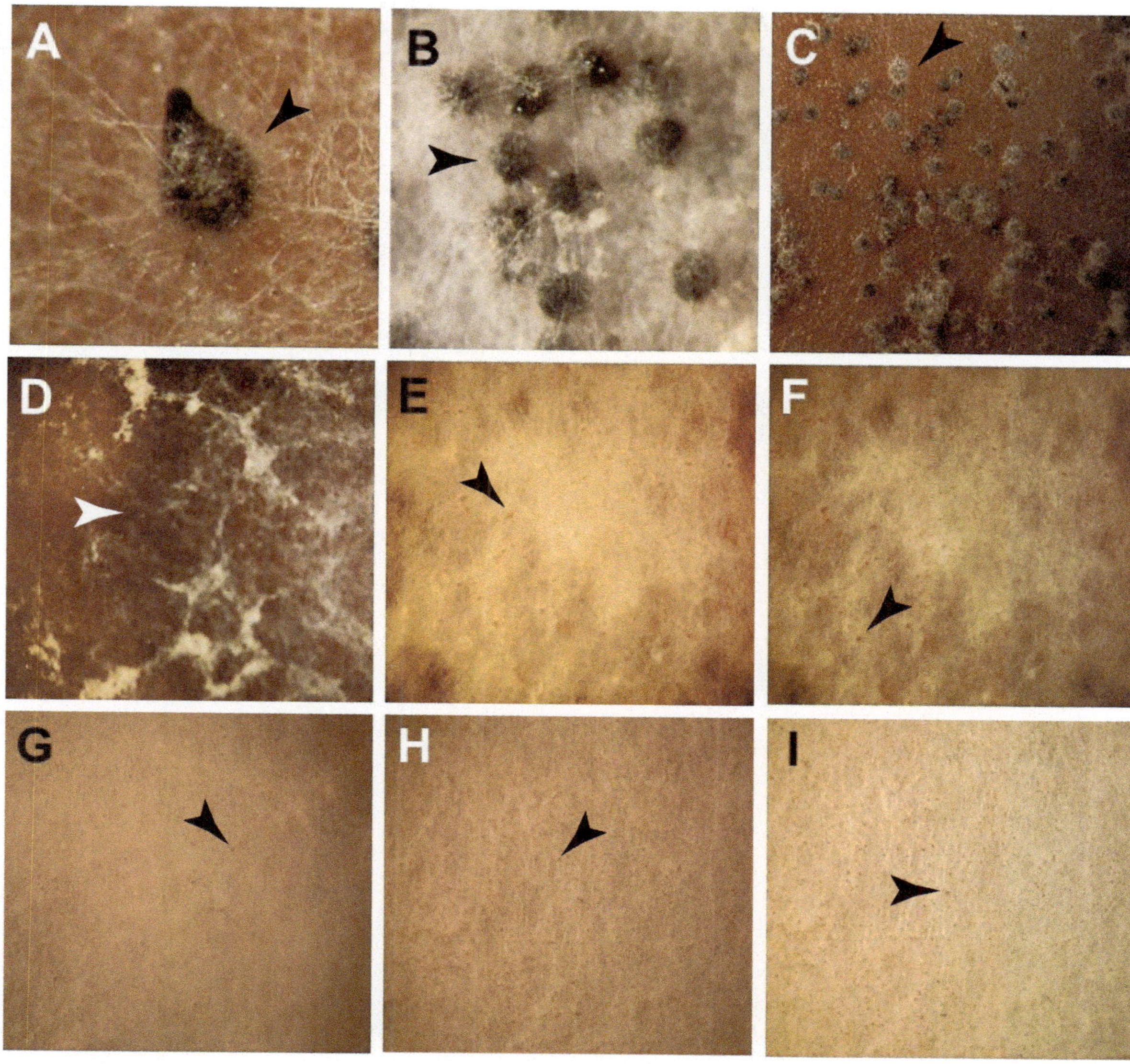

Figure 4. Phenotypes of perithecia from crosses in *N. crassa*. (A) WT control, (B) ΔNCU06316: perithecium development was arrested at early stage equally to 48–72 h, (C) ΔNCU07508: perithecium development was arrested at 48–72 h, (D) ΔNCU06874: no perithecia formed, only protoperithecia, though melanin was released into the medium (white arrow heads), (E) ΔNCU05609: no perithecia formed, only protoperithecia, (F) ΔNCU00175: no perithecia formed, only protoperithecia, (G) ΔNCU00427: no perithecia formed, only protoperithecia. (H) ΔNCU02089: protoperithecia failed to develop into perithecia, and (I) ΔNCU09525: protoperithecia only. Perithecia (large, black) and protoperithecia (small, yellowish-gray to gray) are indicated with black arrow heads.

a fluffy white mycelium. Putative functions of the remaining genes that affect fruiting body formation in *N. crassa* were determined by a Blast search [62] with the NCBI Blast server of the translated sequence against all known GenBank non-redundant protein sequences (Table 1). However, the deletion of either gene NCU06316, a putative argonaute siRNA chaperone complex subunit which in fission yeast is required for histone H3 Lys9 (H3-K9) methylation, heterochromatin, assembly and siRNA generation [1], or NCU07508, a putative type-2 protein geranylgeranyltransferase subunit, caused the perithecium development to arrest at an early stage between 48 and 72 h (Fig. 4, panels B and C). In the perithecia arising from these KO strains, we observed no perithecial beak and found that the developing perithecia remained spherical, indicating an arrest in development. Furthermore, the CA of these cultures was stained black along the mating zone, suggesting an increase in melanin biosynthesis (Fig. 4D).

The expression patterns of the genes whose knockouts showed an impact on fruiting body formation were monitored over the time course of sexual development starting from mature protoperithecia before crossing until 144 h, when wild-type cultures showed mature perithecia. The genes NCU06316 or 07508 exhibited a continuous increase in expression starting at 72 h (Fig. 5A). The increased expression correlated with the arrest of the fruiting body formation in an early stage between 48 and 72 h in deletion mutants.

Knockouts of NCU05609 (a proline-rich protein 8), 00175 (a repetitive proline-rich cell wall protein), 00427 (a YjeF_N domain-containing protein), 02089 (a hypothetical protein SMAC) and 09525 (a hypothetical protein SMAC) were arrested at the protoperithecium stage, indicating their importance in the formation of fruiting bodies. NCU09525 encodes a putative secreted protein and has been described in *Nectria haematococca*

Table 1. Putative functions of genes whose deletion impacts fruiting body formation.

ID	Annotation by homology	Max score	Total score	Query cover	E-value	Max iden
NCU06316	argonaute siRNA chaperone complex subunit Arb1 [*Colletotrichum fioriniae*], XP_007589964.1	246	246	76%	8e–72	37%
NCU07508	putative type-2 protein geranylgeranyltransferase subunit beta protein [*Botryotinia fuckeliana*], EMR87581.1	145	145	68%	1e–34	31%
NCU06874	HMG box-containing protein [*Magnaporthe oryzae*], XP_003718106.1	269	269	42%	7e–75	40%
NCU05609	proline-rich protein [*Coccidioides Posadasii*], ACU44647.1	46.6	46.6	64%	0.020	29%
NCU00175	repetitive proline-rich cell wall protein [*Colletotrichum higginsianum*], CCF45612.1	153	205	88%	1e–39	44%
NCU00427	YjeF_N domain-containing protein [*Magnaporthe oryzae*], ELQ39075.1	734	734	97%	0.0	53%
NCU02089	hypothetical protein SMAC_04986 [*Sordaria macrospora*], XP_003352871.1	1276	1276	99%	0.0	88%
NCU09525	Bacterial-type extracellular deoxyribonuclease [N. haematococca], XP_003051650.1	284	284	85%	2e–93	69%

as bacterial-type extracellular deoxyribonuclease [63]. NCU09525 and 00175 were highly expressed between 48 and 120 h, while expression of NCU05609 peaked at 48 and 96 h after crossing. NCU06874 and 00427 showed an increased expression 96 h after crossing. NCU02089 peaked at 48 h after crossing and remained fairly constant across the remainder of sexual development (Fig. 5C–D). Although only protoperithecia were observed in mutant crosses, the culture medium for the knockout of NCU06874 blackened, indicating the secretion of melanin, and implying that melanin biosynthesis remained functional despite the fact that no perithecia were formed. Genes encoding enzymes in the melanin synthesis pathway showed a highly similar pattern across sexual development within each *Neurospora* species. While melanin synthesis genes were highly expressed in *N. crassa* protoperithecia samples, they were expressed at a lower level in *N. tetrasperma* and *N. discreta* protoperithecial samples than in late perithecial samples (Table S1). Further research is needed to understand function of melanin synthesis early sexual development in different *Neurospora* species. The protein encoded by NCU06874 contains a High Mobility Group (HMG) box, a structure often found in proteins involved in the regulation of DNA-dependent processes and DNA repair. Recent studies have demonstrated that deletion of the ortholog of NCU06874 in *Podospora anserina* and *Fusarium graminearum* had no effect on perithecium development [64,65]. However, an effect on the distribution of perithecia in *P. anserina* was observed [64,65].

The gene NCU02089 belongs to cluster 5. In *N. crassa*, this gene is expressed fairly uniformly, showing only a slight peak in expression at 48 h. The corresponding ortholog in *N. tetrasperma* shows a similar expression pattern. In *N. discreta*, in contrast, the expression of NCU02089 remains uniform until 96 h, then increases until the end of the sexual development.

The genes NCU06316, 07508, 05609, 00427, 09525, and 06874 belong to cluster 6. In *N. crassa*, the genes NCU06316 and 07508 are highly expressed after 72 h. In *N. tetrasperma*, as in *N. crassa*, the ortholog of NCU07508 also increases in expression after 72 h, while in *N. discreta* the expression peaks at 48 h, and increases again after 96 h. The expression of the ortholog NCU06316 in *N. crassa* differs from the one in *N. tetrasperma* and *N. discreta*. Over the time course of sexual development, expression of this gene is very low. The gene NCU05609 shows an increasing expression over the time course of the sexual development, peaking at 48 and 96 h. In contrast, the corresponding ortholog in *N. tetrasperma* does not show any changes in gene expression, while in *N. discreta* the expression decreases until 24 h, then remains steady until the end of the sexual development. In *N. crassa*, the expression of the gene NCU00427 remains constant until 96 h after crossing and then increases towards the end of the development. The expression of the corresponding ortholog in *N. tetrasperma* peaks at 24 and 120 h, while in *N. discreta*, this gene remains constant over the time course of fruiting body formation. The expression of the gene NCU09525 peaks between 48 and 120 h in *N. crassa*, while in *N. tetrasperma* and *N.*

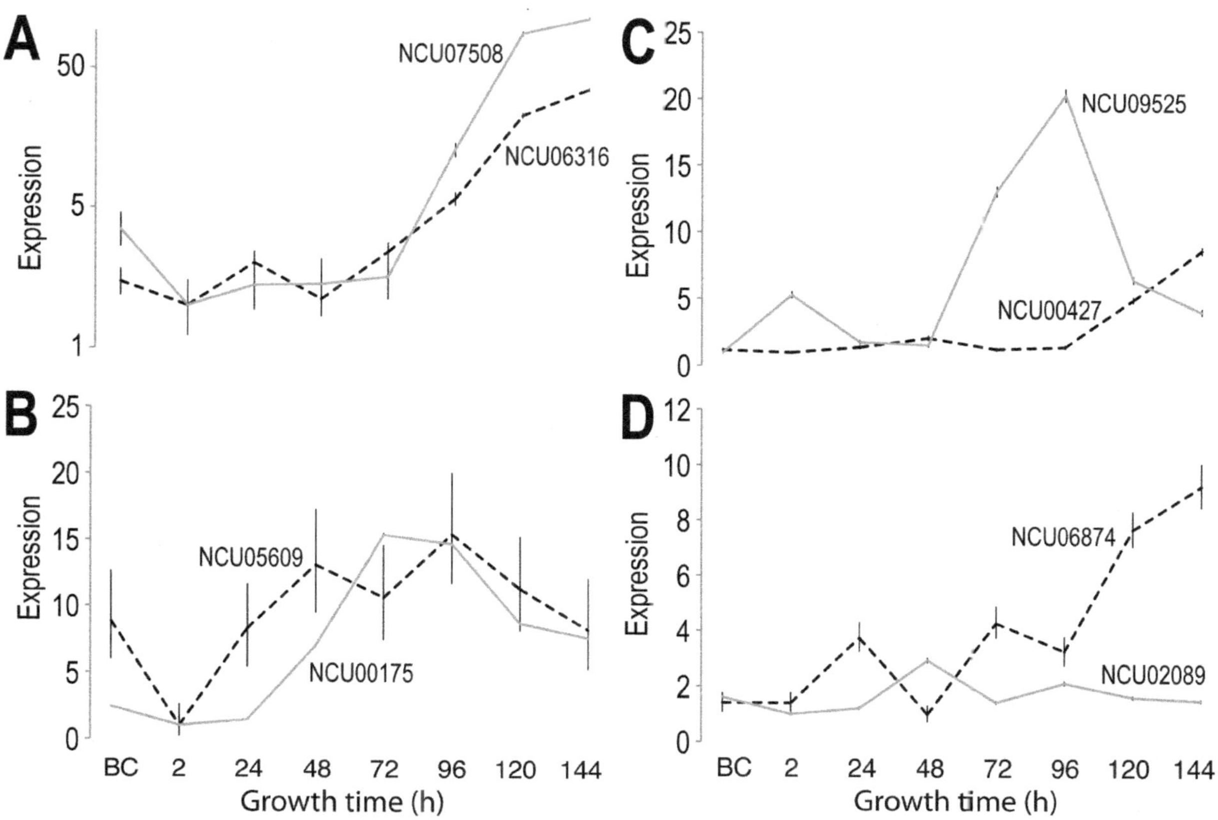

Figure 5. Expression patterns of eight genes with impact on successful perithecium formation in *N. crassa*. A. Expression of NCU06316 (solid line) and NCU07508 (dashed line), knockouts of which lead to early arrest of perithecial development at 48–72 h. Both genes exhibit a continuous increase of expression after 72 h. B. Expression of NCU00175 (solid line) and NCU05609 (dashed line), up-regulated during intermediate stages of perithecial development. C. Expression of NCU09525 (solid line) and NCU00427 (dashed line), both changing significantly in late perithecial development. D. Dynamic changes in expression of NCU02089 (solid line) and NCU06874 (dashed line) across perithecial development. Error bars indicate the inferred 95% credible interval.

discreta, its expression peaks initially at 2 h, potentially as a consequence of mycelial disruption caused by the spreading of conidia and mating of the strains, and then increases until 96 h after crossing followed by a slight decrease towards the end of the development. The gene NCU06874 shows peaks in gene expression at 24 and 72 h, and an increase of expression from 96 h until the end of perithecium development. In contrast, in both *N. tetrasperma* and *N. discreta*, the expression of this gene remains fairly constant over the time course of the sexual development.

The gene NCU00175 belongs to cluster 8 and shows an increased gene expression between 24 and 120 h in *N. crassa*. A similar pattern was observed for *N. tetrasperma*, where the expression increased at 48 and 120 h. In *N. discreta* however, the gene expression increased steadily until 96 h after crossing, remained constant.

Key differences in gene expression patterns of meiosis-related genes between *N. tetrasperma* and *N. crassa* as well as *N. discreta* were observed (Fig. 6). Many RNA processing genes and meiosis- and mitosis- specific genes have experienced multiple duplications. Only a few single copy meiosis-specific gene homologs have been confirmed with our methods. Among these genes, we observed differences in gene expression between homothallic *N. tetrasperma* and heterothallic *N. crassa* and *N. discreta*. The meiotic chromosome segregation protein 3 homologs, *N. tetrasperma* Neute1draft_73667 (JGI identifier for the *mat A* genome), *N. crassa* NCU01858 (Broad Identifier), and *N. discreta* Ndisc8579_91937 (JGI identifier) showed a consistent gene expression pattern in *N. crassa* and *N. discreta*: two up-regulations at 2 h and 72 h (Fig. 6A). In contrast, the expression of the ortholog of chromosome segregation protein 3 in *N. tetrasperma* exhibited a steep down-regulation after strains were crossed and only a slight up-regulation 24 h after crossing. Furthermore, the translation factor pelota (Neute1draft_80736, NCU09161, and Ndisc8579_129376), which functions in meiotic cell division, exhibited a generally consistent expression across perithecial development in *N. crassa* and *N. discreta*, but a dynamic change in expression in *N. tetrasperma* during early perithecial development (Fig. 6B). A dramatic up-regulated expression of the meiosis-specific gene *spo11* (Neute1draft_98113 m, NCU01120, Ndisc8579_51404) started early in *N. tetrasperma* 24 h after crossing, but began in *N. crassa* and *N. discreta* 96 h after crossing (Fig. 6C). For sporulation-related genes, including *asm-1* (ascospore maturation-1), Neute1draft_115825, NCU01414, Ndisc8579_123166), *Rsp* (round spore, Neute1draft_126958, NCU02764, Ndisc8579_94666), *asd-3* (ascus development-3, Neute1draft_103712, NCU05597, Ndisc8579_126512), we also observed up-regulation of expression at the same stages between *N. crassa* and *N. discreta* (Fig. 6D–F). Up-regulation of homologs of *N. crassa asm-1* and *rsp* in *N. tetrasperma* occurred later than in *N. crassa* and *N. discreta*, but interestingly the homolog of *N. crassa asd-3* exhibited an early up-regulation for *N. tetrasperma* at

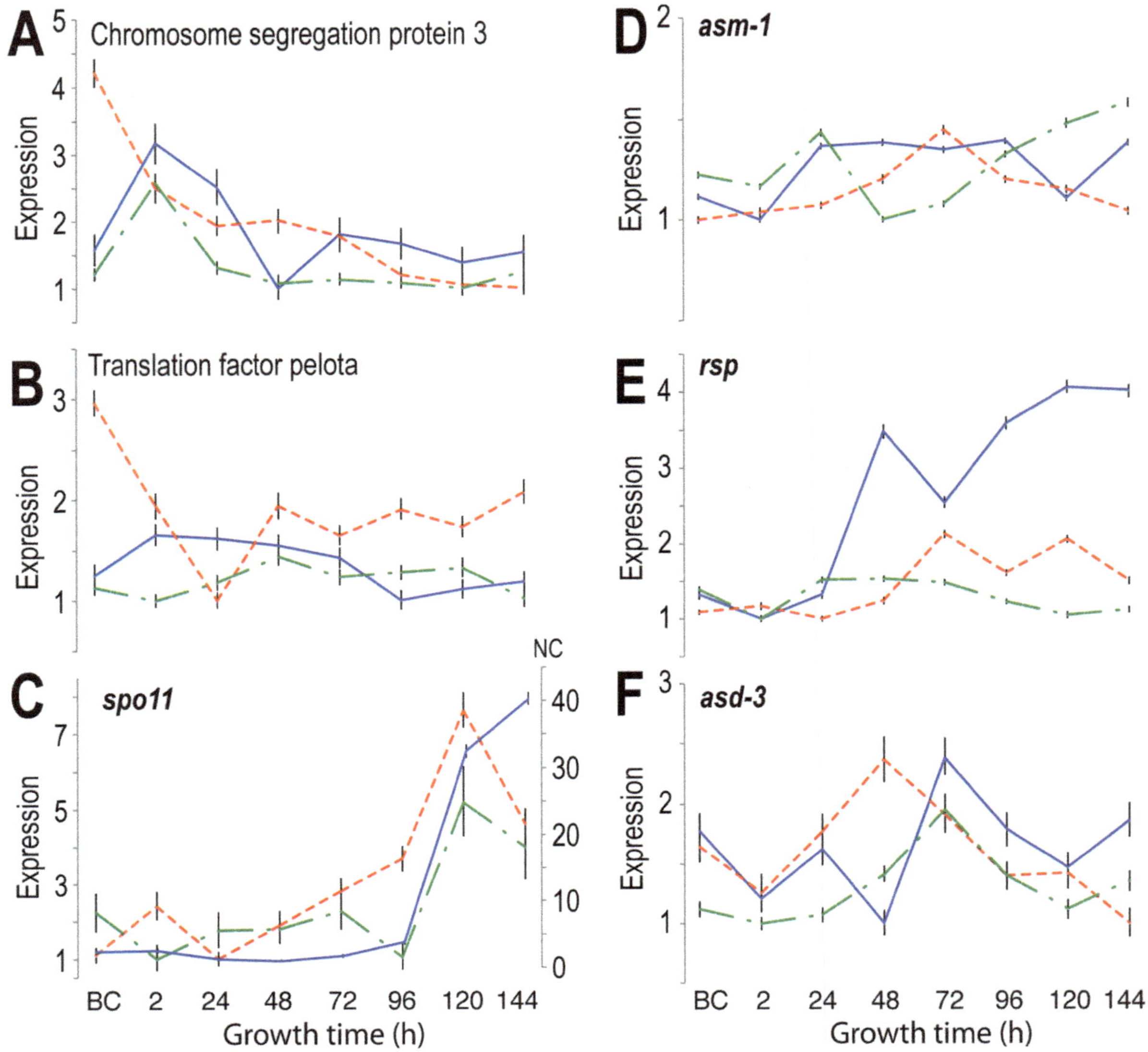

Figure 6. Comparative analysis of meiosis-related genes, exhibiting differences in gene expression between *N. tetrasperma* (red dashed line), *N. crassa* (blue solid line) and *N. discreta* (green dash-dotted line). A) Expression of the *N. crassa* gene encoding chromosome segregation protein 3 was up-regulated during early sexual development in both *N. crassa* and in the orthologous gene in *N. discreta*, but the *N. tetrasperma* ortholog was down-regulated across sexual development, B) Expression of the gene encoding translation factor pelota in *N. crassa* and expression of its ortholog in *N. discreta* was consistent, but the ortholog in *N. tetrasperma* was dramatically and dynamically differentially expressed. C) Expression of meiosis specific gene *spo-11* and its orthologs was up-regulated for all three species from 96 h after crossing, but the up-regulation started early for the *N. tetrasperma* ortholog. D) Expression of *asm-1* (ascus maturation) in *N. crassa* and its orthologin *N. discreta* exhibited the same two-peaked pattern, featuring a peak of expression during early development of the perithecium and a second peak at a later stage of perithecial development, but the ortholog in *N. tetrasperma* exhibited only one peak 72 h after crossing. E) Expression of *N. crassa rsp* (round spore) and its ortholog in *N. discreta* exhibited up-regulation preceding 48 h, but upregulation in *N. tetrasperma* preceded 72 h. F). Expression of *N. crassa asd-3* (ascospore development) peaked at 72 h in *N. crassa,* as did its ortholog in *N. discreta*, but expression of the ortholog in *N. tetrasperma* peaked at 48 h. Error bars indicate the inferred 95% credible interval.

48 h. Similar up-regulation of *asd-3* for *N. crassa* and its homolog in *N. discreta* occurred at 72 h after crossing. The genes NCU01858 and 09161 belong to cluster 1, the genes NCU01120, 02764 and 05597 to cluster 6 and NCU01414 to cluster 7. Except NCU01858 and 09161, all genes shown in Fig. 6 belong to the candidate list.

3.4 Phenotypes of seven knockout mutants cosegregate with hygromycin resistance

To validate the linkage between the insertional mutation and the phenotype, we followed a strategy developed previously for *N. crassa* KO strains [56], and backcrossed progeny (ascospores) were selected and germinated. Twelve to twenty single ascospore progeny displayed hygromycin resistance, and complete cosegregation of hygromycin resistance and identified phenotypes was observed for seven out of eight investigated genes (Table S6). The exception was the knockout of NCU09525, for which only a KO strain of *mat A* was available. No perithecia were produced in this mutant-wild type cross, preventing us from assessing segregation of the cross.

Discussion

Here, we compared the sexual development of three closely related *Neurospora* species, *N. crassa*, *N. tetrasperma*, and *N. discreta,* focusing on characterizing expression patterns for genes involved in sexual development and identifying new genes or new functions of annotated genes in regulating sexual reproduction, taking advantages of the distinct differences among these otherwise highly similar and closely related species. The comparison of gene expression levels across these species during fruiting body formation revealed eight genes that were shown to be crucial for the successful development of perithecia in *N. crassa*: their knockouts were unable to produce mature perithecia, and seven of the eight exhibited cosegregation of the phenotype and a hygromycin marker. We identified different expression patterns for meiosis-related genes between *N. tetrasperma* and *N. crassa* and *N. discreta* that correspond with observed differences in meiosis and sexual spore development between pseudohomothallic *N. tetrasperma* and heterothallic *N. crassa* and *N. discreta*. Since meiosis gene sets and sporulation machinery are largely conserved in presence as well as sequence within these genomes, consistent differences among functionally related genes in these species would be strong evidence of dependent associations for the reconstruction of gene networks and thus for understanding the genetic basis of meiosis and sexual sporulation in *Neurospora* and ascomycetes. Additionally, species-specific expression in later stages of development was observed when clustering the ortholog genes. Future efforts should expand analysis to identify a more complete set of genes involved in meiosis and sporulation based on genetics and reverse genetics on the model *N. crassa*.

We distinguished eight gene clusters based on similarity of overall expression patterns across sexual development in three *Neurospora* species. During early sexual development, genes involved in metabolism are enriched, and clusters that appear later during sexual development are enriched with genes involved in metabolism, energy, transcription, cell cycle, and DNA processing. The expression profiles of these clusters correlated with the morphological changes observed during sexual development: at first metabolic genes are upregulated, then transcription and cell cycle related genes are activated, and eventually protein synthesis and transport genes are transcribed to deliver proteins needed to their destination in order to form the complex three dimensional fruiting body.

Our study revealed eight genes that are required for the successful formation of fruiting bodies in *N. crassa*. Knockouts of two genes, NCU06316 and 07508 resulted in the arrest of perithecium formation between 48 and 72 h after crossing (Fig. 4, panels B and C). In correlation with the gene expression data (Fig. 6A), these genes are of great importance in the later stage of development when structures such as the perithecial beak or ascospores are developed. The developing perithecia for NCU06316 (a putative argonaute siRNA chaperone complex subunit) and 07508 (a putative type-2 protein geranylgeranyltransferase subunit) remain round-shaped without formation of a beak. Six crosses of *N. crassa* mutant strains showed only protoperithecia and no formation of fruiting bodies. These results correlate with the gene expression observed over the time course of the sexual development in *N. crassa*. Here, an up-regulation of gene expression during the later stages of the development was observed. Our observations indicate that there are several genes that are required for the successful formation of perithecia.

Conclusions

We have used comparative transcriptomics as a tool to identify genes that are required for the successful formation of fruiting bodies. The impact of the deletion of the candidate genes was demonstrated in phenotypes observed in crosses exhibiting impaired perithecium formation at different stages of the development. Our findings shed light on the developmental process of fruiting body formation, evolution, and spore development in *Neurospora* species, thus establishing the foundation for future research particularly related to closely related pathogenic fungi. We suggest the potential utility of future research on the eight genes that we have discovered as essential contributors to fruit body development and as candidate genes for targets in the development of fungicides for control of plant pathogens.

Supporting Information

Table S1 Expression results of all genes of all three Neurospora species.
(XLSX)

Table S2 Level of Expression (LOX) data of *N. crassa, N. tetrasperma* and *N. discreta* orthologues. LOX estimates the Level Of gene eXpression from high-throughput-expressed sequence datasets with multiple treatments or samples. The tables includes the corresponding upper and lower confidence intervals (CI) across developmental stages.
(XLSX)

Table S3 Comparative gene expression analysis of *N. crassa, N. tetrasperma,* and *N. discreta*. The table includes the results from the calculations such as the differences of confidence intervals.
(XLSX)

Table S4 Comparison of species using the IF statement function. In each comparison the highest value across all time points for each gene was selected (MAX) and the maximum values were then sorted in descending order.
(XLSX)

Table S5 Prioritization of the maximum value across the time course of the sexual development for each gene in speecies comparisons.
(XLSX)

Table S6 Cosegregation of hygromycin resistance and identified phenotypes in KO strains of interest.
(DOCX)

Acknowledgments

We thank Dr. Joseph Wolenski and Barry Piekos for providing assistance in the Yale MCDB microscopy facility.

Author Contributions

Conceived and designed the experiments: JPT FT. Performed the experiments: NL ZW. Analyzed the data: NAL ZW NL FL-G. Contributed to the writing of the manuscript: NL ZW DAH FT JPT.

References

1. Busch S, Braus GH (2007) How to build a fungal fruit body: from uniform cells to specialized tissue. Mol Microbiol 64: 873–876.
2. Busch S, Schwier EU, Nahlik K, Bayram O, Helmstaedt K, et al. (2007) An eight-subunit COP9 signalosome with an intact JAMM motif is required for fungal fruit body formation. Proc Natl Acad Sci U S A 104: 8089–8094.

3. Coppin E, Berteaux-Lecellier V, Bidard F, Brun S, Ruprich-Robert G, et al. (2012) Systematic deletion of homeobox genes in *Podospora anserina* uncovers their roles in shaping the fruiting body. PLoS One 7: e37488.
4. Debuchy R, Berteaux-Lecellier V, Silar P (2010) Mating systems and sexual morphogenesis in Ascomycetes. In: Borkovich KA, Ebbole DJ, editors. Cellular and Molecular Biology of Filamentous Fungi. Washington, DC: ASM Press. pp. 501–535.
5. Harting R, Bayram O, Laubinger K, Valerius O, Braus GH (2013) Interplay of the fungal sumoylation network for control of multicellular development. Mol Microbiol 90: 1125–1145.
6. Jamet-Vierny C, Debuchy R, Prigent M, Silar P (2007) IDC1, a pezizomycotina-specific gene that belongs to the PaMpk1 MAP kinase transduction cascade of the filamentous fungus Podospora anserina. Fungal Genet Biol 44: 1219–1230.
7. Pöggeler S, Nowrousian M, Kück U (2006) Fruiting-body development in Ascomycetes, in The Mycota I. Growth, Differentiation and Sexuality. In: Kües U, Fischer R, editors. The Mycota I. (2nd edition). Growth, Differentiation and Sexuality. Berlin, Heideberg: Springer-Verlag. pp. 325–355.
8. Trail F, Gardiner DM (2014) Applicatioin of genomics to the study of pathogenecity and development in Fusarium. In: Nowrousian M, Esser K. editors. The Mycota: A comprehensive treatise on fungi as experimental systems for basic and applied research Fungal Genomics XIII. Second edition. Berlin, Heidelberg: Springer-Verlag. pp. 267–300.
9. Trail F (2013) Sex and fruiting in *Fusarium*. In: Brown D, Proctor R, editors. *Fusarium*: genomics, molecular and cellular biology. Norwich, UK: Horizon Scientific Press and Caister Academic Press.
10. Coppin E, Debuchy R, Arnaise S, Picard M (1997) Mating types and sexual development in filamentous ascomycetes. Microbiol Mol Biol Rev 61: 411–428.
11. DeLange AM, Griffiths AJ (1980) Meiosis in *Neurospora crassa*. I. The isolation of recessive mutants defective in the production of viable ascospores. Genetics 96: 367–378.
12. Eckert SE, Kubler E, Hoffmann B, Braus GH (2000) The tryptophan synthase-encoding trpB gene of *Aspergillus nidulans* is regulated by the cross-pathway control system. Mol Gen Genet 263: 867–876.
13. Eckert SE HB, Wanke C, Braus GH. (1999) Sexual development of *Aspergillus nidulans* in tryptophan auxotrophic strains. Arch Microbiol 172: 157–166.
14. Fischer R, Kües U (2003) Developmental processes in filamentous fungi. Genomics of Plants and Fungi. In: Prade RA, Bohnert HJ, editors. New York: Marcel Dekker. pp. 41–118.
15. Johnson TE (1979) A *Neurospora* mutation that arrests perithecial development as either male or female parent. Genetics 92: 1107–1120.
16. Leslie JF, Raju NB (1985) Recessive mutations from natural populations of *Neurospora crassa* that are expressed in the sexual diplophase. Genetics 111: 759–777.
17. Nowrousian M, Ringelberg C, Dunlap JC, Loros JJ, Kuck U (2005) Cross-species microarray hybridization to identify developmentally regulated genes in the filamentous fungus *Sordaria macrospora*. Mol Genet Genomics 273: 137–149.
18. Dodge BO (1927) Nuclear phenomena associated with heterothallism and homothallism in the Ascomycete *Neurospora*. Journal of Agricultural Research 35: 289–305.
19. Mao-Draayer Y, Galbraith AM, Pittman DL, Cool M, Malone RE (1996) Analysis of meiotic recombination pathways in the yeast *Saccharomyces cerevisiae*. Genetics 144: 71–86.
20. Lumbsch HT (2000) Phylogeny of filamentous ascomycetes. Naturwissenschaften 87: 335–342.
21. Samils N, Gioti A, Karlsson M, Sun Y, Kasuga T, et al. (2013) Sex-linked transcriptional divergence in the hermaphrodite fungus *Neurospora tetrasperma*. Proc Royal Soc B: Biol Sci 280: 20130862.
22. Wang Z, Kin K, Lopez-Giraldez F, Johannesson H, Townsend JP (2012) Sex-specific gene expression during asexual development of *Neurospora crassa*. Fungal Genet Biol 49: 533–543.
23. Perkins D, Raju NB (1986) *Neurospora discreta*, a new heterothallic species defined by its crossing behavior. Experimental Mycology 10: 323–338.
24. Turner E, Jacobson DJ, Taylor JW (2010) Reinforced postmating reproductive isolation barriers in *Neurospora*, an Ascomycete microfungus. J Evol Biol 23: 1642–1656.
25. Borkovich KA, Alex LA, Yarden O, Freitag M, Turner GE, et al. (2004) Lessons from the genome sequence of *Neurospora crassa*: tracing the path from genomic blueprint to multicellular organism. Microbiol Mol Biol Rev 68: 1–108.
26. Beadle GW, Tatum EL (1941) Genetic Control of Biochemical Reactions in *Neurospora*. Proc Natl Acad Sci U S A 27: 499–506.
27. Dunlap JC, Loros JJ (2004) The *Neurospora* circadian system. J Biol Rhythms 19: 414–424.
28. Metzenberg RL, Glass NL (1990) Mating type and mating strategies in *Neurospora*. Bioessays 12: 53–59.
29. Kronstad JW, Staben C (1997) Mating type in filamentous fungi. Annu Rev Genet 31: 245–276.
30. Raju N (1992) Genetic control of the sexual cycle in *Neurospora*. Mycol Res 96: 241–262.
31. Raju NB, Perkins DD (1994) Diverse programs of ascus development in pseudohomothallic species of *Neurospora*, *Gelasinospora*, and *Podospora*. Dev Genet 15: 104–118.
32. Dettman JR, Jacobson DJ, Taylor JW (2003) A multilocus genealogical approach to phylogenetic species recognition in the model eukaryote *Neurospora*. Evolution 57: 2703–2720.
33. Beatty NP, Smith ML, Glass NL (1994) Molecular characterization of mating-type loci in selected homothallic species of *Neurospora*, *Gelasinospora* and *Anixiella*. Mycol Res 98: 1309–1316.
34. Casselton LA (2008) Fungal sex genes-searching for the ancestors. Bioessays 30: 711–714.
35. Klix V, Nowrousian M, Ringelberg C, Loros JJ, Dunlap JC, et al. (2010) Functional characterization of MAT1-1-specific mating-type genes in the homothallic ascomycete *Sordaria macrospora* provides new insights into essential and nonessential sexual regulators. Eukaryot Cell 9: 894–905.
36. Nygren K, Strandberg R, Wallberg A, Nabholz B, Gustafsson T, et al. (2011) A comprehensive phylogeny of *Neurospora* reveals a link between reproductive mode and molecular evolution in fungi. Mol Phylogenet Evol 59: 649–663.
37. Paoletti M, Seymour FA, Alcocer MJ, Kaur N, Calvo AM, et al. (2007) Mating type and the genetic basis of self-fertility in the model fungus *Aspergillus nidulans*. Curr Biol 17: 1384–1389.
38. Strandberg R, Nygren K, Menkis A, James TY, Wik L, et al. (2010) Conflict between reproductive gene trees and species phylogeny among heterothallic and pseudohomothallic members of the filamentous ascomycete genus *Neurospora*. Fungal Genet Biol 47: 869–878.
39. Whittle CA, Nygren K, Johannesson H (2011) Consequences of reproductive mode on genome evolution in fungi. Fungal Genet Biol 48: 661–667.
40. Yun SH, Berbee ML, Yoder OC, Turgeon BG (1999) Evolution of the fungal self-fertile reproductive life style from self-sterile ancestors. Proc Natl Acad Sci U S A 96: 5592–5597.
41. Kim H, Wright SJ, Park G, Ouyang S, Krystofova S, et al. (2012) Roles for receptors, pheromones, G proteins, and mating type genes during sexual reproduction in *Neurospora crassa*. Genetics 190: 1389–1404.
42. Bistis GN (1981) Chemotropic Interactions between trichogynes and conidia of opposite mating-Type in *Neurospora crassa*. Mycologia 73: 959–975.
43. Harris JL, Howe HB Jr, Roth IL (1975) Scanning electron microscopy of surface and internal features of developing perithecia of *Neurospora crassa*. J Bacteriol 122: 1239–1246.
44. Raju N (1980) Meiosis and ascospore genesis in *Neurospora*. Eur J Cell Biol 23: 208–223.
45. Palma-Guerrero J, Hall CR, Kowbel D, Welch J, Taylor JW, et al. (2013) Genome wide association identifies novel loci involved in fungal communication. PLoS Genet 9: e1003669.
46. McCluskey K, Wiest A, Plamann M (2010) The Fungal Genetics Stock Center: a repository for 50 years of fungal genetics research. J Biosci 35: 119–126.
47. Hallen HE, Huebner M, Shiu SH, Guldener U, Trail F (2007) Gene expression shifts during perithecium development in *Gibberella zeae* (anamorph *Fusarium graminearum*), with particular emphasis on ion transport proteins. Fungal Genet Biol 44: 1146–1156.
48. Wang Z, Lehr N, Trail F, Townsend JP (2012) Differential impact of nutrition on developmental and metabolic gene expression during fruiting body development in *Neurospora crassa*. Fungal Genet Biol 49: 405–413.
49. Clark TA, Guilmette JM, Renstrom D, Townsend JP (2008) RNA extraction, probe preparation, and competitive hybridization for transcriptional profiling using *Neurospora crassa* long-oligomer DNA microarrays. Fungal Genet Rep 55: 18–28.
50. Trapnell C, Pachter L, Salzberg SL (2009) TopHat: discovering splice junctions with RNA-Seq. Bioinformatics 25: 1105–1111.
51. Galagan JE, Calvo SE, Borkovich KA, Selker EU, Read ND, et al. (2003) The genome sequence of the filamentous fungus *Neurospora crassa*. Nature 422: 859–868.
52. Grigoriev IV, Nordberg H, Shabalov I, Aerts A, Cantor M, et al. (2012) The genome portal of the Department of Energy Joint Genome Institute. Nucleic Acids Res 40: D26–32.
53. Zhang Z, Lopez-Giraldez F, Townsend JP (2010) LOX: inferring Level Of eXpression from diverse methods of census sequencing. Bioinformatics 26: 1918–1919.
54. Poptsova MS, Gogarten JP (2007) BranchClust: a phylogenetic algorithm for selecting gene families. BMC Bioinformatics 8: 120.
55. Zhang Z, Townsend JP (2010) The filamentous fungal gene expression database (FFGED). Fungal Genet Biol 47: 199–204.
56. Wang Z, Lopez-Giraldez F, Lehr N, Farre M, Common R, et al. (2014) Global gene expression and focused knockout analysis reveals genes associated with fungal fruiting body development in *Neurospora crassa*. Eukaryot Cell 13: 154–169.
57. Ruepp A, Zollner A, Maier D, Albermann K, Hani J, et al. (2004) The FunCat, a functional annotation scheme for systematic classification of proteins from whole genomes. Nucleic Acids Res 32: 5539–5545.
58. Colot HV, Park G, Turner GE, Ringelberg C, Crew CM, et al. (2006) A high-throughput gene knockout procedure for *Neurospora* reveals functions for multiple transcription factors. Proc Natl Acad Sci U S A 103: 10352–10357.
59. Dunlap JC, Borkovich KA, Henn MR, Turner GE, Sachs MS, et al. (2007) Enabling a community to dissect an organism: overview of the *Neurospora* functional genomics project. Adv Genet 57: 49–96.
60. Fu C, Iyer P, Herkal A, Abdullah J, Stout A, et al. (2011) Identification and characterization of genes required for cell-to-cell fusion in *Neurospora crassa*. Eukaryot Cell 10: 1100–1109.

61. Seo J, Shneiderman B (2002) Interactively exploring hierarchical clustering results. Computer 35: 80–86.
62. Altschul SF, Gish W, Miller W, Myers EW, Lipman DJ (1990) Basic local alignment search tool. J Mol Biol 215: 403–410.
63. Coleman JJ, Rounsley SD, Rodriguez-Carres M, Kuo A, Wasmann CC, et al. (2009) The genome of *Nectria haematococca*: contribution of supernumerary chromosomes to gene expansion. PLoS Genet 5: e1000618.
64. Ait Benkhali J, Coppin E, Brun S, Peraza-Reyes L, Martin T, et al. (2013) A network of HMG-box transcription factors regulates sexual cycle in the fungus *Podospora anserina*. PLoS Genet 9: e1003642.
65. Son H, Seo YS, Min K, Park AR, Lee J, et al. (2011) A phenome-based functional analysis of transcription factors in the cereal head blight fungus, *Fusarium graminearum*. PLoS Pathog 7: e1002310.

Comparative Analysis and Modeling of the Severity of Steatohepatitis in DDC-Treated Mouse Strains

Vikash Pandey[1], Marc Sultan[1], Karl Kashofer[2], Meryem Ralser[1], Vyacheslav Amstislavskiy[1], Julia Starmann[3], Ingrid Osprian[4,5], Christina Grimm[6], Hendrik Hache[1], Marie-Laure Yaspo[1,7], Holger Sültmann[3], Michael Trauner[8], Helmut Denk[2], Kurt Zatloukal[2], Hans Lehrach[1,7], Christoph Wierling[1]*

1 Max Planck Institute for Molecular Genetics, Department Vertebrate Genomics, Berlin, Germany, **2** Institute of Pathology, Medical University of Graz, Graz, Austria, **3** German Cancer Research Center, Heidelberg, Germany, **4** BIOCRATES Life Sciences AG, Innsbruck, Austria, **5** LKH Wagna, Department of Internal Medicine, Wagna, Austria, **6** Rheumatology and Clinical Immunology, Charité-University Medicine, Berlin, Germany, **7** Dahlem Centre for Genome Research and Medical Systems Biology, Berlin, Germany, **8** Hans Popper Laboratory of Molecular Hepatology, Division of Gastroenterology and Hepatology, Department of Medicine, Medical University of Vienna, Vienna, Austria

Abstract

Background: Non-alcoholic fatty liver disease (NAFLD) has a broad spectrum of disease states ranging from mild steatosis characterized by an abnormal retention of lipids within liver cells to steatohepatitis (NASH) showing fat accumulation, inflammation, ballooning and degradation of hepatocytes, and fibrosis. Ultimately, steatohepatitis can result in liver cirrhosis and hepatocellular carcinoma.

Methodology and Results: In this study we have analyzed three different mouse strains, A/J, C57BL/6J, and PWD/PhJ, that show different degrees of steatohepatitis when administered a 3,5-diethoxycarbonyl-1,4-dihydrocollidine (DDC) containing diet. RNA-Seq gene expression analysis, protein analysis and metabolic profiling were applied to identify differentially expressed genes/proteins and perturbed metabolite levels of mouse liver samples upon DDC-treatment. Pathway analysis revealed alteration of arachidonic acid (AA) and S-adenosylmethionine (SAMe) metabolism upon other pathways. To understand metabolic changes of arachidonic acid metabolism in the light of disease expression profiles a kinetic model of this pathway was developed and optimized according to metabolite levels. Subsequently, the model was used to study *in silico* effects of potential drug targets for steatohepatitis.

Conclusions: We identified AA/eicosanoid metabolism as highly perturbed in DDC-induced mice using a combination of an experimental and *in silico* approach. Our analysis of the AA/eicosanoid metabolic pathway suggests that 5-hydroxyeicosatetraenoic acid (5-HETE), 15-hydroxyeicosatetraenoic acid (15-HETE) and prostaglandin D2 (PGD2) are perturbed in DDC mice. We further demonstrate that a dynamic model can be used for qualitative prediction of metabolic changes based on transcriptomics data in a disease-related context. Furthermore, SAMe metabolism was identified as being perturbed due to DDC treatment. Several genes as well as some metabolites of this module show differences between A/J and C57BL/6J on the one hand and PWD/PhJ on the other.

Editor: Manlio Vinciguerra, University College London, United Kingdom

Funding: This work was funded by a grant of the Austrian Nationalstiftung and the Austria Wirtschaftsservice GmbH in the framework of the IMGuS research program (Institute for Medical Genome Research and Systems Biology, Vienna) and by the Max Planck Society. Co-author Ingrid Osprian was employed by BIOCRATES Life Sciences AG during the runtime of the IMGuS project where she did the metabolic analysis presented in this manuscript. BIOCRATES Life Sciences AG provided support in the form of salary for author Ingrid Osprian, but did not have any additional role in the study design, data collection and analysis, decision to publish, or preparation of the manuscript. The specific role of this author is articulated in the 'author contributions' section.

Competing Interests: Co-author Ingrid Osprian was employed by BIOCRATES Life Sciences AG during the runtime of the IMGuS project where she did the metabolic analysis presented in this manuscript. There are no patents, products in development or marketed products to declare.

* Email: wierling@molgen.mpg.de

Introduction

Nonalcoholic fatty liver disease (NAFLD) is a major cause of chronic liver damage in western countries and dependent on genetic and environmental factors. NAFLD can be considered as the hepatic manifestation of the metabolic syndrome and is linked to visceral obesity and has a higher prevalence among people with hyperlipidemia, hypertension, hyperglycemia and insulin resistance [1,2]. Advanced stages of NAFLD, such as steatohepatitis, ultimately result in fibrosis and cirrhosis and can finally lead to liver failure or hepatocellular carcinoma (HCC) [3]. The importance of nonalcoholic steatohepatitis (NASH) in public health in the western world is demonstrated by the massive increase of obesity and type 2 diabetes mellitus, which are key components of the metabolic syndrome [4].

The molecular basis of complex diseases, such as NASH, is poorly understood and its analysis requires a detailed understanding of the underlying metabolic and regulatory processes on the molecular level. Steatohepatitis is characterized by alterations in the liver, such as steatosis, ballooning of hepatocytes, apoptosis, protein aggregates in hepatocytes (Mallory-Denk bodies), pericellular fibrosis, and predominantly polymorphonuclear granulocytic inflammation [3]. Certain features of steatohepatitis can be reproduced to a variable extent in different mouse models by various treatments like chronic intoxication with DDC (5-diethoxycarbonyl-1,4-dihydrocollidine), methionine- and choline-deficient diet, alcohol or high fat diet [5,6].

Metabolites can be seen as end points of perturbations occurring at the gene level, so that changes of gene expression might to some extent also relate to changes in metabolite concentrations. The metabolic activity changes can be justified from transcript profiles based on the fact that mRNA is translated into a protein, e.g., working as an enzyme, thus changing metabolic flux of its catalyzed reaction [7]. Analysis of experimental data from cellular levels, i.e., transcriptomics, proteomics, fluxomics, and metabolomics show that there is not a high overall correlation between the abundance of RNA and its encoded protein, and between enzyme abundance and their respective catalyzed flux [7]. However, for regulated metabolic pathways the changes of RNA and protein abundances were in accordance with changes in reaction fluxes [7]. Hence, models integrating metabolic pathways and gene expression data may be used for *in silico* studies to predict changes in metabolite levels. Several tools exist to perform pathway analysis of expression and/or metabolic data that provide reasonable candidate pathways for subsequent modeling [8,9]. Past studies already used combined analysis of gene expression and metabolite data for the identification of, e.g., a genetic network of liver metabolism [10], biomarkers of type 2 diabetes [11], and disease related active pathways [12].

Herein, we studied transcriptomics and metabolomics data of mice developing NASH-like phenotypic features. We fed three genetically different mouse strains A/J, C57BL/6J, and PWD/PhJ (henceforth AJ, B6 and PWD) a DDC-supplemented diet. These mouse strains belong to two different subspecies and therefore cover a broad genetic variety. AJ and B6 are classical laboratory mouse strains mainly of *Mus musculus domesticus* origin, whereas PWD is a wild-derived strain mainly of *Mus musculus musculus* origin. Liver samples of DDC-treated and untreated mouse strains were analyzed by RNA sequencing (RNA-Seq) providing comprehensive mRNA expression profiles. Furthermore, we quantified selected metabolites by mass spectrometry and some relevant proteins by reverse-phase protein array (RPPA). Pathway analysis of these data identified deregulated pathways such as, nucleotide, beta-alanine and histidine metabolism. Furthermore, strain-specific deregulation was found in the metabolism of S-adenosyl-methionine (SAMe). Moreover, the metabolism of arachidonic acid was identified to be perturbed due to DDC-treatment irrespective of the strain. For further analysis and correlation of transcriptomics, protein and metabolite data, a kinetic model of the arachidonic acid metabolism was developed and fed with expression data to understand the metabolic changes. The arachidonic acid metabolic model was further used in an *in silico* study to obtain potential drug targets.

Materials and Methods

Mouse experiments and liver preparation

A/J, C57BL/6J and PWD/PhJ (abbreviated AJ, B6, and PWD) animals were obtained from Jackson Laboratories (The Jackson Laboratory, Maine, USA) and bred in the animal facility of the Medical University of Graz under specific pathogen free conditions. Eight weeks old male animals of each strain were fed either a standard (control) diet (Ssniff Spezialdiäten GmbH, Soest, Germany) or the standard diet supplemented with 0.1% DDC (5-diethoxycarbonyl-1,4-dihydrocollidine, Sigma-Aldrich, Vienna, Austria) for eight weeks under constant health monitoring.

All animal experiments were conducted according to the Austrian Animal Welfare Act. All experiments were approved by the Austrian ministry of science and research after review by the Austrian animal research committee under animal license number BMBWK-66.010/0047-BrGT/2005. Austrian law does not require individual institutional animal care and use committees for universities as all animal experimentation has to be pre-approved by the Austrian ministry of science and research.

Up to 4 animals were kept in individually ventilated cages at a monitored temperature of 20–24°C, with humidity between 40–70%, a constant 12 hour light and dark cycle and at least 50–60 air changes per hour. Animals were provided with water and rodent chow ad libitum. Animal health was monitored by daily visual cage inspection and weekly weight check of all experimental animals. Animals losing more than one third of body weight or showing other signs of distress due to DDC feeding were excluded from analysis and immediately sacrificed. All animals were sacrificed by cervical dislocation after anaesthesia by isoflurane inhalation. Liver tissues were harvested and samples of liver tissue were frozen in methyl-butane cooled by liquid nitrogen and subsequently stored in liquid nitrogen. RNA was prepared from frozen liver samples using the RNeasy Mini Kit (Qiagen GmbH, Hilden, Germany) according to manufacturers instructions.

RNA-Seq experiments and data analysis

We performed transcriptome analysis of healthy and DDC-treated livers of three mouse strains AJ, B6, and PWD each, and three biological replicates per strain and condition. In total, 18 paired-end RNA-Seq libraries were prepared from 10 μg of total RNA using a strand-specific strategy and following the protocol described in [13]. The following ligation adapters and PCR primers were used: PE Adapter OligoMix (Cat.#1001782), PCR Primer PE 1.0 (Cat.#1001783), PCR Primer PE2.0 (Cat.#1001784), IndexPE Adapter Oligo Mix Cat.#1005711), PCR PrimerInPE 1.0, (Cat.#100571), PCR PrimerInPE 2.0, (Cat.#1005713). Sequencing was carried out on the GAIIx platform (Illumina) by running 2×51 cycles according to the manufacturer instructions. Sequencing reads were aligned to the mm9 assembly of the mouse reference genome using BWA [14]. Gene levels were then quantified in reads per kilobase of exon model per million mapped reads [15] and using the Ensembl v.53 (*Mus musculus*) annotation. Identification of differentially expressed genes (DEGs) was performed with the R-package "edgeR" [16]. For identification of DEGs three biological replicates of each mouse strain AJ, B6, and PWD for control and DDC-treated states were used. DEGs were identified in two ways: (i) "strain-wise" and (ii) "irrespective of strains". For strain-wise identification DEGs between DDC-treated and control states were computed separately for each mouse strain AJ, B6, and PWD, while for the identification irrespective of strains control and DDC-treated states were analyzed across all three mouse strains together. Using edgeR p-values were calculated and adjusted for multiple testing using the Benjamini-Hochberg procedure implemented in edgeR. Principal components analysis (PCA) was performed to investigate any grouping of strain-wise DEGs.

Metabolomics analysis

Acylcarnitines, Sphingomyelins, Hexoses, Glycerophospholipids (FIA-MS/MS). To determine the concentration of acylcarnitines, sphingomyelins, hexoses and glycerophospholipids in liver homogenate the AbsoluteIDQ kit p150 (Biocrates Life Sciences AG) was prepared as described in the manufacturer's protocol. In brief, 10 µl of liver homogenate was added to the centre of the filter on the upper 96-well kit plate, and the samples were dried using a nitrogen evaporator (VLM Laboratories). Subsequently, 20 µl of a 5% solution of phenyl-isothiocyanate was added for derivatization. After incubation, the filter spots were dried again using an evaporator. The metabolites were extracted using 300 µl of a 5 mM ammonium acetate solution in methanol. The extracts were obtained by centrifugation into the lower 96-deep well plate followed by a dilution step with 600 µl of kit MS running solvent. Mass spectrometric analysis was performed on an API4000 QTrap tandem mass spectrometry instrument (Applied Biosystems/MDS Analytical Technologies) equipped with an electro-sprayionization (ESI)-source using the analysis acquisition method as provided in the AbsoluteIDQ kit. The standard FIA-MS/MS method was applied for all measurements with two subsequent 20 µl injections (one for positive and one for negative mode analysis). Multiple reaction monitoring (MRM) detection was used for quantification applying the spectra parsing algorithm integrated into the MetIQ software (Biocrates Life Sciences AG).

Prostanoids, oxidized fatty acids (LC-MS/MS). Prostanoids – a term summarizing prostaglandins (PG), thromboxanes (TX) and prostacylines – and oxidised fatty acid metabolites were analyzed by LC-ESI-MS/MS [17] by online solid phase extraction (SPE)-LC-MS/MS with an API4000 QTrap tandem mass spectrometry instrument (Applied Biosystems/MDS Analytical Technologies) in negative MRM detection mode. In brief, filter spots in a 96 well plate were spiked with internal standard; 20 µl of sample were added and extracted with aqueous methanol, the individual extracts then were analysed. Data of prostanoids and oxidized fatty acids were quantified with Analyst 1.4.2 software (Applied Biosystems) and finally exported for statistical analysis.

Protein preparation

Proteins were isolated from fresh-frozen tissue using T-Per extraction reagent (Pierce Biotechnology, Inc., Rockford, IL, USA) according to the manufacturer's recommendations, with the addition of the following inhibitors: complete mini protease inhibitor cocktail, staurosporin and PhosStop (Roche, Mannheim, Germany). For protein extraction from tissue (30–40 mg) the Qiagen Tissuelyser (Qiagen, Hilden, Germany) was used. Samples were isolated in 10 µl protein lysis buffer per 10 mg sample. Protein concentrations were determined using the bicinchoninic acid assay reagents (Pierce Biotechnology, Inc.). Protein lysates were stored at −80°C.

Reverse-phase protein array (RPPA) processing and data analysis

Protein lysates were diluted using protein lysis puffer (2 µg/µl). After adding Tween 20 (0.05%, v/v) protein lysates were printed onto nitro-cellulose coated glass slides (Oncyte Nitrocellulose Film Slides, Grace Bio-Labs, Blend, OR, USA) using the Aushon 2470 solid-pin tool arrayer (Aushon Biosystems, Billerica, MA, USA). Antibody incubation and antibody-mediated signal amplification were performed as described [18]. Slides were scanned with the Odyssey NIR scanner (LI-COR Biosciences, Bad Homburg, Germany). Image analysis was carried out with GenePix-Pro 6.0 (Axon Instruments, Sunnyvale, USA). Data sets were analyzed using the RPPAnalyzer package [19]. Quantification results were normalized to the Fast Green FCF staining of total proteins as well as to the median antibody binding signal levels [20].

Antibody validation for RPPA

To verify antibody specificities, a pool of protein lysates from cancer and benign tissues was analyzed by SDS-PAGE followed by Western Blotting. 25 µg of the protein lysate was used and a standard near-infrared detection was applied as described in [18]. Antibodies with a predominant single band in the expected size range were selected for further RPPA analysis. Antibodies were purchased from Cell Signaling Technology (Danvers, MA, USA).

Model Initialization

An example of mapping RNA-Seq or RPPA data to enzyme concentration is described below. Assume a mass action kinetic equation,

$$v = k_{cat} \times E \times S$$

where v is the flux, k_{cat} is the enzymatic turnover number, and E and S are the enzyme and the substrate concentrations, respectively. The aforementioned equation is transformed in equation,

$$v = v_{max} \times u \times S$$

where $v_{\max}$ is the maximum rate and u is a unitless quantity representing a ratio. For the initialization of the scaling factor u with RNA-Seq or RPPA data we use 1 for the control state and the ratio between treatment and control for the DDC-supplemented state. The ratio is computed based on the experimental transcriptomics and proteomics data form mice fed 8 weeks with a DDC-supplemented diet and their respective controls. The state after 8 weeks is assumed as a quasi steady state also regarding the quantified metabolite concentrations.

Modelling and kinetic parameter optimization

The mathematical model of the arachidonic acid (AA)/eicosanoid pathway was developed in PyBioS [21,22] based on information from KEGG and a survey of literature [23–26]. Parameter optimization of the mathematical model was performed with COPASI [27]. COPASI provides a number of different algorithms for optimization of a predefined objective function. We used the genetic algorithm in our analysis to minimize the objective function

$$f_p = \sqrt{\sum_{x \in M} \left([x]_e - [x]_{s,p} \right)^2}$$

where x is a given metabolite of a set of metabolites M of our metabolic model, $[x]_e$ and $[x]_s$ are concentration values of a metabolite x from experiments and simulations, respectively, and $\mathbf{p}$ is a set of kinetic parameters. $[x]_{s,p}$ is the steady state concentration of metabolite x during simulation which depends on $\mathbf{p}$. $\min f_p$ is an optimization problem which minimize f over $\mathbf{p}$.

Results

Liver samples of DDC-treated and untreated mouse strains (AJ, B6 and PWD) were phenotypically characterized and analyzed by RNA-Seq providing comprehensive mRNA expression profiles, by mass spectrometry of selected metabolites, and by RPPA analysis of relevant proteins.

Phenotypic characterization

Phenotypes of liver samples of all three mouse strains after eight weeks of a DDC-supplemented diet were characterized by histological examination and formation of Mallory-Denk bodies using immunohistochemical analysis of ubiquitin and p62 (Fig. 1a). Steatosis was observed to a medium degree in AJ and PWD while it was absent in B6. Ballooning of the hepatocytes and the occurrence of ubiquitin and p62 containing protein aggregates was most pronounced in AJ. Based on the degree of steatohepatic features AJ and B6 were categorized as high and low susceptible, respectively, while PWD is resistant.

Identification of differentially expressed genes using RNA-Seq data

Using RNA-Seq data differentially expressed genes (DEGs) in response to DDC treatment were computed (i) "strain-wise" and (ii) "irrespective of strains" as described in the material and methods section. The total number of DEGs from strain-specific analysis was very similar between the strains with 1,645, 1,657 and 1,373 for AJ, B6 and PWD, respectively (Fig. 1b). In total, 516 genes were deregulated in all three strains in response to DDC feeding, whereas 471 DEGs were found exclusively in AJ (Fig. 1b, Table S1). We call these 471 genes susceptibility genes since this strain shows the most pronounced steatohepatitis phenotype. In addition, to address the DDC treatment effect a principle component analysis (PCA) was performed based on 2,813 genes (Table S2), which were differentially expressed in at least one mouse strain due to DDC treatment (Fig. 1c). While healthy mice show strain-specific expression profiles in PCA, all mouse profiles from DDC treatment group together, implying a similar overall response due to DDC treatment irrespective of the strain (Fig. 1c). Besides the strain-specific analysis, DEGs were also computed irrespective of strains. This analysis yields in total 4,215 DEGs in response to DDC treatment (Table S3).

Analysis of proteomics data

For a screening of changes of the proteome some specific proteins were also analyzed by reverse phase protein arrays (RPPA; Table S4). Notably, the protein expression levels of the liver-type fatty acid-binding protein (Fabp1) are decreased in all strains after DDC-feeding. This observation is in agreement with data from human samples showing an underexpression in mild and progressive stages of NASH but paradoxically an overexpression in simple steatosis [28]. Furthermore, keratin 8 (Krt8) and keratin 18 (Krt18) were found to be upregulated after DDC treatment. Both proteins are major components of the hepatocyte cytoskeleton. Other proteins that were found to be upregulated are glutathione S-transferase alpha (Gsta1) and glutathione S-transferase mu (Gstm1) that are both involved in detoxification processes and catalyze the conjugation of reduced glutathione to xenobiotics. Notably, the level of glutathione S-transferase alpha upregulation in AJ and B6 mice is much higher compared to PWD. Moreover, the protein expression level of the cytochrome Cyp2e1 level is found to be down-regulated in PWD, while it is not changed in B6 and AJ mice. In addition to general protein amounts, also phosphorylated states of the key signaling proteins Erk1 (Mapk3), Akt1 and Stat3 were measured reflecting the activity states of their respective signaling pathways. All of them were found to be induced due to DDC treatment.

Pathway analysis of metabolic profiles and gene expression data

Besides gene expression and proteomics data, also 44 metabolites were identified and quantified by mass spectrometry. Differences in metabolite concentrations due to DDC treatment were judged by t-test (Table S5). A pathway analysis of metabolic profiles and gene expression data was performed with ConsensusPathDB to identify steatohepatitis-specific pathways [29]. Pathway over-representation analysis of the aforementioned 471 susceptibility genes obtained only in AJ identified nucleotide, histidine, beta-alanine, purine metabolism, apoptosis and steroid hormone biosynthesis upon the top-ranked deregulated pathways (Table S6). Within the beta-alanine pathway we found *Srm* encoding spermidine synthase (SPDS) as being upregulated in AJ (2.44-fold in AJ compared to 0.77-fold and 0 75-fold in B6 and PWD, respectively, Tab. S2) as well as the related metabolites spermidine (SPD) and spermine (SPM) that form a module of the hepatic S-adenosylmethionine (SAMe) metabolism (Fig. 1d, [30]). SAMe is needed for methylation of DNA, RNA and lipids, and synthesis and catabolism of SAMe is tightly regulated and changes in SAMe level might lead to fatty liver disease and the development of HCC [30].

Gene expression and metabolite concentrations of hepatic SAMe metabolism were found to be affected in the DDC-treated state (Fig. 1e and 1f). Expression of the genes methionine adenosyltransferase 1 alpha (*Mat1a*), spermine synthase (*Sms*), DNA methyltransferase 1 (*Dnmt1*), adenosylhomocysteinase (*Ahcy*), betaine-homocysteine methyltransferase (*Bhmt*), and spermidine synthase (*Srm*) and concentration changes of the related metabolites spermidine, spermine and putrescine show a different behavior between AJ, B6 and PWD. Gene expression and metabolic profiles of hepatic SAMe metabolism imply that PWD has an opposite response compared to AJ and B6 after feeding a DDC-supplemented diet which resembles the differences in steatohepatitis phenotypes of the strains. The enzyme SAMe decarboxylase (SAMDC) is activated by putrescine and upregulation of spermidine synthase (SPDS) and spermine synthase (SPMS) might explain the increased concentration of spermidine (SPD) and spermine (SPM) in AJ. A higher concentration of putrescine in AJ and B6 lowers the K_m of SAMDC activating polyamine synthesis [30] that may affect the concentration of SAMe. This might affect methylation of various substrates such as DNA, RNA and lipids that might be one reason in NASH disease development. However, it remains difficult to judge the difference in degree of steatoheapatitis since the expression data and metabolic profiles of AJ and B6 show a similar response to the DDC-supplemented diet.

NAFLD has been defined as a metabolic disease associated with the insulin-resistance syndrome [1]. A genome-scale metabolic network of the mouse comprising 3,724 reactions, 2,774 metabolites, and 1,415 enzyme coding genes was used as a reference to identify metabolic pathways of related genes [31]. Irrespective of strain-specific effects, a group of 288 genes (out of 4,215 DEGs) coding for metabolic enzymes of the genome-scale metabolic network were identified as differentially expressed (edgeR p-value $<$1e-6 over all three strains) after DDC treatment (Table S7). This set of 288 genes was subsequently used in a pathway analysis with ConsensusPathDB to identify affected metabolic pathways due to DDC treatment (Table S8). Furthermore, a set of 19 differentially regulated metabolites (t-test p-value $<$0.05, Table S5) was used for

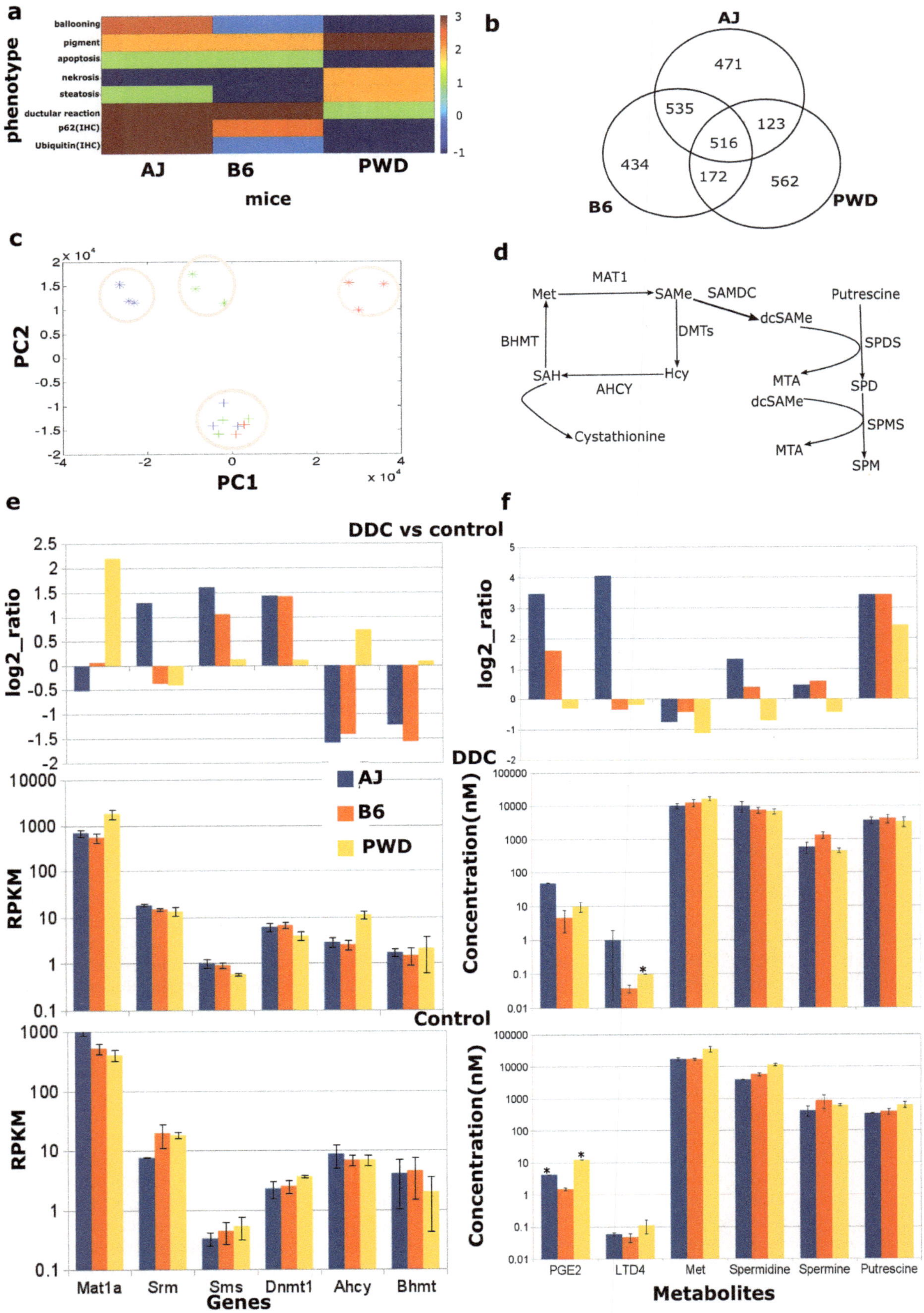
a
phenotype
ballooning
pigment
apoptosis
nekrosis
steatosis
ductular reaction
p62(IHC)
Ubiquitin(IHC)
AJ
B6
PWD
mice
b
AJ
471
535
123
516
434
172
562
B6
PWD
c
PC2
PC1
d
Met
MAT1
SAMe
SAMDC
dcSAMe
Putrescine
BHMT
DMTs
SPDS
SAH
AHCY
Hcy
MTA
SPD
dcSAMe
SPMS
Cystathionine
MTA
SPM
e
f
DDC vs control
log2_ratio
DDC
RPKM
AJ
B6
PWD
Concentration(nM)
Control
RPKM
Mat1a
Srm
Sms
Dnmt1
Ahcy
Bhmt
Genes
PGE2
LTD4
Met
Spermidine
Spermine
Putrescine
Metabolites

Figure 1. Analysis of phenotypic and omics data. a) Qualitative scoring of histological phenotypes of the mouse liver samples. Score -1, absent; score 0, minimal; score 1, mild; score 2, moderate; score 3, severe changes compared to healthy liver tissue. Immunohistochemistry, IHC. **b)** Venn diagram of differentially expressed genes due to DDC treatment in AJ, B6, and PWD mice. **c)** Principle component analysis (PCA) of 2813 genes that were found differentially expressed for at least one mouse strain due to DDC-treatment. * and + indicate control and DDC mice, respectively; red, green and blue represent AJ, B6, and PWD mice, respectively. Principle component 1 (PC1) explains 43% and PC2 29% of the data. **d)** S-adenosylmethionine (SAMe) metabolism. Methionine (Met) is converted to SAMe by the enzyme methionine adenosyltransferase (MAT1). SAMe is converted into S-adenosylhomocysteine (SAH) by DNA-methyltransferase (DMTs) and SAH hydrolase (AHCY) with homocysteine (Hcy) as an intermediate. SAH is substrate for Met formation by betaine-homocysteine methyltransferase (BHMT). SAMe can also be converted into spermine (SPM) via decarboxylated SAMe (dcSAMe) and spermidine (SPD) catalyzed by SAMe decarboxylase (SAMDC), SPD synthase (SPDS), and SPM synthase (SPMS). This pathways is regulated by putrescine, which activates SAMDC. **e)** Arithmetic mean values of RPKM values of aforementioned genes for liver samples of control and DDC-treated mice. Error-bars indicate standard deviations. The bar chart shows log2-ratios of RPKM values of DDC-treated vs control. The genes *Mat1a, Srm, Sms, Dnmt1, Ahcy,* and *Bhmt* encode the enzymes MAT, SPDS, SPMS, DMTs, AHCY and BHMT, respectively. **f)** Bar chart of median concentrations of the metabolites prostaglandin D2 (PGD2), leukotriene D4 (LTD4), methionine (Met), spermidine, spermine, and putrescine. Error-bars indicate median absolute deviations. * indicates samples without a replicate.

a metabolite-based over-representation analysis with ConsensusPathDB (Table S9). Both, the pathway analysis of the transcriptomics as well as the metabolic data set, yield the arachidonic acid pathway upon others as being deregulated after DDC treatment (p-value 2.6e-4 and 1.3e-4, respectively, Table S8 and S9). Analysis of metabolites of the arachidonic acid/eicosanoid pathway yield also significant changes due to DDC treatment of some metabolites of this pathway. The abundances of four metabolites of this pathway are significantly altered due to DDC-treatment: PGD2, 5-HPETE, 15-HETE, and 15-HPETE (p-value$<$0.05, Table S4). An increased activity of the arachidonic acid pathway could also be observed in the metabolic data. Based on these observations we selected the arachidonic acid metabolism as a candidate for further *in silico* analysis.

Model of the arachidonic acid/eicosanoid metabolic pathway

Arachidonic acid (AA) is a fatty acid usually coming from dietary animal sources or being synthesized from dietary linoleic acid. AA is present in cell membranes as a part of phospholipids. AA is released from phospholipids by phospholipase A2 (PLA2) and subsequently it acts as a precursor of prostaglandins and their related compounds, the prostacyclins, thromboxanes and leukotrienes. Several physiological effects of prostaglandins are described in the literature, such as inflammatory response, pain, fever, blood pressure, blood clotting, and regulation of sleep/wake cycle [32,33]. Cyclooxygenase-1, also known as prostaglandin H2 synthase 1 (Ptgs1), catalyzes the reaction of arachidonate to prostaglandin H2 (PGH2), which is the precursor of other prostaglandins, prostacyclins, and thromboxanes. The enzyme 5-lipoxygenase (Alox5) produces 5-HPETE, which is a precursor of leukotrienes. The metabolites PGD2, 5-HPETE, 15-HETE and 15-HPETE have been observed to be significantly ($p<0.05$) changed in concentration due to DDC treatment in the liver samples of all mouse strains (Table S5).

There are only a few kinetic models of the AA metabolism available in the literature to study anti-inflammatory drugs of human polymorphonuclear leukocytes [23] and macrophage cells [24]. Here, an *in silico* model of the AA metabolism in mouse liver was developed based on information from KEGG [26] and a biochemical text book [25] to study the regulation of the perturbed metabolites using the related transcriptome and protein data. The structure of the model is depicted in Fig. 2.

The downstream synthesis of prostaglandins and leukotrienes is initiated by signaling and subsequent release of AA catalyzed by phospholipase A2 (PLA2). In endothelial cell-pericyte co-cultures of rat cells it has been shown that PLA2 is induced through the activation of the MAPK/ERK cascade [34]. PLA2 is involved in inflammation driven liver fibrosis as a key feature of progressive NASH and macrophage PLA2 deficiency prevented activation of hepatic stellate cells and infiltration of F4/80-positive macrophages [35]. 15-lipoxygenase (ALOX15) is an important regulator of inflammation and apoptosis and its expression is regulated by a cytosolic signaling complex with protein kinase C delta (PKCD) and phosphorylated STAT3 [36]. Thus, to describe the activity of PLA2 and PKCD, respectively, RPPA data of phosphorylated ERK and STAT3 was used subsequently for the modeling (Table S4).

The eicosanoid pathway is regulated by several feedback mechanisms (see Fig. 2). The release of AA is controlled by an inhibitory link between AA and phospholipase A2 (PLA2, R1 in Fig. 2, [37]). PLA2 is activated by phosphorylated ERK (p-ERK). Presence of both PLA2 and p-ERK is necessary to drive the metabolic conversion of phosphatidylcholine (PC) into AA (see R1 in Fig.2). To reflect the inhibition of PLA2 by AA an inhibitory feedback link is added to R1. The activity of 5-lipoxygenase (ALOX5) catalyzing the production of 5-HPETE and leukotriene A4 (LTA4) is controlled by product inhibition and by the ALOX5-activating protein ALOX5AP (R2, R4, [38,39]). To convert AA into 15-HPETE the presence of both ALOX15 and a complex with PKCD and STAT3 are necessary. The production of 15-HPETE is also controlled by an inhibitory link between 15-HPETE and ALOX15 which is added to R6 (Fig. 2, [23]). Synthesis of 15-HPETE and PGH2 are controlled by feedback inhibition and glutathione peroxide and PTGS1, respectively (R7 and R9, [23,40]).

We hypothesize that changes of metabolite concentrations of the AA pathway can be explained by changes in mRNA expression. To address this hypothesis, changes in V_{max} values of enzymatic reactions were approximated by fold changes of RNA-Seq expression of the respective enzymes due to DDC treatment. The fold change of PLA2 was approximated by *Pla2g4a* which was expressed at a low level ($<$1 RPKM, [41]) but not differentially (Table S10). Therefore, we hypothesize that PLA2 is at least present to trigger AA release upon its activity state, but it is not differentially expressed due to DDC treatment. Similarly, ALOX5 and ALOX15 were approximated by *Alox5* (p-value = 0.13) and *Alox15* (p-value = 0.62), respectively, which were expressed at low level ($<$1 RPKM) and we hypothesize that the expression of these genes were not affected due to DDC-treatment. PTGDS is approximated by *Ptgds* expression, but it was not significantly changed due to DDC treatment (p-value = 0.64). If there are several isoenzymes that can catalyse the same reaction the most significantly differentially expressed gene was chosen. For instance, GPX can be approximated by *Gpx1, Gpx2, Gpx3, Gpx6* and *Gpx7*. Since *Gpx3* was significantly affected (p = 2.66e-41) and highly expressed it was used to initialize the GPX expression in the model. Fold changes of the enzymes PTGS1, ALOX5AP and PKCD were approximated by expression

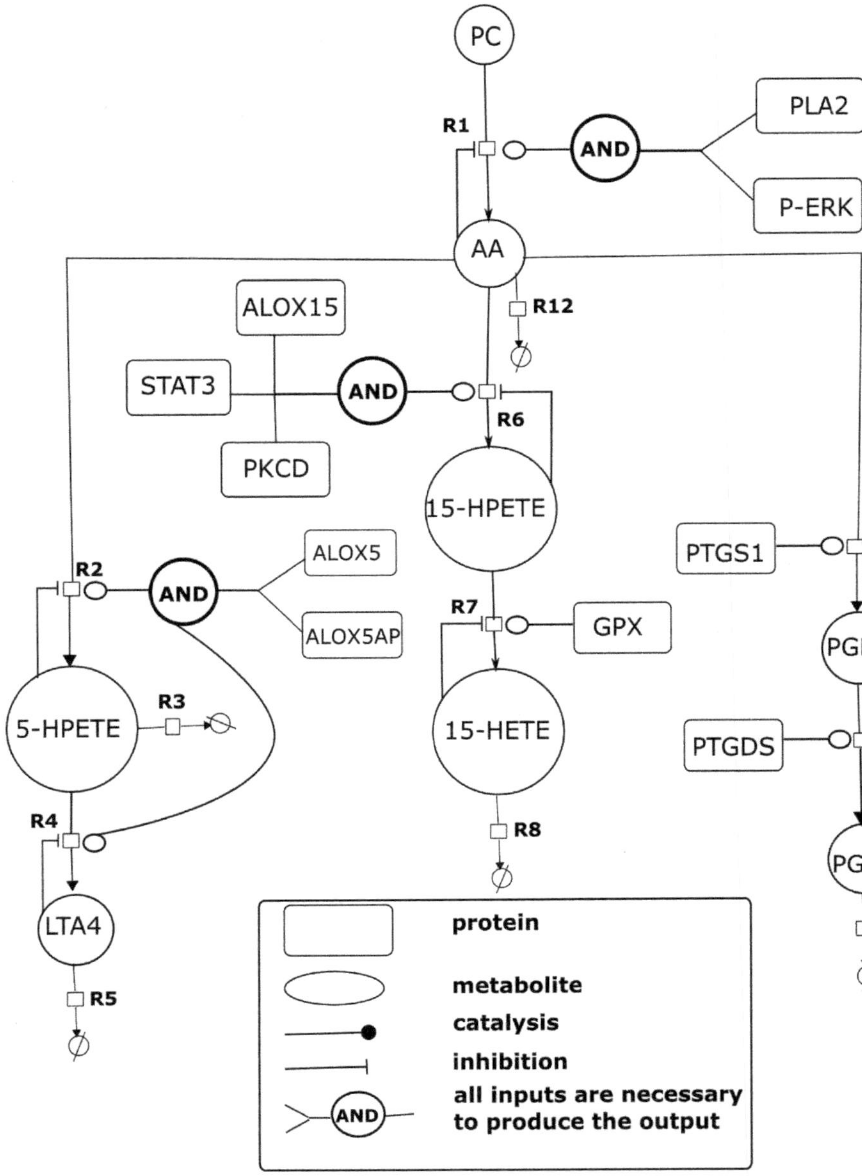

Figure 2. Model of arachidonic acid/eicosanoid metabolism. The model has three branches. Through the first branch arachidonic acid (AA) forms leukotriene A4. In the second branch 15-HPETE and 15-HETE are derived from AA, while in the third branch AA forms prostaglandin D2. See the results section for a detailed description of the model inhibitory links. Phosphatidylcholine, PC; arachidonic acid, AA; phospholipase A2, PLA2; phosphorylated ERK, P-ERK; cyclooxygenase-1, also known as prostaglandin G/H synthase 1, PTGS1; protein kinase C delta, PKCD; signal transducer and activator of transcription 3, STAT3; arachidonate 15-lipoxygenase, ALOX15; arachidonate 5-lipoxygenase, ALOX5; arachidonate 5-lipoxygenase-activating protein, ALOX5AP; prostaglandin H2, PGH2; prostaglandin D2, PGD2; prostaglandin D2 synthase, PTGDS; 15-hydroxyeicosatetraenoic acid, 15-HETE; 5- and 15-hydroperoxyeicosatetraenoic acid, 5- and 15-HPETE; leukotriene A4, LTA4; glutathione peroxidase, GPX.

values of *Ptgs1*, *Alox5ap* and *Prkcd*, respectively. Within our experiments the protein amount of phosphorylated ERK (p-ERK) and phosphorylated STAT3 (p-STAT3) were measured for both control and DDC-treated mice using the reverse phase protein array (RPPA) technology. For this we made use of respective proteomics data and initialized the model components p-ERK and p-STAT3 based on RPPA data. Since p-ERK and p-STAT3 were upregulated due to DDC treatment (Table S4; fold changes of p-ERK and p-STAT3 in AJ, B6, and PWD are 1.90, 2.09, 1.98, and 1.41, 1.42, 2.03, respectively) the effect of p-ERK on PLA2 is modeled by the RNA-Seq expression value of *Pla2g4a* for the DDC-treatment simulation and the p-ERK ratio. Similarly, the effect of the complex formation of PKCD and p-STAT3 on the ALOX15 activity was described by the RNA-Seq expression value of ALOX15 of the DDC-treatment state and the fold change of p-STAT3 and PKCD, respectively. Phosphatidylcholine (PC) was modeled as a fixed component and as its initial concentration the measured concentration of phosphatidylcholine C33:2 was used (Table 1). The experimentally identified fold changes of all enzymes are summarized in Table 1 and were used to approximate the kinetic parameters of the respective reactions. The kinetic parameters of the model were optimized by an objective function integrating the experimental data (see parameter optimization in section Material and Methods). Using data of

Table 1. Initial values of the arachidonic acid/eicosanoid metabolic model.

Symbol	AJ, control	AJ, DDC	B6, control	B6, DDC	PWD, control	PWD, DDC
PTGS1	1.00	2.13	1.00	1.41	1.00	2.09
ALOX5	1.00	1.00	1.00	1.00	1.00	1.00
PTGDS	1.00	1.00	1.00	1.00	1.00	1.00
STAT3	1.00	1.41	1.00	1.42	1.00	2.03
PKCD	1.00	4.30	1.00	4.48	1.00	3.85
GPX	1.00	6.51	1.00	11.70	1.00	19.42
PLA2	1.00	1.90	1.00	2.09	1.00	1.98
ALOX5AP	1.00	4.81	1.00	6.30	1.00	14.23
ALOX15	1.00	1.00	1.00	1.00	1.00	1.00
PC*	269.00	206.00	249.00	209.00	266.00	255.00

The enzyme concentrations of the control state were always initialized with 1.0 nM. The DDC initial value always reflects the fold change between the DDC-treated and the control expression value. The absolute concentration of phosphatidylcholine C33:2 (PC*) comes from MS-analysis and is measured in μM for control and DDC-treatment conditions of AJ, B6, and PWD mice.

DDC treatment *vs.* control we built a quantitative model to explain changes of metabolites of the eicosanoid pathway due to treatment by DDC that mimics the NASH phenotype.

Model Optimization

Model parameters were optimized using experimentally determined metabolite concentrations of the model components. Metabolite concentrations were measured after 8 weeks of DDC treatment and interpreted as steady state concentrations for the model. The metabolites PGD2, 5-HPETE, 15-HPETE and 15-HETE have been found as being affected due to DDC treatment. These metabolites as well as upstream AA were used for model optimization. Therefore, the model was simulated into its steady state and the euclidean distance between simulated steady state concentrations and experimental metabolite data of the aforementioned metabolites were used in an objective function to optimize the kinetic parameters using COPASI. Parameter optimization was done using the metabolic concentrations of B6 control and B6 DDC mice. Kinetic parameters, i.e., the maximum reaction rates (V_{max}), Michaelis-Menten constants (K_m) and inhibition constants (K_i) were optimized based on experimental data. The following five metabolites of the AA model were used to formulate an objective function: AA, 5-HPETE, and 15-hydroperoxyeicosatetraenoic acid (15-HPETE), 15-Hydroxyeicosatetraenoic (15-HETE), and prostaglandin D2 (PGD2). Let aforementioned metabolites be in a set M. The objective function f reads

$$f_p = \sqrt{\left(\sum_{x \in M}\left(\frac{\left([x]_{ec} - [x]_{sc,p}\right)}{[x]_{ec}}\right)^2 + \left(\frac{\left([x]_{ed} - [x]_{sd,p}\right)}{[x]_{ed}}\right)^2\right)}$$

where, **p** is a set of kinetic parameters of the AA model. *ec* and *sc* represent experimental and simulated steady concentration in control condition, while *ed* and *sd* represent experimental and simulated steady concentration in the DDC-treated condition. The objective function f was minimized to obtain optimal kinetic parameters of the AA model using the genetic algorithm of COPASI with a population size of 50 and a generation size of 500. The identified kinetic parameters are listed in Table 2. The model is available in SBML format (suppl. File S1).

Results of the fitted model for the metabolites AA, PGD2, 5-HPETE, 15-HPETE, and 15-HETE using the trained B6 model were compared with the experimental results (Fig. 3; C57Bl6_sim vs. C57Bl6_exp). Except of AA all of the aforementioned metabolites were up-regulated (>1.5 fold) in both, experimental and simulated data, due to DDC in B6 mice (Fig. 3c). AA was not found altered in the experimental data set, but was slightly up-regulated in the simulation (1.35-fold).

For validation of the model we predicted DDC-induced metabolite changes for AJ and PWD using the trained model. Simulation results predict an up-regulation (>1.5-fold) of 5-HPETE, 15-HPETE and 15-HETE in both strains, which was in line with the experimental data. The concentration of PGD2 was found increased (>1.5-fold) between experimental and simulated data for PWD (Fig. 3c). In AJ our modeling approach predicts a 2-fold increase of PGD2 concentration in response to the DDC-treatment, but the experimental data showed no changes (0.9 fold). This disagreement can be strain-specific because one can expect changes in the concentration of PGD2 due to an up-regulated gene expression of *Ptgs1* which is located upstream in the metabolic pathway. AA was up-regulated (>=1.5-fold) in DDC-PWD mice, which is concordant with the experimental data, while for AJ AA is not found as being changed (0.9 fold) in the experimental data, whereas a minor up-regulation (1.2-fold) was predicted by the simulation (Fig. 3c).

Key regulatory enzymes of the DDC mouse model

To identify key regulatory enzymes of the AA/eicosanoid pathway in DDC treated mice, a kind of sensitivity analysis of the trained model was performed. Starting with DDC-treated mice of the AJ strain, each enzyme or enzyme combinations were reverted to its/their activity in normal, untreated condition. This analysis was performed for ALOX5AP, GPX, PKCD, PTGS1, pERK, and pSTAT3 whose activity was found to be perturbed due to DDC-treatment. A reference state was defined where the activity of all enzymes is equal to the normal condition. Results of this analysis are depicted in Fig. 4a. Reverting a single enzyme or combinations of two enzymes were not sufficient to bring the DDC-treated state back to normal because it affects only some branches of the model, as for example a change in the activity of ALOX5 has an effect on the regulation of the downstream metabolites 5-HPETE and LTA4 (cf. Fig. 2). We found that the

Table 2. Kinetic equations and their parameters of the arachidonic acid/eicosanoid metabolism model.

R1: PC → AA	$\frac{V_{max}\times A_1\times A_2\times S}{\left(S+K_m\times\left(1+\frac{I}{K_i}\right)\right)}$	$V_{max}=20.56\ nM^2s^{-1}$, $K_m=2500$ nM [BRENDA], $K_i=100$ nM, $A_1=\{PLA2\}$, $A_2=\{PERK\}$, $I=[AA]$ nM, $S=[PC]$ nM
R2: AA → 5-HPETE	$\frac{V_{max}\times A_1\times A_2\times S}{\left(S+K_m\times\left(1+\frac{I}{K_i}\right)\right)}$	$V_{max}=0.01\ nM^2s^{-1}$, $K_m=0.0107$ nM, $K_i=8.603$ nM, $A_1=\{ALOX5\}$, $A_2=\{ALOX5AP\}$, $I=[5\text{-}HPETE]$ nM, $S=[AA]$ nM
R3: 5-HPETE →	$K_{cat}\times S$	$S=[5\text{-}HPETE]$, $K_{cat}=0.0012s^{-1}$
R4: 5-HPETE → LTA4	$\frac{V_{max}\times A_1\times A_2\times S}{\left(S+K_m\times\left(1+\frac{I}{K_i}\right)\right)}$	$V_{max}=9.7953\ nM^2s^{-1}$, $K_m=99.913$ nM, $K_i=0.709$ nM, $A_1=\{ALOX5\}$, $A_2=\{ALOX5AP\}$, $I=[LTA4]$ nM, $S=[5\text{-}HPETE]$ nM
R5: LTA4 →	$K_{cat}\times S$	$S=[LTA4]$ nM, $K_{cat}=0.0012\ s^{-1}$
R6: AA → 15-HPETE	$\frac{V_{max}\times A_1\times A_2\times A_3\times S}{\left(S+K_m\times\left(1+\frac{I}{K_i}\right)\right)}$	$V_{max}=0.003\ nM^3s^{-1}$, $K_m=0.067$ nM, $K_i=1.004$ nM, $A_1=\{ALOX15\}$, $I=[15\text{-}HPETE]$ nM, $A_2=\{PKCD\}$, $A_3=\{PSTAT3\}$, $S=[AA]$ nM
R7: 15-HPETE → 15-HETE	$\frac{V_{max}\times A\times S}{\left(S+K_m\times\left(1+\frac{I}{K_i}\right)\right)}$	$V_{max}=1.098\ nMs^{-1}$, $K_m=1.58$ nM, $K_i=0.0106$ nM, $A_1=ALOX15$, $I=[15\text{-}HETE]$ nM, $S=[15\text{-}HPETE]$ nM
R8:15-HETE →	$K_{cat}\times S$	$S=[15\text{-}HETE]$, $K_{cat}=0.00127\ s^{-1}$
R9: AA → PGH2	$\frac{V_{max}\times A\times S}{\left(S+K_m\times\left(1+\frac{I}{K_i}\right)\right)}$	$V_{max}=0.168\ nMs^{-1}$, $K_m=3.876$ nM, $K_i=0.013$ nM, $A=\{PTGS1\}I=[PGH2]$, $S=[AA]$ nM
R10: PGH2 → PGD2	$V_{max}\times S\times A$	$V_{max}=0.067\ nMs^{-1}$, $A=\{PTGDS\}$, $S=[PGH2]$
R11:PGD2 →	$K_{cat}\times S$	$S=[PGD2]$ nM, $K_{cat}=0.052\ s^{-1}$
R12: AA →	$K_{cat}\times S$	$S=[AA]$ nM, $K_{cat}=0.00096\ s^{-1}$

A, A_1, A_2 and A_3 are the ratios of gene expression or protein levels of the respective enzymes between DDC-treated vs. control mice. S is the substrate of the corresponding reaction. Squared brackets refer to concentration values and curly braces indicate fold changes of DDC treatment vs. control. The K_m value of PLA2 is taken from BRENDA [55], and other parameters of the table were fitted using the AJ mice metabolic concentrations (see results section).

combinations of the enzymes ALOX5AP, PKCD and PTGS1 with either pERK or pSTAT3 can bring the DDC-treated metabolic state back to normal, with the exception of 15-HETE and AA. Only the combination ALOX5AP, PERK, PKCD, STAT3, and PTGS1 or the combination of all six enzymes was able to bring back the DDC-treated state to normal.

Simulating drug effects in the DDC mouse model

To study potential drug targets for steatosis or inflammation *in silico* drug tests were performed. For instance, cyclooxygenase-2 (PTGS2/COX-2) is a frequent target of anti-inflammatory drugs [42]. In the computer simulation, the activity of each enzyme of the AJ DDC model was inhibited by 3-, 6- and 9-fold, respectively. The calculated effects of those enzyme inhibitions on the regulation of metabolites are shown in Fig. 4b. The inhibition of PTGS1 and ALOX5AP leads to a down-regulation of the respective downstream components (either PGH2 and PGD2 or 5-HPETE and LTA4). ALOX15 is activated by PKCD and pSTAT3, and the inhibition of PKCD or pSTAT3 leads to a down-regulation of the downstream components 15-HPETE and 15-HETE. It can be expected that the inhibition of glutathione peroxidase (GPX) leads to a down-regulation of the downstream component 15-HETE, but the model predicts an up-regulation of 15-HPETE and no change in 15-HETE. This is due to the complex regulation of ALOX15, i.e., an activation of the complex PKCD and pSTAT3 and an inhibition of GPX by 15-HETE.

Discussion

Using a DDC-based mouse model of steatohepatitis we analyzed three different mouse strains AJ, B6, and PWD covering a broad range of genetic variations. AJ, B6, and PWD mice respond with different degrees of steatohepatitis ranging from high, low to resistant, respectively. Based on gene expression and metabolic data we identified differences in hepatic SAMe metabolism in respect to the different steatohepatitis phenotypes across the individual mouse strains. SAMe metabolism might explain susceptibility of AJ mice. SAMe is a key methyl-group donor for phosphatidylcholine synthesis that is required, e.g., for the export of very-low-density lipoproteins (VLDL) and triglycerides from the liver [43]. Furthermore, VLDL synthesis has been found to be impaired by MAT1A-knockout that could be recovered by SAMe administration in MAT1A deficient mice [43]. SAMe and methionine metabolism was found to be perturbed in NASH patients [44] and it may play an important role in development of NAFLD, such as NASH [30]. A study suggests depletion of hepatic anti-oxidants (e.g. reduced glutathione and SAMe) promotes oxidative stress and may induce cellular alterations typical for steatohepatitis [6].

We identified AA metabolism as being the most affected one upon all of the three mouse strains and we propose a dynamic model of AA metabolism of NASH-like phenotypes in mouse induced by DDC. The fitted model provides good predictions between the experimental and simulated metabolic data using the respective gene expression data.

To determine regulatory roles of important enzymes in the upregulation of metabolic levels of the AA/eicosanoid pathway due to DDC-treatment, a sensitivity analysis was performed by reverting to the activity of the enzymes to normal, untreated conditions. The simulated combination of the enzymes ALOX5AP, PERK, PKCD, STAT3, and PTGS1 is able to revert the DDC-treated metabolic state to normal. This multi-targets inhibition might likely become a future approach in respect to individualized medicine as it has already been suggested for cancer [45].

Furthermore, we did *in silico* drug testing of several model enzymes. Inhibition of PTGS1 and ALOX5AP caused down-regulation of PGH2 and PGD2 or 5-HPETE and LTA4, respectively. These enzymes are known targets to treat inflammation. PTGS1 (COX-1), for instance, can be inhibited by

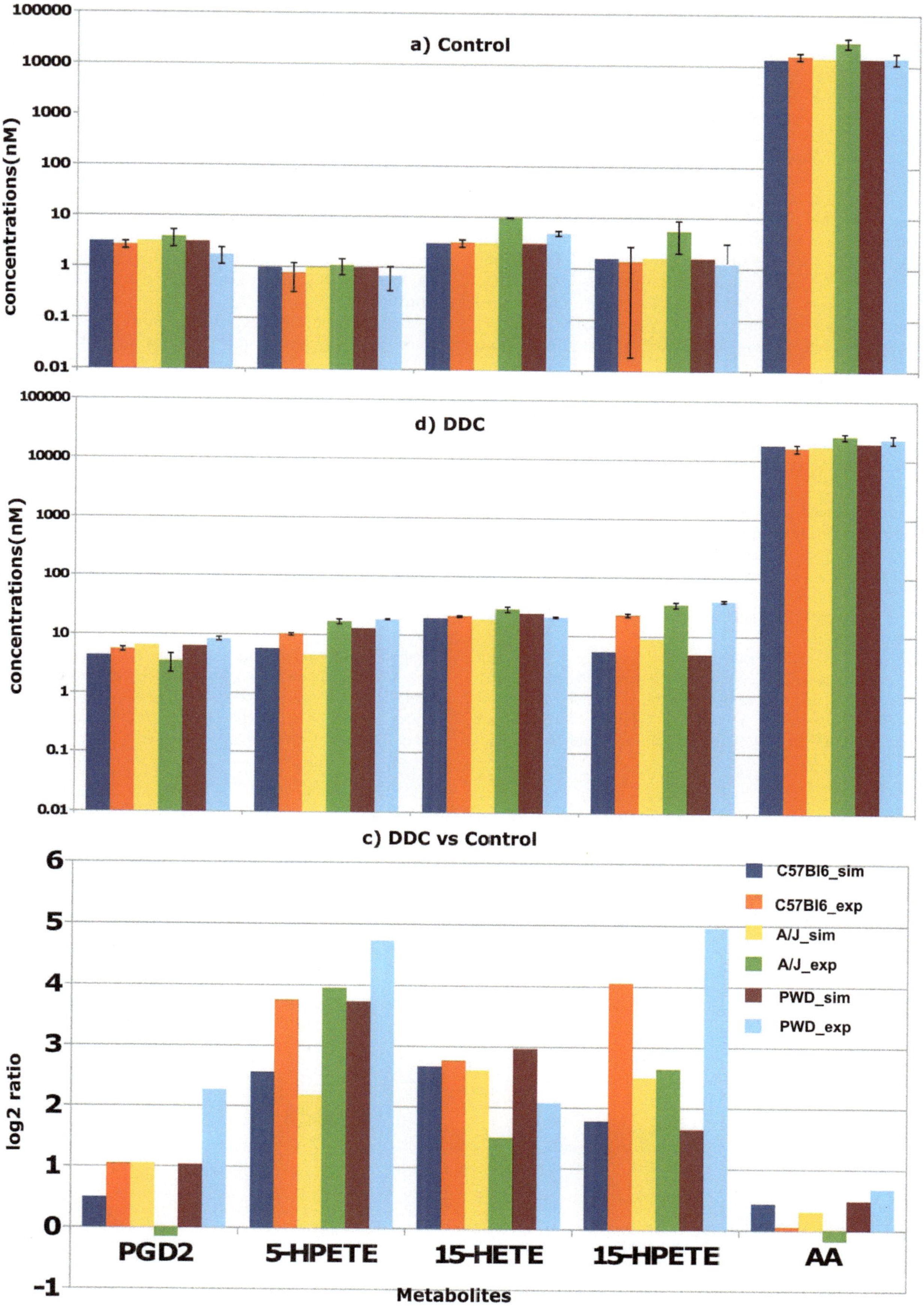
a) Control
concentrations(nM)
100000
10000
1000
100
10
1
0.1
0.01
d) DDC
concentrations(nM)
100000
10000
1000
100
10
1
0.1
0.01
c) DDC vs Control
log2 ratio
6
5
4
3
2
1
0
-1
C57Bl6_sim
C57Bl6_exp
A/J_sim
A/J_exp
PWD_sim
PWD_exp
PGD2
5-HPETE
15-HETE
15-HPETE
AA
Metabolites

Figure 3. Comparison of simulated steady state and experimental metabolite concentrations. Quantitative data of simulated and experimental metabolite concentrations of all three mouse strains AJ, B6, and PWD for **a**) control **b**) DDC treatment, and **c**) their respective ratios of the metabolites prostaglandin D2, PGD2; 5- and 15-hydroperoxyeicosatetraenoic acid, 5- and 15-HPETE; 15-hydroxyeicosatetraenoic acid, 15-HETE; and arachidonic acid, AA.

mofezolac, SC-560, and other drugs [46]. Inhibition of ALOX5AP also known as 5-lipoxygenase activating protein, or FLAP may be useful in the prevention of hepatotoxin-induced necro-inflammatory injury [47]. Drug molecules can interact with multiple targets to alter the state and function of the associated biological network. Licofelone is a novel 5-LOX/COX-inhibitor which inhibits two enzymes to avoid side effects [48]. Overall, the dynamics of the AA model can be used for *in silico* drug studies to test multiple drugs and potential drug targets. The model confirmed the upregulation of PGD2 due to DDC treatment found experimentally. This could be linked to the key transcription factors/ligand-activated nuclear receptors such as PPARδ which has been implicated as a key regulator of energy homeostasis and may represent future research avenues to study the interaction of metabolic and signaling pathways [49].

Prostaglandin E2 (PGE2) was found to be upregulated in AJ and B6 mice but not in PWD (Fig. 1f). PGE2 promotes inflammation after binding to prostaglandin E receptor 2 (EP2) [50] and an inflammatory marker tumor necrosis factor alpha (TNFα) regulates NASH development in a diet-induced mouse model [51]. To explain inflammation phenotypes through signaling by PGE2-EP2-TNFα we observed an upregulation of *Tnf* expression (35-fold) in AJ and B6 mice. Upregulation of PGE2 and TNFα in AJ and B6 mice may explain inflammation of the NASH phenotype.

The spectrum of NAFLD can be characterized by specific alterations in hepatic lipid composition. A comprehensive analysis of plasma lipids and eicosanoids in human revealed a stepwise increase in lipoxygeneasе metabolites 5-HETE, 8-HETE and 15-HETE in NAFLD [52]. This correlates with our observations in the experimental and simulated data, where concentrations of 5-HETE and 15-HETE were increased in DDC mice. Another study reported overexpression of cyclooxygenase-2 (COX-2) in hepatocellular carcinoma (HCC) patients [53]. Using immunohistochemistry they studied COX-2 overexpression in different chronic liver diseases including NASH, chronic hepatitis, and liver cirrhosis. In our study, we detected overexpression of PTGS1 (COX-2) in DDC mice, which is a key regulator of prostaglandin formation.

Martínez-Clemente et al. demonstrated that hyperlipidemia-prone apolipoprotein E-deficient (ApoE(−/−)) mice exhibit hepatic steatosis and increased susceptibility to hepatic inflammation and advanced fibrosis [54]. They found in an experimental model the proinflammatory 5-lipoxygenase (5-LO) pathway to be up-regulated and thus causing liver inflammation and fibrogenesis. They also found that the inhibition of the 5-LO pathway results in a significant reduction in liver inflammation. Our data supports an up-regulation of ALOX5AP through 5-LO pathway due to DDC treatment in AJ, B6, and PWD, leading to an upregulation of the downstream component 5-HPETE in the model that is supported by our experimental data.

In conclusion, mRNA expression data combined with mathematical modeling of metabolic systems provides a useful tool to better understand cellular metabolism although the correlation between transcripts and proteins can deviate depending on cellular location, biological function, and organism [13]. The development of NAFLD is characterized by broad changes on the molecular level. Detailed analysis of three differently susceptible mouse strains, which reflect genetic diversity in humans, showed major deregulation of arachidonic acid metabolism. Detailed modeling of

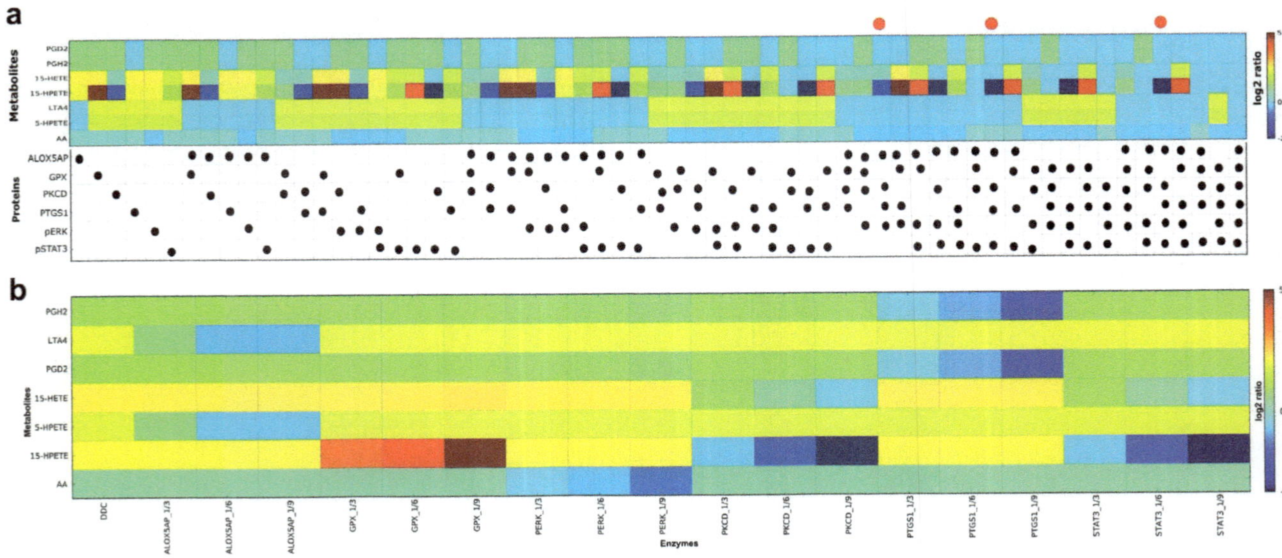

Figure 4. Identification of key regulatory enzymes (a) and drug testing (b) of the arachidonic acid/eicosanoid metabolism model. a) Key enzymes or enzyme combinations of the DDC model condition of AJ were reverted to control conditions of this strain to judge the effect on the change of the metabolite state. Black dots indicate enzymes or enzyme combinations that were reverted to control conditions. Red dots indicate those enzyme combination that are able to bring back the DDC-treated state to control conditions. **b)** *In silico* drug testing of the model by simulating down-regulation of individual enzyme concentrations as given by their respective expression value by 1/3rd-, 1/6th- and 1/9th of the DDC-treated state of AJ.

the arachidonic acid metabolism and model predictions of metabolic levels are in good agreement with experimental profiles when the model is initialized by the measured gene and protein expression data. The study identified deregulated genes, proteins, metabolites and affected pathways of NAFLD etiology in mouse and serves as an integrated resource of omics data for the development of computational models of the disease, such as AA and SAMe metabolism.

Supporting Information

Table S1 Susceptible genes (first column: only differentially expressed in AJ) and all differentially expressed genes of AJ, B6 and PWD mice due to DDC treatment.
(XLS)

Table S2 RNA-Seq expression data (RPKM values) of all three mice that were used for the principle component analysis (PCA analysis).
(XLS)

Table S3 Differentially expressed genes irrespective of strain.
(XLS)

Table S4 Results from RPPA analysis of a set of selected proteins.
(XLS)

Table S5 Metabolic profiles of a set of metabolites and t-test results for identification of metabolic differences.
(XLS)

Table S6 Results of pathway over-representation analysis of susceptibility genes using ConsensusPathDB.
(XLS)

Table S7 288 differentially expressed metabolic genes of the genome-scale metabolic network that were identified by edgeR over all three strains that were subsequently used in pathway over-representation analysis.
(XLS)

Table S8 Pathway over-representation analysis based on suppl. Table S7 using ConsensusPathDB.
(XLS)

Table S9 Pathway over-representation analysis based on metabolic profiles using ConsensusPathDB.
(XLS)

Table S10 Identification of differentially expressed genes between control and DDC-treatment over all strains of all genes of the AA model.
(XLS)

File S1 SBML file of the arachidonic acid/eicosanoid metabolic model.
(SBML)

Acknowledgments

We thank Peter Abuja for his contribution to the scientific discussion and Gintare Siaulyte for excellent technical assistance and performance of the animal experiments.

Author Contributions

Conceived and designed the experiments: MS KK JS IO MLY HS MT KZ HL CW. Performed the experiments: VP MS KK JS IO. Analyzed the data: VP MS HD KZ KK MR VA JS IO CW. Contributed reagents/materials/analysis tools: MS KK JS IO HH CW. Wrote the paper: VP MS CG CW.

References

1. Marchesini G, Bugianesi E, Forlani G, Cerrelli F, Lenzi M, et al. (2003) Nonalcoholic fatty liver, steatohepatitis, and the metabolic syndrome. Hepatol Baltim Md 37: 917–923. doi:10.1053/jhep.2003.50161.
2. Marra F, Gastaldelli A, Baroni GS, Tell G, Tiribelli C (2008) Molecular basis and mechanisms of progression of non-alcoholic steatohepatitis. Trends Mol Med 14: 72–81. doi:10.1016/j.molmed.2007.12.003.
3. Brunt EM, Neuschwander-Tetri BA, Oliver D, Wehmeier KR, Bacon BR (2004) Nonalcoholic steatohepatitis: histologic features and clinical correlations with 30 blinded biopsy specimens. Hum Pathol 35: 1070–1082.
4. Anderson N, Borlak J (2008) Molecular Mechanisms and Therapeutic Targets in Steatosis and Steatohepatitis. Pharmacol Rev 60: 311–357. doi:10.1124/pr.108.00001.
5. Denk H, Stumptner C, Zatloukal K (2000) Mallory bodies revisited. J Hepatol 32: 689–702.
6. Anstee QM, Goldin RD (2006) Mouse models in non-alcoholic fatty liver disease and steatohepatitis research. Int J Exp Pathol 87: 1–16. doi:10.1111/j.0959-9673.2006.00465.x.
7. Hoppe A (2012) What mRNA Abundances Can Tell us about Metabolism. Metabolites 2: 614–631. doi:10.3390/metabo2030614.
8. Kamburov A, Pentchev K, Galicka H, Wierling C, Lehrach H, et al. (2011) ConsensusPathDB: toward a more complete picture of cell biology. Nucleic Acids Res 39: D712–717. doi:10.1093/nar/gkq1156.
9. Kamburov A, Cavill R, Ebbels TMD, Herwig R, Keun HC (2011) Integrated pathway-level analysis of transcriptomics and metabolomics data with IMPaLA. Bioinforma Oxf Engl 27: 2917–2918. doi:10.1093/bioinformatics/btr499.
10. Ferrara CT, Wang P, Neto EC, Stevens RD, Bain JR, et al. (2008) Genetic Networks of Liver Metabolism Revealed by Integration of Metabolic and Transcriptional Profiling. PLoS Genet 4: e1000034. doi:10.1371/journal.pgen.1000034.
11. Connor SC, Hansen MK, Corner A, Smith RF, Ryan TE (2010) Integration of metabolomics and transcriptomics data to aid biomarker discovery in type 2 diabetes. Mol Biosyst 6: 909–921. doi:10.1039/b914182k.
12. Jauhiainen A, Nerman O, Michailidis G, Jörnsten R (2012) Transcriptional and metabolic data integration and modeling for identification of active pathways. Biostat Oxf Engl. Available: http://www.ncbi.nlm.nih.gov/pubmed/22699861. Accessed 11 September 2012.
13. Ghazalpour A, Bennett B, Petyuk VA, Orozco L, Hagopian R, et al. (2011) Comparative Analysis of Proteome and Transcriptome Variation in Mouse. PLoS Genet 7: e1001393. doi:10.1371/journal.pgen.1001393.
14. Li H, Durbin R (2010) Fast and accurate long-read alignment with Burrows-Wheeler transform. Bioinforma Oxf Engl 26: 589–595. doi:10.1093/bioinformatics/btp698.
15. Mortazavi A, Williams BA, McCue K, Schaeffer L, Wold B (2008) Mapping and quantifying mammalian transcriptomes by RNA-Seq. Nat Methods 5: 621–628. doi:10.1038/nmeth.1226.
16. Robinson MD, McCarthy DJ, Smyth GK (2010) edgeR: a Bioconductor package for differential expression analysis of digital gene expression data. Bioinformatics 26: 139–140. doi:10.1093/bioinformatics/btp616.
17. Unterwurzacher I, Koal T, Bonn GK, Weinberger KM, Ramsay SL (2008) Rapid sample preparation and simultaneous quantitation of prostaglandins and lipoxygenase derived fatty acid metabolites by liquid chromatography-mass spectrometry from small sample volumes. Clin Chem Lab Med CCLM FESCC 46: 1589–1597. doi:10.1515/CCLM.2008.323.
18. Brase J, Mannsperger H, Fröhlich H, Gade S, Schmidt C, et al. (2010) Increasing the sensitivity of reverse phase protein arrays by antibody-mediated signal amplification. Proteome Sci 8: 36. doi:10.1186/1477-5956-8-36.
19. Mannsperger HA, Gade S, Henjes F, Beissbarth T, Korf U (2010) RPPanalyzer: Analysis of reverse-phase protein array data. Bioinformatics 26: 2202–2203. doi:10.1093/bioinformatics/btq347.
20. Loebke C, Sueltmann H, Schmidt C, Henjes F, Wiemann S, et al. (2007) Infrared-based protein detection arrays for quantitative proteomics. PROTEOMICS 7: 558–564. doi:10.1002/pmic.200600757.
21. Wierling C, Herwig R, Lehrach H (2007) Resources, standards and tools for systems biology. Brief Funct Genomic Proteomic 6: 240–251. doi:10.1093/bfgp/elm027.
22. Klipp E, Liebermeister W, Wierling C, Kowald A, Lehrach H, et al. (2009) Systems Biology - A Textbook. Wiley-VCH, Weinheim.
23. Yang K, Ma W, Liang H, Ouyang Q, Tang C, et al. (2007) Dynamic simulations on the arachidonic acid metabolic network. PLoS Comput Biol 3: e55. doi:10.1371/journal.pcbi.0030055.
24. Gupta S, Maurya MR, Stephens DL, Dennis EA, Subramaniam S (2009) An integrated model of eicosanoid metabolism and signaling based on lipidomics flux analysis. Biophys J 96: 4542–4551. doi:10.1016/j.bpj.2009.03.011.

25. Michal G, editor (1998) Biochemical Pathways: An Atlas of Biochemistry and Molecular Biology. 1st ed. Wiley-Spektrum. 277 p.
26. Kanehisa M, Goto S, Sato Y, Furumichi M, Tanabe M (2012) KEGG for integration and interpretation of large-scale molecular data sets. Nucleic Acids Res 40: D109–114. doi:10.1093/nar/gkr988.
27. Hoops S, Sahle S, Gauges R, Lee C, Pahle J, et al. (2006) COPASI–a COmplex PAthway SImulator. Bioinformatics 22: 3067–3074. doi:10.1093/bioinformatics/btl485.
28. Charlton M, Viker K, Krishnan A, Sanderson S, Veldt B, et al. (2009) Differential expression of lumican and fatty acid binding protein-1: New insights into the histologic spectrum of nonalcoholic fatty liver disease. Hepatology 49: 1375–1384. doi:10.1002/hep.22927.
29. Kamburov A, Pentchev K, Galicka H, Wierling C, Lehrach H, et al. (2011) ConsensusPathDB: toward a more complete picture of cell biology. Nucleic Acids Res 39: D712–717. doi:10.1093/nar/gkq1156.
30. Mato JM, Martínez-Chantar ML, Lu SC (2013) S-adenosylmethionine metabolism and liver disease. Ann Hepatol 12: 183–189.
31. Sigurdsson MI, Jamshidi N, Steingrimsson E, Thiele I, Palsson BØ (2010) A detailed genome-wide reconstruction of mouse metabolism based on human Recon 1. BMC Syst Biol 4: 140. doi:10.1186/1752-0509-4-140.
32. Miller SB (2006) Prostaglandins in Health and Disease: An Overview. Semin Arthritis Rheum 36: 37–49. doi:10.1016/j.semarthrit.2006.03.005.
33. Hayaishi O (1991) Molecular mechanisms of sleep-wake regulation: roles of prostaglandins D2 and E2. FASEB J Off Publ Fed Am Soc Exp Biol 5: 2575–2581.
34. Anfuso CD, Lupo G, Romeo L, Giurdanella G, Motta C, et al. (2007) Endothelial cell-pericyte cocultures induce PLA2 protein expression through activation of PKCalpha and the MAPK/ERK cascade. J Lipid Res 48: 782–793. doi:10.1194/jlr.M600489-JLR200.
35. Ishihara K, Miyazaki A, Nabe T, Fushimi H, Iriyama N, et al. (2012) Group IVA phospholipase A2 participates in the progression of hepatic fibrosis. FASEB J Off Publ Fed Am Soc Exp Biol 26: 4111–4121. doi:10.1096/fj.12-205625.
36. Bhattacharjee A, Xu B, Frank DA, Feldman GM, Cathcart MK (2006) Monocyte 15-Lipoxygenase Expression Is Regulated by a Novel Cytosolic Signaling Complex with Protein Kinase C δ and Tyrosine-Phosphorylated Stat3. J Immunol 177: 3771–3781.
37. Geddis MS, Tornieri K, Giesecke A, Rehder V (2004) PLA2 and secondary metabolites of arachidonic acid control filopodial behavior in neuronal growth cones. Cell Motil Cytoskeleton 57: 53–67. doi:10.1002/cm.10156.
38. Lepley RA, Fitzpatrick FA (1994) Irreversible inactivation of 5-lipoxygenase by leukotriene A4. Characterization of product inactivation with purified enzyme and intact leukocytes. J Biol Chem 269: 2627–2631.
39. Aharony D, Redkar-Brown DG, Hubbs SJ, Stein RL (1987) Kinetic studies on the inactivation of 5-lipoxygenase by 5(S)-hydroperoxyeicosatetraenoic acid. Prostaglandins 33: 85–100.
40. Mahipal SVK, Subhashini J, Reddy MC, Reddy MM, Anilkumar K, et al. (2007) Effect of 15-lipoxygenase metabolites, 15-(S)-HPETE and 15-(S)-HETE on chronic myelogenous leukemia cell line K-562: Reactive oxygen species (ROS) mediate caspase-dependent apoptosis. Biochem Pharmacol 74: 202–214. doi:10.1016/j.bcp.2007.04.005.
41. Gan Q, Schones DE, Eun SH, Wei G, Cui K, et al. (2010) Monovalent and unpoised status of most genes in undifferentiated cell-enriched Drosophila testis. Genome Biol 11: R42. doi:10.1186/gb-2010-11-4-r42.
42. Jüni P, Rutjes AWS, Dieppe PA (2002) Are selective COX 2 inhibitors superior to traditional non steroidal anti-inflammatory drugs? BMJ 324: 1287–1288.
43. Cano A, Buqué X, Martínez-Uña M, Aurrekoetxea I, Menor A, et al. (2011) Methionine adenosyltransferase 1A gene deletion disrupts hepatic very low-density lipoprotein assembly in mice. Hepatol Baltim Md 54: 1975–1986. doi:10.1002/hep.24607.
44. Kalhan SC, Edmison J, Marczewski S, Dasarathy S, Gruca LL, et al. (2011) Methionine and protein metabolism in non-alcoholic steatohepatitis: evidence for lower rate of transmethylation of methionine. Clin Sci Lond Engl 1979 121: 179–189. doi:10.1042/CS20110060.
45. Broekman F, Giovannetti E, Peters GJ (2011) Tyrosine kinase inhibitors: Multi-targeted or single-targeted? World J Clin Oncol 2: 80–93. doi:10.5306/wjco.v2.i2.80.
46. Perrone MG, Scilimati A, Simone L, Vitale P (2010) Selective COX-1 inhibition: A therapeutic target to be reconsidered. Curr Med Chem 17: 3769–3805.
47. Titos E, Clària J, Planagumà A, López-Parra M, González-Périz A, et al. (2005) Inhibition of 5-lipoxygenase-activating protein abrogates experimental liver injury: role of Kupffer cells. J Leukoc Biol 78: 871–878. doi:10.1189/jlb.1204747.
48. Ulbrich H, Soehnlein O, Xie X, Eriksson EE, Lindbom L, et al. (2005) Licofelone, a novel 5-LOX/COX-inhibitor, attenuates leukocyte rolling and adhesion on endothelium under flow. Biochem Pharmacol 70: 30–36. doi:10.1016/j.bcp.2005.03.031.
49. Wang D, DuBois RN (2010) Eicosanoids and cancer. Nat Rev Cancer 10: 181–193. doi:10.1038/nrc2809.
50. Yao C, Sakata D, Esaki Y, Li Y, Matsuoka T, et al. (2009) Prostaglandin E2-EP4 signaling promotes immune inflammation through Th1 cell differentiation and Th17 cell expansion. Nat Med 15: 633–640. doi:10.1038/nm.1968.
51. Tosello-Trampont A-C, Landes SG, Nguyen V, Novobrantseva TI, Hahn YS (2012) Kuppfer cells trigger nonalcoholic steatohepatitis development in diet-induced mouse model through tumor necrosis factor-α production. J Biol Chem 287: 40161–40172. doi:10.1074/jbc.M112.417014.
52. Puri P, Wiest MM, Cheung O, Mirshahi F, Sargeant C, et al. (2009) The plasma lipidomic signature of nonalcoholic steatohepatitis. Hepatol Baltim Md 50: 1827–1838. doi:10.1002/hep.23229.
53. Giannitrapani L, Ingrao S, Soresi M, Florena AM, Spada EL, et al. (2009) Cyclooxygenase-2 Expression in Chronic Liver Diseases and Hepatocellular Carcinoma. Ann N Y Acad Sci 1155: 293–299. doi:10.1111/j.1749-6632.2009.03698.x.
54. Martínez-Clemente M, Ferré N, González-Périz A, López-Parra M, Horrillo R, et al. (2010) 5-lipoxygenase deficiency reduces hepatic inflammation and tumor necrosis factor alpha-induced hepatocyte damage in hyperlipidemia-prone ApoE-null mice. Hepatol Baltim Md 51: 817–827. doi:10.1002/hep.23463.
55. Scheer M, Grote A, Chang A, Schomburg I, Munaretto C, et al. (2011) BRENDA, the enzyme information system in 2011. Nucleic Acids Res 39: D670–676. doi:10.1093/nar/gkq1089.

16

Proteomic Evidences for Rex Regulation of Metabolism in Toxin-Producing *Bacillus cereus* ATCC 14579

Sabrina Laouami[1,2], Géremy Clair[1,2,3], Jean Armengaud[3], Catherine Duport[1,2]*

1 Avignon Université/INRA, SQPOV UMR408, Avignon, France, **2** INRA, SQPOV UMR408, Avignon, France, **3** Laboratoire de Biochimie des Systèmes Perturbés, CEA Marcoule, DSV-iBEB-SBTN-LBSP, Bagnols-sur-Cèze, France

Abstract

The facultative anaerobe, *Bacillus cereus*, causes diarrheal diseases in humans. Its ability to deal with oxygen availability is recognized to be critical for pathogenesis. The *B. cereus* genome comprises a gene encoding a protein with high similarities to the redox regulator, Rex, which is a central regulator of anaerobic metabolism in *Bacillus subtilis* and other Gram-positive bacteria. Here, we showed that *B. cereus rex* is monocistronic and down-regulated in the absence of oxygen. The protein encoded by *rex* is an authentic Rex transcriptional factor since its DNA binding activity depends on the NADH/NAD$^+$ ratio. *Rex* deletion compromised the ability of *B. cereus* to cope with external oxidative stress under anaerobiosis while increasing *B. cereus* resistance against such stress under aerobiosis. The deletion of *rex* affects anaerobic fermentative and aerobic respiratory metabolism of *B. cereus* by decreasing and increasing, respectively, the carbon flux through the NADH-recycling lactate pathway. We compared both the cellular proteome and exoproteome of the wild-type and Δ*rex* cells using a high throughput shotgun label-free quantitation approach and identified proteins that are under control of Rex-mediated regulation. Proteomics data have been deposited to the ProteomeXchange with identifier PXD000886. The data suggest that Rex regulates both the cross-talk between metabolic pathways that produce NADH and NADPH and toxinogenesis, especially in oxic conditions.

Editor: Alessandro Giuffrè, National Research Council of Italy (CNR), Italy

Funding: This work was supported by the university of Avignon, the Institut National de la Recherche Agronomique, the Commissariat à l'Energie atomique and the French Region of Provence-Alpes-Côtes d'Azur. The funders had no role in study design, data collection and analysis, decision to publish, or preparation of the manuscript.

Competing Interests: The authors have declared that no competing interests exist.

* Email: catherine.duport@univ-avignon.fr

Introduction

Bacillus cereus is a Gram-positive, facultative-anaerobe, rod-shaped endospore-forming human pathogen. Most of the reported illnesses involving *B. cereus* are food-borne intoxications, classified as emetic and diarrheal syndromes [1,2]. Diarrheal disease is due to vegetative outgrowth and secretion of various extracellular factors, including enterotoxins [3]. The most extensively studied diarrheal enterotoxins are hemolysin BL (Hbl), nonhemolytic enterotoxin (Nhe), and cytotoxin K (CytK) [4,5]. These enterotoxins are secreted via the Sec translocation pathway [6]. Although Hbl, Nhe and CytK are currently considered as the etiologic agents of diarrheal syndrome, other toxins, such as EntA, EntB and EntC, may also contribute to the pathogenicity of *B. cereus* [7,8]. To grow and produce virulence factors in the human intestine, *B. cereus* must adapt its metabolism, and regulates its proteome [9] in response to changes in oxygen availability. Indeed, *B. cereus* encounters oxic conditions in zones adjacent to the mucosal surface [10] and anoxic condition in the intestinal lumen [11]. Changes in oxygen availability can influence the relative levels of the dinucleotide, NAD$^+$ and NADH, in the cell, and such changes are sensed by the transcriptional regulator, Rex, in *B. subtilis* [12] as in other Gram-positive bacteria [13,14,15,16,17,18,19]. Depending on the cellular NAD$^+$/NADH ratio, Rex regulators modulate the expression of genes involved in fermentative metabolism, biofilm formation, and oxidative stress [18,20,21]. Structural studies of Rex proteins have identified dinucleotide-binding pockets in the C-terminal domain of the protein. NADH binding in this region leads to a conformational change in the Rex homodimer, triggering a displacement of Rex from its recognition sites on DNA, and thus leading to de-repression of the downstream genes [15,19]. A Rex homologue has been detected in the cellular proteome of *B. cereus* [9]. Furthermore, we found a canonical Rex binding motif [15] overlapping ResD and Fnr binding motifs [22,23,24] in the *ldhA* promoter region [25]. In *B. cereus*, ResD, Fnr and LdhA regulate both catabolism and enterotoxin production under both aerobic and anaerobic growth conditions, probably through a regulatory complex [26]. Rex could be, thus, a regulator of both catabolism and toxinogenesis in *B. cereus*. A role of Rex in toxinogenesis has not yet been reported in *B. cereus* or in any other organism.

In this study, we show that Rex from *B. cereus*, like its orthologues, is a transcriptional factor capable of interacting with DNA in an NADH/NAD$^+$-responsive manner. We demonstrate that *B. cereus* Rex is a key regulator of anaerobic fermentation, aerobic respiration, resistance against external reactive oxygen species, and toxinogenesis, by modulating the cellular and extracellular proteome in an oxygen-dependent manner. This study provides the most comprehensive experimental information on proteins whose synthesis was changed in the presence of Rex. All together, our results offer new information about the metabolic events that maximize *B. cereus* growth in environments with varying oxygen conditions.

Materials and Methods

Bacterial strains, media, and growth conditions

Escherichia coli TOP 10 (Invitrogen) was used as the host for cloning experiments, and *E. coli* SCS110 (Stratagene, La Jolla, CA) was used to prepare DNA for *B. cereus* transformation. *B. cereus* ATCC 14579 [27] was used as the parent strain for the construction of the *rex* deletion mutant. *E. coli* strains were grown at 37°C, with agitation, in Luria broth (LB). *B. cereus* strains were cultured in batches (three independent cultivations per strain) at two oxygen availabilities, i.e. $pO_2 = 0\%$ and $pO_2 = 100\%$ [8,9]. Each batch culture was inoculated with a subculture grown overnight at an initial optical density at 560 nm (OD_{560}) equal to 0.02. For *B. cereus* cultivation, the minimal MOD medium was supplemented with 30 mM glucose as the carbon source [28]. Anaerobic and aerobic batch cultures were performed at 37°C in a 2 L bioreactor (BioFlo/CelliGen 115, New Brunswick), and the working volume was maintained at 1.8 L. The pH was kept at a controlled value of 7.2 by automatic addition of 5 M KOH. *B. cereus* growth was monitored spectrophotometrically at 560 nm and calibrated with cell dry-weight measurements as previously described [25]. *B. cereus* cells were harvested by centrifugation when they reached their maximal growth rate ($\mu = \mu_{max}$) and immediately frozen until proteomic analysis. Supernatants were kept at −20°C for glucose and glucose-by-product assays and exoproteomic analysis.

Construction of the *B. cereus* Δ*rex* mutant strain

The deletion mutant, ATCC 14579 Δ*rex*, was constructed as follows. Two DNA fragments encompassing the 5′untranslated region (UTR) and 3′UTR of *rex* (BC 0291) were generated by PCR using primer pairs, rex1F (5′-GCCATGTTAATGTTTC-GATGTCT-3′) and rex2R (5′-CCCGGGATCTTTTAG-CAGTGGCTTGTGG-3′, *Sma*I restriction site is underlined), and rex3F (5′-CCCGGGGTTTACTTTTTGAAAAACTATC-CACAA-3′, *Sma*I restriction site is underlined) and rex4R (5′-TGCATTATGTATCGTGCTTTGG-3′), respectively. The resulting 840 bp 3′*SmaI* and 827 bp 5′*SmaI* DNA fragments were cloned into the TA cloning vector, pCR4-TOPO (Invitrogen, La Jolla, CA), generating plasmids, pCR4mut*rex*1 and pCR4mut*rex*2, respectively. The 840 bp DNA fragment encompassing the 5′UTR region of *rex* was isolated from pCR4mut*rex*1 with *Pst*I and *Sma*I, and subcloned into pCR4mut*rex*2 to generate pCR4mut*rex*3. A 1.5 kbp *Sma*I fragment containing the entire spectinomycin gene, *spc* [29], was purified from pDIA [25]. This purified fragment was ligated into *Sma*I-digested pCR4mut*rex*3. The resulting plasmid, pCR4mut*rex*4, was digested with *EcoR*I and the resulting 3067 bp 5′rexUTR-*spc*-3′rexUTR was subsequently inserted into the *EcoR*I site of pMAD [30]. The resulting plasmid was introduced into *B. cereus* cells by electroporation. The *rex* ORF was deleted and replaced with *spc* via a double-crossover event [30]. Chromosomal allele exchanges were confirmed by PCR with oligonucleotide primers located upstream and downstream of the DNA regions used for allelic exchange. To complement the *rex* gene in *trans*, a DNA fragment of 932 bp encompassing the *rex* ORF (630 bp) and its promoter region (219 bp) was first PCR amplified using the primer pairs, rex-compF (5′-GGATCCCGTTCGAAAGCGCGTTTACTTG-3′; the *BamH*I restriction site is underlined) and rexcompR (5′-GAGCTCGATTTTAATTTGGCACTTCGCC-3′; the *Sac*I restriction site is underlined), and then cloned into the pCRXL-TOPO plasmid (Invitrogen). The PCR fragment was then cut with *BamH*I and *Sac*I and ligated into pHT304 [31], digested with the same restriction enzymes. The integrity of the recombinant vector (pHT304*rex*) insert was verified by sequencing.

Phenotypic characterization of *B. cereus* Δ*rex* using the API-50CHB testsystem

The carbohydrate metabolism of the wild-type strain (WT) transformed or not with pHT304, Δ*rex* mutant transformed or not with pHT304 and Δ*rex* mutant transformed with pHT304*rex* was examined using API 50 CHB strips (BioMérieux SA, France). The results showed that WT and WT(pHT304) did not ferment turanose. In contrast, Δ*rex* and Δ*rex* (pHT304) showed a typical positive reaction in turanose test fermentation. Transformation of Δ*rex* mutant with pHT304*rex* inhibited the capacity of Δ*rex* to ferment turanose.

Measurement of glucose and by-product concentrations

Enzymatic test kits from Diffchamb (Lyon, France), R-Biopharm (Saint-Didier au Mont-d'Or, France), and Roche (Meylan, France) were used to analyze the glucose, lactate, ethanol, formate, acetate, and succinate concentrations in the supernatants of 4 mL cell cultures obtained after centrifugation at 10,000×*g* for 5 min (4°C). The specific glucose consumption rate, defined as the differential change in glucose concentration with time, was calculated from the equation, $q_{glucose} = \mu/Y_x$, where μ is the specific growth rate (h^{-1}) and Y_x is the biomass yield (g.mol carbon substrate^{-1}).

Gene expression analysis by RT-PCR

RT-PCR was performed using SYBR Green technology on a Lightcycler instrument (Roche applied Science) as described previously [32]. The primers used in this study have been described previously [8].

Purification of Rex

The *rex* ORF was amplified by PCR from *B. cereus* ATCC 14579 using the oligonucleotides, pET101*rex*F (5′-ACCATG-GATCAGCAAAAGATTCCA-3′) and pET101*rex*R (5′-TTGTGGATAGTTTTTCAAAAAGTAAAC-3′). The amplicon was introduced as a blunt-end PCR product into pET101/D-TOPO (Invitrogen). The integrity of the inserted sequence was confirmed by DNA sequencing. The resulting construct was transformed into *E. coli* BL21-CodonPlus(DE3)-RIL strain (Stratagene) for protein production. BL21-CodonPlus(DE3)-RIL cells carrying the pET101-*rex* expression plasmid were grown in 1 L LB medium containing 100 μg.mL^{-1} ampicillin, at 30°C with agitation (200 rpm) until the cell density reached an OD_{600} of about 0.5. After this, 0.5 mM isopropyl-β-D-thiogalactopyranoside (IPTG) was added to the culture, and growth was continued for four hours. The cells were harvested by centrifugation (6500 rpm for 10 min at 4°C), washed, and resuspended in 20 mL ice-cold extraction buffer consisting of 50 mM TRIS buffered at pH 8.0, 0.3 M NaCl, 10% glycerol, and a protease inhibitor cocktail (one tablet of Complete Mini, EDTA free, Roche). Cells were then incubated with 0.5 mg/mL lysozyme for 45 min under gentle agitation at 4°C and disrupted by sonication for 5 min at 4°C using a Vibra Cell ultrasonicator (Fisher Bioblock Scientific). The lysates were clarified by centrifugation at 9000 rpm for 30 min at 4°C and loaded onto a 2 mL

Co^{2+}-immobilized TALON metal affinity chromatography column (Clontech) equilibrated with the extraction buffer. The column was washed with 30 mL of extraction buffer containing 20 mM imidazole and Rex was eluted with 3 mL extraction buffer containing 200 mM imidazole. The eluted fraction was desalted by dialysis, concentrated using Nanosep 30 kDa-molecular-mass cutoff devices (Omega disc membrane; Pall filtron), and stored at −80°C until analysis. The purity of the protein was estimated to be above 90% by Coomassie blue-stained SDS-PAGE. The protein concentration was determined using a Bradford assay (Interchim) with bovine serum albumin as the reference.

Electrophoretic mobility shift assays (EMSAs)

Nucleic acid fragments containing the promoter regions of *ldhA* and *rex* were PCR-amplified using biotinylated forward primers, LdhAF (5′-ACCTGCTAATCCGATGATTG-3′) and RexF (5′-CAAGAATCGTTTCTGCACCG-3′), and nonbiotinylated reverse primers, LdhAR (5′-GGATCCAACTAATCCAGTAC-3′) and RexR (5′-GCTTACCAGAAAGAGATAAG-3′). The DNA used as negative control was a fragment of the ssuRNA BC0007 sequence (NC_004722), which was amplified with the biotinylated ssubioF (5′-GGTAGTCCACGCCGTAAACG-3′) and ssuR (5-GACAACCATGCACCACCTG-3′) primer pair. The 5′-labeled amplicons were purified using the High Pure PCR Product Purification Kit (Roche). Binding reactions were performed for 30 min at 37°C by incubating biotin-labeled DNA fragments (2 nM per reaction) with different amounts of Rex in 10 mM Tris-HCl buffered at pH 7.5, and containing 50 mM KCl, 2.5% glycerol, 5 mM $MgCl_2$ and 5 mg/L poly(dI−dC). The samples were resolved by electrophoresis on a 6% nondenaturing polyacrylamide gel and electrotransferred onto Hybond N+ Nylon membranes (Amersham). Biotin-labeled DNA was detected using the LightShift Chemiluminescent EMSA Kit (Pierce).

Proteomic sample preparation and nanoLC-MS/MS analysis of tryptic peptides

Three independent biological replicates were harvested for each of the two conditions (aerobiosis and anaerobiosis) and two strains (Δ*rex* mutant and its parent strain, ATCC 14579). The extracellular proteins of the 12 cultures were extracted by trichloroacetic acid precipitation [8]. The cellular proteins from the 12 samples were obtained as previously described [9]. The 24 resulting samples were subjected to SDS-PAGE, and then identified after trypsin proteolysis by nanoLC-MS/MS tandem mass spectrometry with an LTQ-Orbitrap XL mass spectrometer as previously described [8,9]. A total of 131 nanoLC-MS/MS runs were carried out to acquire the whole dataset. The MS/MS spectra were assigned to tryptic peptide sequences with the Mascot Daemon software (version 2.3.2; Matrix Science) with mass tolerances of 5 ppm on the parent ion and 0.5 Da on the MS/MS, fixed modification for carbamidomethylated cysteine, and variable modification for methionine oxidation. Mascot results were parsed with a *p*-value threshold below 0.05 for peptide identification and proteins were validated when at least two peptides were detected. The number of MS/MS spectra per protein recorded by nanoLC-MS/MS was extracted for each sample. In each condition, proteins were further considered for comparison only if peptides were seen in at least two of the three replicates. The resulting datasets were normalized taking into account the total protein concentration of the corresponding pellet and supernatant. Protein concentrations in *B. cereus* were determined using the Reducing agent Compatible Detergent Compatible (RCDC) protein assay (Bio-Rad) following the supplier's instructions. MS/MS spectral counts were compared with the TFold method using the PatternLab software program 2.0.0.13 [33] using a *p*-value cut-off set at 0.05. Log_2 (fold-change) were then calculated for comparisons and only the proteins for which the *p*-value was lower than the 0.05 cut-off were considered. The mass spectrometry proteomics data have been deposited to the ProteomeXchange Consortium (http://www.proteomexchange.org) via the PRIDE partner repository [34] with the dataset identifier PXD000856 and DOI 10.6019/PXD000856.

Exposure of bacteria to H_2O_2 and viability assays

Oxidative stress resistance of the *B. cereus* Δ*rex* mutant and the parent strain, ATCC 14579, was assessed by exposing aerobically grown (OD~0.4) and anaerobically grown (OD~0.2) cells to 20 and 5 mM H_2O_2, respectively. Samples were taken prior to oxidative stress (time zero) and after 20 min. Aliquots (100 μl) of the samples were diluted in H_2O, appropriate dilutions of the culture were plated onto LB agar, and after overnight incubation at 37°C the colony forming units (CFUs) were counted. All the experiments were performed at least in triplicate, and at least two technical replicates from each dilution step were carried out to determine the number of CFUs.

Results

Expression analysis of *rex* in *B. cereus* ATCC 14579 at two different pO_2

We identified BC_0291 in the *B. cereus* ATCC 14579 genome [27] as the homologue to the *Bacillus subtilis* transcriptional repressor, Rex [12,35,36], with 90% sequence identities. We established by 5′RACE PCR (Fig. S1) that a transcriptional start site (G) was located 41 bp from the translational start site (ATG) of the BC_0291 open reading frame. Upstream of this transcriptional start site, we identified a potential housekeeping σA-type promoter, TATACAN(17)TAAACT. The stop codon (TAA) overlapped an inverted repeat (AAAACGCAGAGG(N6)C-CTCTGCGTTTT; $\Delta G = -23.2$ kcal/mol) that may be a transcriptional terminator, suggesting that *rex* was monocistronically transcribed. Using real-time RT-PCR, we investigated the expression of *B. cereus rex* under aerobiosis and anaerobiosis in early-growing ATCC 14579 cells. The results indicated that *rex* expression was significantly higher under aerobiosis than under anaerobiosis (log_2 fold-change = 1.1, p-value<0.05).

Rex regulates glucose catabolism in *B. cereus*

Given that Rex orthologues function as NAD^+/NADH-responsive regulators of metabolism in several Gram-positive bacteria (such as *B. subtilis*, *S. aureus* and *S. coelicolor* [12,13,14,15,36,37]), we sought to determine if *B. cereus rex* contributed to anaerobic fermentative and aerobic respiratory growth. As shown in Table 1, *rex* deletion slightly increased the *B. cereus* growth rate under both anaerobiosis and aerobiosis without a significant change in final biomass. Under anaerobiosis, the increase of growth rate was not related to glucose consumption, suggesting perturbation of the fermentative pathways. In accordance with this hypothesis, the spectra of fermentation end products were significantly modified by the *rex* deletion (Table 1). We observed that formate, acetate and ethanol production were promoted at the expense of lactate production. In addition, the ethanol-to-acetate ratio increased in Δ*rex* mutant, while NADH recovery levels remained unchanged and ATP yields tended to increase. Under aerobiosis, respiratory Δ*rex* cells secreted ~8-fold higher amounts of lactate compared with the wild-type cells, while producing lower levels of acetate. We conclude that *rex* deletion affects the carbon flux through the NADH recycling lactate pathway under both anaerobiosis and aerobiosis (Fig. 1). Complementation for growth of the non-polar *rex* mutant in *B. cereus* ATCC 14579 was not possible using a pHT304-based plasmid. This is probably due to the presence of multiple copies of the *rex* promoter region and/or overexpression of *rex* [38]. Therefore, to validate the role of Rex in aerobic respiration and anaerobic fermentation in *B. cereus*, we deleted the *rex* gene of *B cereus* F4430/73 [39] and cultured the mutant strain under both aerobiosis and anaerobiosis. The results showed that the growth phenotype of Δ*rex* was the same in strains ATCC 14579 and F4430/73 (Table S1).

Rex regulates hydrogen peroxide resistance of *B. cereus* cells

To evaluate the impact of Rex deficiency on *B. cereus* resistance to oxidative stress, aerobically and anaerobically grown ATCC 14579 cells were exposed to hydrogen peroxide (20 and 5 mM H_2O_2, respectively). Figure 2 shows that aerobically grown Δ*rex* cells were less susceptible to H_2O_2 harmful effects than WT cells, while the anaerobically grown Δ*rex* cells were more susceptible. These data indicate that Rex restricts the resistance to oxidative stress of aerobically grown *B. cereus* cells, while sustaining a high resistance to oxidative stress of anaerobically growing cells. To validate the role of Rex in the *B. cereus* resistance to external H_2O_2, we repeated the experiments using the F4430/73 Δ*rex* strain (Fig. S2). The results showed that Rex deficiency similarly impacts F4430/73 and ATCC 14579 cells while the survival of F4430/73 anaerobically grown cells was strongly higher than the survival of ATCC 14579 cells.

DNA binding activity of *B. cereus* Rex

Previous studies have shown that proteins belonging to the Rex family function as dimers that bind to promoter regions containing DNA motifs with the 5′-TTGTGAAnnnnTTCACAA-3′ consensus sequence [21]. This typical binding motif is missing in the *rex* regulatory region of *B. cereus*, as reported for *S. aureus* and *B. subtilis* [12,15]. However, a putative binding motif with two mismatches compared with the known consensus motif was found upstream of the transcription start site of *ldhA* [25] in *B. cereus*, as in *S. aureus* and *B. subtilis*. To test whether *B. cereus* Rex binds to the *ldhA* promoter region, we overexpressed *B. cereus rex* in *E. coli*, purified the dimeric His6-tagged recombinant protein (Fig. S3), and performed electrophoretic mobility shift assays (EMSA). As a negative control, we assessed the binding of the Rex protein to the *ssu* DNA fragment (Fig. 3A). The results showed that a concentration of 0.6 μM Rex led to a complete shift of the *ldhA* fragment (2 nM). As expected, no shift was observed with the *rex* DNA fragment. The specificity of the binding was demonstrated by the absence of DNA shift with the negative control. The second EMSA experiment (Fig. 3B) showed that 10 mM NADH impaired Rex binding to the *ldhA* promoter region while 100 mM NAD^+ did not interfere with Rex-DNA complex formation. An NAD^+/NADH ratio of five led to an incomplete shift of the *ldhA* fragment. We conclude that in *B. cereus*, as in *B. subtilis* and *S. aureus* [15], Rex binding activity depends on the NADH/NAD^+ ratio.

Global comparative proteomics of *B. cereus* Δ*rex* and wild-type cells

To investigate the regulatory role of Rex in aerobic respiratory and anaerobic fermentative growth of *B. cereus* and decipher its role in toxinogenesis, we carried out a comparative proteome

Table 1. Results from controlled batch cultures of Δ*rex* mutant and its parent strain, *B. cereus* ATCC 14579[a].

	Anaerobic fermentative growth		Aerobic respiratory growth	
	WT[b]	Δ*rex*	WT[b]	Δ*rex*
Maximal specific growth rate (μ_{max}) (h^{-1})	0.84±0.04	0.90±0.02*	1.22±0.11	1.41±0.04*
Final biomass (g.liter^{-1})	0.78±0.01	0.83±0.06	2.38±0.12	2.31±0.12
$Y_{glucose}$ (g of cells. mol of glucose^{-1})	26±1	28±1	79±1	77±1
Maximal specific glucose consumption(mmol.g^{-1}.h^{-1})	32±2	32±2	15±1	18±2
Yields of end products (mol.mol glucose^{-1})[c]				
Lactate ($Y_{lactate}$)	1.33±0.01	1.25±0.02*	0.05±0.02	0.40±0.01*
Acetate ($Y_{acetate}$)	0.32±0.01	0.36±0.01*	1.05±0.01	0.69±0.03*
Formate ($Y_{formate}$)	0.23±0.01	0.27±0.01*	0.01±0.01	0.01±0.01
Ethanol ($Y_{ethanol}$)	0.13±0.01	0.18±0.01*	NZ[d]	NZ
Succinate ($Y_{succinate}$)	0.01±0.01	0.01±0.01	0.05±0.01	0.04±0.01
Ethanol versus Acetate	0.40	0.50*		
ATP yield[e]	2.10	2.15	ND[f]	ND
NADH recovered[g]	1.2	1.2	ND	ND

[a]Cells were grown under anaerobiosis (pO_2 = 0%) and full aerobiosis (pO_2 = 100%). Data are the means of triplicate measures obtained from three independent cultures.
[b]WT, wild-type parent strain *B. cereus* ATCC 14579.
[c]Yields of end products were calculated at the stationary phase.
[d]NZ, yield was below 0.01 mol.mol glucose^{-1}.
[e]ATP yield was calculated as moles of ATP produced per mole of consumed glucose, and was equal to $Y_{lactate}+Y_{ethanol}+2*Y_{acetate}$.
[f]ND, not determined.
[g]NADH recovery was calculated as the ratio of pathways producing NADH versus those consuming NADH (producing NAD^+), and was equal to (lactate+2×acetate+2×ethanol−formate)/(lactate+2×ethanol).
*$p<0.05$ *versus* WT in Student's t-test.

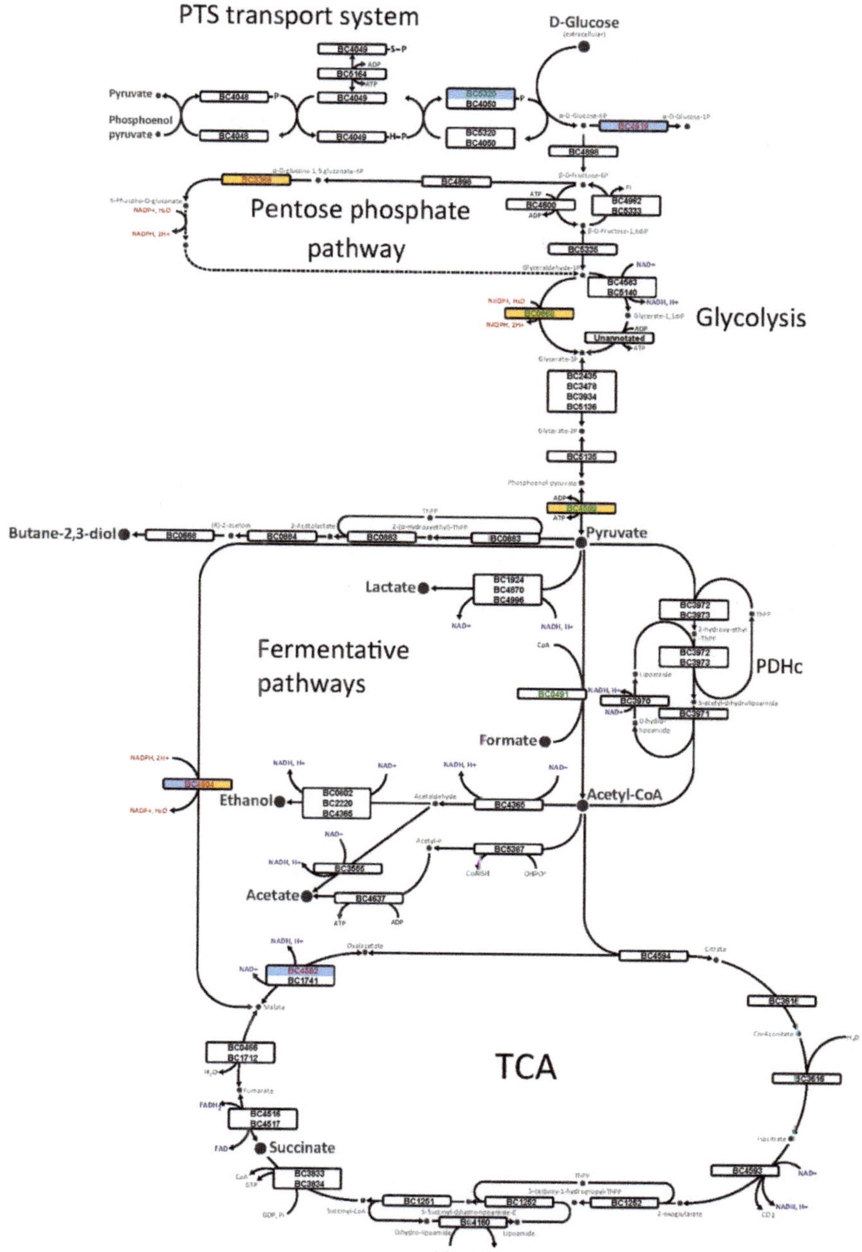

Figure 1. An overview of the main anaerobic and aerobic glucose catabolic pathways utilized by *B. cereus* ATCC 14579. The proteins detected in this study are indicated by their BC number. Protein names and functions are listed in Table S6. The form filling indicates fold-change values that satisfied the Student's *t*-test statistical criteria (*p*-value<0.05) in anaerobiosis (blue) and aerobiosis (yellow). Red and green BC numbers indicated significant increase and decrease, respectively, of abundance level of the proteins in Δ*rex* mutant compared with wild-type. The nicotinamide nucleotides are indicated in red ($NADP^+$/NADPH) or blue (NAD^+/NADH).

analysis from cells harvested in exponential growth phase ($\mu = \mu_{max}$). We compared the cellular and extracellular proteomes of aerobically and anaerobically grown Δ*rex* with those of its parental strain, with biological triplicates for each condition. A total of 1,620 proteins were identified in the whole study (Tables S2, S3, S4 and S5), with 1,450 proteins specifically detected in the whole-cell shotgun analysis and 170 further proteins observed in the exoproteome study. These proteins were quantified by spectral count as previously described [8,9]. The Rex-dependent changes in protein abundance, expressed as $\log_2$ of the ratio of a protein's abundance at a given condition (aerobiosis or anaerobiosis) relative to the wild-type, are presented in Tables S6 and S7, where the proteins with significant abundance changes based on the PatternLab Tfold student t-test (*p*-value below 0.05) are highlighted while proteins with non-significant changes are in grey color.

(i) Anaerobiosis. Among the whole cellular proteome (Table S6), 145 proteins showed significant abundance level changes in Δ*rex* cells relative to WT (*p*-values<0.05): 89 proteins were significantly up-regulated and 56 proteins were significantly down-regulated in Δ*rex* cells. In the exoproteome (Table S7), 41 proteins showed significant abundance level changes in Δ*rex* cells relative

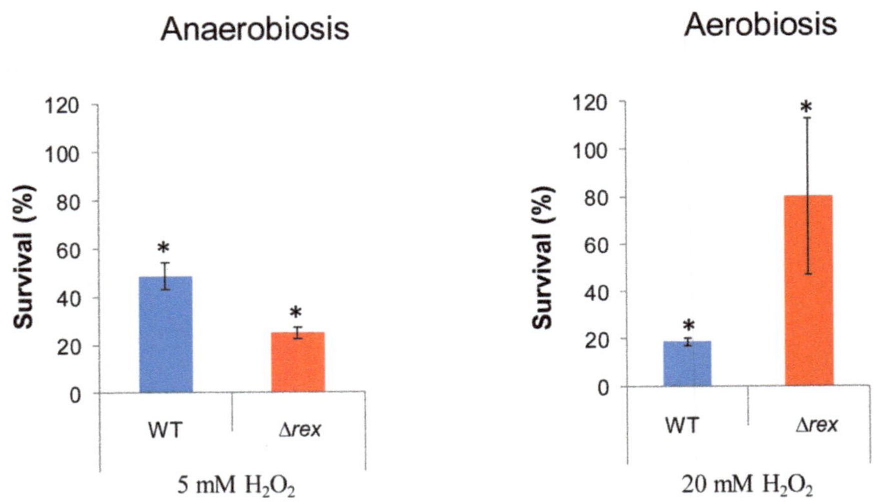

Figure 2. Survival of *B. cereus* cells towards external hydrogen peroxide under aerobiosis and anaerobiosis. Cells were grown in liquid cultures to the mid-exponential growth phase either aerobically or anaerobically, and subjected to 20 or 5 mM H_2O_2, respectively. Samples were taken at 0 and 20 min after H_2O_2 addition. Colony forming units per mL were counted and expressed as (N/No)×100. Error bars represented the standard deviation from three independent measures. Significant differences (p-value<0.05) between mutant and wild-type strains are indicated with asterisks.

to WT: 17 proteins were significantly up-regulated and 24 proteins were significantly down-regulated in Δ*rex* cells.

(ii) Aerobiosis. Table S6 indicates that 132 proteins showed significant abundance level changes in Δ*rex* cellular proteome (p-values<0.05): 82 proteins were significantly up-regulated and 50 proteins were significantly down-regulated in Δ*rex* cells. Table S7 indicates that 70 proteins showed significant abundance level changes in Δ*rex* cellular exoproteome (p-values<0.05): 35 were significantly up-regulated and 35 were significantly down-regulated in Δ*rex* cells.

(iii) Global analysis (Fig S4). Only 8 cellular proteins showed similar behavior under both aerobiosis and anaerobiosis (4 were up-regulated and 4 were down-regulated in Δ*rex* cells). Only 3 extracellular proteins were down-regulated under both anaerobiosis and aerobiosis. However, all the extracellular proteins that were up-regulated under anaerobiosis were also up-regulated under aerobiosis. Taken together, the data indicate that Rex (i) differently modulates the cellular proteome of *B. cereus* anaerobically and aerobically grown cells in terms of the identity of proteins that show significant abundance level changes, (ii) acts mainly as a repressor at the cellular proteome level under both

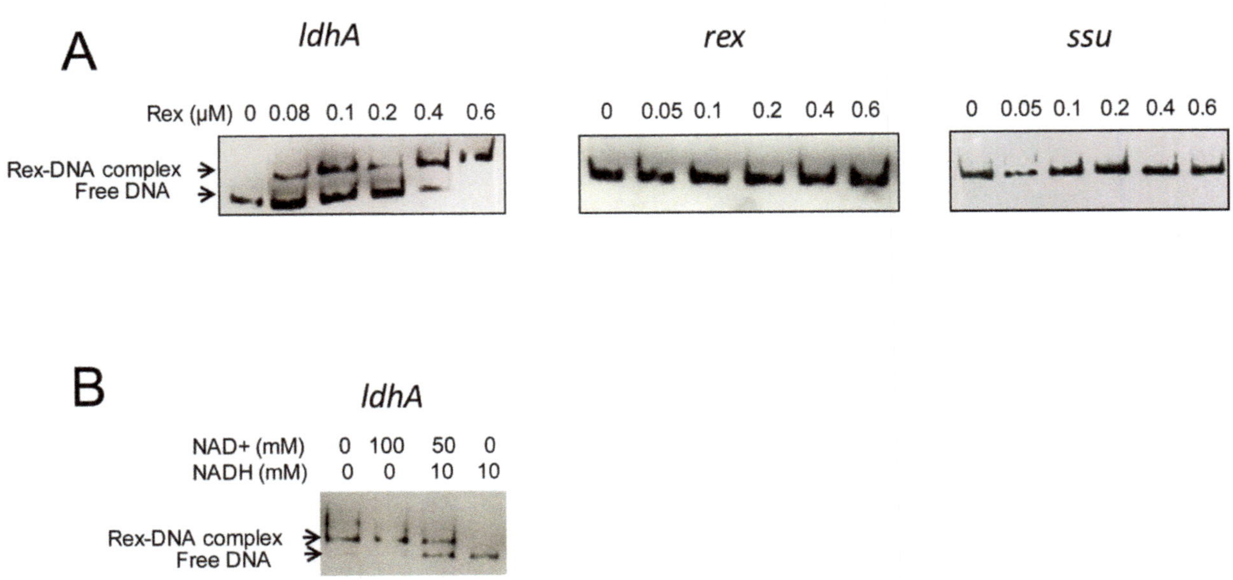

Figure 3. *In vitro* binding of Rex to *ldhA* promoter region determined by EMSA. Panel A. Biotin-labeled PCR products (2 nM) corresponding to *ldhA* and *rex* promoter regions and a negative control (a fragment of ssuRNA BC_0007 sequence) were mixed with increasing concentrations of purified Rex. **Panel B.** PCR product corresponding to the *ldhA* promoter region was incubated with 0.6 μM purified Rex and different concentrations of NAD^+ and NADH, as indicated.

Table 2. Cellular abundance level changes of cell membrane, transport, stress response, transcriptional regulation, virulence and phage-related proteins in Δ*rex* mutant relative to wild-type under anaerobiosis and aerobiosis.

				Abundance level change and significance			
				Anaerobiosis		Aerobiosis	
Accession no (NP)	Protein name	Gene number	Functionnal annotation	log_2 (fold-change)[a]	*p*-value	log_2 (fold-change)	*p*-value
			Cell wall, membrane and cell surface proteins				
NP_833630.1	-	BC_3910	N-acetylglucosaminyl transferase	**1,561**	**0,018**	−1,515	0,197
NP_835080.1	TagA	BC_5419	N-acetyl-beta-D-mannosaminyltransferase	**1,263**	**0,042**	−0,377	0,382
NP_832677.1	-	BC_2929	Peptidoglycan N-acetylglucosamine deacetylase	**0,832**	**0,019**	0,799	0,183
NP_835088.1	-	BC_5427	Cell wall biosynthesis glycosyltransferase	**−1,120**	**0,046**	−0,889	0,070
NP_834932.1	AmsK	BC_5269	Amylovoran biosynthesis AmsK	−0,713	0,112	**1,595**	**0,001**
NP_834404.1	-	BC_4698	Choline kinase	**0,949**	**0,032**	0,214	0,199
NP_834827.1	Lgt	BC_5163	Prolipoprotein diacylglyceryl transferase	**1,007**	**0,012**	**−0,811**	**0,004**
NP_834936.1	-	BC_5273	UDP-bacillosamine synthetase	**1,322**	**0,033**	0,632	0,110
NP_835093.1	-	BC_5432	Bactoprenol glucosyl transferase	**−1,218**	**0,032**	0,536	0,306
NP_835019.1	-	BC_5358	Collagen adhesion protein	0,918	0,088	**1,170**	**0,045**
NP_830682.1	-	BC_0896	S-layer protein	−0,234	0,180	**−0,286**	**0,017**
			Transport-related proteins				
NP_829996.1	SecE	BC_0115	Preprotein translocase subunit	**−0,971**	**0,044**	0,411	0,252
NP_834447.1	-	BC_4743	ABC transporter ATP-binding protein	**1,561**	**0,018**	−0,029	0,485
NP_834539.1	-	BC_4839	ABC transporter ATP-binding protein	**1,585**	**0,031**	0,856	0,115
NP_830141.1	Uup	BC_0290	ABC transporter ATP-binding protein	0,151	0,382	**0,705**	**0,045**
NP_830187.1	-	BC_0348	Methionine ABC transporter	**−1,690**	**0,004**	2,160	0,064
NP_830967.1	OppD	BC_1182	Oligopeptide transport ATP-binding protein	**0,651**	**0,002**	−0,044	0,421
NP_830660.1	-	BC_0874	Arginine ABC transporter ATP-binding protein	−0,396	0,349	**−1,218**	**0,024**
NP_832817.1	CutC	BC_3071	Copper tansport	**0,903**	**0,007**	−0,358	0,341
NP_830432.1	-	BC_0615	Di-/tripeptide transporter	**−0,837**	**0,047**	0,422	0,232
NP_830429.1	-	BC_0612	L-lactate permease	**0,748**	**0,026**	−0,644	0,261
			Stress response-related proteins				
NP_831381.1	-	BC_1603	Cold shock protein	**−1,120**	**0,037**	−0,474	0,277
NP_830286.1	TelA	BC_0447	Tellurite resistance protein	0,084	0,432	**0,840**	**0,008**
NP_833675.1	-	BC_3956	GTP-binding protein TypA/BipA	**0,310**	**0,049**	0,043	0,419
NP_833972.1	-	BC_4258	Hydroxyacylglutathione hydrolase	**0,824**	**0,028**	−0,578	0,213
NP_833576.1	-	BC_3855	Alkaline-shock protein	0,014	0,483	**0,740**	**0,026**
NP_834187.1	OhrA	BC_4475	Organic hydroxyperoxide protein A	0,189	0,340	**2,029**	**0,015**
NP_829959.1	Hsp15	BC_0062	Heat shock protein 15	0,799	0,082	**1,667**	**0,007**
NP_830216.1	AhpC	BC_0377	Alkyl hydroperoxide reductase C22	−0,136	0,426	**−1,029**	**0,007**
NP_830941.1	KatE	BC_1155	Catalase	**1,111**	**0,014**	−0,074	0,466
NP_834345.1	-	BC_4639	Thiol peroxidase	−0,044	0,463	**−0,737**	**0,034**
			Transcriptional Regulation				
NP_833888.1	ArgR2	BC_4174	Arginine biosynthesis repressor	**1,511**	**0,030**	1,269	0,079
NP_833771.1	-	BC_4053	Transcriptional regulator, GntR family	**0,895**	**0,019**	0,227	0,399
NP_833799.1	-	BC_4081	MarR family transcriptional regulator	0,029	0,464	**1,903**	**0,001**
NP_834861.1	-	BC_5197	MarR family transcriptional regulator	**0,345**	**0,029**	0,604	0,123
NP_832113.1	-	BC_2351	MerR family transcriptional regulator	0,485	0,276	**1,170**	**0,045**
NP_830532.1	TenA	BC_0742	Transcriptional activator	**0,791**	**0,046**	−0,621	0,193
NP_834137.1	-	BC_4425	Transcriptional regulator	**0,632**	**0,044**	−1,029	0,051
NP_831310.1	-	BC_1531	Transcriptional regulatory protein	−0,029	0,471	**1,480**	**0,004**
NP_831256.1	ResD	BC_1477	Transcriptional regulatory protein	**0,356**	**0,008**	−0,599	0,119
NP_830596.1	-	BC_0806	BigG family transcription antiterminator	0,485	0,276	**1,922**	**0,000**

Table 2. Cont.

Accession no (NP)	Protein name	Gene number	Functionnal annotation	Abundance level change and significance			
				Anaerobiosis		Aerobiosis	
				log2 (fold-change)[a]	*p*-value	log2 (fold-change)	*p*-value
			Virulence-related proteins				
NP_834769.1	Hlyl	BC_5101	Hemolysin I	−0,152	0,438	**2,611**	**0,001**
NP_832488.1	Npr2	BC_2735	Bacillolysin	−1,089	0,193	**−2,120**	**0,013**
NP_833332.1	HhoA	BC_3600	Protease	0,485	0,098	**−1,690**	**0,026**
NP_831147.1	-	BC_1366	SSEB protein	**1,163**	**0,036**	1,269	0,055
			Phage-related proteins				
NP_831667.1	-	BC_1894	Phage protein	**3,023**	**0,014**	−1,358	0,286
NP_831632.1	-	BC_1859	Phage protein	**2,526**	**0,039**	−0,644	0,261
NP_831635.1	-	BC_1862	Phage protein	**1,714**	**0,027**	−0,761	0,302
NP_831673.1	-	BC_1901	Phage protein	**1,727**	**0,012**	−1,889	0,205
NP_831676.1	-	BC_1904	Phage protein	**1,975**	**0,038**	−0,515	0,367

Only changes satisfying statistical criteria (p-value<0.05) at least in one growth condition are shown.
[a]The relative amount of each protein was determined using PatternLab software. Numbers in bold indicate that satisfied statistical criteria. See Table S6 for details on all the proteins showing significant abundance level changes in Δ*rex* cells.

aerobiosis and anaerobiosis and, has a stronger impact on the exoproteome of aerobically grown than anaerobically grown cells.

Insights into the cellular proteome of the Δ*rex* mutant

The list of cellular proteins showing significant abundance changes (p-values<0.05) due to the absence of Rex includes proteins related to glucose metabolism, amino acid, nucleotide and lipid metabolism, protein folding, response to stress, virulence, cell wall/membrane assembly, transport, and a number of proteins of unknown function under both aerobiosis and anaerobiosis (see Table S6 for details on protein characteristics, the fold-changes and p-values in each comparison).

(i) Impact of Rex on glucose metabolism-related proteins under anaerobic fermentative conditions. Figure 1 shows an overview of the glucose catabolic pathways of *B. cereus*. As illustrated in Fig. 1, the abundance level of key fermentative enzymes, such L-lactate dehydrogenases (Ldh; BC_1924, BC_4870, BC_4996), pyruvate formate lyase (PflB; BC_0491), alcohol dehydrogenase (BC_4365, BC_2220, BC_0802) and acetate kinase (BC_4634) were not significantly different in the mutant compared with wild-type cells. However, the lack of Rex resulted in a decrease of the abundance of the IIA component of the glucose-specific phosphotransferase (PTS) system (BC_5320, $\log_2 = -0.4$; p-value = 0.021). A significant decrease in abundance was also observed for the pyruvate-formate-lyase-activating enzyme, PflA (BC_0492; $\log_2 = -1$; p-value = 0.048). PflA is the sole enzyme able to activate the oxygen-sensitive PflB (BC_0491), which converts pyruvate to acetyl-CoA and formate (Fig. 1) [40]. In contrast, anaerobically grown Δ*rex* cells sustain a higher level of phosphoglucomutase (Pgm; BC_4919) than the wild-type cells ($\log_2 = 0.6$; p-value = 0.036) and thus, could sustain a higher carbon flux through gluconeogenesis than glycolysis (Fig. 1). As for Pgm, the malate dehydrogenase (Mdh; BC_4592) abundance level was increased in the absence of Rex under anaerobiosis ($\log_2 = 1.7$; p-value = 0.017). Mdh is one of the enzymes involved in the NADH-dependent reduction of oxaloacetate to malate, which is the first reaction in the succinate pathway (Fig. 1). The increase of Mdh abundance in the Δ*rex* mutant compared with wild-type was associated with the increase of NADPH-producing malic enzyme (ME; BC_4604; $\log_2 = 0.5$; p-value = 0.033), which converts malate into pyruvate. Taken together, our data suggest that Rex may modulate the abundance level of key metabolic enzymes to control carbon flow through fermentative pathways, possibly at the expense of the pathway that couples NADH-recycling Mdh with NADPH-producing ME.

(ii) Impact of Rex on glucose metabolism-related proteins under aerobic respiratory conditions. Figure 1 shows that ME was the sole enzyme showing an abundance level change in both aerobically and anaerobically grown mutant cells ($\log_2 = 0.4$; p-value = 0.01 under aerobiosis). 6-Phosphogluconolactonase (Pgl; BC_3368) showed a higher increase than ME ($\log_2 = 1.3$; p-value = 0.047) in Δ*rex* aerobically grown cells. The increase of the Pgl abundance level may have an impact on the activity of the NADPH-generating pentose phosphate pathway (PPP) in the Δ*rex* mutant (Fig. 1). Two glycolytic proteins showed significant abundance decreases in aerobically grown mutant cells: the NADPH-producing glyceraldehyde-3-phosphate-dehydrogenase (NADPH-GAPDH; BC_0868; ($\log_2 = -0.5$; p-value = 0.012) and pyruvate kinase (PK; BC_4599; ($\log_2 = -0.4$; p-value = 0.032), which catalyzes the last steps of glycolysis. The decrease of the NADPH-GAPDH level could increase the carbon flow through the NADH-producing GAPDH (BC_4583, BC_5333) and thus stimulate flux through the upstream glycolysis pathways, especially through the NADH-recycling lactate pathway (Fig. 1). Finally, Rex may modulate glycolytic flux by controlling the abundance level of key enzymes of glycolysis and pentose phosphate pathway when cells are grown under aerobiosis in glucose-containing medium.

(iii) Impact of Rex on other relevant cellular pathways. *Cell-wall and cell surface related-proteins.* Among the 20 proteins classified in the cell-wall/surface functional group, 8 showed significant abundance level changes (p<0.05) in anaerobically grown Δ*rex* mutant cells while 4 proteins showed significant abundance level changes in aerobically grown cells

(Table 2). A protein annotated as a putative prolipoprotein diacylglyceryl transferase (Lgt; BC_5163), which may be required for lipid modification of the cysteine residue present within the lipobox of prolipoproteins [41], showed a significant change of abundance level under both conditions: it was more highly detected in anaerobically grown cells and less detected in aerobically grown cells when Rex was absent.

Transport-related proteins. The largest change in transport-related protein abundance was observed in anaerobically grown Δ*rex* cells: 8 proteins showed significant abundance level changes ($p<0.05$). This suggests that the Rex-dependent control of the abundance pattern of transport related-proteins was stronger in anaerobic fermentative conditions than in aerobic respiratory conditions.

Transcriptional regulation-related proteins. Δ*rex* up-regulated the synthesis of 6 and 4 transcriptional regulators under anaerobiosis and aerobiosis, respectively. Among these, we found regulators of the Mer and Mar families, which control many cellular processes including oxidative stress response and virulence [43,44]. Interestingly, we noted that ResD was significantly up-regulated in anaerobically grown Δ*rex* cells but remained unchanged in aerobically grown Δ*rex* cells. ResD is involved in both metabolism and toxinogenesis in *B. cereus* [22,32].

Stress-response related proteins. Among the stress response-related proteins, the antioxidant protein, OhrA [7,9], was significantly up-regulated in aerobically grown Δ*rex* cells and its abundance remained constant under anaerobiosis (Table 2). Like OhrA, the ribosome-associated heat-shock protein, Hsp15 (BC_0062), showed, under aerobiosis, a significant change in abundance level. This protein could be involved in the recycling of free 50 S ribosomal subunits [45]. Interestingly, we noted significant changes in the abundance patterns of 50 S ribosomal subunits in aerobically grown Δ*rex* cells (Table S6). Unlike aerobically grown cells, anaerobically grown Δ*rex* cells exhibited higher levels of catalase, KatE (BC_1155), lower levels of the BC1603 protein, which is functionally annotated as a cold-shock protein and no significant level change of alkyl hydroperoxide reductase (AhpC, BC_0377). Clearly, Rex controls the abundance pattern of stress-related proteins in an oxygen-dependent manner.

Phage-related proteins. Five phage-related proteins were strongly up-regulated in anaerobically grown Δ*rex* cells ($\log_2>1.5$, p-value<0.05) and their abundance remained constant in aerobically grown Δ*rex* cells. The functions of these proteins are unknown but they are considered to be potentially beneficial to *B. cereus* in facing adverse environmental conditions [42].

Virulence-related proteins. Table 2 shows that *rex* deletion has a significant impact on the abundance levels of hemolysin I (Hly I), neutral protease Npr2 (BC_2735), protease HhoA (BC_3600), and SSEB (BC_1366) proteins under aerobiosis, all of which are predicted to be secreted virulence factors [4,46].

Insights into the extracellular proteome of the Δ*rex* mutant

(i) Focus on toxin-related proteins. Although the intracellular abundance level of HlyI was higher in the aerobically grown Δ*rex* mutant strain, the extracellular level of this protein was not significantly different in the mutant compared with wild-type cells (Table S7). This was also the case for EntB, a putative enterotoxin [9]. However, 12 other toxin-related proteins showed a significantly higher abundance level in the exoproteome of aerobically grown Δ*rex* cells than wild-type cells (Table 3). Some of these proteins also showed a higher abundance level under anaerobiosis, but to a lesser extent, *i.e.* the three components of Nhe (NheA, NheB and NheC), EntA, EntC and the L1 component of Hbl. From these data, we conclude that Rex regulates the toxinogenic profile of the *B. cereus* exoproteome in response to oxygen availability.

(ii) Other predicted extracellular proteins. Like the toxin-related proteins, many degradative enzymes, flagella components and proteins that are released from the cell wall, are more abundant in the exoproteome of the Δ*rex* mutant strain than wild-type cells, especially under aerobiosis (Table 4). All of these proteins are predicted to be extracellular proteins and, except for flagella components, all of them encompass signal peptides in their N-terminal primary structure and/or transmembrane helices. One metabolism-related protein, the subunit α of F0F1 ATP synthase (AtpA), and three proteins of unknown function (BC_1894, BC_5360 and BC_4062) also showed a higher abundance level in the exoproteome of aerobically grown Δ*rex* cells. Except for BC_1894, which was annotated as a prophage protein and thus could be secreted via holin(-like) pathway [47], these proteins contain a signal peptide, and thus are predicted to be secreted by classical export pathways.

(iii) Predicted cytoplasmic proteins. In contrast to proteins predicted to be classically secreted, many proteins predicted to be cytoplasmic proteins, including metabolic proteins, stress response-related proteins, translation-related proteins, and other proteins were significantly less abundant in the exoproteome of the Δ*rex* mutant than the parental strain (Table 4). We noted that most of the cytoplasmic proteins that showed a significant abundance level decrease under aerobiosis did not show significant abundance level changes under anaerobiosis, and vice-versa (Table 4 and Table S7). Finally, the percentage of predicted cytoplasmic proteins significantly decreased in the Δ*rex* exoproteome compared with the wild-type exoproteome in the presence of oxygen, while the percentage of typically secreted proteins (toxin-related proteins and degradative enzymes and adhesins) was significantly higher (Fig. 4).

The absence of Rex does not affect the mRNA level of toxin-related proteins

An explanation for the increased abundance levels of toxin-related proteins in the Δ*rex* mutant exoproteome could be a change in their mRNA levels. We thus tested whether transcription of genes encoding these proteins was altered in the Δ*rex* mutant compared with wild-type. Table 3 indicates that *rex* deletion significantly down-regulated *entC* and *entFM* under aerobiosis, but did not significantly change transcription of all other toxin-related genes. These transcriptomic results are not concordant with the proteomic data. This suggests that toxin-related protein abundance may be mainly modulated at the post-transcriptional level when Rex is lacking.

Discussion

B. cereus Rex is an authentic Rex transcriptional factor because (i) it is able to bind to promoter fragments *in vitro*, indicating that at least in some cases the effect of Rex is mediated by direct interaction with the promoters, (ii) it impacts the cellular metabolism and oxidative stress tolerance, and (iii) it acts mainly as a repressor at the proteomic level [12,13,14,15,20,48]. However, *B. cereus* Rex also plays an important role in fully aerobic respiratory metabolism and it modulates the toxinogenic profile of exoproteome.

When oxygen was supplied to pH-regulated batch cultures at $pO_2 = 100\%$, *B. cereus* metabolizes glucose to carbon dioxide by oxidation of glycolytic pyruvate in the tricarboxylic acid (TCA) cycle. This reaction produces NADH, which then fuels oxidative

Table 3. Changes in mRNA and protein abundance of toxins and putative toxins induced by Δ*rex* mutation under anaerobic and aerobic growth conditions[#].

Protein name	Gene	Log_2(fold-change) in *B. cereus* Δ*rex* vs wild-type[a]					
		Anaerobiosis			Aerobiosis		
		mRNA	Protein		mRNA	Protein	
			cellular	extracellular	s	cellular	extracellular
CytK	BC_1110	NS	NS[a]	NS	NS	NS	**1,72 (*p*=0.001)**
EntA	BC_5239	NS	NS	**0,29 (*p*=0.035)**	NS	NS	**0,53 (*p*=0.001)**
EntB	BC_2952	NS	NS	NS	NS	NS	NS
EntC	BC_0813	NS	NS	**0,37 (*p*=0.042)**	**−1,29 (p<0.05)**	NS	**0,75 (*p*=0.002)**
EntFM	BC_1953	NS	NS	NS	**−1,49 (p<0.05)**	NS	**0,68 (*p*=0.000)**
HblB	BC_3102	NS	NS	NS	NS	NS	**1,38 (*p*=0.001)**
HblB′	BC_3101	NS	NS	NS	NS	NS	**1,28 (*p*=0.002)**
HblL1	BC_3103	NS	NS	**0,32 (*p*=0.039)**	NS	NS	**1,09 (*p*=0.006)**
HblL2	BC_3104	NS	NS	NS	NS	NS	**1,29 (*p*=0.000)**
HlyI	BC_5101	NS	NS	NS	NS	**2,61(*p*=0.001)**	NS
HlyII	BC_3523	NS	NS	NS	NS	NS	**2,45 (*p*=0.000)**
NheA	BC_1809	NS	NS	**0,47 (*p*=0.003)**	NS	NS	**1,35 (*p*=0.002)**
NheB	BC_1810	NS	NS	**0,42 (*p*=0.008)**	NS	NS	**1,34 (*p*=0.001)**
NheC	BC_1811	NS	NS	**0,83 (*p*=0.001)**	NS	NS	**1,70 (*p*=0.001)**

[a]Concerning mRNA levels, each log_2(fold-change) represents the mean level of mRNA in the Δ*rex* mutant samples relative to the mean level in the wild-type sample. The mean values were obtained from three measurements done on triplicate independent cultures. Only significant log_2 ratios are indicated in bold. Concerning protein levels, the relative amount of each protein in Δ*rex* mutant compared to its parental strain was determined using PatternLab software. Numbers in bold indicate data satisfied statistical criteria (*p*-value<0.05). The *p*-values were indicated in brackets. Plus and minus indicate increased and decreased abundance levels, respectively. NS: no significant change was observed. For details, see Tables S6 and S7.

Table 4. Changes in protein abundances in Δ*rex* exoproteome compared to wild-type exoproteome under anaerobiosis and aerobiosis.

Protein name	Gene	Accession no. (NP)	Functional annotation	TM domain	Export Signal peptide	Log_2 fold-change and significance[a]			
						Anaerobiosis		Aerobiosis	
						log_2-fold-change	*p*-value	log_2-fold-change	*p*-value
Degradative enzymes and adhesins									
ColC	BC_0556	NP_830373	Collagenase	N	Y	0,4	0,03	1,9	0,00
NprB	BC_5351	NP_835012	Bacillolysin	N	Y	0,67	0,00	1,97	0,00
NprP2	BC_2735	NP_832488	Bacillolysin	N	Y	1,58	0,00	2,6	0,01
HhoA	BC_3600	NP_833332	Protease	N	Y	0,176	0,241	0,6	0,03
Tgc	BC_1991	NP_831760	Putative murein endopeptidase	N	Y	0,45	0,01	1,71	0,00
PlcB	BC_0670	NP_830483	phospholipase C	N	Y	0,6	0,00	2,25	0,00
PlcA	BC_3761	NP_833485	Phosphatidylinositol phosphodiesterase	N	Y	0,48	0,02	2,55	0,00
InhA2	BC_0666	NP_830479	Immune inhibitor A precursor	N	Y	1,59	0,01	3,43	0,00
PgdA	BC_2929	NP_832677	Peptidoglycan deacetylase	N	Y	0,585	0,055	0,87	0,04
Sph	BC_0671	NP_830484	Sphingomyelin phosphodiesterase	N	Y	0,82	0,03	2,82	0,00
Flagella components									
FlaA	BC_1659	NP_831436	Flagellin	N	N	0,516	0,117	0,6	0,02
FlgK	BC_1636	NP_831414	Flagellar hook-associated protein	N	N	0,96	0,01	1,04	0,05
FlgL	BC_1637	NP_831415	Flagellar hook-associated protein	N	N	0,78	0,04	1,28	0,00
FliD	BC_1638	NP_831416	Flagellar capping protein	N	N	0,526	0,063	1,22	0,00
FlgB	BC_1641	NP_831419	Flagellar basal body rod protein	N	N	0,111	0,411	0,9	0,01
Cell wall and membrane-related proteins									
FtsI	BC_4270	NP_833984	Penicillin-binding protein	Y	N	0,189	0,145	1,73	0,02
YvgJ1	BC_2932	NP_832680	Phosphoglycerol transferase	Y	N	−0,03	0,462	1,25	0,03
YvgJ2	BC_5232	NP_834895	phosphoglycerol transferase	Y	N	−0,03	0,463	0,71	0,00
CwlD	BC_5196	NP_834860	N-acetylmuramoyl-L-alanine amidase	N	Y	0,485	0,022	0,67	0,00
CwlB	BC_0902	NP_830688	N-acetylmuramoyl-L-alanine amidase	N	Y	1,091	0,187	−1,1	0,04
Metabolism									
Eno	BC_5135	NP_834803	Phosphopyruvate hydratase	N	N	−1,2	0,01	−0,12	0,385
PfkA	BC_4600	NP_834306	6-Phosphofructokinase	N	N	−0,9	0,04	−0,18	0,377
PykA	BC_4599	NP_834305	Pyruvate kinase	N	N	−0,15	0,429	−4,6	0,01
Pgi	BC_4898	NP_834571	Glucose-6-phosphate isomerase	N	N	−1	0,04	0,014	0,493
AtpA	BC_3230	NP_832971	F0F1 ATP synthase subunit alpha	N	Y	−0,03	0,448	1,75	0,00
GapA	BC_5140	NP_834805	Glyceraldehyde-3-phosphate dehydrogenase	N	N	−0,62	0,144	−2,9	0,00
PtA	BC_5387	NP_835048	Phosphotransacetylase	N	N	−1,6	0,01	−2,1	0,03

Table 4. Cont.

Protein name	Gene	Accession no. (NP)	Functional annotation	TM domain	Export Signal peptide	Log_2 fold-change and significance[a]			
						Anaerobiosis		Aerobiosis	
						log_2-fold-change	*p*-value	log_2-fold-change	*p*-value
GapN	BC_0868	NP_830654	NADP-dependent glyceraldehyde-3-phosphate dehydrogenase	N	N	−3,6	0,00	−1,32	0,054
FbA	BC_5335	NP_834997	Fructose-bisphosphate aldolase	N	N	−0,76	0,298	−3,6	0,03
Hpr	BC_4049	NP_833767	Phosphocarrier protein HPr	N	N	−2,4	0,03	−1,18	0,079
Bdh	BC_0868	NP_830481	(R,R)-butanediol dehydrogenase	N	N	−3,32	0,06	−3,6	0,02
PtsA	BC_4048	NP_833766	Phosphotransferase	N	N	−2,12	0,058	−3,2	0,00
SfcA	BC_4604	NP_834310	NAD-dependent malic enzyme	N	N	−0,86	0,187	−1,6	0,04
GlmM	BC_0188	NP_830056	Phosphoglucosamine mutase	N	N	−1,32	0,058	−3,8	0,00
GdhA	BC_4162	NP_833877	Leucine dehydrogenase	N	N	−1,8	0,04	−2,4	0,081
-	BC_4366	NP_834078	Cystathionine beta-lyase	N	N	0,151	0,447	−2	0,03
AdsS	BC_5468	NP_835123	Adenylosuccinate synthetase	N	N	−1,8	0,04	−1,15	0,202
-	BC_3799	NP_833521	Aspartate-semialdehyde dehydrogenase	N	N	−2,56	0,079	−3,2	0,01
ThiG	BC_0749	NP_830539	Thiazole synthase	N	N	ND	ND	−2,2	0,01
-	BC_0071	NP_829966	Phosphoribosyltransferase	N	N	−0,23	0,425	−3,5	0,00
DeoD	BC_1463	NP_831242	Purine nucleoside phosphorylase	N	N	−2,6	0,02	−1,06	0,185
-	BC_0015	NP_829919	Pyridoxine biosynthesis protein	N	N	−1,8	0,01	−3,6	0,00
-	BC_3981	NP_833700	tetrahydrodipicolinate -acetyltransferase	N	N	ND	ND	−3,3	0,00
MoaB	BC_4758	NP_834462	Molybdenum cofactor	N	N	−0,94	0,187	−2,3	0,02
FrvB	BC_4571	NP_834277	Deblocking aminopeptidase	N	N	−3,2	0,00	−0,92	0,215
PepT	BC_4143	NP_833858	Peptidase T	N	N	−1,6	0,04	−1,94	0,061
Stress-related proteins									
AhpC	BC_0377	NP_830216	Alkyl hydroperoxide reductase C22	N	N	−0,94	0,154	−2,2	0,00
GrpE	BC_4313	NP_834025	Chaperon	N	N	−1,4	0,04	−0,38	0,335
GroEL	BC_0295	NP_830146	Chaperon	N	N	−0,69	0,214	−2,3	0,01
Dps2	BC_5044	NP_834714	Non-specific DNA-binding protein	N	N	−2,4	0,00	−1	0,165
Dps1	BC_2011	NP_831779	Non-specific DNA-binding protein	N	N	−1	0,04	0	0,499
Sod	BC_5445	NP_835106	Superoxide dismutase [Mn]	N	N	−2,4	0,03	−0,92	0,226
AhpF	BC_0376	NP_830215	Alkyl hydroperoxide reductase subunit F	N	N	ND	ND	−2,4	0,02
TerD	BC_0443	NP_830282	tellurium resistance protein	N	N	−2,7	0,00	−0,97	0,064
Translation									
RpsF	BC_5476	NP_835129	30 S ribosomal protein S6	N	N	−2,3	0,01	−2,32	0,073
RpsG	BC_0126	NP_830007	30 S ribosomal protein S7	N	N	0,566	0,2	−1,8	0,01

Table 4. Cont.

Protein name	Gene	Accession no. (NP)	Functional annotation	TM domain	Export Signal peptide	Log$_2$ fold-change and significance[a]			
						Anaerobiosis		Aerobiosis	
						log$_2$-fold-change	*p*-value	log$_2$-fold-change	*p*-value
RpsH	BC_0145	NP_830025	30 S ribosomal protein S8	N	N	1,709	0,059	−1,8	0,00
RpsJ	BC_0130	NP_830010	30 S ribosomal protein S10	N	N	ND	ND	−2,7	0,01
RplJ	BC_0119	NP_830000	50 S ribosomal protein L10	N	N	−0,58	0,16	−4,3	0,00
RplK	BC_5075	NP_834743	50 S ribosomal protein L11	N	N	−0,94	0,187	−2,4	0,02
RplL	BC_0120	NP_830001	50 S ribosomal protein L7/L12	N	N	−1,7	0,03	−1,1	0,03
RpsT	BC_4320	NP_834032	30 S ribosomal protein S20	N	N	ND	ND	−2	0,03
RplX	BC_0142	NP_830022	50 S ribosomal protein L24	N	N	−2,2	0,03	−0,79	0,154
RpmD	BC_0149	NP_830029	50 S ribosomal protein L30	N	N	−1,03	0,095	−2,4	0,02
Frr	BC_3822	NP_833543	Ribosome recycling factor	N	N	−1,6	0,04	−0,23	0,405
RaiA	BC_5190	NP_834854	SSU ribosomal protein S30P	N	N	−2,3	0,04	−2,6	0,02
EftS	BC_3824	NP_833545	Elongation factor Ts	N	N	1,263	0,204	−3,1	0,01
Transcriptional Regulation									
AbrB	BC_0042	NP_829939	Transcription state regulator	N	N	−1,15	0,082	−3,3	0,00
CodY	BC_3826	NP_833547	Transcriptional repressor	N	N	ND	ND	−2,6	0,02
-	BC_3728	NP_833453	DNA-binding protein HU	N	N	−0,23	0,422	−2,5	0,00
Other									
-	BC_1984	NP_831667	Phage protein	N	N	3,113	0,134	4,77	0,01
-	BC_1012	NP_830798	unknown	N	N	−0,81	0,291	−3,5	0,00
-	BC_5360	NP_835021	unknown	Y	Y	0,299	0,134	2,38	0,00
-	BC_0002	NP_829890	unknown	N	N	−4,1	0,00	−3,1	0,01
-	BC_2077	NP_831845	unknown	N	N	−4,6	0,00	−0,01	0,477
-	BC_4062	NP_833780	unknown	Y	N	0,31	0,363	0,9	0,01
-	BC_2705	NP_832458	unknown	N	N	−1,89	0,084	−3,6	0,00

Only changes satisfying statistical criteria (*p*-value<0.05) at least in one growth condition are shown.

[a]Each log$_2$ fold-change value represents the mean protein level of the Δ*rex* sample in relation to the wild-type sample. The relative amount of each protein was determined using PatternLab software. Plus and minus symbols indicate up-and down regulation of the protein, respectively. Numbers in bold indicate data that satisfied statistical criteria (*p*-value<0.05). ND: not detected. For details, see Table S7.

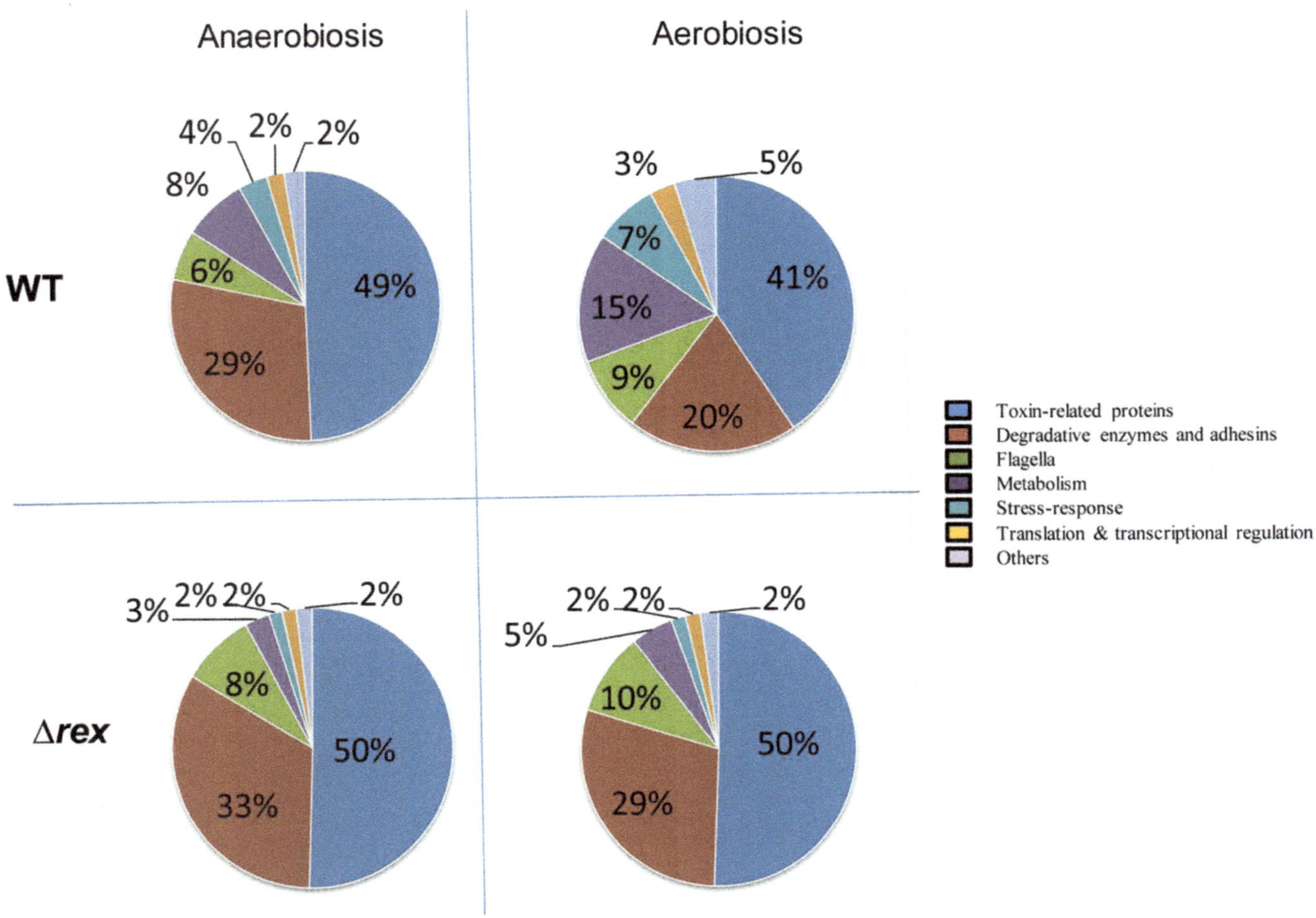

Figure 4. Functional distribution of proteins identified in Δ*rex* and wild-type *B. cereus* exoproteomes under aerobiosis and anaerobiosis. The diagrams represent the average proportion of each functional protein group based on total spectral counts (Table S5). For readability reasons, standard deviations (below 10%) are not shown.

phosphorylation to produce ATP, with minimal production of lactate and elimination of carbon excess through excretion of acetate. Rex deficiency facilitates the entry of carbon flow into the NADH-recycling lactate pathway at the expense of pyruvate oxidation into acetyl-CoA (Fig. 1). This promotes lactate production at the expense of acetate without impairment of TCA capacity, since acetate was always produced as a waste [28]. In the absence of oxygen or other external electron acceptors, ATP synthesis occurs at the level of substrate phosphorylation and NADH is reoxidized in terminal step fermentative reactions from pyruvate (Fig. 1). When grown in pH-controlled anaerobic batch cultures ($pO_2 = 0\%$), *B. cereus* produces lactate as the main glucose by-product to satisfy the demand for redox balance. Rex deficiency favored the production of more reduced metabolites, as evidenced by the increase in the ethanol to acetate ratio and the decrease of lactate production. This indicates that Rex regulates the carbon flow distribution at the pyruvate node by favoring carbon flow through the NADH-recycling lactate pathway at the expense of Pfl-dependent fermentative pathways under anaerobiosis. A key question is how Rex controls the carbon flow into the NADH-recycling lactate pathway in *B. cereus* cells grown in fully aerobic respiratory and anaerobic fermentative conditions? This control process could involve interplay of transcriptional and/or post-transcriptional regulation of key glycolytic enzymes (such as Pgm, BC_4919 under anaerobiosis and PK, BC_4599 under aerobiosis) and post-translational regulation of lactate dehydrogenase. Evidently, further intensive work is required to unravel the complexity of the regulatory network involving Rex. Finally, by controlling the entry of carbon flow into the lactate pathway, Rex controls the availability of glycolytic intermediates for macromolecular synthesis as well as supporting NADPH production under both aerobiosis and anaerobiosis.

In contrast to anaerobic fermentation, aerobic respiration is a major source of reactive oxygen species (ROS) generation [49]. *B. cereus*, like other facultative anaerobes, uses scavenging systems to control the level of ROS [50]. Sequestration of ROS leads to oxidative deactivation of these scavenging systems, which are then reactivated directly or indirectly depending on NADPH level [51]. By modulating the distribution pattern of antioxidant enzymes (Tables 2 and 4), and possibly NADPH production, Rex may thus modulate the effectiveness of antioxidative defense systems. Specifically, Rex may maximize antioxidant system activity under anaerobiosis while restricting this activity under aerobiosis; this could explain why the resistance of *B. cereus* cells towards external H_2O_2, when Rex is absent, is lower under anaerobiosis and higher under aerobiosis.

Except for HlyI, the abundance levels of classical extracellular proteins (such as toxins) were enhanced in *B. cereus* exoproteome when Rex was absent, especially under aerobiosis. In *B. cereus*, as in other Gram-positive bacteria, most extracellular proteins are synthesized as precursors, with typical N-terminal signal peptides, and exported from the cytoplasm by the Sec-dependent pathway [52]. This is the case for toxins [6]. However, unlike other toxins, HlyI contains a putative N-terminal twin arginine motif that could target the protein to the Tat secretion pathway instead of the Sec-dependent pathway [53,54]. Besides classical export routes, *B. cereus* possesses specialized protein export systems such the flagellar type III system [5,55], which translocates most flagellar components across the cytoplasmic membrane, and possibly the ESAT-6 secretion system, which mediates the secretion of virulence factors belonging to the WXG100 protein family, such as the protein BC_5360 [56,57]. A feature of these two specialized secretion systems is the lack of a classical signal peptide to direct the substrate protein for secretion. However, like the Sec-dependent pathway and unlike the TAT-dependent pathway, these protein export systems require chaperones to prevent premature folding, aggregation and degradation by cytoplasmic proteases [58,59]. Rex may thus influence protein-assisted export by modulating the distribution pattern of the components of export systems as observed in our proteomic data (Table 2, [60]). In addition, it has been reported that chaperones may be targets for intracellular ROS [61]. Keeping the functional integrity of these chaperones, specifically in the context of secretion, may thus depend on Rex-dependent NADPH availability. The role and contribution of NADPH in classical and specialized secretory mechanisms remains an open question that deserves further experiments.

Our results show that many cytoplasmic proteins predicted to be involved in active growth, either directly (such as metabolic enzymes, translational and post-translational-related proteins) or indirectly by preventing inappropriate gene expression (such as AbrB and CodY under aerobiosis [62]), were less abundant in *B. cereus* exoproteome when Rex was absent. We assume that cell lysis cannot account for changes in the abundance level of these cytoplasmic proteins since: (i) cell viability count did not change (data not shown), (ii) EF-Tu did not accumulate in extracellular medium (Table S7, [63]), and (iii) the abundance pattern of cytoplasmic proteins was dependent on growth conditions and did not always correlate with intracellular abundance changes (Table 2 and 4). Among the cytoplasmic proteins identified in *B. cereus* exoproteome, several have already been identified in the exoproteome of other bacteria, and many have been described as extracellular moonlighting components playing a role in bacterial virulence [64,65]. The non classical mechanism of secretion that could explain their release remains to be firmly described [47,66]. One hypothesis is that signal-less predicted cytoplasmic proteins might be secreted within outer membrane vesicles [63]. Such nanovesicles have been observed in *Bacillus anthracis* [67]. Interestingly, ATP is the primary metabolic factor involved in secretory vesicle movements. ATP generation is lower in fermenting cells than in non-fermenting cells, according to metabolism. In addition, Rex promotes ATP generation by enhancing NADH-producing pathways at the expense of NADPH pathways. Therefore, if ATP availability is a triggering factor for export of cytoplasmic proteins, this could explain the higher export of cytoplasmic proteins in aerobically grown cells compared to anaerobically grown cells, as well as the substantial decrease of cytoplasmic protein export under aerobiosis when Rex was absent.

In conclusion, Rex in *B. cereus* plays a pivotal role in controlling the cross-talk between the metabolic networks that control growth, oxidative defense machinery and extracellular accumulation of toxins. In the context of *B. cereus* natural environment, Rex may thus help the pathogen to (i) maintain its host alive until transmission to the next host can be achieved, by controlling its growth and production of its toxins [68], and (ii) better survive stress conditions under anaerobiosis, albeit to the detriment of maximizing its survival when oxygen is present [10,69,70].

Supporting Information

Figure S1 Genetic organization of *rex* chromosomal region of *B. cereus* ATCC 14579. Panel A. Large arrows represent open reading frames identified in strain ATCC 14579, and their orientation shows the transcriptional direction. **Panel B.** Nucleotide sequence of *rex* ORF (black capital letters), as well as promoter and terminator regions (blue letters), are shown. Transcriptional start site (+1), putative −35 and −10 motifs and putative terminator are underlined. Start and stop translation codons are in bold.
(TIF)

Figure S2 Survival of *B. cereus* F4430/73 cells towards external hydrogen peroxide. Cells were grown in liquid cultures to mid-exponential growth phase either aerobically or anaerobically, and subjected to 20 and 10 mM H_2O_2 stress, respectively. Samples were taken at 0 and 20 min. Colony forming units per mL were counted and expressed as (N/No)×100. Error bars represented the standard deviation from three independent measures. Significant differences ($p<0.05$) between mutant and wild-type strains are indicated with asterisks.
(TIF)

Figure S3 SDS-PAGE analysis of overproduced and purified *B. cereus* Rex. Rex purification fractions were analyzed by electrophoresis on an SDS-12% polyacrylamide gel after Coomassie Brillant Blue staining (Panel A) or Imperial Protein Stain (Thermo scientific, Panel B). Position and molecular weights (kDa) of markers (lanes 1) are given on the left. **Panel A**: Lane 1, standard proteins; Lane 2, soluble whole cell extract from *E. coli*. **Panel B**: Lane 1, standard proteins; Lane 2, purified protein after Co^{2+} IMAC.
(TIF)

Figure S4 Venn diagrams summarizing proteins that showed significant abundance level changes in the cellular and extracellular proteomes of Δ*rex* compared with wild-type *B. cereus* cells under anaerobiosis (blue) and aerobiosis (red). Proteins up- and down-regulated by Δ*rex* were distinguished.
(TIF)

Table S1 Results from controlled batch cultures of Δ*rex* mutants and its parent strain, *B. cereus* F4430/73.
(DOCX)

Table S2 List of assigned MS/MS spectra from the cellular proteomes.
(XLSX)

Table S3 List of assigned MS/MS spectra from the exoproteomes.
(XLSX)

Table S4 List of proteins identified in cellular proteomes and their normalized spectral count.
(XLSX)

Table S5 List of proteins identified in exoproteomes and their normalized spectral count.
(XLSX)

Table S6 Fold changes of cellular proteins and their statistical significance.
(XLSX)

Table S7 Fold changes of exoproteins and their statistical significance.
(XLSX)

Acknowledgments

S.L. held a fellowship from the Institut National de la Recherche Agronomique and the French Region of Provence Alpes Côtes d'Azur and G.C. held a fellowship from the Ministère de la Recherche et de l'Enseignement Supérieur.

Author Contributions

Conceived and designed the experiments: SL JA CD. Performed the experiments: SL GC CD. Analyzed the data: SL JA CD. Contributed reagents/materials/analysis tools: SL GC JA CD. Contributed to the writing of the manuscript: CD.

References

1. Bottone EJ (2010) *Bacillus cereus*, a volatile human pathogen. Clin Microbiol Rev 23: 382–398.
2. Stenfors Arnesen LP, Fagerlund A, Granum PE (2008) From soil to gut: *Bacillus cereus* and its food poisoning toxins. FEMS Microbiol Rev 32: 579–606.
3. Ceuppens S, Boon N, Uyttendaele M (2013) Diversity of *Bacillus cereus* group strains is reflected in their broad range of pathogenicity and diverse ecological lifestyles. FEMS Microbiol Ecol. 84: 433–50.
4. Ramarao N, Sanchis V (2013) The pore-forming haemolysins of *Bacillus cereus*: a review. Toxins (Basel) 5: 1119–1139.
5. Senesi S, Ghelardi E (2010) Production, secretion and biological activity of *Bacillus cereus* enterotoxins. Toxins (Basel) 2: 1690–1703.
6. Fagerlund A, Lindback T, Granum PE (2010) Bacillus cereus cytotoxins Hbl, Nhe and CytK are secreted via the Sec translocation pathway. BMC Microbiol 10: 304.
7. Clair G, Lorphelin A, Armengaud J, Duport C (2013) OhrRA functions as a redox-responsive system controlling toxinogenesis in *Bacillus cereus*. J Proteomics 94: 527–539.
8. Clair G, Roussi S, Armengaud J, Duport C (2010) Expanding the known repertoire of virulence factors produced by *Bacillus cereus* through early secretome profiling in three redox conditions. Mol Cell Proteomics 9: 1486–1498.
9. Clair G, Armengaud J, Duport C (2012) Restricting Fermentative Potential by Proteome Remodeling: An adaptative strategy evidenced In *Bacillus cereus*. Mol Cell Proteomics 11: M111 013102.
10. Marteyn B, West NP, Browning DF, Cole JA, Shaw JG, et al. (2010) Modulation of *Shigella* virulence in response to available oxygen in vivo. Nature 465: 355–358.
11. Moriarty-Craige SE, Jones DP (2004) Extracellular thiols and thiol/disulfide redox in metabolism. Annu Rev Nutr 24: 481–509.
12. Larsson JT, Rogstam A, von Wachenfeldt C (2005) Coordinated patterns of cytochrome bd and lactate dehydrogenase expression in *Bacillus subtilis*. Microbiology 151: 3323–3335.
13. Brekasis D, Paget MS (2003) A novel sensor of NADH/NAD+ redox poise in *Streptomyces coelicolor* A3(2). EMBO J 22: 4856–4865.
14. Gyan S, Shiohira Y, Sato I, Takeuchi M, Sato T (2006) Regulatory loop between redox sensing of the NADH/NAD(+) ratio by Rex (YdiH) and oxidation of NADH by NADH dehydrogenase Ndh in *Bacillus subtilis*. J Bacteriol 188: 7062–7071.
15. Pagels M, Fuchs S, Pane-Farre J, Kohler C, Menschner L, et al. (2010) Redox sensing by a Rex-family repressor is involved in the regulation of anaerobic gene expression in Staphylococcus aureus. Mol Microbiol 76: 1142–1161.
16. Sickmier EA, Brekasis D, Paranawithana S, Bonanno JB, Paget MS, et al. (2005) X-ray structure of a Rex-family repressor/NADH complex insights into the mechanism of redox sensing. Structure 13: 43–54.
17. Somerville GA, Proctor RA (2009) At the crossroads of bacterial metabolism and virulence factor synthesis in *Staphylococci*. Microbiol Mol Biol Rev 73: 233–248.
18. Vesic D, Kristich CJ (2013) A Rex family transcriptional repressor influences H2O2 accumulation by *Enterococcus faecalis*. J Bacteriol 195: 1815–1824.
19. Wang E, Ikonen TP, Knaapila M, Svergun D, Logan DT, et al. (2011) Small-angle X-ray scattering study of a Rex family repressor: conformational response to NADH and NAD+ binding in solution. J Mol Biol 408: 670–683.
20. Bitoun JP, Nguyen AH, Fan Y, Burne RA, Wen ZT (2011) Transcriptional repressor Rex is involved in regulation of oxidative stress response and biofilm formation by Streptococcus mutans. FEMS Microbiol Lett 320: 110–117.
21. Ravcheev DA, Li X, Latif H, Zengler K, Leyn SA, et al. (2012) Transcriptional regulation of central carbon and energy metabolism in bacteria by redox-responsive repressor Rex. J Bacteriol 194: 1145–1157.
22. Esbelin J, Armengaud J, Zigha A, Duport C (2009) ResDE-dependent regulation of enterotoxin gene expression in *Bacillus cereus*: evidence for multiple modes of binding for ResD and interaction with Fnr. J Bacteriol 191: 4419–4426.
23. Esbelin J, Jouanneau Y, Armengaud J, Duport C (2008) ApoFnr binds as a monomer to promoters regulating the expression of enterotoxin genes of *Bacillus cereus*. J Bacteriol 190: 4242–4251.
24. Zigha A, Rosenfeld E, Schmitt P, Duport C (2007) The redox regulator Fnr is required for fermentative growth and enterotoxin synthesis in *Bacillus cereus* F4430/73. J Bacteriol 189: 2813–2824.
25. Laouami S, Messaoudi K, Alberto F, Clavel T, Duport C (2011) Lactate dehydrogenase A promotes communication between carbohydrate catabolism and virulence in *Bacillus cereus*. J Bacteriol 193: 1757–1766.
26. Esbelin J, Jouanneau Y, Duport C (2012) *Bacillus cereus* Fnr binds a [4Fe-4 S] cluster and forms a ternary complex with ResD and PlcR. BMC Microbiol 12: 125.
27. Ivanova N, Sorokin A, Anderson I, Galleron N, Candelon B, et al. (2003) Genome sequence of *Bacillus cereus* and comparative analysis with *Bacillus anthracis*. Nature 423: 87–91.
28. Rosenfeld E, Duport C, Zigha A, Schmitt P (2005) Characterization of aerobic and anaerobic vegetative growth of the food-borne pathogen *Bacillus cereus* F4430/73 strain. Can J Microbiol 51: 149–158.
29. Murphy E (1985) Nucleotide sequence of a spectinomycin adenyltransferase AAD(9) determinant from *Staphylococcus aureus* and its relationship to AAD(3″) (9). Mol Gen Genet 200: 33–39.
30. Arnaud M, Chastanet A, Debarbouille M (2004) New vector for efficient allelic replacement in naturally nontransformable, low-GC-content, gram-positive bacteria. Appl Environ Microbiol 70: 6887–6891.
31. Arantes O, Lereclus D (1991) Construction of cloning vectors for *Bacillus thuringiensis*. Gene 108: 115–119.
32. Duport C, Zigha A, Rosenfeld E, Schmitt P (2006) Control of enterotoxin gene expression in *Bacillus cereus* F4430/73 involves the redox-sensitive ResDE signal transduction system. J Bacteriol 188: 6640–6651.
33. Carvalho PC, Fischer JS, Chen EI, Yates JR, 3rd, Barbosa VC (2008) PatternLab for proteomics: a tool for differential shotgun proteomics. BMC Bioinformatics 9: 316.
34. Vizcaino JA, Cote RG, Csordas A, Dianes JA, Fabregat A, et al. (2013) The PRoteomics IDEntifications (PRIDE) database and associated tools: status in 2013. Nucleic Acids Res 41: D1063–1069.
35. Wang Q, Ou MS, Kim Y, Ingram LO, Shanmugam KT (2010) Metabolic flux control at the pyruvate node in an anaerobic *Escherichia coli* strain with an active pyruvate dehydrogenase. Appl Environ Microbiol 76: 2107–2114.
36. Wang E, Bauer MC, Rogstam A, Linse S, Logan DT, et al. (2008) Structure and functional properties of the *Bacillus subtilis* transcriptional repressor Rex. Mol Microbiol 69: 466–478.
37. Schau M, Chen Y, Hulett FM (2004) *Bacillus subtilis* YdiH is a direct negative regulator of the cydABCD operon. J Bacteriol 186: 4585–4595.
38. Sprynski N, Felix C, O'Callaghan D, Vergunst AC (2012) Restoring virulence to mutants lacking subunits of multiprotein machines: functional complementation of a *Brucella virB5* mutant. FEBS Open Bio 2: 71–75.
39. Spira WM, Goepfert JM (1975) Biological characteristics of an enterotoxin produced by *Bacillus cereus*. Can J Microbiol 21: 1236–1246.
40. Garrigues C, Goupil-Feuillerat N, Cocaign-Bousquet M, Renault P, Lindley ND, et al. (2001) Glucose metabolism and regulation of glycolysis in *Lactococcus lactis* strains with decreased lactate dehydrogenase activity. Metab Eng 3: 211–217.
41. Wichgers Schreur PJ, Rebel JM, Smits MA, van Putten JP, Smith HE (2011) Lgt processing is an essential step in *Streptococcus suis* lipoprotein mediated innate immune activation. PLoS One 6: e22299.
42. Wang X, Kim Y, Ma Q, Hong SH, Pokusaeva K, et al. (2010) Cryptic prophages help bacteria cope with adverse environments. Nat Commun 1: 147.
43. Holley TA, Stevenson CE, Bibb MJ, Lawson DM (2013) High resolution crystal structure of Sco5413, a widespread actinomycete MarR family transcriptional regulator of unknown function. Proteins 81: 176–182.
44. Newberry KJ, Brennan RG (2004) The structural mechanism for transcription activation by MerR family member multidrug transporter activation, N terminus. J Biol Chem 279: 20356–20362.
45. Jiang L, Schaffitzel C, Bingel-Erlenmeyer R, Ban N, Korber P, et al. (2009) Recycling of aborted ribosomal 50 S subunit-nascent chain-tRNA complexes by the heat shock protein Hsp15. J Mol Biol 386: 1357–1367.
46. Gohar M, Faegri K, Perchat S, Ravnum S, Okstad OA, et al. (2008) The PlcR virulence regulon of *Bacillus cereus*. PLoS One 3: e2793.
47. Tjalsma H, Antelmann H, Jongbloed JD, Braun PG, Darmon E, et al. (2004) Proteomics of protein secretion by *Bacillus subtilis*: separating the "secrets" of the secretome. Microbiol Mol Biol Rev 68: 207–233.
48. Bitoun JP, Liao S, Yao X, Xie GG, Wen ZT (2012) The redox-sensing regulator Rex modulates central carbon metabolism, stress tolerance response and biofilm formation by *Streptococcus mutans*. PLoS One 7: e44766.
49. Seaver LC, Imlay JA (2001) Alkyl hydroperoxide reductase is the primary scavenger of endogenous hydrogen peroxide in *Escherichia coli*. J Bacteriol 183: 7173–7181.
50. Mols M, Abee T (2011) Primary and secondary oxidative stress in *Bacillus*. Environ Microbiol. 13: 1387–1394.

51. Mailloux RJ, Lemire J, Appanna VD (2011) Metabolic networks to combat oxidative stress in *Pseudomonas fluorescens*. Antonie Van Leeuwenhoek 99: 433–442.
52. Schneewind O, Missiakas D (2013) Sec-secretion and sortase-mediated anchoring of proteins in Gram-positive bacteria. Biochim Biophys Acta.
53. Yuan J, Zweers JC, van Dijl JM, Dalbey RE (2010) Protein transport across and into cell membranes in bacteria and archaea. Cell Mol Life Sci 67: 179–199.
54. Goosens VJ, Monteferrante CG, van Dijl JM (2013) The Tat system of Gram-positive bacteria. Biochim Biophys Acta.
55. Senesi S, Salvetti S, Celandroni F, Ghelardi E (2010) Features of *Bacillus cereus* swarm cells. Res Microbiol 161: 743–749.
56. Burts ML, Williams WA, DeBord K, Missiakas DM (2005) EsxA and EsxB are secreted by an ESAT-6-like system that is required for the pathogenesis of Staphylococcus aureus infections. Proc Natl Acad Sci U S A 102: 1169–1174.
57. Pallen MJ (2002) The ESAT-6/WXG100 superfamily – and a new Gram-positive secretion system? Trends Microbiol 10: 209–212.
58. Alonzo F, 3rd, Xayarath B, Whisstock JC, Freitag NE (2011) Functional analysis of the *Listeria monocytogenes* secretion chaperone PrsA2 and its multiple contributions to bacterial virulence. Mol Microbiol 80: 1530–1548.
59. Moran-Barrio J, Limansky AS, Viale AM (2009) Secretion of GOB metallo-beta-lactamase in *Escherichia coli* depends strictly on the cooperation between the cytoplasmic DnaK chaperone system and the Sec machinery: completion of folding and Zn(II) ion acquisition occur in the bacterial periplasm. Antimicrob Agents Chemother 53: 2908–2917.
60. Roberts IN, Lam XT, Miranda H, Kieselbach T, Funk C (2012) Degradation of PsbO by the Deg protease HhoA Is thioredoxin dependent. PLoS One 7: e45713.
61. Ezraty B, Aussel L, Barras F (2005) Methionine sulfoxide reductases in prokaryotes. Biochim Biophys Acta 1703: 221–229.
62. Frenzel E, Doll V, Pauthner M, Lucking G, Scherer S, et al. (2012) CodY orchestrates the expression of virulence determinants in emetic *Bacillus cereus* by impacting key regulatory circuits. Mol Microbiol 85: 67–88.
63. Meneses N, Mendoza-Hernandez G, Encarnacion S (2010) The extracellular proteome of *Rhizobium etli* CE3 in exponential and stationary growth phase. Proteome Sci 8: 51.
64. Copley SD (2012) Moonlighting is mainstream: paradigm adjustment required. Bioessays 34: 578–588.
65. Henderson B, Martin A (2013) Bacterial moonlighting proteins and bacterial virulence. Curr Top Microbiol Immunol 358: 155–213.
66. Yang CK, Ewis HE, Zhang X, Lu CD, Hu HJ, et al. (2011) Nonclassical protein secretion by *Bacillus subtilis* in the stationary phase is not due to cell lysis. J Bacteriol 193: 5607–5615.
67. Rivera J, Cordero RJ, Nakouzi AS, Frases S, Nicola A, et al. (2010) *Bacillus anthracis* produces membrane-derived vesicles containing biologically active toxins. Proc Natl Acad Sci U S A 107: 19002–19007.
68. ten Bokum AM, Movahedzadeh F, Frita R, Bancroft GJ, Stoker NG (2008) The case for hypervirulence through gene deletion in *Mycobacterium tuberculosis*. Trends Microbiol 16: 436–441.
69. Jakesevic M, Aaby K, Borge GI, Jeppsson B, Ahrne S, et al. (2011) Antioxidative protection of dietary bilberry, chokeberry and *Lactobacillus plantarum* HEAL19 in mice subjected to intestinal oxidative stress by ischemia-reperfusion. BMC Complement Altern Med 11: 8.
70. Marteyn B, Scorza FB, Sansonetti PJ, Tang C (2010) Breathing life into pathogens: the influence of oxygen on bacterial virulence and host responses in the gastrointestinal tract. Cell Microbiol 13: 171–176.

Characterization of PHB1 and Its Role in Mitochondrial Maturation and Yolk Platelet Degradation during Development of *Artemia* Embryos

Xiang Ye[1], Yang Zhao[2], Ling-Ling Zhao[1], Yu-Xia Sun[1], Jin-Shu Yang[1], Wei-Jun Yang[1]*

1 Key Laboratory of Conservation Biology for Endangered Wildlife of the Ministry of Education and College of Life Sciences, Zhejiang University, Hangzhou, Zhejiang, People's Republic of China, **2** Department of Molecular and Cell Biology, University of California, Berkeley, California, United States of America

Abstract

Background: To cope with harsh environments, crustaceans such as Artemia produce diapause gastrula embryos (cysts) with suppressed metabolism. Metabolism and development resume during post-diapause development, but the mechanism behind these cellular events remains largely unknown.

Principal Finding: Our study investigated the role of prohibitin 1 (PHB1) in metabolic reinitiation during post-diapause development. We found that PHB1 was developmentally regulated via changes in phosphorylation status and localization. Results from RNA interference experiments demonstrated PHB1 to be critical for mitochondrial maturation and yolk degradation during development. In addition, PHB1 was present in yolk platelets, and it underwent ubiquitin-mediated degradation during the proteolysis of yolk protein.

Conclusions/Significance: PHB1 has an indispensable role in coordinating mitochondrial maturation and yolk platelet degradation during development in *Artemia*. This novel function of PHB1 provides new clues to comprehend the roles of PHB1 in metabolism and development.

Editor: Bin He, Baylor College of Medicine, United States of America

Funding: This work was supported by the National Natural Science Foundation of China (31270424), the National Basic Research Program of China (973 Program, 2012CB944903), and 863 programs of China (2012AA10A401). The funders had no role in study design, data collection and analysis, decision to publish, or preparation of the manuscript.

Competing Interests: The authors have declared that no competing interests exist.

* Email: w_jyang@cls.zju.edu.cn

Introduction

Prohibitin 1 (PHB1), a prohibitin family protein, forms a large complex with PHB2 at the inner mitochondrial membrane. PHB1 has emerged as a key regulator of several cellular events, including mitochondrial morphogenesis, cell adhesion, cell cycle regulation, and fat metabolism [1]. For example, PHB1 regulates transcription in the nucleus by interacting with E2F, retinoblastoma protein (Rb), and chromatin-remodeling complexes (Brg-1/Brm). Repression of E2F-mediated [3] transcription by PHB1 requires histone deacetylation and co-repressors (e.g., N-CoR) [2] [4] [5] [6]. PHB1 also inhibits DNA replication by interacting with members of Minichromosome maintenance complex of proteins (MCM2-7) proteins in the nucleus [7]. On the other hand, phosphorylation of PHB1 results in its association with the plasma membrane, activation of the Ras–Raf signaling pathway, which regulates epithelial cell adhesion and migration [8], and metastasis [9]. In light of these findings, PHB1 is a promising target for clinical applications because natural anticancer compounds such as rocaglamide can inhibit the Ras−Raf signaling pathway by binding to PHB1 [10].

Among the many emerging roles of PHB1, its role in the maintenance of mitochondrial integrity is perhaps the most important. At the mitochondrial inner membrane, PHB1 and PHB2 assemble into a large ring-like complex that stabilizes newly synthesized mitochondrial respiratory enzymes [11]. Prohibitins also function in cell proliferation and apoptosis by regulating the processing of the dynamin-like GTPase OPA1, thus control morphogenesis of mitochondrial cristae [12]. PHB1-mediated regulation of mitochondrial metabolism also links with cell senescence [13] [14]. For example, Marta et al. have reported that prohibitins promote longevity in *Caenorhabditis elegans* by affecting mitochondrial function and fat metabolism, which extends life span in certain genetic background [15]. The loss of prohibitin also impairs mitochondrial architecture and results in neurodegeneration in the mouse, which is indicative of aging [16].

PHB1 has many tissue-specific functions. It is strongly expressed by cells such as adipocytes, muscle cells, and β-cells that rely on efficient mitochondrial function and high energy levels. The loss of PHB markedly reduces fat content in adult nematodes and induces mitochondrial damages in β-cells of mammals, which results in the development of diabetes [15] [17]. In addition, PHB1 is critical for adipocyte differentiation, attenuating insulin-stimulated fatty acid and glucose oxidation in adipose tissue [18] [19] [20]. Thus, PHB1 plays an important role in age-related diseases by affecting metabolic processes; however, the molecular mechanism behind aberrant PHB1 function is not known.

PHB1 contains several highly conserved domains, namely, an N-terminal hydrophobic domain, a PHB domain, a coiled-coil domain, and a nuclear export sequence [2] [9] [21]. The N-terminal hydrophobic domain is critical for its attachment to the mitochondrial inner membrane, whereas the coiled-coil domain is important for protein–protein interactions [22]. The PHB domain is an integral membrane domain that is conserved from protozoa to mammals, and Tyr 114 and Ser 121 within this domain are phosphorylated and the O-GlcNAc modification of Ser 121 can be reduced by mutation of Tyr 114 to phenylalanine. Thus both residues have important roles in tyrosine kinase signaling pathways, including insulin, growth factors, and immune receptors signaling [23] [24].

Artemia (brine shrimp) are extremophiles that possess a unique mechanism of reproduction, that is, they are capable of laying encysted gastrula embryos (cysts) to cope with harsh environments such as anoxia, high salinity, high pH, and ion and temperature fluctuations. The cysts maintain an extremely low metabolism rate, remain in a state of obligate dormancy (diapause), and survive for prolonged periods of time. Diapause can be halted by specific environmental stimuli, which allows embryos to enter dormancy and post-embryonic development [25]. Post-embryonic development in *Artemia* cysts is divided into pre-emergent development, emergence, and larval development. During the emergence stage, mitochondrial maturation and yolk platelet degradation in a related manner, which results in the production of metabolites and enzymes that are essential for mitochondrial function [26] [28]. Vallejo et al. have shown that mitochondrial maturation in the absence of proliferation increases mitochondrial activity after the resumption of development in *Artemia* cysts [29]. The unique process of mitochondrial maturation provides us with a unique model in which mitochondrial biogenesis and metabolic regulation can be investigated.

In addition to mitochondria, yolk platelets are also important organelles involved in the regulation of metabolism during the resumption of development in *Artemia* cysts. Yolk platelets, which contain mainly proteins and lipids, are not static storage structures; instead, they have important roles in nauplius development and beyond. Yolk platelet degradation occurs in two phases. In the first non-enzymatic phase, the pH increases from 6.3 to 8.2, whereas in the second phase, yolk platelets fuse with lysosomes to initiate enzymatic degradation. This process releases enzymes and metabolites that are required for development [26]. Hand et al. have identified oxygen and intracellular pH as important factors in metabolic suppression in *Artemia* encysted embryos, which might occur via transcriptional repression in mitochondria [30] [31] [32]. However, the mechanism responsible for metabolic suppression is largely unknown.

In the present study, PHB1 was cloned and its expression during the resumption of development was characterized in *Artemia*. By RNA interference, PHB1 was found to have a role in mitochondrial maturation during post-embryonic development. A role for PHB1 in the degradation of yolk platelets was also found.

Materials and Methods

Animal culture and sample collection

Artemia parthenogenetica from Gahai Lake, China, was a gift from Professor Feng-Qi Liu (Nankai University, Tianjin, China). Specimens were separated into two random groups and cultured under different conditions. The first group was cultured in 8% artificial seawater (Blue Starfish, Hangzhou, Zhejiang, China) with a 5 h light cycle per day. Under these conditions, most *Artemia* reproduced oviparously and released encysted diapause embryos. The second group was cultured in 4% artificial seawater with a 19 h light cycle per day. Almost all Artemia in this group reproduced ovoviviparously and yielded swimming nauplii. Both groups were reared at 28°C and fed with Chlorella powder (Fuqing King Dnarmsa Spirulina Co. Ltd., Fuqing, China) every 2 days.

To obtain post-diapause (activated) embryos, cysts released by adult Artemia were dehydrated in a saturated sodium chloride solution for 24 h and then frozen at −20°C for 3 months. To obtain hatching samples, post-diapause cysts were hydrated for 5 h at 4°C and then incubated in 2% artificial seawater under continuous light at 25°C. Samples were taken at 0, 6, 12, and 24 h. All samples were snap-frozen in liquid nitrogen and stored at −80°C until use.

Molecular cloning of PHB1

Total RNA was extracted using TRIzol reagent (Invitrogen, Carlsbad, CA, USA) according to the manufacturer's instructions. First-strand cDNAs were synthesized from 1 μg of total RNA by M-MLV reverse transcriptase (TaKaRa Bio, Shiga, Japan) in a reaction volume of 12.5 μl. The PHB1-encoding cDNA fragment was obtained through two rounds of polymerase chain reaction (PCR) amplification using degenerate primers (Table S1). To obtain the full-length PHB1 cDNA, 3′ and 5′ rapid amplification of cDNA ends was performed with gene-specific primers (Table S1) using the FirstChoice RLM-RACE kit (Ambion, Austin, TX, USA).

Real-time polymerase chain reaction

Real-time PCR reactions were performed on the Bio-Rad MiniOpticon real-time PCR system using SYBR Premix Ex Taq (TaKaRa Bio, Shiga, Japan), and 200 nM of each specific primer (ATP6F, ATP6R, ND5F, ND5R, CytoBF, and CytoBR; Table S1) was used. Relative transcript levels were presented as fold-change, which was calculated using the comparative CT method as described by Livak and Schmittgen with α-tubulin as the internal reference [33].

Antibody production

The open reading frame of the ArPHB1 gene was cloned (ExF and ExR, Table S1) into the pET-28 vector (Novagen) to generate the 6×His-PHB1 fusion protein. After purification using Ni-NTA resin and the QIAexpressionist kit (Qiagen, Germany), the protein was used to generate a polyclonal antibody against PHB1 (HuaAn, Hangzhou, China). For the phospho-PHB1 antibody, the phosphorated peptides (D(p)YEERVLPSITSEVL and DYEERVL P(p)SITSEVL) were used to generate a polyclonal antibody (HuaAn, Hangzhou, China). Serum was pre-cleared with the corresponding peptide that was not phosphorated before purification. The specificity of the anti-phospho-PHB1 antibody was tested by dot blotting (Fig. S2). In brief, peptides (0.04 and 0.004 μg) were applied directly onto a PVDF (polyvinylidene difluoride) membrane (Millipore Corp., Billerica, MA, USA) as dots and then detected with the corresponding antibody.

Western blot analysis

After resolving proteins by 8% or 10% SDS-PAGE, proteins were transferred onto PVDF membranes. The membranes were incubated with primary antibodies overnight at 4°C, and proteins were visualized using the BM chemiluminescence western blotting kit (Roche Applied Science, Penzberg, Upper Bavaria, Germany). The ubiquitin antibody was purchased from Abcam (ab134953) (Cambridge, England, United kingdom).

RNA interference

The double-stranded RNA was prepared as previously described using gene-specific primers (dsPHBF and dsPHBR; Table S1) and subcloned into the pET-T7 vector between Xho I and EcoR I sites. Plasmids expressing green fluorescent protein (GPF) dsRNA as the negative control were constructed as described. The recombinant plasmids were transformed into *Escherichia coli* HT115, and the dsRNA was produced and purified as described [28]. PHB and GFP dsRNAs (1 μg of each) were injected separately into specimens using the UltraMicro-Pump II system (World Precision Instruments Inc., Sarasota, FL, USA) equipped with a Micro4 microsyringe pump controller (World Precision Instruments Inc.). The injected specimens were cultured under previously described conditions to obtain ovoviviparous nauplii.

Transmission electron microscopy analysis

Nauplii were fixed for 12 h in 2.5% glutaraldehyde prepared in phosphate-buffered saline (PBS). Thereafter, they were washed and post-fixed in 1% osmium tetroxide, dehydrated in a graded acetone series, and embedded in Spurr resin. Sections of 70 nm in thickness were obtained with a Leica EM UC6 microtome (Wetzlar, Germany), stained with 2% uranyl acetate and Reynold's solution (0.2% sodium citrate and 0.2% lead nitrate), viewed by a JEM-1230 transmission electron microscope (JEOL, Akishima, Tokyo, Japan), and photographed at a voltage of 70 kV.

Isolation of yolk platelets and nuclei

Isolation of yolk platelets and nuclei was performed as previously described [34] using small volumes of buffers, which were supplemented with a protein inhibitor complex (Sangon Biotech, ShangHai, China) and 1 mM PMSF (Phenylmethanesulfonyl fluoride). In brief, approximately 100 mg of sample was homogenized using a glass homogenizer in 1 ml HM buffer (10 mM Tris−HCl, 10 mM NaCl, and 10 mM $MgCl_2$, pH 7.5) containing 0.1% NP-40. The homogenate was filtered through a cell strainer (pore size, 38 μm) and centrifuged at 800×*g* for 10 min at 4°C. The pellet was gently resuspended in 200 μl of HM buffer, and the sample was layered onto a 1.2 ml Percoll layer (75% Percoll containing 150 mM NaCl, 10 mM $MgCl_2$, and 10 mM Tris−HCl, pH 7.5) in a 2 ml centrifuge tube. The sample was centrifuged in a fixed-angle rotor at 15000×*g* for 15 min at 4°C. Nuclei accumulating at the sample/Percoll interface and yolk platelets sedimenting at the bottom were collected separately, washed with 5-volumes of HM buffer, and centrifuged at 850×*g* for 10 min at 4°C. Thereafter, yolk platelets were resuspended in RIPA lysis buffer (Beyotime, Shanghai, China), and the sample was maintained at 4°C for 30 min. The sample was then centrifuged at 12000×*g* for 5 min, the supernatant was collected, and stored at −80°C for immunoprecipitation. Nuclei were collected and fixed in 4% paraformaldehyde for immunofluorescence.

Membrane and cytosol fraction

In yolk platelets and nuclei isolation, after 800×*g* centrifugation, the supernatant was subjected to further centrifugation at 20000×g. The pellet and supernatant was taken as membrane and non-membrane fractions respectively.

Nuclear immunofluorescence

Nuclei were fixed in 4% paraformaldehyde overnight and then washed with PBS. Fixed nuclei were incubated in blocking buffer (0.3% Triton X-100 and 5% bovine serum albumin in PBS) for 1 h at room temperature. Thereafter, nuclei were labeled with an anti-PHB antibody in 0.3% Triton X-100 and 3% bovine serum albumin in PBS overnight at 4°C, followed by incubation in a secondary Alexa Fluor 488 donkey anti-rabbit antibody (Invitrogen) for 1 h at room temperature. DNA was stained by DAPI (4′,6-diamidino-2-phenylindole dilactate) for 10 min at room temperature. Then stained nuclei were visualized by a LSM 710 confocal microscope (Jena, Germany).

Immunoprecipitation

For immunoprecipitation, yolk platelet lysate (500–1000 μg) and an anti-PHB1 antibody (10 μg) were incubated with gentle rocking overnight at 4°C. Thereafter, protein A agarose beads (20 μl of a 50% bead slurry, Millipore Corp.) were added, and samples were incubated with gentle rocking for 1 h at 4°C. After centrifugation for 30 sec at 4°C, the pellet was washed five times with 100 μl of RIPA lysis buffer. The immunocomplex was eluted by adding SDS loading buffer (4% SDS, 0.2% bromophenol blue, 20% glycerol, and 200 mM β-mercaptoethanol) and boiled in a water bath for 5 min.

Limited trypsin digestion assay

The immunoprecipitation product, other than eluted by SDS loading buffer,from the protein A agarose beads, was washed twice with PBS. Then the the protein A beads with the immunoprecipitates were incubated with trypsin (0.05% in 0.53 mM EDTA; Corning, VA, USA) at an enzyme to substrate ratio of 1:25 (w/w) at 30°C for 30 min. The digest was then subjected to Western blot analysis as above by using anti-ubiquitin (Abcam, ab134953) or anti-PHB1 antibody. The intact immunoprecipitated products served as controls.

AICAR treatment

As a cell permeable activator of AMP-activated protein kinase (AMPK), 5-Aminoimidazole-4-carboxamide 1-β-D-ribofuranoside Acadesine N1-(β-D-Ribofuranosyl)-5-aminoimidazole-4-carboxamide (AICAR) (Sigma, 3050 Spruce St., St. Louis, Missouri 63103, United States) was used to activate AMPK. In the NM treatment experiment, rehydrated post-diapause embryos were incubated in 2% seawater containing 0.5 and 1 mM for 24 h. After that the nauplii were collected for Western blotting analysis.

Results

To investigate the role of ArPHB1 in the development of *Artemia*, a 988-bp cDNA encoding ArPHB1 was cloned by PCR from post-diapause embryos of *A. parthenogenetica* from Gahai Lake using DF1, DF2, DR1, and DR2 primers for degenerate PCR and 3F1, 3F2, 5R1, and 5R2 primers for the rapid amplification of cDNA ends (Table S1). Sequence analysis revealed the cDNA to contain an open reading frame encoding a protein of 272 amino acids with a predicted mass of 29.9 kDa (Fig. S1). The deduced amino acid sequence showed that ArPHB1 had a conserved prohibitin domain, a coiled-coil domain, an N-terminal hydrophobic sequence, and a nuclear export sequence (NES) (Fig. 1A). ArPHB1 was homologous to PHB1 in other species. For example, sequence alignment showed 78% sequence

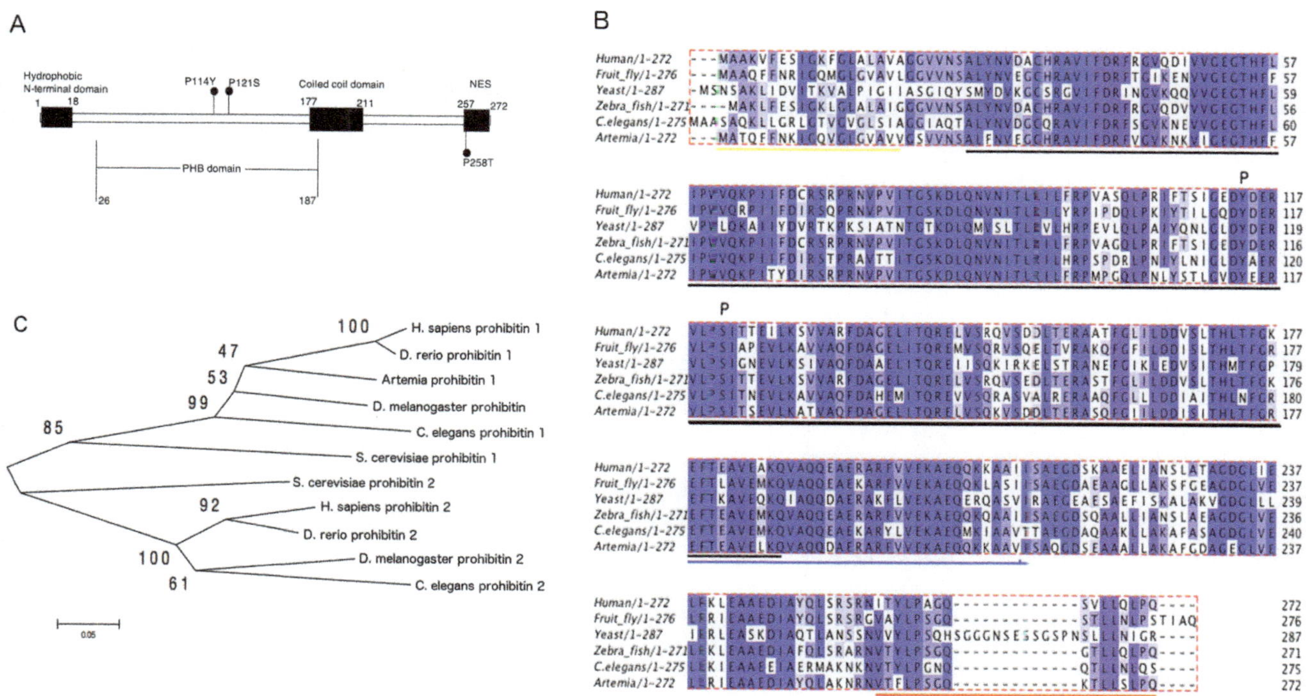

Figure 1. Sequence and phylogenetic analysis of ArPHB1. A, Schematic view of PHB1 architecture. B, Amino acid sequence alignment of ArPHB1 with PHB1 from other species. The sequences used in this alignment and their Genbank/EMBL/DDBJ or SWISS-PROT database accession numbers are listed in Table S2. Gaps inserted to maximize alignment are denoted by hyphens. The amino acid position numbers are shown to the right of the sequences. The phosphorylation sites against which the antibody was generated is marked by P. Conserved domains are marked by lines below sequences (the N-terminal hydrophobic domain by yellow, the PHB domain by black, the coiled-coil domain by blue, and the nuclear export sequence by red). The alignment was performed by MEGA5 using the ClustalW method. C, A phylogenetic tree of the amino acid sequences of prohibitins. The sequences used in this analysis and their GenBank/EMBL/DDBJ or SWISS-PROT database accession numbers are listed in Table S1. The phylogenetic tree was constructed using the Neighbor-joining method. Bootstrap percentage values for 1000 replicate analysis are shown at branching points. The bar at the bottom shows the branch length, and it corresponds to the mean number of differences (0.05) per residue along each branch.

identity with fruit fly PHB1, 74% with zebrafish PHB1, 68% with nematode PHB1, 74% with human PHB1, and 50% with yeast PHB1 (Fig. 1B). Phylogenetic analysis confirmed ArPHB1 to derive from PHB1 (Fig. 1C).

To characterize ArPHB1, an antibody was produced using full-length ArPHB1 after the protein was expressed in *E. coli*. Western blot analysis showed two bands of 30 kDa (predicted molecular weight) for PHB1-30 and 75 kDa for PHB1-75. The major protein band, PHB1-30, showed no significant changes in expression (Fig. 2A), while PHB1-75 was observed during pre-emergent development (6 and 12 h incubation) (Fig. 2B) when the embryos were preparing for energy- and nutrient- demanding processes (i.e., emergence and larval development). Through subcellular fraction we found that PHB1-75 existed as a non-membrane form of PHB1 in *Artemia* (Fig. 2C), which is consistent with previous report that the PHB1-75 existed in non-membrane fraction [27].We also produced antibodies against phosphor-Tyr-114 and phosphor-Ser-121 using respective peptides, and their specificities were tested by dot blotting (Fig. S2). The phosphorylation status of Tyr-114 and Ser-121, which are thought to be involved in metabolic control, were investigated. Both PHB1-114Y (35 kDa) and PHB1-121S (38 kDa) were phosphorylated in embryos during pre-emergent development (0, 6, and 12 h after incubation), but not in the nauplius (24 h after incubation) (Fig. 2B). These results indicate that PHB1 phosphorylation has roles in cell proliferation and metabolic control during post-embryonic development of *Artemia*.

To explore whether the nuclear localization of PHB1 was also developmentally regulated, we localized PHB1 by immunofluorescence. PHB1 was found in the nuclear matrix during pre-emergent development (0, 6, and 12 h after incubation), but not 24 h after incubation (Fig. 3A), which was further confirmed by Western blotting analysis (Fig. 3B). Although PHB1 was exported out of the nuclear matrix, it remained at the nuclear membrane (Fig. 3A), illustrating that the localization of PHB1 was developmentally-related. These changes may be related to cell cycle progression or metabolic remodeling during the emergence stage.

RNAi was performed by microinjecting dsRNA into female embryos, which had oocytes in the oviducts. PHB1 knockdown resulted in the death of embryos or the production of swimming nauplii (Fig. 4A). Swimming nauplii released by the PHB1 knockdown group were sluggish compared to nauplii released by the control group (Video S1, S2), indicating that there were defects in energy homeostasis. Nauplii released by the PHB1 knockdown group were also weaker than those released by the control group (Fig. 4A). Northern and western blotting analyses showed successful RNAi knockdown (Fig. 4B).

To investigate if PHB1 affected energy homeostasis, nauplii released at 1–1.5 day were collected. Western blot analysis illustrated an increase in AMPK activity (Fig. 4C). Thereafter, we investigated the role of PHB1 in mitochondrial function by RNAi. PHB1 knockdown adversely affected the morphology of mitochondria (Fig. 5A), which were smaller and more compact than those in nauplii released by the control group. RT-PCR analysis

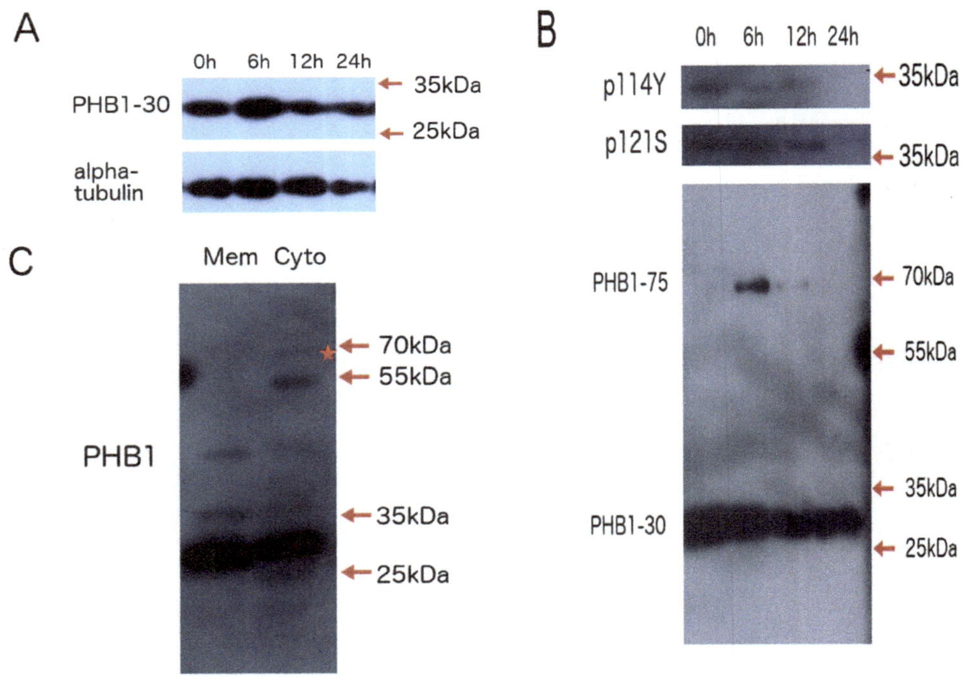

Figure 2. Characterization of ArPHB1 during post-embryonic development. A, Western blot analysis of PHB1. 0 h, 0 h-incubated cysts; 6 h, 6 h-incubated cysts; 12 h, 12 h-incubated cysts; 24 h, 24 h-incubated nauplius. B, Western blot analysis of phosphorylated PHB1 and PHB1 modifications. The same samples were loaded in B as in A. C, PHB1-75 existed as a non-membrane form of PHB1 in *Artemia*. Mem, membrane fraction obtained by 20000×g centrifugation after nuclei isolation (800×g, 10 min); Cyto, cytosol fraction obtained after 20000×g centrifugation. PHB1-75 was marked by an asterisk.

showed downregulation of mtDNA-encoded mRNAs in the PHB1 knockdown group (Fig. 5B), confirming the morphological defects in mitochondria. Moreover, western blot analysis revealed that F-ATPase β subunit, the enzyme for ATP synthesis, was also downregulated after PHB1 knockdown (Fig. 5C). These results indicate that PHB1 maintains normal mitochondrial biogenesis during the development of *Artemia*.

Yolk platelets, which contain mainly yolk protein, are important organelles in metabolism. The degradation of yolk proteins is critical for the normal development of *Artemia* because the process produces nutrients and necessary enzymes [26]. Because PHB1 knockdown yielded a nutrient-deficient phenotype, we investigated whether PHB1 knockdown could affect the degradation of yolk proteins. We used an antibody against peptide located at the N-terminal of vitellogenin (anti-VGN), which was previously used by our lab to analyze the enzyme cleavage pattern of yolk proteins [35], to examine yolk platelets degradation. In this study, we isolated yolk platelets from post-embryonic developmental samples (0, 6, 12, and 24 h after incubation) (Fig. 6A). Yolk proteins in yolk platelets showed different bands at 12 h after incubation which may suggest initiation of enzymic degradation of yolk proteins in yolk platelets. Western blot analysis showed that PHB1 existed in yolk platelets, and the amount of PHB1 gradually increased during pre-emergent development and decreased in nauplii (Fig. 6B). In addition, several proteins bands were observed after using the anti-PHB1 antibody. Two of these protein bands electrophoresed at 130 and 170 kDa at 12 h after incubation (Fig. 6B), indicating that PHB1 has a role in yolk protein degradation during post-embryonic development. And the degradation of yolk protein is consistent with the amount of ArPHB1 in yolk platelets (Fig. 6C). In agreement with these results, PHB1 knockdown resulted in the accumulation of yolk proteins in nauplii (Fig. 6D). TEM analysis found that yolk platelet degradation was blocked in nauplii after PHB1 knockdown, which might have caused nutrient deficiency (Fig. 6E). These results illustrate that PHB1 is indispensable for yolk platelet degradation during the early development of *Artemia*.

Attachment of ubiquitin molecules can target organelles (peroxisomes) to lysosomes for degradation [36]. To determine whether the two additional protein bands at 12 h after incubation were ubiquitinated PHB1, PHB1 in yolk platelet lysates was immunoprecipitated with an anti-PHB1 antibody. The 130 kDa protein band was found to be ubiquitinated PHB1 (Fig. 6F). These results illustrate that PHB1 may mediates yolk platelets degradation via ubiquitination.

Discussion

In response to unfavorable environmental conditions, *Artemia* produce encysted gastrula embryos in diapause, a state of obligate dormancy, in which complex metabolic processes are downregulated to minimize cellular activities and conserve resources. Organisms in diapause remain hypometabolic, even under conditions that would normally promote active metabolism and development. In this state, organisms in diapause are more tolerant of environmental stresses. In addition, a greater metabolic arrest usually associates with a longer dormant state [37]. During post-embryonic development, metabolic suppression ceases to meet energy and nutrient needs for normal development, providing us with a unique model in which metabolic remodeling can be studied.

The control of energy storage in organisms in diapause is an emerging area of research, which may provide insights on the aging process and age-related diseases because there are crosstalk between diapause and aging. A recent study has shown the

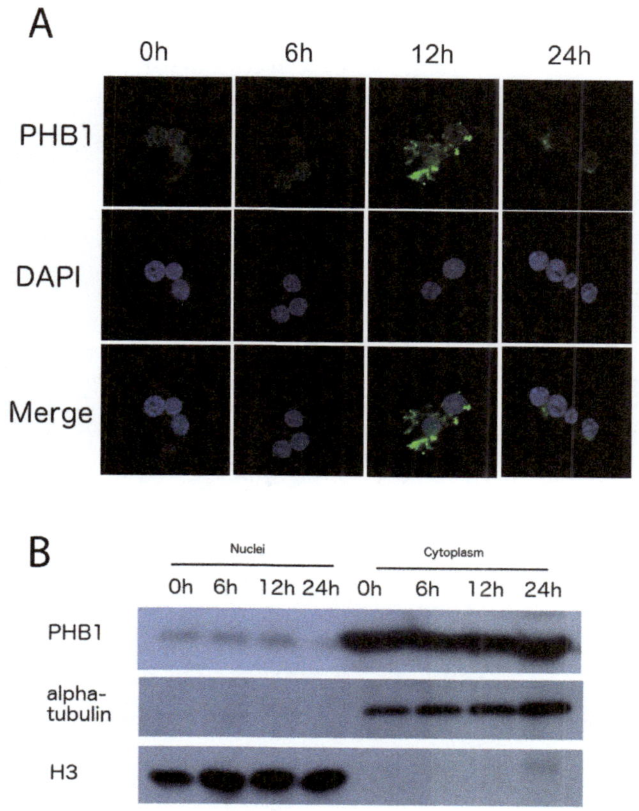

Figure 3. The localization of PHB1 in nuclei during *Artemia* post-embryonic development. A, Immunofluorescence analysis of the nucleus. Nuclei from 0, 6, 12, and 24 h-incubated nauplii were isolated and stained with an anti-PHB1 antibody. DNA was stained with DAPI (4,6-diamidino-2-phenylindole dilactate). B, Nuclei from the respective developmental stages were isolated and analyzed by Western blotting (anti-PHB1).

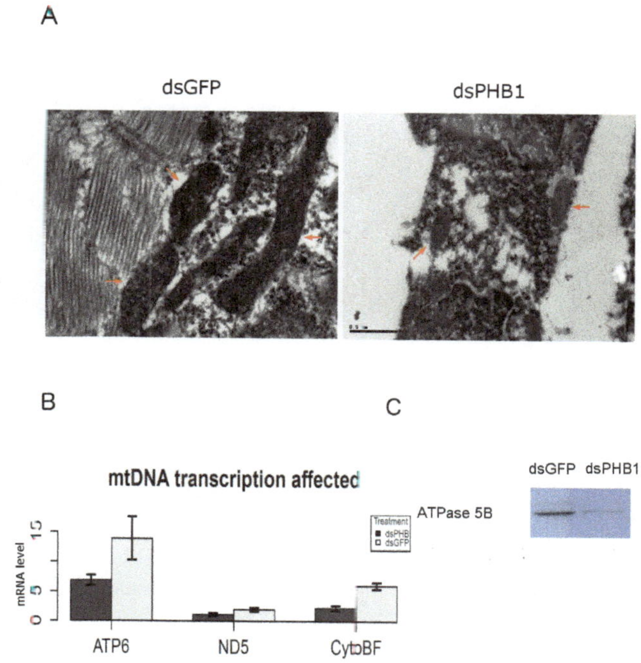

Figure 5. ArPHB1 knockdown resulted in aberrant mitochondrial function. A, mitochondrial morphology by TEM (magnification, 24000×; scale bar = 0.5 μm). Mitochondria are denoted by red arrows. B, Real-time PCR of mitochondrial genes (ATP6, ATP synthase F0 subunit 6; ND5, NADH dehydrogenase 5; and Cyto B, cytochrome b). Results were normalized to α-tubulin. Means ± SD are plotted. $P<0.05$. The primers used for real-time PCR are listed in Table S1. C, Western blot analysis of ATPase β subunit.

metabolic and stress sensor, AMP-activated protein kinase (AMPK), directly suppress adipose triglyceride lipase (ATGL-1) activity by phosphorylation, which reserves lipids in the dauer of *C. elegans* [38]. And appropriate level of ROS was also important for reserving lipids in the dauer of *C. elegans* through hypoxia

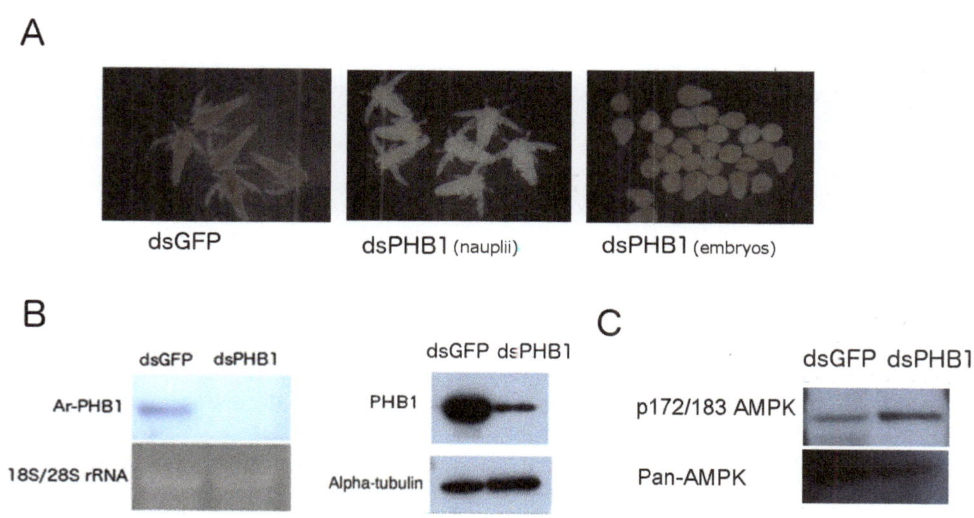

Figure 4. ArPHB1 knockdown resulted in the production of metabolic stress. A, Released nauplii were fixed in absolute ethanol and photographed under a dissecting microscope. dsGFP, dsGFP injected as control; dsPHB1(nauplii), nauplii released after ArPHB1 knockdown; dsPHB1(embryos), embryos released after ArPHB1 knockdown. B, Northern and western blot analyses to verify the efficiency of PHB1 knockdown. C, AMPK activity (p172/183) was investigated by western blot analysis in PHB1 knockdown and control groups.

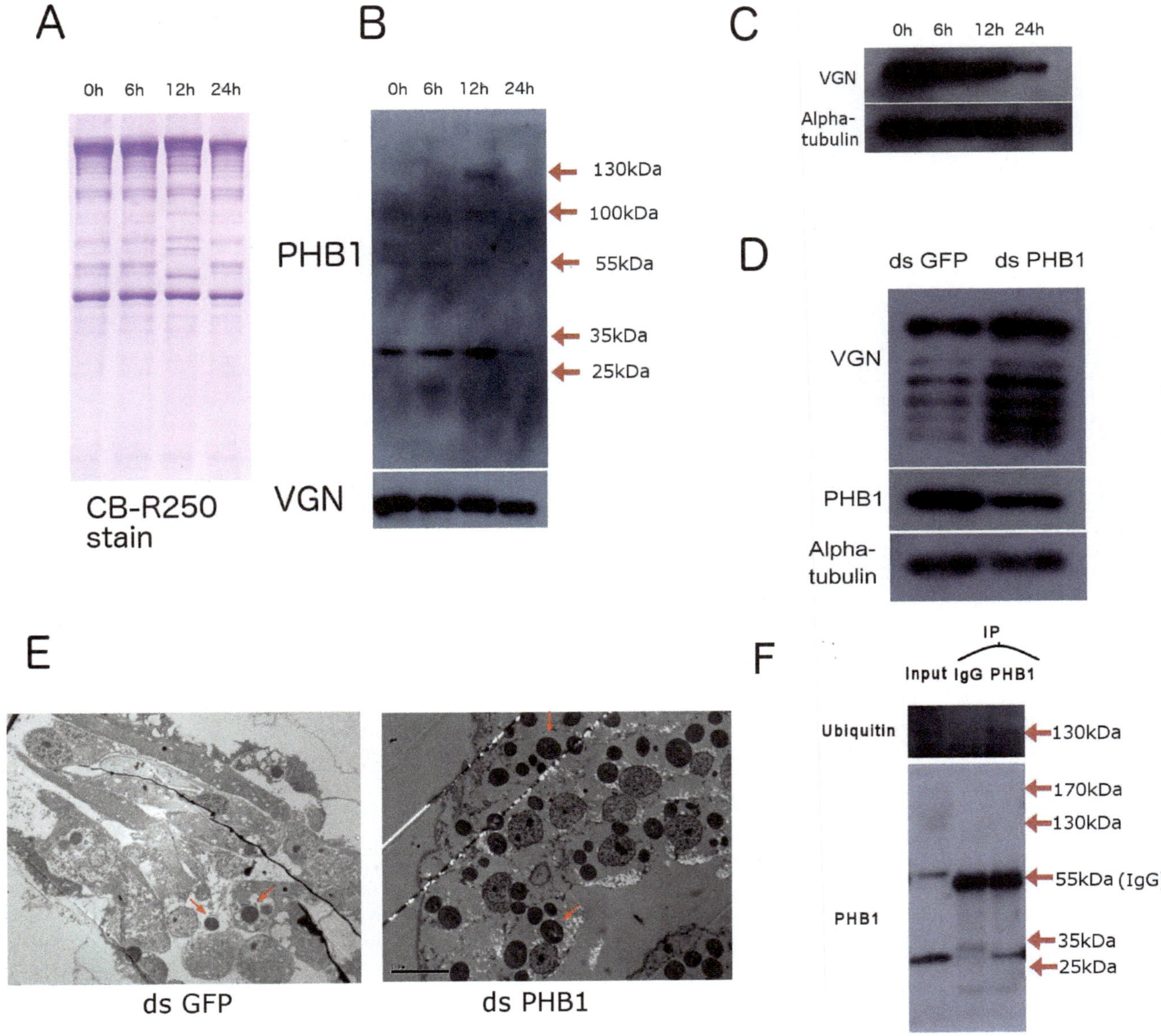

Figure 6. Involvement of PHB1 in the degradation of yolk platelets. A, SDS-PAGE analysis of isolated yolk platelet lysates. Proteins were stained by Coomassie brilliant blue R-250. 0 h, 0 h after incubation; 6 h, 6 h after incubation; 12 h, 12 h after incubation; 24 h, 24 h after incubation. B, Yolk platelets were isolated by density centrifugation and probed with an anti-PHB1 antibody. An anti-VGN antibody was used as the control. 0 h, 0 h after incubation of cysts; 6 h, 6 h after incubation of cysts; 12 h, 12 h after incubation of cysts; 24 h, 24 h after incubation of nauplii. C. Yolk protein degradation during embryonic development. 0 h, 0 h after incubation of cysts; 6 h, 6 h after incubation of cysts; 12 h, 12 h after incubation of cysts; 24 h, 24 h after incubation D, Western blot analysis of yolk protein degradation in PHB1 knockdown nauplii. E, TEM analysis showed aberrant yolk platelet degradation in the PHB1 knockdown group (magnification 1250×, scale bar = 1 μm). F, Immunoprecipitation with an anti-PHB1 antibody using isolated yolk platelet lysates (12 h after incubation of cysts). The precipitates were then probed with an anti-ubiquitin antibody.

inducible factor-1 (HIF-1) signalling [39]. It is reported that PHBs modulate fat content and mitochondrial metabolism in *C. elegans* [15]. According to previous report that the ubiquitin of protein and metabolic state in dormancy cyst is reversibly regulated in response to anoxia [25] [40], our results here suggests that keep ArPHB1 from ubiquitination may be a mechanism to reserve the yolk platelet in dormancy cysts.

In this study, we investigated the expression of ArPHB1 during post-embryonic development of *Artemia* when encysted embryos resume their metabolism to meet developmental requirements [26]. The expression of ArPHB1 during post-embryonic development was relatively constant because ArPHB1 is an important scaffolding protein involved in several biological processes. In characterization of PHB1 expression during embryonic development, a band of 75 kDa, which was analysed in a previous paper as a non-membrane form of PHB1 (27), was detected after 6 h of hatching and gradually disappeared after 12 h incubation. After 6 h incubation the encysted embryos have resumed respiration activities, the appear of a non-membrane-attached form of PHB1 may respresent the tranportation of ArPHB1 in the cell, which might lead to the increase of ArPHB1 in yolk platelet. There is a strong association between the regulation of PHB1, and its modifications and cellular localizations. There are changes in the phosphorylation status and localization of ArPHB1 during emergence, which marks the initiation of cell proliferation and the increased activities of many enzymes [26] [28]. Given that cell proliferation requires amino acids, nucleotides, and energy, PHB1 may mediate metabolic remodeling to meet the needs of cell proliferation and development. It is also possible that ArPHB1 directly regulates cell proliferation during post-embryonic development.

ArPHB1 knockdown increased AMPK activity in nauplii. Thus, we used AICAR, an activator of AMPK, to activate AMPK; however, AICAR did not affect swimming or cold resistance in nauplii from the knockdown group (data not shown). By contrast AICAR treatment accelerated yolk degradation when the anti-VGN antibody was used for western blot analysis (Fig. S3). Therefore, we conclude that the increase in AMPK activity was not responsible for the defects in nauplii caused by ArPHB1 knockdown. The increase in AMPK activity after ArPHB1 knockdown was most likely caused by damaged mitochondria. Although an increase in AMPK activity usually associates with the degradation of yolk protein, ArPHB1 knockdown blocked this effect.

Yolk degradation is critical in the cascade of events leading to emergence [26]. It has been reported that AMPK homolog in *C. elegans* (aak-2) affected the role of PHB1 in fat metabolism [15]. However, this does not apply to the degradation of yolk proteins. Our results showed that PHB1 participate directly in the degradation of yolk platelets independent of AMPK activity. Due to the highly conserved amino acid and functional domains, the involvement of PHB1 in yolk degradation is likely to be conserved in other organisms. And PHB1 may serve as substrates of ubiquitination in yolk platelets that target the yolk platelets for selective autophagy. Besides, a tryptic digestion of the anti-PHB1 IP product resulted in a band of about 37 kDa, indicating that the 130-kDa molecule was the ubiquitinated form of PHB1 (Fig. S4).

In the ArPHB1 knockdown experiment, 50% of the embryos were released as swimming nauplii and the remaining 50% were released as dead embryos (Fig. 4A). The blockage of degradation of yolk platelets caused malnutrition may be the cause of the weaker nauplii in the ArPHB1 knockdown group. The different fates of the embryos may be due to PHB1 knockdown at different stages of oocyte development. For example, the dsRNA was injected when oocytes were in the oviduct, and oocytes were not developmentally synchronized. These results show that ArPHB1 has other lethal roles in embryonic development.

Our results illustrate that PHB1 is critical for post-embryonic and early larval development of *Artemia*. PHB1 regulates mitochondrial maturation and mediates yolk platelet degradation, which support high energy cellular activities during post-embryonic and larval development. Our results also suggest that PHB1 functions in yolk platelet degradation as substrates of ubiquitination in yolk platelet.

Supporting Information

Figure S1 The ArPHB1 cDNA sequence and its deduced amino acid sequence.
(TIF)

Figure S2 Antibody specificities for phosphor-Tyr-114 and phosphor-Ser-121 of PHB1. The peptides were applied directly onto PVDF membranes as dots and then detected with the corresponding antibody.
(TIF)

Figure S3 The effect of AICAR, an AMPK activator, on the degradation of yolk platelet proteins. The encysted embryos were incubated at 2% seawater containing 0, 0.5, and 1 mM AICAR for 24 h. For each lane, 7 nauplius were homogenized in SDS loading buffer (4% SDS, 0.2% bromophenol blue, 20% glycerol, and 200 mM β-mercaptoethanol) and loaded.
(TIF)

Figure S4 Tryptic digestion and Western blotting analysis of PHB1 IP product. 1 and 2, Western blotting of intact (1) and digested (2) IP product by anti-ubiquitin; 1′ and 2′, Western blotting of intact (1′) and digested (2′) IP product by anti-PHB1. The 130- and 37-kDa bands are indicated by red and blue asterisks respectively.
(TIF)

Table S1 Nucleotide sequences of primers used in polymerase chain reactions.
(DOCX)

Table S2 Protein sequences used in the alignment of sequences and phylogenetic analysis.
(DOCX)

Video S1 Nauplii reproduced by the adults, into which fragment of GFP dsRNA was microinjected as control.
(AVI)

Video S2 Nauplii reproduced by the adults, into which fragment of ArPHB1 dsRNA was microinjected.
(AVI)

Acknowledgments

We thank Prof. Feng-Qi Liu of the Nankai University, Tianjin, China, for the kind gift of cysts of *A. parthenogenetica*.

Author Contributions

Conceived and designed the experiments: WJY XY. Performed the experiments: XY YZ LLZ YXS. Analyzed the data: WJY XY. Contributed reagents/materials/analysis tools: JSY. Wrote the paper: WJY XY JSY LLZ.

References

1. Back JW, Sanz MA, De Jong L, De Koning LJ, Nijtmans LGJ, et al. (2002) A structure for the yeast prohibitin complex: Structure prediction and evidence from chemical crosslinking and mass spectrometry. Protein Science 11: 2471–2478.
2. Mishra S, Murphy LC, Nyomba BLG, Murphy LJ (2005) Prohibitin: a potential target for new therapeutics. Trends in Molecular Medicine 11: 192–197.
3. Faller DV, Zhang BH, Sheng W (2002) Prohibitin recruits the chromatin-remodeling ATPases Brg-1 and Brm for repression of E2F-driven transcription and cell proliferation. Blood 100: 738A–738A.
4. Fusaro G, Wang S, Chellappan S (2002) Differential regulation of Rb family proteins and prohibitin during camptothecin-induced apoptosis. Oncogene 21: 4539–4548.
5. Wang S, Nath N, Fusaro G, Chellappan S (1999) Rb and prohibitin target distinct regions of E2F1 for repression and respond to different upstream signals. Molecular and Cellular Biology 19: 7447–7460.
6. Wang S, Fusaro G, Padmanabhan J, Chellappan SP (2002) Prohibitin co-localizes with Rb in the nucleus and recruits N-CoR and HDAC1 for transcriptional repression. Oncogene 21: 8388–8396.
7. Rizwani W, Alexandrow M, Chellappan S (2009) Prohibitin physically interacts with MCM proteins and inhibits mammalian DNA replication. Cell Cycle 8: 1621–1629.
8. Rajalingam K, Wunder C, Brinkmann V, Churin Y, Hekman M, et al. (2005) Prohibitin is required for Ras-induced Raf-MEK-ERK activation and epithelial cell migration. Nature Cell Biology 7: 837–843.
9. Chiu CF, Ho MY, Peng JM, Hung SW, Lee WH, et al. (2013) Raf activation by Ras and promotion of cellular metastasis require phosphorylation of prohibitin in the raft domain of the plasma membrane. Oncogene 32: 777–787.
10. Polier G, Neumann J, Thuaud F, Ribeiro N, Gelhaus C, et al. (2012) The Natural Anticancer Compounds Rocaglamides Inhibit the Raf-MEK-ERK Pathway by Targeting Prohibitin 1 and 2. Chemistry & Biology 19: 1093–1104.

11. Nijtmans LGJ, de Jong L, Sanz MA, Coates PJ, Berden JA, et al. (2000) Prohibitins act as a membrane-bound chaperone for the stabilization of mitochondrial proteins. Embo Journal 19: 2444–2451.
12. Merkwirth C, Dargazanli S, Tatsuta T, Geimer S, Loewer B, et al. (2008) Prohibitins control cell proliferation and apoptosis by regulating OPA1-dependent cristae morphogenesis in mitochondria. Genes & Development 22: 476–488.
13. Coates PJ, Jamieson DJ, Smart K, Prescott AR, Hall PA (1997) The prohibitin family of mitochondrial proteins regulate replicative lifespan. Current Biology 7: 607–610.
14. Coates PJ, Nenutil R, McGregor A, Picksley SM, Crouch DH, et al. (2001) Mammalian prohibitin proteins respond to mitochondrial stress and decrease during cellular senescence. Experimental Cell Research 265: 262–273.
15. Artal-Sanz M, Tavernarakis N (2009) Prohibitin couples diapause signalling to mitochondrial metabolism during ageing in C-elegans. Nature 461: 793–U747.
16. Merkwirth C, Martinelli P, Korwitz A, Morbin M, Bronneke HS, et al. (2012) Loss of Prohibitin Membrane Scaffolds Impairs Mitochondrial Architecture and Leads to Tau Hyperphosphorylation and Neurodegeneration. Plos Genetics 8.
17. Supale S, Thorel F, Merkwirth C, Gjinovci A, Herrera PL, et al. (2013) Loss of Prohibitin Induces Mitochondrial Damages Altering beta-Cell Function and Survival and Is Responsible for Gradual Diabetes Development. Diabetes 62: 3488–3499.
18. Ande SR, Xu Z, Gu Y, Mishra S (2012) Prohibitin has an important role in adipocyte differentiation. International Journal of Obesity 36: 1236–1244.
19. Liu D, Lin YM, Kang T, Huang B, Xu W, et al. (2012) Mitochondrial Dysfunction and Adipogenic Reduction by Prohibitin Silencing in 3T3-L1 Cells. Plos One 7.
20. Vessal M, Mishra S, Moulik S, Murphy LJ (2006) Prohibitin attenuates insulin-stimulated glucose and fatty acid oxidation in adipose tissue by inhibition of pyruvate carboxylase. Febs Journal 273: 568–576.
21. Yurugi H., K . Rajalingam (2013) A role for prohibitin in mast cell activation: location matters. Sci Signal **6**(292): pe29.
22. Joshi B, et al. (2003) A putative coiled-coil domain of prohibitin is sufficient to repress E2F1-mediated transcription and induce apoptosis. Biochem Biophys Res Commun **312**(2): 459–466.
23. Ande SR, Y Gu, BL Nyomba and S Mishra (2009) Insulin induced phosphorylation of prohibitin at tyrosine 114 recruits Shp1. Biochim Biophys Acta **1793**(8): 1372–1378.
24. Ande S R, S . Moulik and S . Mishra (2009) Interaction between O-GlcNAc modification and tyrosine phosphorylation of prohibitin: implication for a novel binary switch. PLoS One 4(2): e4586.
25. TH J Abatzopoulos, J . Beardmore, JS . Clegg, P . Sorgeloos (2002) Artemia: Basic and Applied Biology. Dordrecht, the Netherlands, Kluwer Academic Publishers.
26. Robert AB, Patrick S., Clive NAT (1991) Artemia Biology. United States, CRC Press.
27. Sripathi SR, W He, CL Atkinson, JJ Smith, Z Liu, et al. (2011) Mitochondrial-nuclear communication by prohibitin shuttling under oxidative stress. Biochemistry 50(39): 8342–8351.
28. Dai JQ, Zhu XJ, Liu FQ, Xiang JH, Nagasawa H, et al. (2008) Involvement of p90 ribosomal S6 kinase in termination of cell cycle arrest during development of Artemia-encysted embryos. Journal of Biological Chemistry 283: 1705–1712.
29. Vallejo CG, Lopez M, Ochoa P, Manzanares M, Garesse R (1996) Mitochondrial differentiation during the early development of the brine shrimp Artemia franciscana. Biochemical Journal 314: 505–510.
30. Kwast KE, Hand SC (1996) Oxygen and pH regulation of protein synthesis in mitochondria from Artemia franciscana embryos. Biochemical Journal 313: 207–213.
31. Eads BD, Hand SC (1999) Regulatory features of transcription in isolated mitochondria from Artemia franciscana embryos. American Journal of Physiology-Regulatory Integrative and Comparative Physiology 277: R1588–R1597.
32. Eads BD, Hand SC (2003) Transcriptional initiation under conditions of anoxia-induced quiescence in mitochondria from Artemia franciscana embryos. Journal of Experimental Biology 206: 577–589.
33. Schmittgen TD, Livak KJ (2008) Analyzing real-time PCR data by the comparative C-T method. Nature Protocols 3: 1101–1108.
34. Van Breukelen F, Maier R, Hand SC (2000) Depression of nuclear transcription and extension of mRNA half-life under anoxia in Artemia franciscana embryos. J Exp Biol 203(Pt 7): 1123–1130.
35. Chen S, Chen DF, Yang F, Nagasawa H, Yang WJ (2011) Characterization and processing of superoxide dismutase-fused vitellogenin in the diapause embryo formation: a special developmental pathway in the brine shrimp, Artemia parthenogenetica. Biol Reprod 85(1): 31–41.
36. Kim PK, Hailey DW, Mullen RT, Lippincott-Schwartz J (2008) Ubiquitin signals autophagic degradation of cytosolic proteins and peroxisomes. Proc Natl Acad Sci USA 105(52): 20567–20574.
37. Hand SC (1998) Quiescence in Artemia franciscana embryos: reversible arrest of metabolism and gene expression at low oxygen levels. J Exp Biol 201(Pt 8): 1233–1242.
38. Narbonne P, Roy R (2009) Caenorhabditis elegans dauers need LKB1/AMPK to ration lipid reserves and ensure long-term survival. Nature 457(7226): 210–214.
39. Xie M, Roy R (2012) Increased levels of hydrogen peroxide induce a HIF-1-dependent modification of lipid metabolism in AMPK compromised C. elegans dauer larvae. Cell Metab 16(3): 322–335.
40. Anchordoguy T, Hand SC (1995) Reactivation of ubiquitination in Artemia franciscana embryos during recovery from anoxia-induced quiescence. J Exp Biol 198(Pt 6): 1299–1305.

18

The Potential of a High Protein-Low Carbohydrate Diet to Preserve Intrahepatic Triglyceride Content in Healthy Humans

Eveline A. Martens*, **Blandine Gatta-Cherifi**¤, **Hanne K. Gonnissen**, **Margriet S. Westerterp-Plantenga**

Department of Human Biology, Maastricht University, Maastricht, The Netherlands

Abstract

Background: Protein supplementation has been shown to reduce the increases in intrahepatic triglyceride (IHTG) content induced by acute hypercaloric high-fat and high-fructose diets in humans.

Objective: To assess the effect of a 12-wk iso-energetic high protein-low carbohydrate (HPLC) diet compared with an iso-energetic high carbohydrate-low protein (HCLP) diet on IHTG content in healthy non-obese subjects, at a constant body weight.

Design: Seven men and nine women [mean ± SD age: 24±5 y; BMI: 22.9±2.1 kg/m^2] were randomly allocated to a HPLC [30/35/35% of energy (En%) from protein/carbohydrate/fat] or a HCLP (5/60/35 En%) diet by stratification on sex, age and BMI. Dietary guidelines were prescribed based on individual daily energy requirements. IHTG content was measured by ^{1}H-magnetic resonance spectroscopy before and after the dietary intervention.

Results: IHTG content changed in different directions with the HPLC (CH_2H_2O: 0.23±0.17 to 0.20±0.10; IHTG%: 0.25±0.20% to 0.22±0.11%) compared with the HCLP diet (CH_2H_2O: 0.34±0.20 vs. 0.38±0.21; IHTG%: 0.38±0.22% vs. 0.43±0.24%), which resulted in a lower IHTG content in the HPLC compared with the HCLP diet group after 12 weeks, which almost reached statistical significance ($P = 0.055$).

Conclusions: A HPLC vs. a HCLP diet has the potential to preserve vs. enlarge IHTG content in healthy non-obese subjects at a constant body weight.

Trial Registration: Clinicaltrials.gov NCT01551238

Editor: Vincent Wong, The Chinese University of Hong Kong, Hong Kong

Funding: This project has received funding from the European Union's Seventh Framework Programme for research, technological development and demonstration under grant agreement n° 266408. BG-C was supported by grants from the French Society of Endocrinology. The funders had no role in study design, data collection and analysis, decision to publish, or preparation of the manuscript.

Competing Interests: The authors have declared that no competing interests exist.

* Email: eap.martens@maastrichtuniversity.nl

¤ Current address: Group Energy Balance and Obesity, INSERM U862, Neurocentre Magendie, Bordeaux, France

Introduction

Energy-restricted high-protein diets are commonly applied as interventions aiming to induce body weight loss and related improvements in metabolic profile [1,2]. On the other hand, high carbohydrate-high fat diets may increase the susceptibility for overeating, which may result in the development of metabolic disturbances [3,4,5]. Visceral adipose tissue (VAT) volume has been linked to the metabolic disturbances associated with obesity, such as a diminished insulin sensitivity and dyslipidemia [6]. However, high ectopic lipid content, especially intrahepatic triglyceride (IHTG) content, and not VAT volume, is an independent risk factor for these metabolic disturbances [7,8,9]. In obese subjects, increases in BMI, total adipose tissue volume, or VAT volume were only associated with increases in insulin resistance and disturbances in very-low-density lipoprotein (VLDL)-TG metabolism when IHTG content was increased simultaneously [9]. Furthermore, insulin sensitivity was lower and VLDL secretion was higher in obese subjects with a high IHTG content (>5.5% of liver volume), matched on VAT volume [8]. No differences were observed between subjects with a high VAT volume (>1 100 cm^3) but similar IHTG content [8]. Therefore, measuring IHTG content as a marker of metabolic function in dietary intervention studies is highly relevant.

Several human studies suggest that high protein intake may modulate IHTG content. A 4-wk uncontrolled study showed that protein supplementation was associated with a reduction in IHTG content in obese subjects [10]. In two short-term dietary intervention studies, protein supplementation reduced the increases in IHTG content induced by hypercaloric high-fat and high-fructose diets in healthy subjects [11,12]. However, the longer-term effects of dietary protein content on IHTG content, thereby excluding the possible effects of metabolic disturbances and changes in body weight are unknown. Therefore, in the context of prevention of metabolic disturbances, the aim of this dietary intervention study was to assess the effect of a 12-wk iso-energetic high protein-low carbohydrate (HPLC) diet compared with an iso-energetic high carbohydrate-low protein (HCLP) diet on IHTG content in healthy non-obese subjects, at a constant body weight.

Subjects and Methods

The Medical Ethical Committee of Maastricht University approved the study, and all subjects gave written informed consent. The study was registered on clinicaltrials.gov with Identifier: NCT01551238. The protocol for this trial and supporting CONSORT checklist are available as supporting information; see Checklist S1 and protocol S1.

Study subjects

Based on the study by Thetaz et al. [12], power analysis showed that with an α of 0.05 and a β of 0.80, at least 7 subjects per group were needed to show a difference in IHTG content between the dietary interventions of this study.

Twenty-four subjects were recruited by advertisements in local newspapers and on notice boards at the university. Data from measurements of four subjects were unreliable, and were therefore excluded from analysis. Four subjects had to be excluded from the data analysis because of non-compliance with the designated protein intake, as shown by the urinary nitrogen biomarker. The baseline characteristics of the excluded subjects did not differ from the total sample. Overall, 16 subjects (7 men and 9 women) were included in the final data analysis (**Figure 1**).

Subjects underwent a screening that included anthropometric measurements and the completion of questionnaires eliciting information about health, smoking behaviour, use of medications, and alcohol consumption. Subjects were aged between 19 and 31 y had a BMI between 19.0 and 25.3 kg/m^2. Age and BMI were equally divided between males and females. Subjects were non-smoking, not using more than moderate amounts of alcohol (>10 drinks/wk), were weight stable (body weight change <3 kg during the last 6 mo and had no planned weight change during the study period), and were not using medication or supplements except for oral contraceptives in women.

The validated Dutch translation of the Baecke Activity Questionnaire was used to measure habitual physical activity scores, which were converted into a standardized individual physical activity level [13,14].

Dietary intervention

The study used a single-blind parallel design in which subjects were randomly allocated to the HPLC [30/35/35% of energy (En%) from protein/carbohydrate/fat] or the HCLP (5/60/35 En%) diet following stratification on sex, age and BMI. Detailed dietary guidelines were prescribed based on individual daily energy requirements, calculated as the basal metabolic rate determined with the formula of Harris and Benedict [15] times the physical activity level [16]. This was carried out in order to keep body weight constant. Moreover, subjects were told that they were expected to maintain their body weight and physical activity level. The dietary guidelines included a variety of recipes for breakfasts, lunches, dinners and snacks, consisting of commercially available food items and the amounts that had to be consumed to reach the intended diet compositions. The guidelines included the prescription to consume additional protein [HPLC: whey with α-lactalbumin (Hiprotal Whey Protein Alpha, FrieslandCampina Domo EMEA, Amersfoort, The Netherlands)] or carbohydrate [HCLP: maltodextrin (Fantomalt, Nutricia, Zoetermeer, The Netherlands)] by shakes twice daily (**Table 1**).

Body weight and biomarker of protein intake

At baseline and at weeks 5, 9 and 12, body weight was measured by using a digital balance in order to check whether it remained constant, and 24-h urinary nitrogen excretion was determined as biomarker of protein intake throughout the intervention period.

IHTG content

MRI and proton magnetic resonance spectroscopy (^{1}H-MRS) measurements were performed using a 1.5-T whole-body scanner (Intera; Philips Healthcare, Best, The Netherlands). Sagittal, coronal and transversal images through the right lobe in the lower third of the liver were obtained by MRI (slice thickness 8 mm; repetition time/echo time, 500/15 ms; matrix 180×96) in order to gain anatomical and positional information of the liver, ^{1}H-MRS enabled to measure IHTG content by distinguishing hepatic water from a spectrum of IHTG. Fatty acid chains that represent different proton moiety were separated by the relative magnitude of resonance [17]. A 20×20×20 mm volume of interest was positioned, avoiding the lateral margin of the liver parenchyma, major blood vessels, and intrahepatic bile ducts. The position of the volume of interest was the same at the baseline measurement and at the measurement after the intervention. Following shimming (10×10×10 cm volume), eight spectra were acquired using a Q-body coil [18]. Point-resolved spectroscopy sequence (PRESS) was used for spatial localization and signal acquisition using a protocol set up by Schrauwen-Hinderling et al (repetition time/echo time, 4000/33 ms; number of signals averaged fat/water, 64/16) [19]. Water signal was suppressed using 150 Hz frequency-selective prepulses. The unsuppressed water signal was measured in the same volume of interest under the same shimming conditions, and was used as a reference signal [19]. Spectra were fitted using the Java-based magnetic resonance user interface (jMRUI) package fitted in the domain using a non-linear least-squares algorithm (AMARES) [19,20,21]. Areas of resonance from protons of methylene [-$(CH_2)_n$-] and methyl (CH_3) in the fatty acid chains, relative to the area of protons of H_2O, were used as a measure for the IHTG content. A more extensive description of the applied post processing procedure has been published previously [18,19,21,22,23].

Statistical analysis

All statistical analyses were performed by using SPSS version 20 for Macintosh OS X (SPSS Inc.). Differences in subject characteristics and anthropometrics between the diet groups were assessed using Factorial ANOVA. Mann-Whitney tests were used to test for differences in IHTG content between the diet groups.

Factorial ANOVAs with repeated measures were used to test whether nitrogen excretion, body weight, and BMI changed over time within the diet groups, and to test whether these effects differed between the diet groups. Data on IHTG content were log transformed before these statistical analyses were performed. Differences were regarded statistically significant if $P<0.05$. Data are presented as mean ± SD.

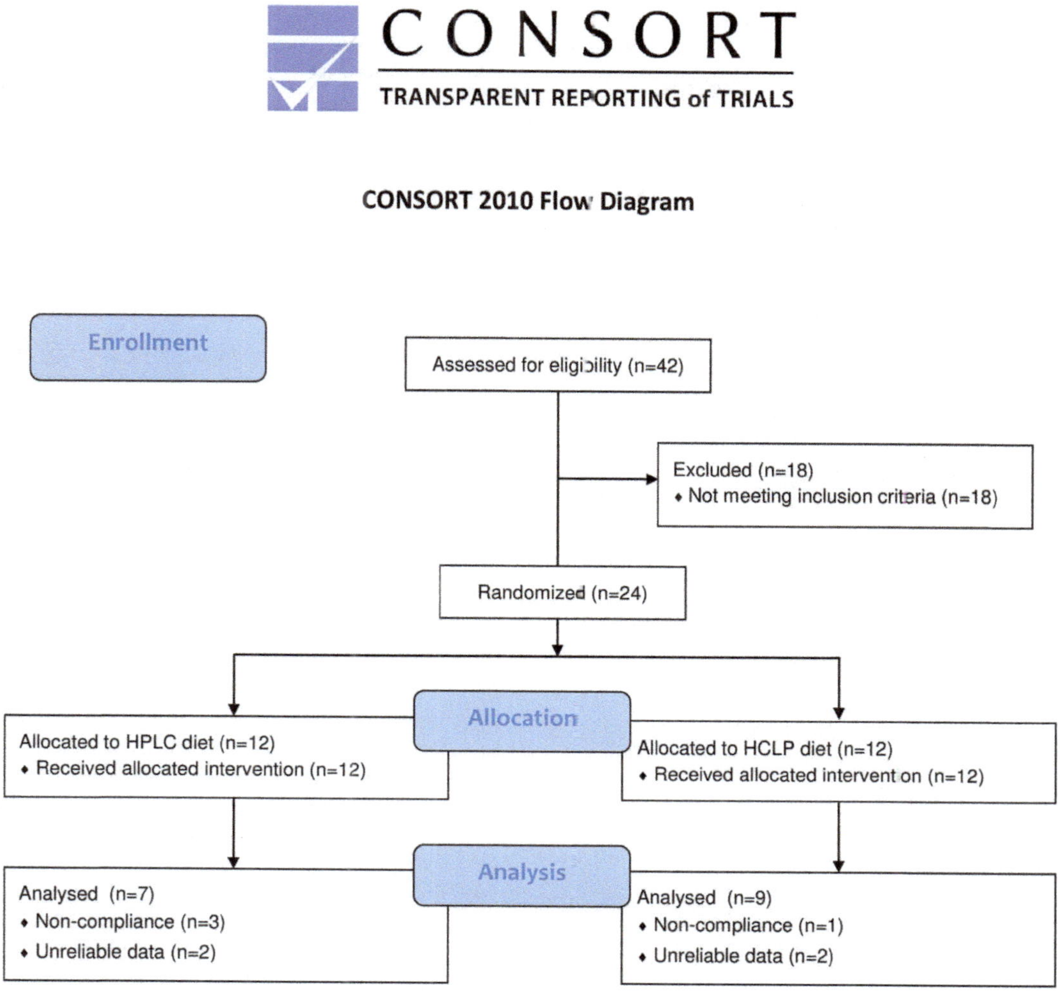

Figure 1. Flow diagram (CONSORT). HCLP, high-carbohydrate low-protein; HPLC, high-protein low-carbohydrate.

Results

Subject characteristics

Subject characteristics did not differ between the diet groups at baseline (**Table 2**). No adverse events were reported.

Body weight and biomarker of protein intake

The condition of stable body weight was met in both arms. Baseline nitrogen excretion did not differ significantly between the HPLC (12.2±3.6 g/d) and HCLP (11.4±3.4 g/d) diets. Nitrogen excretion increased significantly in the HPLC (23.3±4.4 g/d, $P<0.05$), and decreased in the HCLP (6.5±1.3 g/d, $P=0.001$) diet group compared with baseline. Significant differences in protein intake between the diet groups were confirmed in that nitrogen excretion differed significantly between the diet groups throughout the intervention period ($P=0.001$).

IHTG content

IHTG content decreased with the HPLC (CH_2H_2O: 0.23±0.17 to 0.20±0.10; IHTG%: 0.25±0.20% to 0.22±0.11%) and increased with the HCLP diet (CH_2H_2O: 0.34±0.20 vs. 0.38±0.21; IHTG%: 0.38±0.22% vs. 0.43±0.24%), although not reaching statistical significance (Table 2). These changes of IHTG content in different directions resulted in a lower IHTG content in the HPLC compared with the HCLP diet group after 12 weeks, which almost reached statistical significance ($P=0.055$).

Discussion

Protein supplementation has been shown to reduce the increases in IHTG content induced by acute hypercaloric high-fat and high-fructose diets in humans. In the context of prevention of metabolic disturbances, we assessed the longer-term effects of iso-energetic diets with large contrasts in relative protein content on IHTG content in healthy non-obese subjects, thereby excluding the possible effects of changes in body weight. The present study showed that IHTG content changed in different directions with the HPLC compared with the HCLP diet, which resulted in a lower IHTG content in the HPLC compared with the HCLP diet group after 12 weeks, which almost reached statistical significance. The observation that the condition of stable body weight was met in both diet groups confirmed that this observation was not affected by changes in body weight. The successful implementation of the dietary intervention in each diet group was confirmed by urinary nitrogen concentrations.

A proposed mechanism for the modulation of IHTG content by a high protein intake is the stimulation of hepatic lipid oxidation due to the high energetic demand for amino acid catabolism and

Table 1. Sample menu for the HPLC and HCLP diets.

	HPLC	HCLP
Breakfast	Low-fat yoghurt	Whole-grain juice with forest fruits
	Breakfast cereals	Breakfast cereals
	Kiwi	Apple
	Semi-skimmed milk	Orange juice
Lunch	Whole-wheat bread	Whole-wheat bread
	Margarine	Margarine
	Cheese spread	Apple spread
	Chicken breast	Hazelnut paste
Snack	Protein shake	Carbohydrate shake
Dinner	White rice	White rice
	French beans	French beans
	Chicken breast	Red cabbage with apple
	Walnuts	Red pepper
	Olive oil	Olive oil
	Savoury sauce	Savoury sauce
Snack	Protein shake	Carbohydrate shake
	Cereal bar	Corn crisps
	Apple	Pear

The amounts that were prescribed to consume were based on individual daily energy requirements.

ketogenesis [24,25]. Hepatic lipid oxidation may also be stimulated by an increased bile acid production in response to high protein intake, a process that may also inhibit lipogenesis [26]. Furthermore, protein-induced glucagon secretion inhibits de novo lipogenesis and stimulates hepatic ketogenesis [27,28].

On the other hand, previous studies showed that hyper- and eucaloric high-carbohydrate diets stimulated the synthesis of fatty acids from dietary glucose via de novo lipogenesis [29,30,31,32]. The carbohydrate-induced de novo lipogenesis correlated with plasma TG concentrations [30,31]. Increased IHTG deposition was shown to be the result of hepatic TG-synthesis exceeding the rate of its secretion into VLDL [29,30]. The relationship between high carbohydrate intake and increased VLDL-TG concentrations, possibly caused by stimulation of VLDL-TG production and inhibition of VLDL-TG clearance [33], may result in increases in hepatic TG concentrations, and subsequently in IHTG content [31]. Protein supplementation has been observed to blunt the increase in IHTG content induced by short-term hypercaloric high-fructose intake in healthy subjects [12]. Therefore, it is likely that the observed trend for a difference in IHTG content after the consumption of the HPLC compared with the HCLP diet may be the result of combined effects involving changes in protein and carbohydrate intake.

The strength of the present study was the application of an isoenergetic macronutrient intervention for a period that was long enough to reliably measure changes in IHTG content, thereby excluding the possible effects of metabolic disturbances and changes in body weight. Our healthy, non-obese subjects had a low IHTG content at baseline, which relates to a healthy metabolic function. This may exhibit a strong resistance to metabolic changes in response to dietary interventions when stimuli such as a high-fat or a high-fructose intake are absent [11,12]. This indicates that high protein-low carbohydrate diets may be biologically less relevant for inducing a significant decrease in IHTG content in healthy subjects. Nevertheless, the observations of this study are relevant by suggesting that high protein-low carbohydrate diets may be beneficial for the prevention of metabolic disturbances in healthy subjects. Measurements of energy expenditure, body composition, body fat distribution including ectopic fat deposition, and metabolic function are needed to assess the translation of high-protein diets as a treatment strategy into a preventive measure. By addressing ectopic fat deposition in response to dietary interventions, this study adds to the discussion of the clinical relevance of dietary protein for the prevention of obesity. It would also be relevant to implement the measurement of IHTG content in studies assessing the effects of high-protein diets during body weight loss and weight maintenance in patients with metabolic disturbances.

To conclude, a HPLC vs. a HCLP diet has the potential to preserve vs. enlarge IHTG content in healthy non-obese subjects at a constant body weight.

Table 2. Subject characteristics, anthropometrics and IHTG content in the HPLC and HCLP diet groups at baseline and after 12 weeks.

	Baseline		After 12 weeks	
	HPLC	HCLP	HPLC	HCLP
No. of subjects (M/F)	7 (4/3)	9 (3/6)		
Age (y)	23±5	25±5		
Baecke total score	9.4±1.0	8.9±1.1		
Height (cm)	173±8	170±9		
BW (kg)	67.6±6.6	64.5±8.2	67.8±6.2	64.3±7.5
BMI (kg/m²)	22.6±2.0	22.3±2.1	22.7±2.3	22.2±2.0
IHTG content				
CH_2H_2O	0.23±0.17	0.34±0.20	0.20±0.10*	0.38±0.21*
IHTG (%)	0.25±0.20	0.38±0.22	0.22±0.11*	0.43±0.24*

Values represent numbers or means ± SD.
*Trend ($P=0.055$) for factorial ANOVA with repeated measures between diet groups.
BW, body weight; HCLP, high-carbohydrate low-protein; HPLC, high-protein low-carbohydrate; IHTG, intrahepatic triglyceride.

Supporting Information

Checklist S1 CONSORT Checklist.
(DOC)

Protocol S1 Study Protocol.
(PDF)

Acknowledgments

We acknowledge Prof. Dr. Joachim E. Wildberger for giving us the opportunity to perform the MRI and ^{1}H-MRS measurements at the Department of Radiology. We thank Dr. Vera B. Schrauwen-Hinderling for providing us a validated protocol for the determination of IHTG content by using MRI and ^{1}H-MRS, and for offering assistance with the post processing procedure, and Siti N. Wulan for her help in performing part of the studies.

Author Contributions

Conceived and designed the experiments: MSW-P EAM BG-C HKG. Performed the experiments: EAM HKG BG-C. Analyzed the data: EAM BG-C. Wrote the paper: EAM BG-C. Contributed to the interpretation of the data and reviewed the manuscript: MSW-P.

References

1. Acheson KJ (2013) Diets for body weight control and health: the potential of changing the macronutrient composition. Eur J Clin Nutr 67: 462–466.
2. Wycherley TP, Moran LJ, Clifton PM, Noakes M, Brinkworth GD (2012) Effects of energy-restricted high-protein, low-fat compared with standard-protein, low-fat diets: a meta-analysis of randomized controlled trials. Am J Clin Nutr 96: 1281–1298.
3. Westerterp KR, Verboeket-van de Venne WP, Westerterp-Plantenga MS, Velthuis-te Wierik EJ, de Graaf C, et al. (1996) Dietary fat and body fat: an intervention study. Int J Obes Relat Metab Disord 20: 1022–1026.
4. Westerterp KR (2006) Perception, passive overfeeding and energy metabolism. Physiol Behav 89: 62–65.
5. Stubbs RJ, Mazlan N, Whybrow S (2001) Carbohydrates, appetite and feeding behavior in humans. J Nutr 131: 2775S–2781S.
6. Despres JP, Lemieux I (2006) Abdominal obesity and metabolic syndrome. Nature 444: 881–887.
7. Lettner A, Roden M (2008) Ectopic fat and insulin resistance. Curr Diab Rep 8: 185–191.
8. Fabbrini E, Magkos F, Mohammed BS, Pietka T, Abumrad NA, et al. (2009) Intrahepatic fat, not visceral fat, is linked with metabolic complications of obesity. Proc Natl Acad Sci U S A 106: 15430–15435.
9. Magkos F, Fabbrini E, Mohammed BS, Patterson BW, Klein S (2010) Increased whole-body adiposity without a concomitant increase in liver fat is not associated with augmented metabolic dysfunction. Obesity (Silver Spring) 18: 1510–1515.
10. Bortolotti M, Maiolo E, Corazza M, Van Dijke E, Schneiter P, et al. (2011) Effects of a whey protein supplementation on intrahepatocellular lipids in obese female patients. Clin Nutr 30: 494–498.
11. Bortolotti M, Kreis R, Debard C, Cariou B, Faeh D, et al. (2009) High protein intake reduces intrahepatocellular lipid deposition in humans. Am J Clin Nutr 90: 1002–1010.
12. Theytaz F, Noguchi Y, Egli L, Campos V, Buehler T, et al. (2012) Effects of supplementation with essential amino acids on intrahepatic lipid concentrations during fructose overfeeding in humans. Am J Clin Nutr 96: 1008–1016.
13. Baecke JA, Burema J, Frijters JE (1982) A short questionnaire for the measurement of habitual physical activity in epidemiological studies. Am J Clin Nutr 36: 936–942.
14. Philippaerts RM, Matton L, Wijndaele K, Balduck AL, De Bourdeaudhuij I, et al. (2006) Validity of a physical activity computer questionnaire in 12- to 18-year-old boys and girls. Int J Sports Med 27: 131–136.
15. Harris JA, Benedict FG (1918) A Biometric Study of Human Basal Metabolism. Proc Natl Acad Sci U S A 4: 370–373.
16. Westerterp KR, Wouters L, van Marken Lichtenbelt WD (1995) The Maastricht protocol for the measurement of body composition and energy expenditure with labeled water. Obes Res 3 Suppl 1: 49–57.
17. Hamilton G, Yokoo T, Bydder M, Cruite I, Schroeder ME, et al. (2011) In vivo characterization of the liver fat (1)H MR spectrum. NMR Biomed 24: 784–790.
18. Szczepaniak LS, Nurenberg P, Leonard D, Browning JD, Reingold JS, et al. (2005) Magnetic resonance spectroscopy to measure hepatic triglyceride content: prevalence of hepatic steatosis in the general population. Am J Physiol Endocrinol Metab 288: E462–468.
19. Schrauwen-Hinderling VB, van Loon LJ, Koopman R, Nicolay K, Saris WH, et al. (2003) Intramyocellular lipid content is increased after exercise in nonexercising human skeletal muscle. J Appl Physiol (1985) 95: 2328–2332.
20. Naressi A, Couturier C, Devos JM, Janssen M, Mangeat C, et al. (2001) Java-based graphical user interface for the MRUI quantitation package. MAGMA 12: 141–152.
21. Vanhamme L, van den Boogaart A, Van Huffel S (1997) Improved method for accurate and efficient quantification of MRS data with use of prior knowledge. J Magn Reson 129: 35–43.
22. Boesch C, Slotboom J, Hoppeler H, Kreis R (1997) In vivo determination of intra-myocellular lipids in human muscle by means of localized 1H-MR-spectroscopy. Magn Reson Med 37: 484–493.
23. Schick F, Eismann B, Jung WI, Bongers H, Bunse M, et al. (1993) Comparison of localized proton NMR signals of skeletal muscle and fat tissue in vivo: two lipid compartments in muscle tissue. Magn Reson Med 29: 158–167.
24. Veldhorst MA, Westerterp-Plantenga MS, Westerterp KR (2009) Gluconeogenesis and energy expenditure after a high-protein, carbohydrate-free diet. Am J Clin Nutr 90: 519–526.
25. Westerterp-Plantenga MS, Lemmens SG, Westerterp KR (2012) Dietary protein - its role in satiety, energetics, weight loss and health. Br J Nutr 108 Suppl 2: S105–112.
26. Watanabe M, Houten SM, Wang L, Moschetta A, Mangelsdorf DJ, et al. (2004) Bile acids lower triglyceride levels via a pathway involving FXR, SHP, and SREBP-1c. J Clin Invest 113: 1408–1418.
27. Torres N, Tovar AR (2007) The role of dietary protein on lipotoxicity. Nutr Rev 65: S64–68.
28. Gannon MC, Nuttall JA, Damberg G, Gupta V, Nuttall FQ (2001) Effect of protein ingestion on the glucose appearance rate in people with type 2 diabetes. J Clin Endocrinol Metab 86: 1040–1047.
29. Agius L (2013) High-carbohydrate diets induce hepatic insulin resistance to protect the liver from substrate overload. Biochem Pharmacol 85: 306–312.
30. Hudgins LC, Hellerstein MK, Seidman CE, Neese RA, Tremaroli JD, et al. (2000) Relationship between carbohydrate-induced hypertriglyceridemia and fatty acid synthesis in lean and obese subjects. J Lipid Res 41: 595–604.
31. Schwarz JM, Neese RA, Turner S, Dare D, Hellerstein MK (1995) Short-term alterations in carbohydrate energy intake in humans. Striking effects on hepatic glucose production, de novo lipogenesis, lipolysis, and whole-body fuel selection. J Clin Invest 96: 2735–2743.
32. Hudgins LC, Hellerstein M, Seidman C, Neese R, Diakun J, et al. (1996) Human fatty acid synthesis is stimulated by a eucaloric low fat, high carbohydrate diet. J Clin Invest 97: 2081–2091.
33. Parks EJ, Hellerstein MK (2000) Carbohydrate-induced hypertriacylglycerolemia: historical perspective and review of biological mechanisms. Am J Clin Nutr 71: 412–433.

19

Existence of Molten Globule State in Homocysteine-Induced Protein Covalent Modifications

Tarun Kumar, Gurumayum Suraj Sharma, Laishram Rajendrakumar Singh*

Dr. B. R. Ambedkar Center for Biomedical Research, University of Delhi, Delhi, India

Abstract

Homocysteine thiolactone is a toxic metabolite produced from homocysteine by amino-acyl t-RNA synthetase in error editing reaction. The basic cause of toxicity of homocysteine thiolactone is believed to be due to the adduct formation with lysine residues (known as protein N-homocysteinylation) leading to protein aggregation and loss of enzyme function. There was no data available until now that showed the effect of homocysteine thiolactone on the native state structural changes that led to aggregate formation. In the present study we have investigated the time dependent structural changes due to homocysteine thiolactone induced modifications on three different proteins having different physico-chemical properties (cytochrome-c, lysozyme and alpha lactalbumin). We discovered that N-homocysteinylation leads to the formation of molten globule state—an important protein folding intermediate in the protein folding pathway. We also found that the formation of the molten globule state might be responsible for the appearance of aggregate formation. The study indicates the importance of protein folding intermediate state in eliciting the homocysteine thiolactone toxicity.

Editor: Rizwan H. Khan, Aligarh Muslim University, India

Funding: This work is supported by the grant provided by University of Delhi (DRCH/R&D). LRS, TK and GSS acknowledge Council of Scientific and Industrial Research (File No.: 09/045(1131)/2011-EMR-1) and Department of Science and Technology (Ref No.: SR/SO/BB-0003/2011) for the financial assistance provided in the form of research fellowship. The funders had no role in study design, data collection and analysis, decision to publish, or preparation of the manuscript.

Competing Interests: The authors have declared that no competing interests exist.

* Email: lairksingh@gmail.com

Introduction

Homocysteine is a sulfur containing toxic metabolite produced as a byproduct in methionine metabolism pathway. This toxic homocysteine is known to metabolize to methionine by remethylation or to cystiene by trans-sulfurylation [1]. However, mutations in the homocysteine metabolizing enzymes, cystathionine β-synthase (CBS) or methylene tetrahydrofolate reductase (MTHFR) cause an impaired ability to metabolize the toxic homocysteine resulting in an increased levels of cellular and plasma homocysteine [2,3]. The increased accumulation of plasma homocysteine results in homocystinuria and the symptoms include arteriosclerosis, osteoporosis, mental retardation, thrombosis, dislocated eye lenses and neurodegenerative pathologies such as dementia, Parkinson's and Alzheimer's diseases [4,5,6,7]. The serum level of homocysteine in healthy adults is 5–10 μM, it may range from 15–20 μM in mild form but in case of hyperhomocysteinimia it may rise up to 500 μM [8,9]. The toxic effect of homocysteine has been believed due to the formation of homocysteine thiolactone (HTL), which is synthesized from homocysteine by amino acyl t-RNA synthetase in proofreading or error editing reaction mechanism whenever homocysteine gets incorporated in place of methionine by mistake during translation process [10,11,12,13]. It has been demonstrated that HTL preferentially forms amide bonds with ε-amino group of lysine residues of protein in a non-enzymatic mechanism; a process referred to as "protein N-homocysteinylation" [13] which results in various kinds of toxic effects on macromolecules including loss of enzymatic activity, generation of oxidative stress within the cell and autoimmune response generation against N-homocysteinylated self proteins [14].

It has been known that protein N-homocysteinylation results in the formation of toxic multimers, aggregate or amyloid and therefore, this has been considered to be a risk factor for neurodegeneration [15,16,17,18,19,20,21]. Various studies have also shown that the protein aggregation can also be induced *in vitro* by partially folded; molten globule (MG) states [22,23]. Molten globule states are indeed compact denatured protein folding intermediates that exist between the native and denatured states [24]. MG state has attracted much attention in recent years because it is believed to be involved in many biological processes [25,26,27,28,29,30]. It has the following common structural characteristics: (i) presence of intact secondary structure, (ii) loss of tertiary interactions, (iii) presence of loosely packed hydrophobic core that increases the hydrophobic surface accessibility to solvent environment and (iv) 10–30% increase in the radius of gyration [29,31,32,33]. Studies on covalent modification of native proteins by glycating agents that target lysine residues (similar to HTL induced covalent modifications) have also been shown to result in the formation of molten globule states ultimately leading to aggregate formation [23]. Since covalent modification by HTL has been known to induce aggregate/amyloid formation, it is important to investigate if HTL induced covalent modification of proteins also results in formation of MG states for which the detailed structural consequences that HTL modification has on the native state of proteins have to be understood.

Most studies reported earlier on the structural consequences of homocysteine on proteins were however made at time intervals ranging from 12–72 hours [15,17,19,34,35,36] where proteins have already formed aggregate. To the best of our knowledge there are no reports of the effect of protein N-homocysteinylation on the structural characteristics of native proteins at different time intervals. Therefore, what effects HTL exposure results in the native state conformation of the protein has not been properly understood. Here, we investigated the effect of protein N-homocysteinylation on the conformation of three different proteins having different physico-chemical properties and different lysine contents (cytochrome-c from bovine heart, bovine alpha lactalbumin and lysozyme from chicken egg white) at different time intervals. The present work clearly demonstrates that protein N-homocysteinylation results in the formation of molten globule state- an important protein folding intermediate in case of cytochrome-c (cyt-c) and alpha lactalbumin (α-LA) while there exists no structural changes on lysozyme. We also found that molten globule state might be responsible for the formation of aggregate due to protein N-homocysteinylation. Our results indicate the importance of the protein folding intermediate in eliciting the toxic effect of protein N-homocysteinylation.

Materials and Methods

Materials

Cytochrome c (from bovine heart), lysozyme (from chicken egg white), alpha lactalbumin (from bovine milk), DL-homocysteine thiolactone hydrochloride (HTL) and 8-Anilinonaphthalene-1-sulphonic acid were purchased from Sigma-Aldrich chemical Co. Potassium chloride, di-potassium hydrogen phosphate and potassium di-hydrogen phosphate were purchased from Merck, India. Dithiobis(2-nitrobenzoic acid), the Ellman's reagent was also purchased from Sigma-Aldrich chemical Co. Guanidium hydrochloride (Gdmcl) was purchased from M.P. Biomedicals. Unless otherwise stated, all chemicals were used without further purification and double distilled water was used as aqueous phase.

Analytical Procedures

Lysozyme, cyt-c and α-LA solutions were dialyzed extensively against 0.1 M KCl at pH 7.0 at ~4°C. Protein stock solutions were filtered using 0.22-μm Millipore syringe driven filter. Cytochrome c was oxidized using 0.01% potassium ferrocyanide before dialysis. All the proteins gave a single band during polyacrylamide gel electrophoresis (see Figure S1). Concentration of the protein solutions was determined experimentally using the molar absorption coefficient (ε) values 3.9×10^4 M^{-1} cm^{-1} at 280 nm for Lysozyme [37], 1.06×10^5 M^{-1} cm^{-1} at 409 nm for cyt c [38] and 29210 M^{-1} cm^{-1} at 280 nm for α-LA [39]. All solutions for optical measurements were prepared in the desired degassed buffer (0.05 M phosphate buffer, pH 7.0). Since pH of the protein solution may change on the addition of HTL, pH of each solution was also measured after addition of HTL. We observed no significant change in the pH of the protein solutions after addition of HTL.

Modification of proteins (cyt c, α-LA and lysozyme) by HTL and sulfhydryl estimation using Ellman's reagent

Proteins were treated with different concentrations of HTL (1 mM, 2 mM, 5 mM and 10 mM) and then incubated (18 hours for cyt-c, 4 hours for α-LA and 24 hours for lysozyme) at 25°C in 0.05 M phosphate buffer of pH 7.0. Aliquotes of proteins

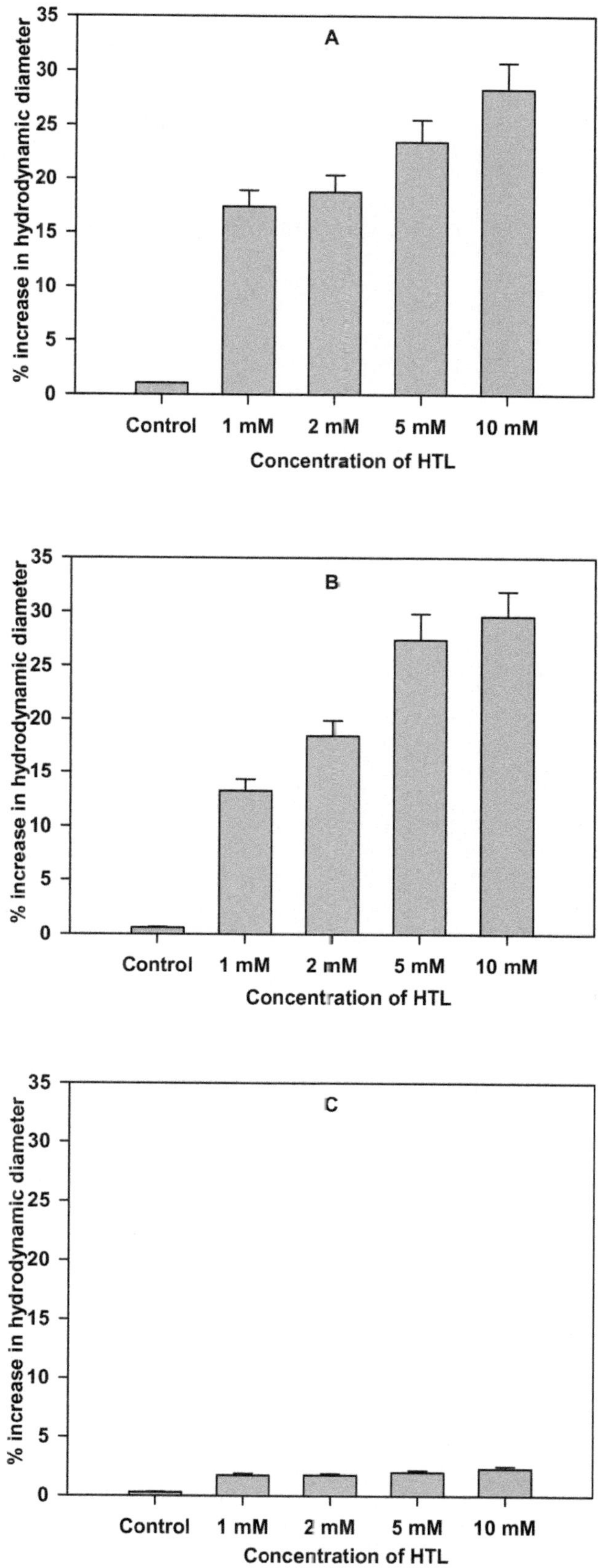

Figure 1. Effect of homocysteine thiolactone (HTL) on the hydrodynamic diameter of the native state of proteins. Percent increase in hydrodynamic diameter of (A) cyt-c (at 18 hours) and (B) α-LA (at 4 hours) and (C) lysozyme (at 24 hours) after modification with different concentrations of HTL.

Table 1. Hydrodynamic diameter of cyt-c, α-LA and lysozyme*.

Time in hours	Hydrodynamic diameter of cyt-c (nm)				
	Native	1 mM HTL	2 mM HTL	5 mM HTL	10 mM HTL
0	2.84	2.81	2.82	2.86	2.79
1	2.81	2.83	2.81	2.84	2.78
2	2.82	2.88	2.85	2.8	2.83
3	2.75	2.8	2.85	2.85	2.84
4	2.88	2.89	2.82	2.83	2.87
6	2.81	2.79	2.72	2.85	2.98
8	2.83	2.71	2.82	2.93	3.07
10	2.78	2.78	2.88	2.98	3.28
12	2.85	2.93	2.91	3.05	3.34
14	2.83	2.95	3.03	3.24	3.41
16	2.8	3.13	3.16	3.48	3.54
18	2.87	3.3	3.35	3.53	3.58, 90.12
20	2.83	3.56	3.4	3.49, 25.26	140.8, 365.13, 831.54
22	2.91	3.45, 18.13	3.58, 20.15	116.7, 260.8, 741.8	160.56, 370.24, 840.34
24	2.88	3.52, 20.12	3.54, 24.65	120.5, 278.9, 801.4	165.45, 369.59, 845.21
Time in hours	**Hydrodynamic diameter of α-LA (nm)**				
	Native	**1 mM HTL**	**2 mM HTL**	**5 mM HTL**	**10 mM HTL**
0	3.42	3.55	3.46	3.54	3.58
1	3.53	3.49	3.49	3.67	3.69
2	3.52	3.56	3.58	3.69	4.14
3	3.51	3.67	3.63	4.02	4.66
4	3.44	3.98	4.1	4.51	4.64
6	3.57	4.02	4.52	6.51, 20.58	6.98, 26.47
8	3.68	6.91, 15.32	6.93, 20.13	6.88, 23.46	6.83, 30.15
Time in hours	**Hydrodynamic diameter of lysozyme (nm)**				
	Native	**1 mM HTL**	**2 mM HTL**	**5 mM HTL**	**10 mM HTL**
0	3.38	3.45	3.44	3.55	3.49
1	3.41	3.55	3.33	3.52	3.52
2	3.38	3.33	3.38	3.52	3.49
3	3.39	3.43	3.32	3.58	3.53
4	3.39	3.52	3.37	3.42	3.53
6	3.38	3.58	3.49	3.42	3.64
8	3.39	3.49	3.55	3.53	3.54
10	3.32	3.51	3.49	3.53	3.54
12	3.47	3.44	3.46	3.55	3.62
14	3.52	3.43	3.49	3.46	3.57
16	3.52	3.55	3.52	3.34	3.61
18	3.44	3.56	3.49	3.46	3.51
20	3.46	3.44	3.45	3.49	3.56
22	3.39	3.53	3.57	3.55	3.47
24	3.39	3.51	3.5	3.62	3.57

*Experimental error in hydrodynamic diameter measurement is in the range of 7–9%.

modified with HTL were first precipitated down with 10% TCA to remove unbound HTL. Protein pellets were collected and then resolubilized in phosphate buffer, pH 7.0. The levels of thiol groups in control and homocysteinylated protein samples were assayed using Dithiobis(2-nitrobenzoic acid), the Ellman's reagent. The absorbance of the samples were measured at 412 nm,

Table 2. Sulfhydryl content measurement of HTL treated cyt-c, α-LA and lysozyme*.

Concentration of HTL	Sulfhydryl content (μM/mg)		
	cyt-c (At 18 hours)	α- LA (At 4 hours)	Lysozyme (At 24 hours)
0 mM	19	36	40
1 mM	753	305	391
2 mM	972	865	819
5 mM	2052	1958	1736
10 mM	2982	3098	2013

*Error in -SH content measurement is in the range of 5–8%.

using a 1 cm path-length cuvette. The amount of 5′-nitrothiobenzoate released was calculated from the molar extinction coefficient of 13,700 M^{-1} cm^{-1} [40,41,42].

Dynamic light scattering (DLS) measurements

Size distribution of the particles present in the protein sample were obtained using a Zetasizer Micro V/ZMV 2000 (Malvern, UK). Measurements were made at a fixed angle of 90° using an incident laser beam of 689 nm. Fifteen measurements were made with an acquisition time of 30 seconds for each sample at sensitivity of 10%. The data was analysed using Zetasizer software provided by the manufacturer to get hydrodynamic diameters, and polydispersity which is a measure of the standard deviation of the size of the particles. The protein concentration was 2.0 mg/ml. All measurements were performed at 25°C.

Circular Dichroism (CD) Measurements

CD measurements were made in a Jasco J-810 spectropolarimeter equipped with peltier controller at 25°C with six accumulations. Protein concentration used for the CD measurements was 0.5 mg/ml. Cells of 0.1 and 1.0 cm path length were used for the measurements of the far- and near-UV CD spectra respectively. Necessary blanks were subtracted. The CD instrument was routinely calibrated with D-10-camphorsulfonic acid. The molar ellipticity in units of Deg cm^2 $dmol^{-1}$ was determined for average residue molecular weight for the proteins used in the study.

Fluorescence Measurements

Fluorescence spectra of the protein samples were measured in a Perkin Elmer LS 55 Spectrofluorimeter in a 3 mm quartz cell, with both excitation and emission slits set at 10 nm. Protein concentration for all the experiments was in the range of 0.06–0.07 mg/ml. For intrinsic fluorescence measurements, cyt-c, α-LA and lysozyme were excited at 295 nm, while the emission spectra were recorded from 300–500 nm. For ANS-protein binding experiments the excitation wavelength was 360 nm, and emission spectra were recorded from 400–600 nm. ANS concentration was kept 16 fold than that of protein concentration.

Absorption spectroscopy measurements

The light scattering intensity of the protein samples were measured by monitoring the absorbance at 500 nm for cyt-c [43], 400 nm for α-LA and 450 nm for lysozyme [44] in a JASCO V-660 spectrophotometer equipped with temperature controlled cuvette holder at 25°C. The protein concentration was 1 mg/ml.

Results

We have investigated the effect of protein N-homocysteinylation on three different proteins (cyt-c, α-LA and lysozyme) having different physico-chemical properties. The hydrophobicity indices, pI values and lysine content of these proteins are 1110, 10.5, 19 for cyt-c, 1050, 4.2, 12 for α-LA and 890, 10.7, 6 for lysozyme respectively. Concentrations of HTL for the treatment used for all the studies were 1 mM, 2 mM, 5 mM and 10 mM.

Hydrodynamic diameter of proteins increases upon treatment with HTL

In the present study, we have initially studied the conformational changes due to N-homocysteinylation at different concentrations of HTL using DLS as a tool at different time intervals by monitoring the hydrodynamic diameter of the proteins (Table 1). We observed significant changes in the hydrodynamic diameter of cyt-c in the time interval 4–18 hours depending on HTL concentrations and α-LA in the time interval 1–4 hours after treatment with HTL beyond which both the proteins tend to get aggregated. However, lysozyme shows no differences in the hydrodynamic diameter due to N-homocysteinylation up to the maximum time of incubation (Table 1) at all HTL concentrations used in this study. Figure 1 also shows that the changes in the maximum increase in hydrodynamic diameter of cyt-c and α-LA is around 13–30% at all HTL concentrations while in the case of lysozyme there is negligible effect on hydrodynamic diameter at all concentrations of HTL. The result suggests that hydrodynamic diameter of different proteins increases at different time intervals depending on the HTL concentration.

Estimation of -SH groups in all the three proteins at specified time periods confirmed protein N-homocysteinylation

To verify if the changes in the conformation of the native states is due to N-homocysteinylation, we have further measured -SH content of the protein samples treated with HTL. Increase in the -SH content has been previously reported to be a signature of the protein N-homocysteinylation [5]. Table 2 shows increase in the sulfhydryl content of all the three proteins due to treatment with HTL at 4 hours for α-LA, 18 hours for cyt-c and 24 hours for lysozyme. It is seen in this table that there is an increase in the -SH groups of the proteins upon treatment with HTL at the specified time periods mentioned in the previous section. The results confirmed that each of the native protein has been N-homocysteinylated at the specified time periods.

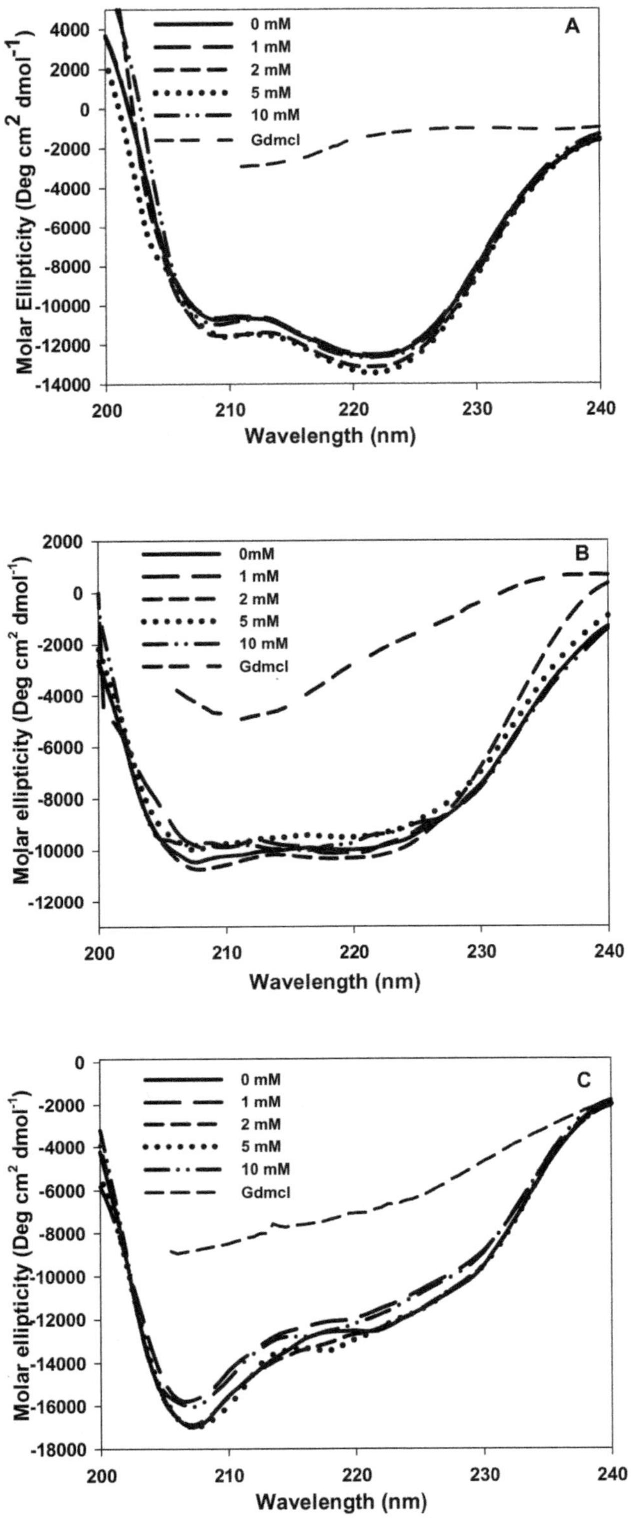

Figure 2. Effect of homocysteine thiolactone (HTL) on the secondary structure of the native state of proteins. Far UV CD spectra of (A) cyt-c (at 18 hours), (B) α-LA (at 4 hours) and (C) lysozyme at (24 hours) modified with different concentrations of HTL. Unfolded state control (7.0 M GdmCl) have also been shown for each protein.

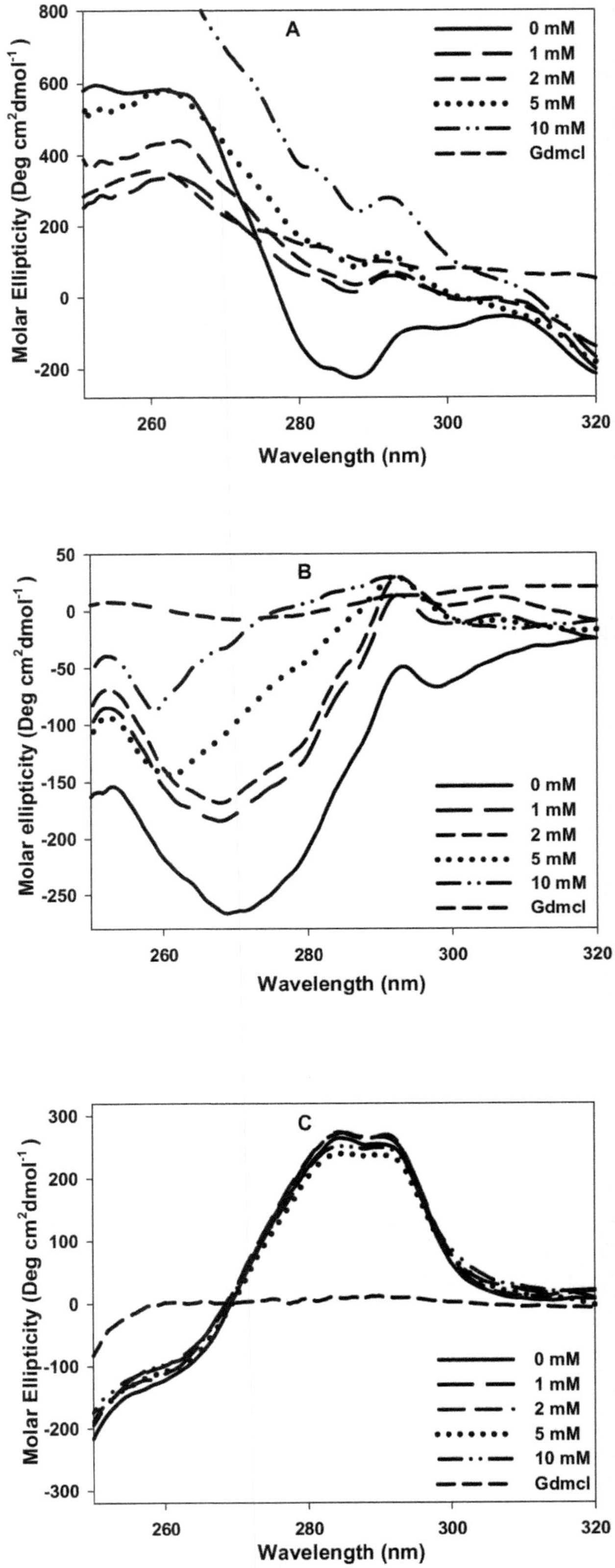

Figure 3. Effect of homocysteine thiolactone (HTL) on the tertiary structure of the native state of proteins. Near UV CD spectra of (A) cyt-c (at 18 hours), (B) α-LA (at 4 hours) and (C) Lysozyme (at 24 hours) modified with different concentrations of HTL. Unfolded state control (7.0 M GdmCl) have also been shown for each protein.

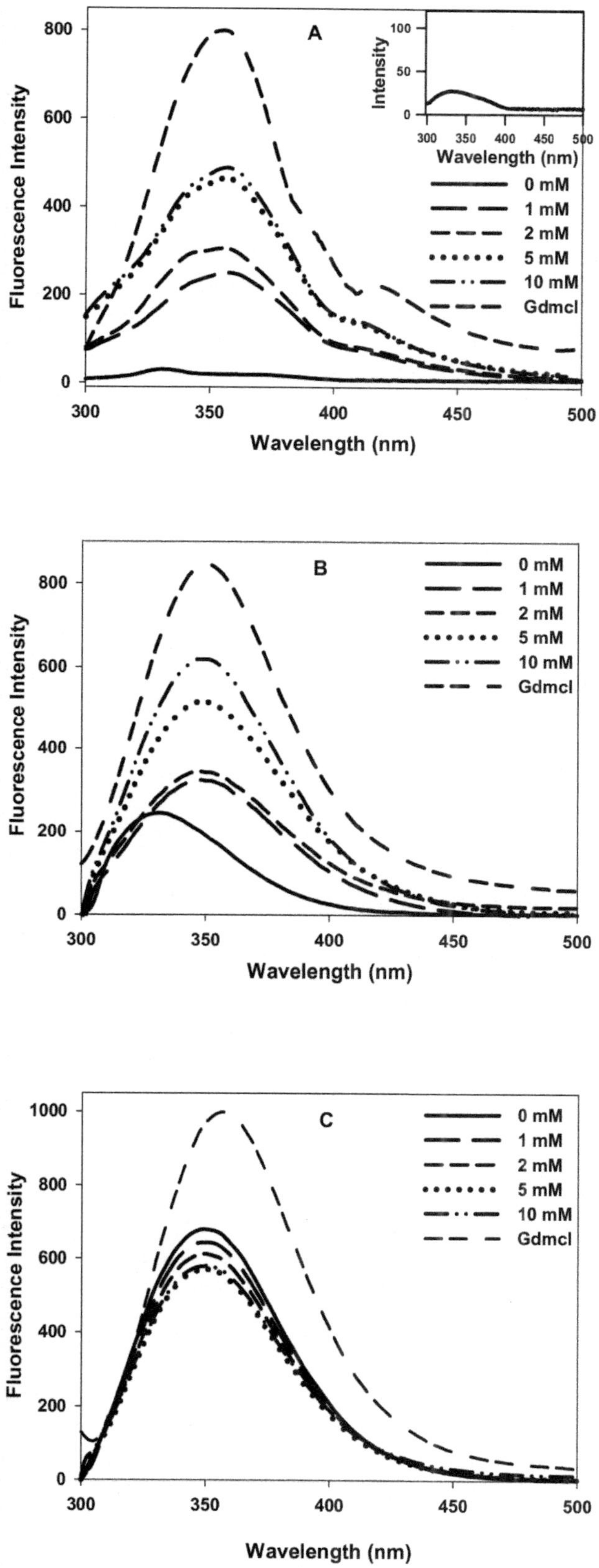

Figure 4. Effect of homocysteine thiolactone (HTL) on the intrinsic fluorescence of the native state of proteins. Intrinsic fluorescence spectra of (A) cyt-c (at 18 hours), (B) α-LA (at 4 hours) and (C) lysozyme (at 24 hours) modified with different concentrations of HTL. Unfolded state control (7.0 M GdmCl) have also been shown for each protein. Inset in Figure 4 (A) shows the enlarged view of the native state emission spectra of cyt-c. Excitation wavelength of 295 nm was used for all the three proteins and emissions were recorded at the wavelength range of 300–500 nm.

N-homocysteinylation results in the formation of molten globule state in cyt-c and α-LA but not in lysozyme

To characterize the structural changes in the proteins due to covalent modification by HTL, we measured the changes in secondary and tertiary structures of the proteins at 4 hours for α-LA, 18 hours for cyt-c and 24 hours for lysozyme by measuring the far-, near-UV CD and intrinsic fluorescence spectra of all the three HTL treated proteins. It may, however, be noted that in the case of cyt-c there appears a non-native species in the presence of 10 mM HTL at 18 hours (Table 1). We have intentionally taken 18 hours incubation time for cyt-c for structural characterization although it has one additional non-native species in our DLS study, as this time interval represents the maximum change in the native state hydrodynamic diameters of protein at all HTL concentrations and the volume fraction of the non-native species observed is negligible (volume distribution less than 1%). Figure 2 shows the secondary structural changes of the proteins treated and incubated for the respective time periods (as shown in figure legend) with different concentrations of HTL. It is seen in this figure that there is (i) neither significant alterations in the spectral properties in terms of the respective peaks of each of the protein (ii) nor the intensity (molar ellipticity) of the homocysteinylated and non-homocysteiylated proteins are changed at different concentrations of HTL. The result leads us to believe that there are no changes in the secondary structure of all the three proteins due to N-homocysteinylation. However, the near-UV CD spectra shown in Figure 3 suggests that the tertiary structures of both cyt-c and α-LA have been unfolded due to N-homocysteinylation at the specified time intervals, while the tertiary structure of lysozyme is not affected due to N-homocysteinylation at different concentrations of HTL. Figure 4 also shows an increase in the tryptophan fluorescence of cyt-c and α-LA along with very prominent red shift in emission maxima (also see Table S1) but in case of lysozyme there was no alteration in the fluorescence properties indicating that N-homocysteinylation (at the respective time periods) disrupts the tertiary interactions of the native states of both cyt-c and α-LA but does not have any perturbing effect on lysozyme. Taken together, the results indicate that N-homocysteinylation leads to different consequences on the native state structure of different proteins. Absence of tertiary structure with intact secondary structure in case of cyt-c and α-LA might indicate a molten globule state. To further verify for this possibility we have intentionally performed ANS binding experiment of these N-homocysteinylated proteins at the respective time periods (Figure 5). It is seen in Figure 5, that there was increase in the intensity of ANS fluorescence and a blue shift in case of both cyt-c and α-LA upon HTL treatment after the respective time of incubations confirming that ANS binds to both the proteins indicating the existence of molten globule state in case of cyt-c and α-LA at all HTL concentrations used in this study. No such ANS binding was observed in case of lysozyme as there was neither an apparent increase in ANS fluorescence intensity nor a blue shift. The results indicate that N-homocysteinylation at the specified time periods results in molten globule state formation in case of cyt-c and α-LA but not in lysozyme.

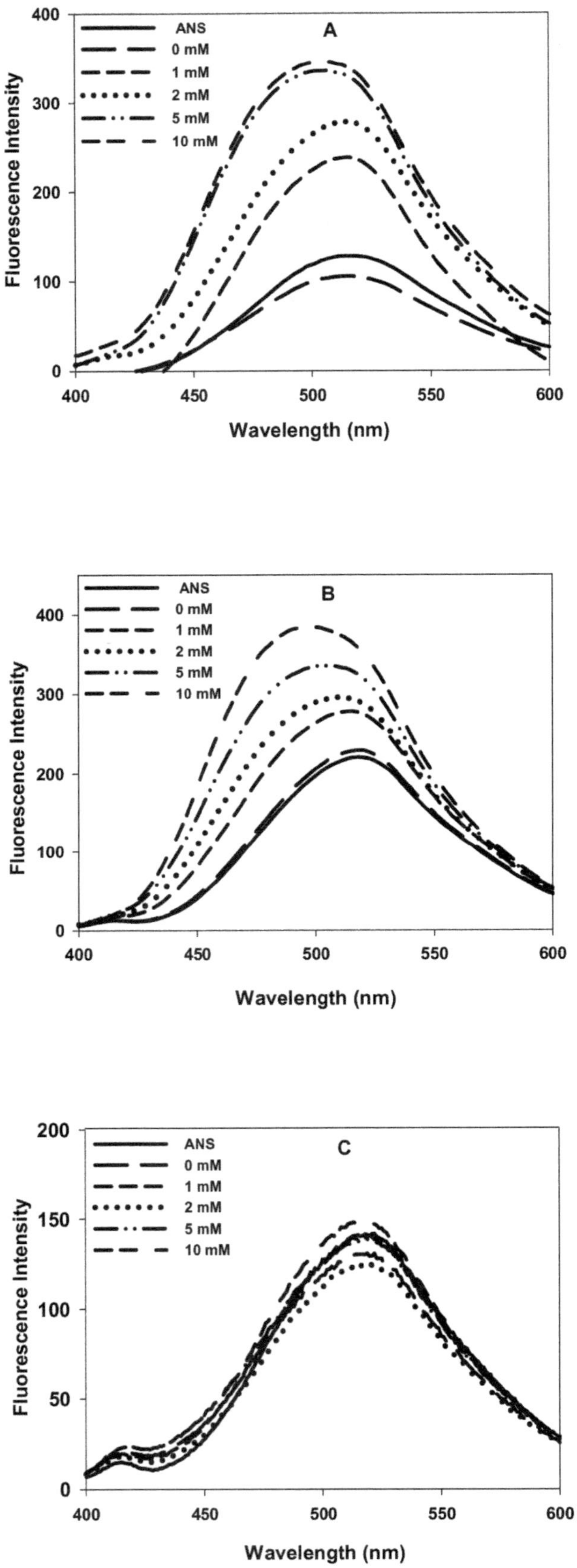

Figure 5. Effect of homocysteine thiolactone (HTL) on the extrinsic fluorescence of the native state of proteins. ANS binding study (A) cyt-c (at 18 hours), (B) α-LA (at 4 hours) and (C) lysozyme (at 24 hours) modified with different concentrations of HTL. Excitation wavelength of 360 nm was used and emissions were recorded at the wavelength range of 400–600 nm.

N-homocysteinylation induces aggregate formation in cyt-c and α-LA but not in lysozyme at 24 hours

In order to study the aggregation propensities of the proteins due to N-homocysteinylation, the turbidity of the modified protein samples was measured using UV-visible spectroscopy. Figure 6 shows the light scattering intensity of the proteins at the maximum time of incubation with HTL (24 hours). It is seen in this figure that both cyt-c and α-LA have been aggregated due to N-homocysteinylation. However, lysozyme was protected against HTL-induced protein aggregation (figure not shown). Since the proteins chosen in the study have different physico-chemical properties, the results indicate that protein N-homocysteinylation has different consequences on the aggregation propensities of the proteins having different physico-chemical properties.

Discussion

To investigate for the possible effect of protein N-homocysteinylation on the native state structure of different proteins, we have first of all measured the change in the hydrodynamic diameter of the proteins at different time intervals. Results shown in Table 1 and Table 2 suggest that HTL has been incorporated to the proteins resulting in the adduct formation that affects hydrodynamic diameter of the native proteins in a time dependent manner. There is increase in the hydrodynamic diameter of the cyt-c (up to 18 hours) and α-LA (up to 4 hours). Beyond these time intervals aggregate formation starts as evidenced by large increase in the hydrodynamic diameter of the HTL modified proteins. There was no significant change in the hydrodynamic diameter of lysozyme till 24 hours. The results indicate that homocysteinylation has different consequences on the native states of different proteins. To investigate for the difference in the native state structure of the proteins due to N-homocysteinylation, we have further measured the secondary and tertiary structures of the native proteins after incubation with HTL for 4 hours in case of α-LA and 18 hours in case of cyt-c. Our results indicate that the secondary structural content (see Figure 2) for the proteins are not at all changed due to N-homocysteinylation while the tertiary structures (Figure 3 and Figure 4) have been modified differently by HTL on the different proteins. In case of lysozyme the tertiary interactions are not significantly changed while in case of cyt-c and α-LA, the tertiary interactions have been disrupted completely at the highest concentrations of the HTL used in the study. Loss of tertiary structure while having no effect on the secondary structure suggests that the native state of both the cyt-c at 18 hours and α-LA at 4 hours of incubation might be a molten globule (MG) state. Indeed MG states are characterized by the (i) presence of intact secondary structure, (ii) the loss of tertiary interactions and (iii) 10–30% increase in the radius of gyration [29,31,32,33]. Interestingly, increase in the hydrodynamic diameter shown in Figure 1 (17–28% for cyt-c and 12–29% for α-LA) further evidenced that the modified proteins might exist in the MG state at the specified time periods. There is no increase in the hydrodynamic diameter of lysozyme suggesting that lysozyme does not lose its native conformation by HTL modification. If the HTL modified native state of cyt-c (at 18 hours) and α-LA (at 4 hours) is really a molten globule state then hydrophobic clusters must be exposed to the solvent due to disruption of tertiary structure. To probe for this, we have further measured the ANS binding to the proteins at the

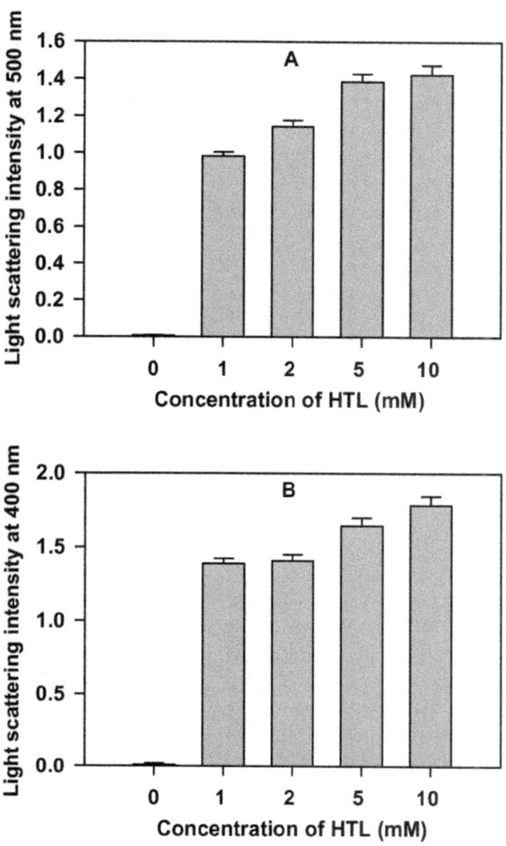

Figure 6. Homocysteine thiolactone (HTL) induced aggregation study of cyt-c and α-LA. Turbidometric study of (A) cyt-c (at 500 nm) and (B) α-LA (at 400 nm) modified with different concentrations of HTL at 24 hours.

given experimental conditions as ANS specifically binds to the hydrophobic clusters exposed to the solvent [45]. Results in Figure 5 suggest that ANS has been bound to both cyt-c and α - LA but not in lysozyme. Taken together the results lead us to believe that N-homocysteinylation results in the formation of MG state in cyt-c and α -LA. It is however seen in case of lysozyme that there is no formation of such MG state since neither the tertiary structure gets unfolded nor there is increase in the hydrodynamic diameter which are prerequisite characteristic features reported for the proteins to be in molten globule state [46]. We therefore conclude that the formation of MG state due to HTL modification is dependent on the amino acid sequence of the proteins. In accordance with our results, glycating agents that target and modify proteins in the same manner by forming covalent adduct with lysine residues, also have been reported to induce MG state formation and hence aggregation [23]. At present we do not have a concrete explanation for having different effects on the native state of different proteins due to HTL induced modification. It might be possible that disruption of the tertiary structure due to N-homocysteinylation might be the limiting step to the conversion of the N-state to the MG state as the tertiary structure of lysozyme could not be disrupted due to the modification. Perhaps the incorporation of HTL at a specific lysine in each of the proteins might be responsible for opening of the tertiary structure which in the case of α-LA and cyt-c is easily accessible while is difficult to target in case of lysozyme. In support to our argument it has been shown earlier in cyt-c that incorporation of HTL in certain lysine residues does not result in significant change in the native structure while there are four lysine residues (Lys 8 or 13, Lys86 or 87, Lys 99, and Lys 100) which are susceptible to N-homocysteinylation resulting in subtle structural changes in the native structure of protein [34]. Interestingly, modification of only one single lysine (Lys 29) residue in bovine pancreatic insulin results in aggregation [15]. It has also been reported that some of the lysine residues in hemoglobin which are directly accessible to solvent environment do not get modified by HTL [35]. Based on these studies we can conclude that the modification of only selective lysine residues results in the perturbation in the structure of proteins.

It has been reported earlier that most of the end products of protein N-homocysteinylation is the formation of aggregates or amyloids [15,16,17,18,19]. We speculate that since MG states are very unstable and have exposed hydrophobic clusters, they might play an important role in the formation of aggregate or amyloid. If our speculation is true, then it is expected that both cyt-c and α - LA should form aggregate while lysozyme should not aggregate (as there are no existence of MG states in lysozyme due to N-homocysteinylation). To pursue for this possibility we have measured the aggregation propensity of the HTL adducted proteins after 24 hours by using light scattering as a tool. It is seen in Figure 6 that there is increase in the light scattering intensity in case of both cyt-c and α-LA while in case of lysozyme the scattering intensity is not at all increased (not shown in Figure). This result is also in agreement with the DLS measurement that there is existence of 2–3 aggregated species at the maximum time of incubation. Thus we conclude that the formation of MG state is responsible for different appearance of aggregates in two different proteins. One of the basic causes of homocysteine toxicity in the cells is the formation of protein aggregates or amyloids and loss of enzyme activity. The results therefore indicate the role of protein folding intermediates in eliciting the HTL toxicity. Indeed such MG state induced by HTL might be more toxic than the aggregates induced by HTL modifications. Further research should focus on these issues.

Conclusions

In summary, we are sure of at least two things (i) N-homocysteinylation results in different consequences on the native state of different proteins: N-homocysteinylation of cyt-c and α-LA induces formation of molten globule state while there is no significant alteration in the native state structure of lysozyme. (ii) The formation of the protein aggregates due to N-homocysteinylation is protein dependent and appears to be related with the formation of molten globule state. Protein N-homocysteinylation has earlier been reported to result in different consequences on the aggregation propensities of different proteins (no aggregation to multimer and aggregate/amyloid formation) and formation of aggregates/amyloids is considered to be the basic cause of neurodegeneration [15,16,17,18,19,20,21]. The study therefore indicates the importance of the MG state as one of the mechanisms in eliciting homocysteine toxicity and neurodegeneration. Studies aimed at preventing the MG state formation induced by HTL modification might yield clinical implications to prevent or delay the onset of homocysteine induced proteopathies and associated disorders.

Supporting Information

Figure S1 SDS-PAGE profile of lysozyme (lane L1), cyt-c (lane L2) and α-LA (lane L3). Lane M represents the protein ladder.
(TIF)

Table S1 Fluorescence emission maxima of HTL treated cyt-c, α-LA and lysozyme.
(DOCX)

Author Contributions

Conceived and designed the experiments: LRS. Performed the experiments: TK GSS. Analyzed the data: TK GSS LRS. Contributed reagents/materials/analysis tools: LRS. Wrote the paper: LRS.

References

1. Brosnan JT, Brosnan ME (2006) The sulfur-containing amino acids: an overview. J Nutr 136: 1636S–1640S.
2. Jakubowski H (1991) Proofreading in vivo: editing of homocysteine by methionyl-tRNA synthetase in the yeast *Saccharomyces cerevisiae*. EMBO J 10: 593–598.
3. Jakubowski H (2002) The determination of homocysteine-thiolactone in biological samples. Anal Biochem 308: 112–119.
4. Seshadri S (2006) Elevated plasma homocysteine levels: risk factor or risk marker for the development of dementia and Alzheimer's disease? J Alzheimers Dis 9: 393–398.
5. Jakubowski H (1999) Protein homocysteinylation: possible mechanism underlying pathological consequences of elevated homocysteine levels. FASEB J 13: 2277–2283.
6. White AR, Huang X, Jobling MF, Barrow CJ, Beyreuther K, et al. (2001) Homocysteine potentiates copper- and amyloid beta peptide-mediated toxicity in primary neuronal cultures: possible risk factors in the Alzheimer's-type neurodegenerative pathways. J Neurochem 76: 1509–1520.
7. Jacobsen DW (1998) Homocysteine and vitamins in cardiovascular disease. Clin Chem 44: 1833–1843.
8. Gellekink H, den Heijer M, Heil SG, Blom HJ (2005) Genetic determinants of plasma total homocysteine. Semin Vasc Med 5: 98–109.
9. Refsum H, Ueland PM, Nygard O, Vollset SE (1998) Homocysteine and cardiovascular disease. Annu Rev Med 49: 31–62.
10. Jakubowski H (1997) Metabolism of homocysteine thiolactone in human cell cultures. Possible mechanism for pathological consequences of elevated homocysteine levels. J Biol Chem 272: 1935–1942.
11. Jakubowski H, Zhang L, Bardeguez A, Aviv A (2000) Homocysteine thiolactone and protein homocysteinylation in human endothelial cells: implications for atherosclerosis. Circ Res 87: 45–51.
12. Jakubowski H, Goldman E (1993) Synthesis of homocysteine thiolactone by methionyl-tRNA synthetase in cultured mammalian cells. FEBS Lett 317: 237–240.
13. Jakubowski H (2000) Calcium-dependent human serum homocysteine thiolactone hydrolase. A protective mechanism against protein N-homocysteinylation. J Biol Chem 275: 3957–3962.
14. Undas A, Perla J, Lacinski M, Trzeciak W, Kazmierski R, et al. (2004) Autoantibodies against N-homocysteinylated proteins in humans: implications for atherosclerosis. Stroke 35: 1299–1304.
15. Jalili S, Yousefi R, Papari MM, Moosavi-Movahedi AA (2011) Effect of homocysteine thiolactone on structure and aggregation propensity of bovine pancreatic insulin. Protein J 30: 299–307.
16. Khazaei S, Yousefi R, Alavian-Mehr MM (2012) Aggregation and fibrillation of eye lens crystallins by homocysteinylation; implication in the eye pathological disorders. Protein J 31: 717–727.
17. Paoli P, Sbrana F, Tiribilli B, Caselli A, Pantera B, et al. (2010) Protein N-homocysteinylation induces the formation of toxic amyloid-like protofibrils. J Mol Biol 400: 889–907.
18. Stroylova YY, Chobert JM, Muronetz VI, Jakubowski H, Haertle T (2012) N-homocysteinylation of ovine prion protein induces amyloid-like transformation. Arch Biochem Biophys 526: 29–37.
19. Stroylova YY, Zimny J, Yousefi R, Chobert JM, Jakubowski H, et al. (2011) Aggregation and structural changes of alpha(S1)-, beta- and kappa-caseins induced by homocysteinylation. Biochim Biophys Acta 1814: 1234–1245.
20. Obeid R, Herrmann W (2006) Mechanisms of homocysteine neurotoxicity in neurodegenerative diseases with special reference to dementia. FEBS Lett 580: 2994–3005.
21. Mattson MP, Shea TB (2003) Folate and homocysteine metabolism in neural plasticity and neurodegenerative disorders. Trends Neurosci 26: 137–146.
22. Rochet JC, Lansbury PT, Jr. (2000) Amyloid fibrillogenesis: themes and variations. Curr Opin Struct Biol 10: 60–68.
23. Iram A, Alam T, Khan JM, Khan TA, Khan RH, et al. (2013) Molten Globule of Hemoglobin Proceeds into Aggregates and Advanced Glycated End Products. PloS one 8: e72075.
24. Rabbani G, Ahmad E, Zaidi N, Fatima S, Khan RH (2012) pH-Induced molten globule state of Rhizopus niveus lipase is more resistant against thermal and chemical denaturation than its native state. Cell Biochem Biophys 62: 487–499.
25. Jennings PA, Wright PE (1993) Formation of a molten globule intermediate early in the kinetic folding pathway of apomyoglobin. Science 262: 892–896.
26. Arai M, Kuwajima K (2000) Role of the molten globule state in protein folding. Advances in Protein Chemistry = Advances in Protein Chemistry 53: 209–282.
27. Hameed M, Ahmad B, Fazili KM, Andrabi K, Khan RH (2007) Different molten globule-like folding intermediates of hen egg white lysozyme induced by high pH and tertiary butanol. J Biochem 141: 573–583.
28. Arai M, Kuwajima K (2000) Role of the molten globule state in protein folding. Adv Protein Chem 53: 209–282.
29. Ptitsyn OB (1995) Molten globule and protein folding. Adv Protein Chem 47: 83–229.
30. Naeem A, Khan RH (2004) Characterization of molten globule state of cytochrome c at alkaline, native and acidic pH induced by butanol and SDS. Int J Biochem Cell Biol 36: 2281–2292.
31. Ansari MA, Zubair S, Atif SM, Kashif M, Khan N, et al. (2010) Identification and characterization of molten globule-like state of hen egg-white lysozyme in presence of salts under alkaline conditions. Protein Pept Lett 17: 11–17.
32. Kuwajima K (2002) The role of the molten globule state in protein folding: the search for a universal view of folding. PROCEEDINGS-INDIAN NATIONAL SCIENCE ACADEMY PART A 68: 333–340.
33. Kuwajima K (1989) The molten globule state as a clue for understanding the folding and cooperativity of globular' protein structure. Proteins: Structure, Function, and Bioinformatics 6: 87–103.
34. Perla-Kajan J, Marczak L, Kajan L, Skowronek P, Twardowski T, et al. (2007) Modification by homocysteine thiolactone affects redox status of cytochrome C. Biochemistry 46: 6225–6231.
35. Zang T, Dai S, Chen D, Lee BW, Liu S, et al. (2009) Chemical methods for the detection of protein N-homocysteinylation via selective reactions with aldehydes. Anal Chem 81: 9065–9071.
36. Yousefi R, Jalili S, Alavi P, Moosavi-Movahedi AA (2012) The enhancing effect of homocysteine thiolactone on insulin fibrillation and cytotoxicity of insulin fibril. Int J Biol Macromol 51: 291–298.
37. Hamaguchi K, Kurono A (1968) Structure of muramidase (lysozyme). I. The effect of guanidine hydrochloride on muramidase. J Biochem 54: 111–122.
38. Margoliash E, Frohwirt N (1959) Spectrum of horse-heart cytochrome c. Biochem J 71: 570–572.
39. Sugai S, Yashiro H, Nitta K (1973) Equilibrium and kinetics of the unfolding of alpha-lactalbumin by guanidine hydrochloride. Biochim Biophys Acta 328: 35–41.
40. Riener CK, Kada G, Gruber HJ (2002) Quick measurement of protein sulfhydryls with Ellman's reagent and with 4,4′-dithiodipyridine. Anal Bioanal Chem 373: 266–276.
41. Butterworth PH, Baum H, Porter JW (1967) A modification of the Ellman procedure for the estimation of protein sulfhydryl groups. Arch Biochem Biophys 118: 716–723.
42. Ellman GL (1959) Tissue sulfhydryl groups. Arch Biochem Biophys 82: 70–77.
43. FedunovÃi D, AntalÃ-k Mn (2006) Prevention of thermal induced aggregation of cytochrome c at isoelectric pH values by polyanions. Biotechnology and bioengineering 93: 485–493.
44. Raman B, Ramakrishna T, Rao CM (1996) Refolding of denatured and denatured/reduced lysozyme at high concentrations. Journal of Biological Chemistry 271: 17067–17072.
45. Rabbani G, Ahmad E, Zaidi N, Khan RH (2011) pH-dependent conformational transitions in conalbumin (ovotransferrin), a metalloproteinase from hen egg white. Cell Biochem Biophys 61: 551–560.
46. Rabbani G, Kaur J, Ahmad E, Khan RH, Jain SK (2014) Structural characteristics of thermostable immunogenic outer membrane protein from Salmonella enterica serovar Typhi. Appl Microbiol Biotechnol 98: 2533–2543.

PERMISSIONS

All chapters in this book were first published in PLOS ONE, by The Public Library of Science; hereby published with permission under the Creative Commons Attribution License or equivalent. Every chapter published in this book has been scrutinized by our experts. Their significance has been extensively debated. The topics covered herein carry significant findings which will fuel the growth of the discipline. They may even be implemented as practical applications or may be referred to as a beginning point for another development.

The contributors of this book come from diverse backgrounds, making this book a truly international effort. This book will bring forth new frontiers with its revolutionizing research information and detailed analysis of the nascent developments around the world.

We would like to thank all the contributing authors for lending their expertise to make the book truly unique. They have played a crucial role in the development of this book. Without their invaluable contributions this book wouldn't have been possible. They have made vital efforts to compile up to date information on the varied aspects of this subject to make this book a valuable addition to the collection of many professionals and students.

This book was conceptualized with the vision of imparting up-to-date information and advanced data in this field. To ensure the same, a matchless editorial board was set up. Every individual on the board went through rigorous rounds of assessment to prove their worth. After which they invested a large part of their time researching and compiling the most relevant data for our readers.

The editorial board has been involved in producing this book since its inception. They have spent rigorous hours researching and exploring the diverse topics which have resulted in the successful publishing of this book. They have passed on their knowledge of decades through this book. To expedite this challenging task, the publisher supported the team at every step. A small team of assistant editors was also appointed to further simplify the editing procedure and attain best results for the readers.

Apart from the editorial board, the designing team has also invested a significant amount of their time in understanding the subject and creating the most relevant covers. They scrutinized every image to scout for the most suitable representation of the subject and create an appropriate cover for the book.

The publishing team has been an ardent support to the editorial, designing and production team. Their endless efforts to recruit the best for this project, has resulted in the accomplishment of this book. They are a veteran in the field of academics and their pool of knowledge is as vast as their experience in printing. Their expertise and guidance has proved useful at every step. Their uncompromising quality standards have made this book an exceptional effort. Their encouragement from time to time has been an inspiration for everyone.

The publisher and the editorial board hope that this book will prove to be a valuable piece of knowledge for researchers, students, practitioners and scholars across the globe.

LIST OF CONTRIBUTORS

Raymond N. Allan
Academic Unit of Clinical and Experimental Sciences, Faculty of Medicine and Institute for Life Sciences, University of Southampton, Southampton, United Kingdom
Southampton NIHR Wellcome Trust Clinical Research Facility, University Hospital Southampton NHS Foundation Trust, Southampton, United Kingdom

Paul Skipp
Centre for Biological Sciences, University of Southampton, Southampton, United Kingdom
Centre for Proteomic Research, Institute for Life Sciences, University of Southampton, Southampton, United Kingdom

Johanna Jefferies
Academic Unit of Clinical and Experimental Sciences, Faculty of Medicine and Institute for Life Sciences, University of Southampton, Southampton, United Kingdom
Public Health England, Southampton, United Kingdom

Stuart C. Clarke
Academic Unit of Clinical and Experimental Sciences, Faculty of Medicine and Institute for Life Sciences, University of Southampton, Southampton, United Kingdom
Public Health England, Southampton, United Kingdom
Southampton NIHR Respiratory Biomedical Research Unit, University Hospital Southampton NHS Foundation Trust, Southampton, United Kingdom

Saul N. Faust
Academic Unit of Clinical and Experimental Sciences, Faculty of Medicine and Institute for Life Sciences, University of Southampton, Southampton, United Kingdom
Southampton NIHR Wellcome Trust Clinical Research Facility, University Hospital Southampton NHS Foundation Trust, Southampton, United Kingdom
Southampton NIHR Respiratory Biomedical Research Unit, University Hospital Southampton NHS Foundation Trust, Southampton, United Kingdom

Jeremy Webb
Centre for Biological Sciences, University of Southampton, Southampton, United Kingdom
Southampton NIHR Respiratory Biomedical Research Unit, University Hospital Southampton NHS Foundation Trust, Southampton, United Kingdom

Luanne Hall-Stoodley
Academic Unit of Clinical and Experimental Sciences, Faculty of Medicine and Institute for Life Sciences, University of Southampton, Southampton, United Kingdom
Southampton NIHR Wellcome Trust Clinical Research Facility, University Hospital Southampton NHS Foundation Trust, Southampton, United Kingdom
Microbial Infection and Immunity, Centre for Microbial Interface Biology, The Ohio State University, Columbus, Ohio, United States of America

Ulfat I. Baig
Department of Microbiology, Abasaheb Garware College, Pune, Maharashtra, India
Department of Biology, Indian Institute of Science Education and Research, Pune, Maharashtra, India

Bharati J. Bhadbhade
Department of Microbiology, Abasaheb Garware College, Pune, Maharashtra, India

Dincy Mariyam and Milind G. Watve
Department of Biology, Indian Institute of Science Education and Research, Pune, Maharashtra, India

Sunjae Lee, Doheon Lee and Seyeol Yoon
Department of Bio and Brain Engineering, KAIST, Yuseong-gu, Daejeon, Republic of Korea

KiYoung Lee
Department of Medical Informatics, School of Medicine, Ajou University, Yeongtong-gu, Suwon-si, Republic of Korea

Jae W. Lee
Neuroscience Section, Papé Family Pediatric Research Institute, Department of Pediatrics, Oregon Health and Science University, Portland, Oregon, United States of America

Gilad W. Vainera and Juliane Kania-Almog
Department of Pathology, Tel-Aviv Sourasky Medical Center, Tel-Aviv, Israel

Ann Saada
Monique and Jacques Roboh Department of Genetic Research, and Department of Genetics and Metabolic Diseases, Hadassah-Hebrew University Medical Center, Jerusalem, Israel

Adir Amartely
Department of Pathology, Hadassah-Hebrew University Medical Center, Jerusalem, Israel

Jacob Bar-Tana and Rachel Hertz
Human Nutrition and Metabolism, Hebrew University Medical School, Jerusalem, Israel

Gilad W. Vainer and Juliane Kania-Almog
Department of Pathology, Tel-Aviv Sourasky Medical Center, Tel-Aviv, Israel

Ann Saada
Monique and Jacques Roboh Department of Genetic Research, and Department of Genetics and Metabolic Diseases, Hadassah-Hebrew University Medical Center, Jerusalem, Israel

Adir Amartely
Department of Pathology, Hadassah-Hebrew University Medical Center, Jerusalem, Israel

Jacob Bar-Tana and Rachel Hertz
Human Nutrition and Metabolism, Hebrew University Medical School, Jerusalem, Israel

Vincent Billard, Alain Ourry, Anne Maillard and Philippe Etienne
Normandie Université, Caen, France
UNICAEN, UMR 950 Ecophysiologie Végétale, Agronomie et nutritions N, C, S, Caen, France INRA, UMR 950 Ecophysiologie Végétale, Agronomie et nutritions N, C, S, Caen, France

Maria Garnica and José-Maria Garcia-Mina
Timac Agro Spain, Poligono de Arazuri-Orcoyen, Orcoyen, Spain

Laurent Coquet and Thierry Jouenne
Plateforme de protéomique PISSARO, UMR6270 CNRS Facultédes Sciences de Rouen, Mont-Saint-Aignan, France

Florence Cruz and Jean-Claude Yvin
Centre de Recherche International en Agroscience, CRIAS-TAI, Groupe Roullier, Dinard, France

Ludovic Jeanson
Service de Génétique et Embryologie Médicales, UnitéMixte de Recherche_Scientifique 933, Institut National de la Santéet de la Recherche Médicale, Université Pierre et Marie Curie - Paris 6, and Assistance Publique - Hôpitaux de Paris, Hôpital Armand Trousseau, Paris, France
Unité_1151, Institut National de la Santéet de la Recherche Médicale, Université Paris Descartes, Paris, France

Ida Chiara Guerrera, Cerina Chhuon and Aleksander Edelman
Unité_1151, Institut National de la Santéet de la Recherche Médicale, Université Paris Descartes, Paris, France
Plateau Proteome Necker, Structure Fédérative de Recherche de Necker, Université Paris Descartes, Paris, France

Jean-François Papon and AndréCoste
Unité Mixte de Recherche_Scientifique 855, Institut National de la Santéet de la Recherche Médicale, Université Paris 12, Faculté de Médecine, Créteil, France
Service d'Otorhinolaryngologie et de chirurgie cervico-faciale, Assistance Publique - Hôpitaux de Paris, Hôpital inter-communal et Groupe Hospitalier Henri Mondor- Albert Chenevier, Créteil, France

Patricia Zadigue
Unité Mixte de Recherche_Scientifique 855, Institut National de la Santéet de la Recherche Médicale, Université Paris 12, Faculté de Médecine, Créteil, France

Virginie Prulière-Escabasse
Service d'Otorhinolaryngologie et de chirurgie cervico-faciale, Assistance Publique - Hôpitaux de Paris, Hôpital inter-communal et Groupe Hospitalier Henri Mondor- Albert Chenevier, Créteil, France

Serge Amselem and Estelle Escudier
Service de Génétique et Embryologie Médicales, UnitéMixte de Recherche_Scientifique 933, Institut National de la Santéet de la Recherche Médicale, Université Pierre et Marie Curie - Paris 6, and Assistance Publique - Hôpitaux de Paris, Hôpital Armand Trousseau, Paris, France

Dimah Z. Habash, Marcela Baudo and David W. Lawlor
Plant Biology and Crop Science, Rothamsted Research, Harpenden, United Kingdom

Matthew Hindle, Stephen J. Powers, Michael Defoin-Platel, Rowan Mitchell, Mansoor Saqi and Chris Rawlings
Computational and Systems Biology, Rothamsted Research, Harpenden, United Kingdom

Kawther Latiri
Laboratoire D'agronomie, National Agricultural Research Institute of Tunisia, Ariana, Tunisia

Jose L. Araus
Dept. of Vegetal Biology, Faculty of Biology, Barcelona, Spain

Ahmad Abdulkader
Biotechnology Department, General Commission for Scientific Agricultural Research, Damascus, Syria

Roberto Tuberosa
Dept. of Agroenvironmental Science and Technology, University of Bologna, Bologna, Italy

Miloudi M. Nachit
Biodiversity and Integrated Gene Management Program, International Center for Agricultural Research in the Dry Areas, Rabat, Morocco

Melissa Dsouza and Michael W. Taylor
Centre for Microbial Innovation, School of Biological Sciences, University of Auckland, Auckland, New Zealand

Susan J. Turner
Centre for Microbial Innovation, School of Biological Sciences, University of Auckland, Auckland, New Zealand
BioDiscovery New Zealand Limited, Parnell, Auckland, New Zealand

Jackie Aislabie
Landcare Research, Hamilton, New Zealand

Judit Marsillach and Clement E. Furlong
Department of Medicine (Division of Medical Genetics), University of Washington, Seattle, Washington, United States
Department of Genome Sciences, University of Washington, Seattle, Washington, United States

Stephanie M. Suzuki and Rebecca J. Richter
Department of Medicine (Division of Medical Genetics), University of Washington, Seattle, Washington, United States

Matthew G. McDonald, Peter M. Rademacher and Allan E. Rettie
Department of Medicinal Chemistry, University of Washington, Seattle, Washington, United States

Michael J. MacCoss and dward J. Hsieh
Department of Genome Sciences, University of Washington, Seattle, Washington, United States

Tingting Liu, Chaofeng Shen, Yi Wang, Canke Huang and Jiyan Shi
Institute of Environmental Science and Technology, College of Environmental and Resource Sciences, Zhejiang University Hangzhou, P.R. China

Deug-Nam Kwon and Jin-Hoi Kim
Department of Animal Biotechnology, Konkuk University, Seoul, South Korea

Byung-Soo Chang
Department of Cosmetology, Hanseo University, Seosan, Chungnam, South Korea

Esther Roselló-Lletí, Estefanía Tarazón, Maria Micaela Molina-Navarro, Ana Ortega and Miguel Rivera
Cardiocirculatory Unit, Health Research Institute Hospital La Fe, Valencia, Spain

María G. Barderas
Department of Vascular Physiopathology, Hospital Nacional de Paraple´ jicos, SESCAM, Toledo, Spain

Manuel Otero, Francisca Lago and Jose Ramó n González-Juanatey
Cellular and Molecular Cardiology Research Unit, Department of Cardiology and Institute of Biomedical Research, University Clinical Hospital, Santiago de Compostela, Spain

Antonio Salvador
Cardiology Service, Hospital La Fe, Valencia, Spain

Manuel Portolés
Cell Biology and Pathology Unit, Health Research Institute Hospital La Fe, Valencia, Spain

Caitlin Pitt, Jeanne A. Robert, Tiffany R. Bonnett and Dezene P. W. Huber
Ecosystem Science and Management Program, University of Northern British Columbia, Prince George, BC, Canada

Christopher I. Keeling and Jörg Bohlmann
Michael Smith Laboratories, University of British Columbia, Vancouver, BC, Canada

Nina A. Lehr, Ning Li and Francesc López-Giráldez
Department of Ecology and Evolutionary Biology, Yale University, New Haven, Connecticut, United States of America

Zheng Wang
Department of Ecology and Evolutionary Biology, Yale University, New Haven, Connecticut, United States of America
Department of Biostatistics, Yale University, New Haven, Connecticut, United States of America

David A. Hewitt
Department of Botany, Academy of Natural Sciences, Philadelphia, Pennsylvania, United States of America
Wagner Free Institute of Science, Philadelphia, Pennsylvania, United States of America

Frances Trail
Department of Plant Biology, Michigan State University, East Lansing, Michigan, United States of America
Department of Plant, Soil and Microbial Sciences, Michigan State University, East Lansing, Michigan, United States of America

Jeffrey P. Townsend
Department of Ecology and Evolutionary Biology, Yale University, New Haven, Connecticut, United States of America
Department of Biostatistics, Yale University, New Haven, Connecticut, United States of America
Program in Computational Biology and Bioinformatics, Yale University, New Haven, Connecticut, United States of America
Program in Microbiology, Yale University, New Haven, Connecticut, United States of America

Karl Kashofer, Helmut Denk and Kurt Zatloukal
Institute of Pathology, Medical University of Graz, Graz, Austria

Vikash Pandey, Marc Sultan, Meryem Ralser, Vyacheslav Amstislavskiy, Hendrik Hache and Christoph Wierling
Max Planck Institute for Molecular Genetics, Department Vertebrate Genomics, Berlin, Germany

Julia Starmann and Holger Sültmann
German Cancer Research Center, Heidelberg, Germany

Ingrid Osprian
BIOCRATES Life Sciences AG, Innsbruck, Austria
LKH Wagna, Department of Internal Medicine, Wagna, Austria

Christina Grimm
Rheumatology and Clinical Immunology, Charite´-University Medicine, Berlin, Germany

Marie-Laure Yaspo and Hans Lehrach
Max Planck Institute for Molecular Genetics, Department Vertebrate Genomics, Berlin, Germany
Dahlem Centre for Genome Research and Medical Systems Biology, Berlin, Germany

Michael Trauner
Hans Popper Laboratory of Molecular Hepatology, Division of Gastroenterology and Hepatology, Department of Medicine, Medical University of Vienna, Vienna, Austria

Sabrina Laouami and Catherine Duport
Avignon Université/INRA, SQPOV UMR408, Avignon, France, 2 INRA, SQPOV UMR408, Avignon, France

Géremy Clair
Avignon Université/INRA, SQPOV UMR408, Avignon, France
INRA, SQPOV UMR408, Avignon, France
Laboratoire de Biochimie des Systèmes Perturbés, CEA Marcoule, DSV-iBEB-SBTN-LBSP, Bagnols-sur-Cèze, France

Jean Armengaud
Laboratoire de Biochimie des Systèmes Perturbés, CEA Marcoule, DSV-iBEB-SBTN-LBSP, Bagnols-sur-Cèze, France

Xiang Ye, Ling-Ling Zhao, Yu-Xia Sun, Jin-Shu Yang and Wei-Jun Yang
Key Laboratory of Conservation Biology for Endangered Wildlife of the Ministry of Education and College of Life Sciences, Zhejiang University, Hangzhou, Zhejiang, People's Republic of China

Yang Zhao
Department of Molecular and Cell Biology, University of California, Berkeley, California, United States of America

Eveline A. Martens, Blandine Gatta-Cherifi, Hanne K. Gonnissen and Margriet S. Westerterp-Plantenga
Department of Human Biology, Maastricht University, Maastricht, The Netherlands

Tarun Kumar, Gurumayum Suraj Sharma and Laishram Rajendrakumar Singh
Dr. B. R. Ambedkar Center for Biomedical Research, University of Delhi, Delhi, India

Index